Current Topics in Behavioral Neurosciences

Volume 67

Current Topics in Behavioral Neurosciences provides critical and comprehensive discussions of the most significant areas of behavioral neuroscience research, written by leading international authorities. Each volume in the series represents the most informative and contemporary account of its subject available, making it an unrivalled reference source. Each volume will be made available in both print and electronic form.

With the development of new methodologies for brain imaging, genetic and genomic analyses, molecular engineering of mutant animals, novel routes for drug delivery, and sophisticated cross-species behavioral assessments, it is now possible to study behavior relevant to psychiatric and neurological diseases and disorders on the physiological level. The Behavioral Neurosciences series focuses on translational medicine and cutting-edge technologies. Preclinical and clinical trials for the development of new diagnostics and therapeutics as well as prevention efforts are covered whenever possible. Special attention is also drawn on epigenetical aspects, especially in psychiatric disorders.

CTBN series is indexed in PubMed and Scopus.

Josh M. Cisler • Kevin M. Crombie •
Thomas G. Adams
Editors

Exercise and Mental Health

Springer

Editors
Josh M. Cisler
Department of Psychiatry and Behavioral Sciences, Dell Medical School
The University of Texas at Austin
Austin, TX, USA

Kevin M. Crombie
Department of Kinesiology
The University of Alabama
Tuscaloosa, AL, USA

Thomas G. Adams
Department of Psychiatry
Yale University
New Haven, CT, USA

ISSN 1866-3370 ISSN 1866-3389 (electronic)
Current Topics in Behavioral Neurosciences
ISBN 978-3-031-68976-5 ISBN 978-3-031-68977-2 (eBook)
https://doi.org/10.1007/978-3-031-68977-2

This Springer imprint is published by the registered company Springer Nature Switzerland AG
The registered company address is: Gewerbestrasse 11, 6330 Cham, Switzerland

Preface

The goal of this volume is to provide a comprehensive overview of the emerging science regarding the impact of physical activity and exercise on constructs related to mental health. Towards this goal, the volume is divided into three sections. The first section describes basic research on the effects of exercise on physiological systems and health. This section provides foundational definitions of exercise and physical activity, an overview of physiological systems engaged during exercise, and basic neurophysiological processes by which exercise enhances resilience towards the general negative effects of stress. This section of basic research on exercise and associated neurophysiology serves as a foundation for understanding of the basic processes and mechanisms by which exercise influences constructs related to mental health.

Subsequently, the second section discusses laboratory studies of the impact of exercise on mood and cognition. This section explores different facets of mood and cognition that serve as basic mechanisms for understanding the pathophysiology and treatment of mental health disorders. As an example, our chapter on the impact of exercise on fear and safety learning describes how fear extinction is a foundational process for understanding the psychological treatment of exposure therapy. Delineating whether, and how, exercise impacts fear extinction then provides a basic science foundation for understanding whether, and how, exercise might be helpful in the treatment of disordered fear and anxiety. Similarly, the chapter discussing the impact of exercise on acute affect and mood provides a foundational understanding for the prescription of exercise in the treatment of mood disorders. Overall, the critical review of basic laboratory studies investigating acute impacts of exercise on various aspects of cognition, learning, memory, and mood and affect fosters an understanding of constructs essential for understanding how exercise can be used to target and treat symptoms of mental health disorders.

Upon establishing a foundation of basic neurophysiology and mechanisms of mood and cognition engaged through exercise, section three describes the use of exercise in the prevention, treatment, and pathophysiology of mental health disorders. There are interesting bidirectional relationships between physical activity and mental health symptoms for most mental health disorders. For example, certain clinical symptoms may be a determinant of alterations in physical activity and exercise (e.g., depression reducing overall activity levels). Reciprocally, physical activity and exercise can directly impact certain clinical symptoms (e.g., an acute bout of moderate intensity exercise improving depressed mood and anhedonia). The use of exercise as either an adjunct to existing evidence-based treatments or as a stand-alone treatment is accordingly accumulating a promising supporting body of data.

It is our hope that this volume is interesting and informative for those working in the intersection of exercise physiology, exercise psychology, and mental health pathophysiology and treatment. We are grateful to all of the authors and contributors whose hard work has resulted in this volume, which we hope will contribute to further integration of exercise science into the fields of mental health research and treatment.

Austin, TX, USA Josh M. Cisler
Tuscaloosa, AL, USA Kevin M. Crombie
New Haven, CT, USA Thomas G. Adams

Contents

Part III Exercise and Clinical Symptoms and Treatment

Part I
Basic Research on Effects of Exercise on Physiological Systems and Health

Foundations of Exercise and Physical Activity Research

Angelique G. Brellenthin and Zoe Sirotiak

Contents

Abstract While recognition of the link between physical activity and mental health dates back nearly two millennia, the academic study of physical activity is a relatively young discipline emerging over the last 50 years. This chapter provides an overview of key terms and measurement techniques in physical activity, exercise, and mental health research. The most common study designs in physical activity research include cross-sectional, cohort, randomized controlled trials, systematic reviews, and meta-analytic studies. Examples from the literature as well as the advantages and disadvantages of various methodological approaches are discussed throughout the chapter in the context of mental health research.

A. G. Brellenthin (✉) and Z. Sirotiak
Department of Kinesiology, Iowa State University, Ames, IA, USA
e-mail: abrellen@iastate.edu

Curr Topics Behav Neurosci (2024) 67: 3–22
https://doi.org/10.1007/7854_2024_488
Published Online: 31 July 2024

Keywords Exercise · Measurement · Physical activity · Study design

1 Origins of Physical Activity Research

The relationship between physical activity and mental health has been acknowledged since antiquity. In 600 BCE, a physician in India named Susruta advised his patients to take moderate exercise every day to not only strengthen the body but also "reduce senility" (Tipton 2014). However, it was not until the epidemiological transition of the twentieth century, which marked a long-term shift in "mortality and disease patterns whereby pandemics of infection [were] gradually displaced by degenerative and man-made diseases as the chief form of morbidity and primary cause of death," that scientific and systematic efforts to quantify the health effects of physical activity truly began (Omran 2005). The introduction of a series of highly effective vaccines and antibiotics beginning in the 1930s contributed to reduced mortality from infectious disease (McKinlay and McKinlay 1977). Reductions in death from infectious diseases and the resulting increase in lifespan led to growing concerns regarding the high rates of death from heart disease, a chronic disease that could be neither prevented nor treated through traditional biomedical approaches. Dr. Jeremy Morris first proposed the link between physical activity and chronic disease prevention in his 1953 *Lancet* paper, "Coronary Heart-Disease and Physical Activity of Work" (Morris et al. 1953). Morris found that among 31,000 male employees of the London Transport system, bus drivers, who sat for the majority of their work shifts, had higher rates of heart attack than bus conductors, who mostly walked and climbed stairs during their shifts (Morris et al. 1953). Morris's seminal findings gave way to a line of inquiry into the associations of physical activity with a wide range of chronic diseases, including many psychological disorders, for which there were few medical treatments at the time. This body of research was summarized in the first physical activity recommendations put forth by the Centers for Disease Control and Prevention and the American College of Sports Medicine in 1995. They advised that adults should accumulate ≥30 min of moderate-intensity aerobic physical activity on most, preferably all, days of the week due to the associations between physical activity and the reduced risk of chronic diseases including anxiety and depression (Pate et al. 1995). The United States Department of Health and Human Services has since updated the *Physical Activity Guidelines for Americans* in 2008 and 2018 as physical activity research expands into new topical areas and as significant methodological advancements in the instruments and approaches used to measure physical activity and mental health-related outcomes have emerged (Physical Activity Guidelines for Americans 2018). The following sections will provide an overview of key terms and definitions, measurement considerations, and study designs commonly used in physical activity, exercise, and mental health research.

2 Physical Activity Terms and Definitions

2.1 *Physical Activity, Exercise, and Fitness*

The *Physical Activity Guidelines for Americans* contains the following specific and standardized definitions for the conceptualization and measurement of physical activity and exercise (Physical Activity Guidelines for Americans 2018). Physical activity refers to "any bodily movement produced by the contraction of skeletal muscle that increases energy expenditure above a basal level." By this definition, physical activity encompasses a wide range of activities from sitting and watching television to a maximal effort sprint to catch a bus. Physical activity occurs throughout the day, including any movement that takes place in work and home settings or as part of recreation or transportation. Exercise is a form of physical activity that is "planned, structured, repetitive, and performed with the goal of improving health or fitness." Exercise is always classified as physical activity, but physical activity is not necessarily exercise. Physical fitness is the "ability to carry out daily tasks with vigor and alertness, without undue fatigue, and with ample energy to enjoy leisure-time pursuits and respond to emergencies." There are many types of physical fitness such as cardiorespiratory fitness, muscular strength, and flexibility. Cardiorespiratory fitness, which represents the body's capacity to use oxygen, is the most commonly assessed type of fitness and serves as a strong predictor of a wide variety of health outcomes (Ross et al. 2016).

2.2 *Parameters of Physical Activity*

Physical activity is prescribed and measured according to the parameters of frequency, intensity, and time (Physical Activity Guidelines for Americans 2018). Frequency is how often a person engages in a certain type of activity, often on a "per week" basis. Intensity describes the level of effort needed to perform an activity. This can be defined in terms of "absolute" intensity or "relative" intensity. Absolute intensity is often quantified with metabolic equivalents (METs). One MET is equivalent to an individual's resting metabolic rate or the amount of energy expended while sitting quietly (Physical Activity Guidelines for Americans 2018). Brisk walking has an approximate MET value of 4, meaning the metabolic effort required to walk briskly is about 4 times the metabolic effort required to sit quietly. The Compendium of Physical Activities is an important research tool that was created to ensure consistency in determining the intensity levels of hundreds of activities (Ainsworth et al. 2011). On an absolute scale, aerobic activities are commonly categorized as sedentary (≤1.5 METs), light (1.6 to 2.9 METs), moderate (3.0 to 5.9 METs), or vigorous (≥6.0 METs) intensity (Physical Activity Guidelines for Americans 2018). Relative intensity allows for physical effort to be quantified relative to individual differences (e.g., baseline muscular strength or baseline cardiorespiratory fitness), and it is often used to prescribe exercise in recreational, clinical, and research settings. Common units used to assess relative intensity include the percentage of maximal heart rate (% HR_{max}), maximal oxygen uptake (% VO_{2max}), or 1-repetition max (% 1RM) as well as ratings of perceived exertion (RPE) (Riebe 2018; Scherr et al. 2013). For example, an activity that requires a sustained heart rate between 64 and 76% of HR_{max} is considered a moderate-intensity activity. A vigorous-intensity activity is between 77 and 95% HR_{max} (Riebe 2018). If HR_{max} is unknown, it can be estimated using the eq. 207-(0.7 × age in years) (Gellish et al. 2007). Time or duration is simply how long a person performs an activity in a given session (e.g., 30 min or 1 h). Frequency, intensity, and time are multiplied together to calculate an overall volume or dose of physical activity.

Physical activity is also classified in terms of its modality or type. The *Physical Activity Guidelines for Americans* provide specific recommendations listed in Table 1 for the two most prevalent modes of physical activity: aerobic physical activity and muscle-strengthening physical activity (Physical Activity Guidelines for Americans 2018). Aerobic physical activity, also known as endurance or cardio activity, involves the sustained rhythmic movement of the body's large muscles. Aerobic physical activity includes brisk walking, running, swimming, and bicycling. The guidelines recommend that adults perform at least 150 min (time) per week (frequency) of moderate-intensity (intensity) aerobic physical activity or 75 min per week of vigorous-intensity physical activity or an equivalent combination of moderate- and vigorous-intensity physical activity per week. This level of physical activity is associated with substantial mental health benefits including decreases in depressed mood and anxiety symptoms and increases in cognitive function, sleep quality, and quality of life (Physical Activity Guidelines Advisory Committee Scientific Report 2018). This volume of aerobic physical activity could be achieved in one session (e.g., a weekend pickup basketball game) or several sessions per week

Table 1 Physical activity guidelines for Americans

• Adults should move more and sit less throughout the day. Some physical activity is better than none
• For substantial physical and mental health benefits, adults should do at least 150 to 300 min a week of moderate-intensity, or 75 to 150 min a week of vigorous-intensity aerobic physical activity, or an equivalent combination of moderate- and vigorous-intensity aerobic activity
• Additional physical health benefits may be gained by engaging in physical activity beyond the equivalent of 300 min of moderate-intensity aerobic activity a week. There is limited evidence on the association between high amounts of physical activity and mental health outcomes
• Adults should also do muscle-strengthening activities of a moderate or greater intensity that involve all major muscle groups on 2 or more days a week for additional health benefits

Adapted from: 2018 Physical Activity Guidelines Advisory Committee. *2018 Physical Activity Guidelines Advisory Committee Scientific Report* and the U.S. Department of Health and Human Services. *Physical Activity Guidelines for Americans, 2nd edition.* Washington, DC: U.S. Washington, DC: U.S. Department of Health and Human Services, 2018 (Physical Activity Guidelines for Americans 2018; Physical Activity Guidelines Advisory Committee Scientific Report 2018) (e.g., a 30-min walk during one's lunch break 5 days per week) (O'Donovan et al. 2017). Individuals can be categorized according to the dose of aerobic physical activity that they regularly achieve. "Inactive" individuals do not engage in any moderate-to-vigorous activity; "insufficiently active" individuals do some but less than the recommended 150 min per week of moderate-to-vigorous intensity activity; "active" individuals engage in 150–300 min per week; and "highly active" individuals exceed 300 min per week of moderate-to-vigorous intensity aerobic activity (Physical Activity Guidelines for Americans 2018).

Muscle-strengthening physical activity focuses on activities that enhance skeletal muscle strength, power, endurance, and mass (Physical Activity Guidelines for Americans 2018). Examples of such activities include strength training, weight lifting, and resistance training as well as heavy gardening or lifting heavy loads in occupational or household tasks. The guidelines recommend that adults perform ≥2 days per week of muscle-strengthening physical activity including movements that strengthen the large muscle groups of the upper and lower body. The *Physical Activity Guidelines for Americans* do not provide specific guidance regarding the intensity or duration of muscle-strengthening activities at this time; however, the American College of Sports Medicine recommends that healthy adults train each major muscle group for 2–4 sets with 8–12 repetitions per set to improve general muscular fitness (Riebe 2018). While aerobic and muscle-strengthening physical activities are by far the most commonly measured and prescribed activities in mental health research, sedentary behavior, light-intensity physical activity, and mind–body physical activities (e.g., yoga, Tai Chi) are emerging areas of research.

2.3 Acute Exercise and Exercise Training

Physical activity's effects on mental health have been researched predominantly in relation to acute or chronic exercise rather than incidental (non-exercise, non-purposeful) or lifestyle-embedded (housework, transportation) physical activity. However, stemming from the increased availability of activity monitors to accurately capture these types of activities, there is growing investigation into the effects of incidental and lifestyle-embedded physical activity on mental health (Tremblay et al. 2007). Acute exercise is a single bout of exercise performed at a specific intensity, duration, and modality. In the case of muscle-strengthening physical activity, it may also be described by the number of sets and repetitions performed of specific exercises. Acute exercise studies may measure outcomes before, during, and after the exercise bout and are particularly useful for mechanistic investigation of transient outcomes such as mood, affect, pain sensitivity, and some aspects of cognition. Exercise training or "chronic exercise" is repeated bouts of acute exercise over a period of time such as several weeks or months. Exercise training studies are designed to determine the accumulated effects of exercise on mental health, particularly on clinically-relevant outcomes such as remission rates. Many studies investigate acute and chronic effects of exercise in the same protocol.

For example, changes in brain-derived neurotrophic factor (BDNF) in response to acute aerobic exercise may be examined to help understand long-term changes in hippocampal volume or memory in response to exercise training (Knaepen et al. 2010; Erickson et al. 2011).

3 Physical Activity Measurement

3.1 Supervised Physical Activity

Physical activity is measured through a variety of approaches, and selection of the best approach often depends on the research question, study design, and available resources. In physical activity and exercise studies, physical activity may be supervised, unsupervised, or a combination of unsupervised and supervised. In supervised studies, participants perform acute or chronic exercise in the presence of researchers. Researchers collect exercise performance data such as time, intensity (e.g., heart rate or ratings of perceived exertion), estimated energy expenditure, or weight lifted. Supervised exercise studies also provide direct data on exercise adherence (how many exercise sessions attended out of how many assigned), compliance (when participants attended, how much of the exercise dose was performed out of how much was prescribed), and attrition (patterns of participant dropout from the exercise intervention over time). While the exact operational definitions of adherence, compliance, and attrition may vary between studies, these data help researchers draw more balanced conclusions regarding both the efficacy (i.e., how well an intervention works under ideal circumstances) and effectiveness (i.e., how well an intervention works in the "real world") of an exercise intervention for mental health outcomes. For example, one clinical trial of aerobic exercise versus pharmacotherapy on major depression reported similar remission rates among supervised aerobic exercise and sertraline therapy groups, yet 20% of supervised exercise patients dropped out of the study compared to 14% of sertraline patients (Blumenthal et al. 2007). Therefore, the similar efficacy of the two treatments on remission rates must be considered alongside the higher dropout and possibly reduced effectiveness of exercise. It is also generally recommended that supervised exercise interventions periodically (e.g., at baseline and follow-up) collect data on unsupervised physical activity from all participants (e.g., control groups and exercise groups). Unsupervised activity occurring outside the laboratory setting should be assessed to account for potential differential patterns and exposures of physical activity between experimental groups that might confound the results.

3.2 Unsupervised Physical Activity

Much of our foundational knowledge regarding the health outcomes associated with physical activity comes, somewhat paradoxically, from observational studies of *unsupervised* (i.e., unobserved) physical activity. These studies have participants recall their recent physical activity behaviors in a questionnaire or wear an activity monitor during their daily activities for a certain period of time to measure unsupervised physical activity. Regardless of the measurement approach, physical activity is not directly witnessed by the study team.

Assessment of unsupervised physical activity can be either objective or subjective. Objective measures include activity monitors such as pedometers and accelerometers as well as gold standard, criterion measures like indirect calorimetry and doubly-labeled water. As their name implies, objective measures tend to produce more unbiased measures of physical activity as they do not rely on participant recall. However, they are not completely bias-free either, since researchers processing these data must make decisions that might affect the findings. For example, many studies ask participants to wear an accelerometer continuously for 7 days to measure weekly physical activity. The participant may forget to wear the monitor for certain periods of time, wear it improperly, or remove it during bathing. The researcher must carefully review the raw data and make decisions about what is considered valid wear time and whether they have enough data to get a reasonably accurate summary of the participant's physical activity behaviors. Even after collecting sufficient data, researchers must further select cut-points for the accelerometer-generated activity counts that categorize physical activity into different intensities (e.g., sleeping versus sitting versus light-intensity activity, etc.). The selection of cut-points can affect results dramatically, yet cut-points are not always consistent between studies and may differ depending on the study population (e.g., children, older adults) (Watson et al. 2014; Kim et al. 2012).

Subjective measures are at the other end of the physical activity measurement spectrum. Subjective measures are based on participant self-report of physical activity in the form of a physical activity interview, diary or log, or a questionnaire. Questionnaires are particularly inexpensive and convenient, but they are prone to poor validity as they rely on participants to accurately and honestly report past physical activity behaviors, which may not be suitable in some clinical populations (e.g., patients with mild cognitive impairment). Short recall surveys are the most popular type of questionnaire. These surveys typically consist of 10 or fewer questions and capture recent physical activity behaviors within the past week or month. They more accurately capture structured exercise rather than incidental daily physical activity since participants may be able to remember and therefore report exercise to a greater extent than other types of physical activity (Colley et al. 2018). These surveys are useful in categorizing individuals based on their activity levels, such as "inactive" or "active" according to the physical activity guidelines. They are also useful for ranking participants by their physical activity levels relative to other participants in the sample, which is often used to examine dose-response relationships between physical activity and the outcome of interest. The *International Physical Activity Questionnaire-Short Form (IPAQ-SF)* is one of the most widely used short recall surveys having undergone extensive reliability and validity testing in a variety of populations. The IPAQ-SF consists of seven questions and assesses vigorous-intensity activity, moderate-intensity activity, walking, and sitting. The

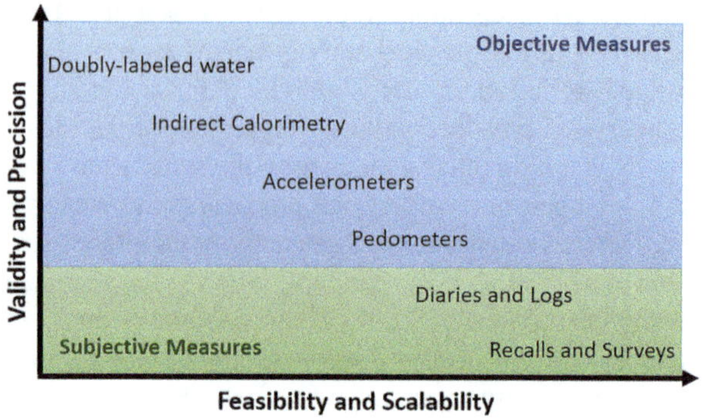

Fig. 1 Tools for measuring unsupervised physical activity. Adapted from Fig. 5A in Welk, G. Morrow, J. Saint-Maurice, P. *Measures Registry User Guide: Individual Physical Activity.* Washington (DC): National Collaborative on Childhood Obesity Research, January 2017 (Welk et al. 2017)

IPAQ-SF is a free resource with a standardized data processing and scoring protocol allowing findings based on the IPAQ to be compared across a wide array of studies (Sjostrom et al. 2005).

In general, objective measures tend to be more accurate than subjective measures. However, objective measures also tend to be more time- and resource-intensive, so they are not the most practical option for many research studies. Furthermore, while an approach like doubly-labeled water may be the most valid method to assess total energy expenditure, information on energy expenditure, without accompanying information on actual physical activity behaviors during the measurement period, does not translate well into practical public health recommendations for physical activity. Therefore, many studies use a combination of assessment tools, such as a pedometer paired with a physical activity diary, to capture physical activity. These approaches are relatively inexpensive and scalable for large sample sizes, while also providing important information on physical activity context and setting, which may be critical factors in mental health research. Finally, objective measures are currently limited in that they are only able to measure aerobic physical activity. Quantification of time spent in muscle-strengthening or mind–body activities can only be assessed through direct supervision or self-report. Methods to measure unsupervised physical activity are summarized in Fig. 1.

In addition to physical activity behaviors, cardiorespiratory fitness and muscular strength are sometimes measured as proxies for recent participation in aerobic and muscle-strengthening physical activity. People who are more physically active generally have higher physical fitness than their less active counterparts. Changes in cardiorespiratory fitness or muscular strength are also often included in aerobic or resistance exercise training trials as secondary outcomes since these fitness parameters are partially reflective of trial fidelity. Among previously untrained participants, exercise groups should generally show a mean improvement in physical fitness compared to control groups if the exercise intervention was delivered as intended. When using physical fitness measures as proxies for recent physical activity, however, it is crucial to account for other factors that heavily influence fitness like age, sex, medications, and genetics (Lee et al. 2010). It is also well documented that cardiorespiratory fitness responses to exercise programs are highly variable, and not everyone who begins an aerobic exercise program will meaningfully improve their cardiorespiratory fitness, despite high adherence and compliance to the exercise protocol (Ross et al. 2019). The choice of the most suitable physical activity assessment method depends on various research priorities, including study size, target population, research objectives, and outcome measures.

4 Study Designs in Physical Activity Research

4.1 Cross-Sectional

Cross-sectional studies are observational studies wherein physical activity behaviors and other variables of interest are assessed without experimental manipulation, and they are the most common study design in physical activity research. Cross-sectional designs collect data at a single time point. For instance, a single questionnaire may include items that ask participants to report current or past (e.g., past week, past month) physical activity as well as items querying the mental health outcome of interest (e.g., current diagnosis or symptom questionnaire).

A study by Harvey et al. (2010) provides an example of a cross-sectional study. This study included over 40,000 participants who completed questionnaires on physical activity and anxiety and depression symptoms. Participants were asked to rate how often they performed light- and high-intensity leisure time and occupational physical activity. They also completed The Hospital Anxiety and Depression Scale to measure depression and anxiety symptoms over the past 2 weeks. The researchers found an inverse relationship between leisure-time physical activity and depression symptoms, with a significant linear trend ($p < 0.001$) for both light- and high-intensity leisure-time activity. No significant relationship was found between workplace activity and depression or anxiety (linear trend p-values: anxiety $= 0.81$; depression $= 0.16$). Those engaging in leisure-time activity are less likely to report symptoms of depression, leading the authors to speculate that social support and engagement may be contributing to mental health among those engaging in activity during leisure (Harvey et al. 2023).

There are several advantages of cross-sectional studies. These studies can be relatively resource-efficient compared to other study designs, as data is only collected at one time. Relatedly, these studies often allow a larger sample size, enhancing the statistical power and generalizability of results. Large, diverse samples also make it possible to efficiently examine relationships between physical activity and mental health outcomes in different sociodemographic groups (e.g., gender, race, age, income), which is useful for public health surveillance and identification of inequities and potential targets for physical activity promotion efforts.

However, there are disadvantages to consider. Cross-sectional studies do not allow for causal determination as variables are measured simultaneously, and it is impossible to determine the temporal sequence of events. Using Harvey et al. (2010) as an example, it could be reasoned that experiencing more depressive symptoms could cause low physical activity participation just as easily as low physical activity participation could cause depressive symptoms (Harvey et al. 2023). There are several sources of potential bias in cross-sectional designs. Many physical activity and mental health questionnaires ask participants to recall their behaviors or symptoms over the past few weeks or months, which is prone to recall bias. Response bias must also be considered, such as the participant answering in a way that they think may please the researcher or providing socially acceptable answers. While cross-sectional studies do not involve manipulating an independent variable, demand characteristics can still be present, in which participants change their reporting or behavior due to being in a study. Cross-sectional studies often attempt to generalize to a population, but selection bias, which occurs when participants are not randomly selected from the population, can limit generalizability. All research studies but especially physical activity studies are vulnerable to healthy response bias whereby research participants, even if randomly selected, are nonetheless volunteers who are likely to be healthier and more active than the general population, which further limits generalizability (Craig et al. 2009).

4.2 Case–Control

Case–control studies are an observational research design in which individuals with a particular health outcome (cases) are compared to individuals without this health outcome (controls). These studies are usually retrospective and consider exposure history between groups to identify associations between specific exposures like physical activity and the development of a health outcome. Ideally, the cases and controls are similar in all attributes except the health outcome, minimizing confounding factors that may influence associations.

Werneck et al. (2020) performed a cross-sectional study of 60,000 adults in Brazil, investigating the relationship between severe mental illnesses, such as schizophrenia, and lifestyle behaviors, such as leisure-time physical activity, tobacco use, and consumption of alcohol. The researchers identified cases of severe mental illness by physician diagnosis. Participants self-reported their engagement in leisure-time physical activity over the past 3 months. Using logistic regression analysis, the researchers found that schizophrenia was associated with lower odds of physical activity in the prior 3 months. The authors concluded that lifestyle factors, including leisure-time physical activity, could be targeted in prevention efforts in the context of severe mental illness (Werneck et al. 2020).

The main advantage of case–control studies is to investigate rare outcomes like schizophrenia. Compared to the large samples needed in other study designs to identify sufficient cases of a rare outcome, investigators using a case–control design can directly recruit participants with the outcome of interest. Case–control designs often involve retrospective data, which can save on resources involved in collecting data. Multiple exposures can be assessed simultaneously, allowing one study to evaluate many contributing factors. Like in Werneck et al. the associations between physical activity and the outcome can be examined in relation to other lifestyle variables such as tobacco or alcohol use (Werneck et al. 2020). In addition, case–control studies often require one contact point with participants, lessening time and resource commitment.

However, there are also disadvantages to case–control studies. Like a cross-sectional design, case–control studies may rely heavily on participant recall, potentially going back several decades, allowing reporting and recall bias to affect results. Relying on participant recall is overtly problematic for some mental health conditions such as dementia. Even with mental disorders that do not affect memory or other cognitive functions, participants recently diagnosed with a rare medical condition may recall their past behaviors differently than someone without a diagnosis. Obtaining medical records, while more objective, may not be readily accessible and may present ethical considerations. In addition, selection bias is an important consideration, as finding a representative sample in rare conditions can be particularly difficult. Selection bias is also a concern when identifying control group participants. Control group participants should be representative of the population from which the cases were obtained, not necessarily from the general population. Since differences in the exposure histories of the control group and the general population are unknown, case–control studies, which compare exposures between controls and cases only, may produce associations that are not reflective of the true relationship between physical activity and the outcome. Temporal uncertainty also limits interpretability. Not only are past exposures recalled at the same time as other data are collected (similar to a cross-sectional study), but the recall period (e.g., past 5 years, past 20 years, lifetime) may not capture the time period between all the possible exposures and disease progression.

4.3 Cohort

Prospective cohort studies are another type of observational study design that recruits and follows a cohort, or group of individuals, over a specific period. Whereas case–control studies focus primarily on identifying the outcome (i.e., cases) and then examining the outcome in relation to a wide variety of potential exposures, prospective cohort studies focus primarily on the exposure and then examine the exposure in relation to a wide variety of outcomes over time. In physical activity research, this method also allows for determining a population attributable fraction for physical activity or the fraction of cases of a disease or disorder that could be prevented if each participant engaged in physical activity.

Harvey et al. (2018) analyzed data from the Health Study of Nord-Trøndelag County (HUNT Study), a prospective cohort study in rural Norway. In this study, over 33,000 adults without mental health disorders at baseline were followed for 11 years. Participants reported the frequency and duration of aerobic exercise at baseline, and depression and anxiety symptoms were assessed at follow-up using The Hospital Anxiety and Depression Scale. Physical activity was associated with a reduced odds of depression at follow-up, though they noted no associations between physical activity and anxiety. The authors also calculated a population attributable fraction, estimating that 12% of cases of depression could have been prevented if each participant had engaged in 1 h of physical activity per week (Harvey et al. 2018).

Compared to other observational research designs, prospective cohort studies are advantageous in that they provide a temporal event sequence. Participants without the outcome of interest at baseline are followed until they either develop the outcome or until the latest study contact. Recall bias is less of a concern with prospective cohort designs as the exposure is measured before an outcome occurs. Likewise, exposures can often be assessed with more detail and accuracy in prospective cohort studies compared to other forms of observational research. Prospective cohort studies can be especially helpful for investigating conditions that develop slowly over time, as both the exposure and disease progression can be measured at many time points if resources allow. In addition, these studies can help calculate incidence rates or the probability that an individual will be diagnosed with a condition within a given period. Like case–control studies, several exposures can be measured simultaneously, saving time and resources.

However, there are disadvantages to a prospective cohort design. Like other designs, prospective cohort studies are limited in generalizability by the degree to which their sample represents the larger population. These studies can require significant time and resources, as participants must be tracked over time, potentially with multiple follow-up visits. Similarly, participant attrition and loss to follow-up are important considerations in a prospective cohort design. If the characteristics of those who drop out significantly differ from those who remain in the study, bias can result. This is particularly true for cohorts assessing physical activity and mental health. Participants who remain in a cohort over many years are generally healthier, both physically and mentally, and more active than participants who are lost to follow-up (Mein et al. 2012). Confounding variables, or spurious variables that can affect the relationship between the main variables of interest, can also be a concern in prospective cohort studies. Finally, while temporal sequence can be established, causal connections cannot.

4.4 Randomized Controlled Trial

Randomized controlled trials (RCTs) are considered the gold-standard study design for determining causality. Researchers frequently use RCTs to evaluate the effectiveness of exercise interventions on mental health outcomes. Randomization is a key feature of RCTs, with participants randomized to either a treatment or a control group. The outcome of interest is measured before and after the trial, and changes in the outcome between the treatment and control groups are analyzed.

A study by Blumenthal et al. (2007) provides an example RCT in their investigation of the effect of aerobic exercise, pharmacotherapy, or a placebo control on depressive symptoms in individuals diagnosed with major depression. In this study, 202 adults with a major depressive disorder diagnosis were randomized to four conditions: supervised group aerobic exercise, home-based aerobic exercise, antidepressant medication, or a placebo pill. Each intervention lasted 16 weeks, after which each participant completed a clinical interview for depression to assess remission. Forty-one percent of participants achieved depression remission, with the two exercise interventions and pharmacotherapy group having greater remission rates than the placebo group. Remission rates by group (aerobic exercise group results combined) are shown in Fig. 2. The authors concluded that aerobic exercise and pharmacotherapy were similarly effective in reducing depression among those with major depressive disorder, offering exercise as an efficacious alternative to pharmaceutical approaches (Blumenthal et al. 2007).

Like observational studies, RCTs provide their own set of advantages and disadvantages. Perhaps the most significant advantage of an RCT design is the ability to determine causality. Randomizing participants allows researchers to balance measured and unmeasured confounding variables between groups, reducing the likelihood that any observed effects are due to these other factors. However, RCTs present limitations and challenges as well. These studies can require significant time and resources both for the research team and participants. In addition, achieving high exercise adherence rates can be difficult, particularly over a long study duration. Blinding is another hallmark feature of RCTs, yet appropriate blinding can be notably challenging, if not impossible, in the context of exercise and mental health. Participants in an exercise intervention cannot be blinded to their condition as participants in a drug trial might be. While exercise and mental health RCTs may involve blinding the researcher collecting data, performing analyses, and drawing conclusions, participants cannot be blinded to their exercise group. Expectation bias,

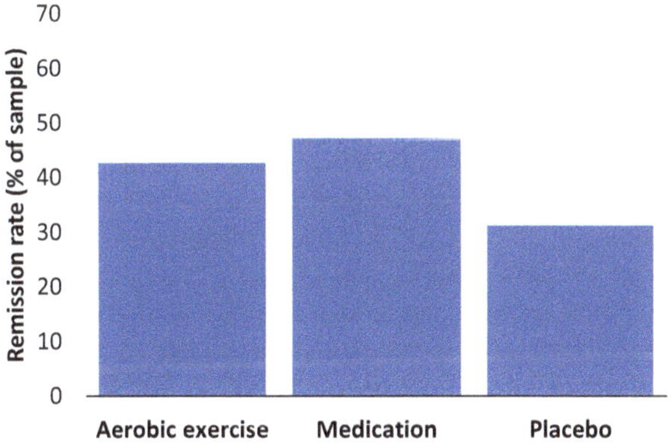

Fig. 2 Remission rates of major depressive disorder by group in Blumenthal et al. (2007). Adapted from Fig. 3 in Blumenthal JA, Babyak MA, Doraiswamy PM, Watkins L, Hoffman BM, Barbour KA, Herman S, Craighead WE, Brosse AL, Waugh R, Hinderliter A, Sherwood A. Exercise and pharmacotherapy in the treatment of major depressive disorder. Psychosom Med. 2007;69 (7): 587–96. doi: https://doi.org/10.1097/PSY.0b013e318148c19a

or when an individual's expectations about a treatment affect their perceptions and behaviors, can be a concern in exercise and mental health trials. Participants may enter an exercise RCT believing that exercise positively impacts mental health, which can contribute to an exaggerated mental health improvement in the exercise group.

Some studies have attempted to address the issue of blinding and outcome expectations by using placebo-control groups (see Chapter "Psychedelics and Drug Checking" for comprehensive overview). Placebos are interventions that offer no expected therapeutic benefit, and they are particularly important when blinding is not possible. Theoretically, placebo groups allow isolation of the effect of the treatment, subtracting the influence of a participant's expectations due to being in a study or being offered a treatment. The placebo effect has been suggested to account for approximately half of the observed psychological benefits of exercise (Lindheimer et al. 2015). In the previously mentioned study by Blumenthal and colleagues, the remission rate of the placebo group was 75% of the remission rate observed among the active treatment groups (Fig. 2) (Blumenthal et al. 2007). This study also used a placebo pill, a convenient and routine approach, since another experimental group was receiving active medication. Though for exercise trials without a medication comparison group, it can be difficult to develop an appropriate placebo control or comparison condition. Relaxation, social activities, and very low-intensity exercise have been used frequently as placebo conditions in exercise and mental health trials (Lindheimer et al. 2015). However, there is substantial heterogeneity in placebo conditions across trials, and many placebos have unclear associations with the outcome, making it difficult to compare these conditions to each other and to the exercise treatment of interest. Ideally, exercise and control groups would be equivalent in every way except the intervention, but this can be difficult to implement in practice. Participants in control conditions may change their exercise or lifestyle habits, and any group could be impacted by poor adherence, affecting the study results. There are also ethical considerations in selecting comparison or control groups, as it is ethically problematic to withhold potentially beneficial treatment from comparison or control groups in patients with mental illness.

Conducting an exercise RCT with sufficient power to find differences between groups can also be challenging. Supervised exercise RCTs are particularly labor-, time-, and resource-intensive trials. Thus, most exercise RCTs are relatively small ($N < 100$) (Abt et al. 2020) compared to typical pharmaceutical trials that might enroll thousands of patients. Furthermore, the effect sizes for exercise on various mental health outcomes can be quite modest, especially after considering placebo effects in comparison groups (Lindheimer et al. 2015). The combination of modest effect sizes and small sample sizes in the exercise science field has led some researchers to conclude that the average power to detect true effects in exercise RCTs is between 40–50% (Mesquida et al. 2022). This means that roughly one out of two studies should report null findings. Yet, over 70% of published studies report that their primary hypothesis was supported by their findings (Büttner et al. 2020). Thus, there are serious concerns of the methodological quality and bias in reporting for much of the existent literature. It should be noted, however, that this problem is not unique to exercise science, and traditional reporting and publication practices are under scrutiny and revision in many scientific fields.

4.5 Systematic Review and Meta-Analysis

Systematic reviews and meta-analyses are summaries of the literature generated after following explicit methodology to comprehensively search, critique, and synthesize together individual studies. A systematic review is a structured review and summary of existing literature on a specific topic or research area. These reviews follow a sequence of forming a research question, identifying relevant literature, assessing the quality of that literature, summarizing evidence based on the accumulation of literature, and interpreting these findings (Khan et al. 2003). The review aims to provide a thorough overview of the current literature on a specific topic. Often combined with a systematic review, a meta-analysis is a statistical method combining data from multiple studies evaluating the same research topic. Relevant data are extracted from the original studies, leading to overall effect size and statistical significance calculations. Meta-analyses offer greater statistical power than individual studies, theoretically leading to a more comprehensive and conclusive assessment of the effects of exercise on mental health.

Stubbs et al. (2017) evaluated literature describing the anxiety-reducing effects of exercise among individuals with anxiety and stress-related disorders. The authors analyzed six RCTs, including a total of 262 adults. The combined data suggested that exercise significantly reduces anxiety symptoms for those with anxiety or stress-related disorders compared to control conditions (standardized mean difference = −0.582; $p = 0.02$) (Stubbs et al. 2017). Meta-analyses provide several advantages, including increasing the sample size leading to increased statistical power to detect differences that exist. More precise effect sizes can be measured with meta-analyses compared to individual studies, and bias in one study may be lessened by including other studies. Systematic reviews and meta-analyses are limited by the quality of the studies they evaluate. Publication bias and the availability of databases and publications also limit these analyses. Articles written in English also receive publication advantages and are more likely to be included in systematic reviews and meta-analyses (Morrison et al. 2012). The heterogeneity of available studies can also provide challenges and difficulty incorporating different measures of variables may limit the ability to combine datasets.

4.6 Qualitative Studies

Qualitative research is an important but less common approach for studying physical activity and mental health. Qualitative work allows for assessing the subjective realities and the lived experience of individuals and groups, analyzing identity, perspectives, and meaning. Qualitative research may allow the identification of beliefs, motivations, and experiences that impact one's perception of physical activity, mental health, and associations between the constructs. Qualitative methods include observation, fieldwork, focus groups, and in-depth interviews. Analysis often involves the identification of common themes, as well as the consideration of findings in the context of existing literature, offering insights related to participant perception and personal impact that are not found with quantitative approaches. Mixed-methods research or combining quantitative and qualitative data analysis in one research project can be a strong research design as it combines the strengths of quantitative and qualitative research and allows for integrated understanding of multiple sources of data.

Mason and Holt (2012) conducted a review of qualitative literature investigating physical activity among those with mental health difficulties. The authors found common concepts noted across articles, noting several themes: social interaction and support, feeling safe, improved symptoms, a sense of meaning, purpose and achievement, identity, and the role of others in the physical activity experience (Mason and Holt, 2012). Qualitative research can offer powerful improvements in understanding, allowing participants to offer their lived experiences and subsequent knowledge rather than being limited to selecting pre-determined options on a survey. Qualitative research allows for participant voice and involvement in the research process, an important consideration in working with individuals living with mental health issues, conditions long associated with stigma and lack of autonomy among patients (Palinkas 2014). Qualitative research can also help develop hypotheses that can then be tested with follow-up quantitative or mixed-methods studies. Figure 3 provides an overview of the types of study designs used in physical activity and mental health research.

5 Summary

This chapter provided an overview of the key concepts and methods used in the study of physical activity and mental health. While there is no perfect or universal approach to examine physical activity and mental health, there are several tools and study designs, each with their own advantages and disadvantages that can be used effectively to conduct rigorous research and generate impactful findings to improve the lives of individuals living with mental illness. Chapters in Sects. 2 and 3 of this volume will present findings from studies that used a variety of these approaches to improve our understanding of: (1) the acute effects of exercise on mood, affect, and cognition, and (2) the acute and chronic effects of physical activity and exercise on preventing and treating mental health disorders including depression, anxiety, eating disorders, cognitive impairment, substance use, schizophrenia, pain, and fatigue.

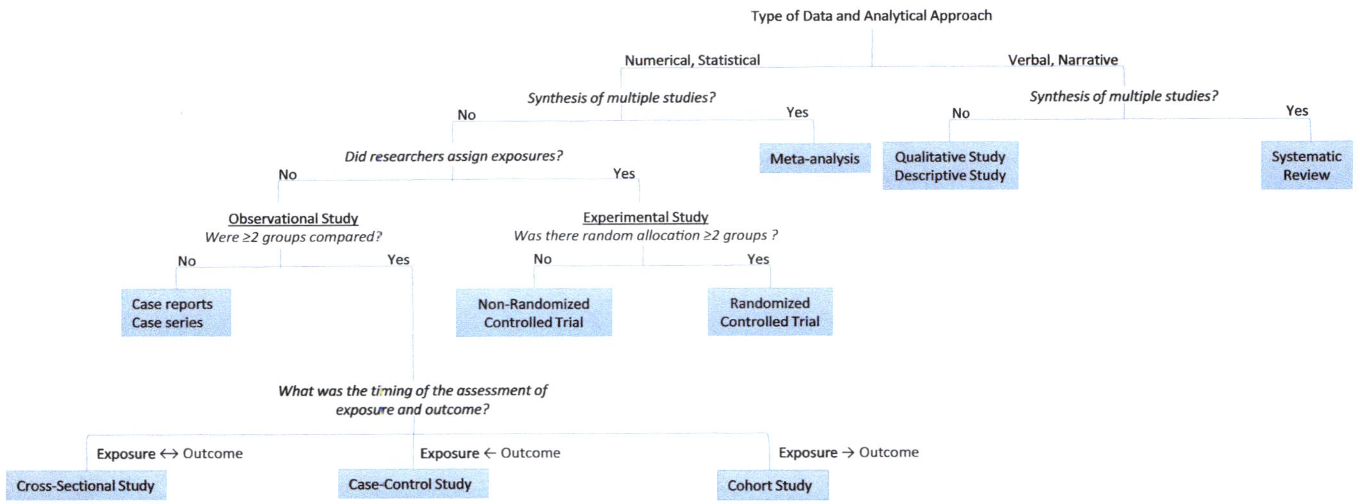

Fig. 3 Flow chart of study designs in physical activity and mental health research

References

Abt G, Boreham C, Davison G et al (2020) Power, precision, and sample size estimation in sport and exercise science research. J Sports Sci 38(17):1933–1935. https://doi.org/10.1080/02640414.2020.1776002

Ainsworth BE, Haskell WL, Herrmann SD et al (2011) 2011 Compendium of Physical Activities: a second update of codes and MET values. Med Sci Sports Exerc 43(8):1575–1581. https://doi.org/10.1249/MSS.0b013e31821ece12

Blumenthal JA, Babyak MA, Murali Doraiswamy P et al (2007) Exercise and Pharmacotherapy in the Treatment of Major Depressive Disorder. Psychosom Med 69(7):587–596. https://doi.org/10.1097/PSY.0b013e318148c19a

Büttner F, Toomey E, McClean S, Roe M, Delahunt E (2020) Are questionable research practices facilitating new discoveries in sport and exercise medicine? The proportion of supported hypotheses is implausibly high. Br J Sports Med 54(22):1365–1371. https://doi.org/10.1136/bjsports-2019-101863

Colley RC, Butler G, Garriguet D, Prince SA, Roberts KC (2018) Comparison of self-reported and accelerometer-measured physical activity in Canadian adults. Health Rep 29(12):3–15

Craig CL, Cameron C, Griffiths J, Bauman A, Tudor-Locke C, Andersen RE (2009) Non-response bias in physical activity trend estimates. BMC Public Health 9:425. https://doi.org/10.1186/1471-2458-9-425

Erickson KI, Voss MW, Prakash RS et al (2011) Exercise training increases size of hippocampus and improves memory. Proc Natl Acad Sci USA 108(7):3017–3022. https://doi.org/10.1073/pnas.1015950108

Gellish RL, Goslin BR, Olson RE, McDonald A, Russi GD, Moudgil VK (2007) Longitudinal modeling of the relationship between age and maximal heart rate. Med Sci Sports Exerc 39(5):822–829. https://doi.org/10.1097/mss.0b013e31803349c6

Harvey SB, Hotopf M, Overland S, Mykletun A (2010) Physical activity and common mental disorders. Br J Psychiatry 197(5):357–364. https://doi.org/10.1192/bjp.bp.109.075176. PMID: 21037212

Harvey SB, Øverland S, Hatch SL, Wessely S, Mykletun A, Hotopf M (2018) Exercise and the Prevention of Depression: results of the HUNT Cohort Study. AJP 175(1):28–36. https://doi.org/10.1176/appi.ajp.2017.16111223

Harvey SB, Hotopf M, Overland S, Mykletun A (2023) Physical activity and common mental disorders. Br J Psychiatry [Internet]. 2010 Nov [cited 2023 Aug 9];197(5). Available from: https://pubmed.ncbi.nlm.nih.gov/21037212/

Khan KS, Kunz R, Kleijnen J, Antes G (2003) Five steps to conducting a systematic review. J R Soc Med 96(3):118–121. https://doi.org/10.1177/014107680309600304

Kim Y, Beets MW, Welk GJ (2012) Everything you wanted to know about selecting the "right" Actigraph accelerometer cut-points for youth, but…: a systematic review. J Sci Med Sport 15(4):311–321. https://doi.org/10.1016/j.jsams.2011.12.001

Knaepen K, Goekint M, Heyman EM, Meeusen R (2010) Neuroplasticity - exercise-induced response of peripheral brain-derived neurotrophic factor: a systematic review of experimental studies in human subjects. Sports Med 40(9):765–801. https://doi.org/10.2165/11534530-000000000-00000

Lee D, Artero EG, Sui X, Blair SN (2010) Mortality trends in the general population: the importance of cardiorespiratory fitness. J Psychopharmacol 24(4 Suppl):27–35. https://doi.org/10.1177/1359786810382057

Lindheimer JB, O'Connor PJ, Dishman RK (2015) Quantifying the placebo effect in psychological outcomes of exercise training: a meta-analysis of randomized trials. Sports Med 45(5):693–711. https://doi.org/10.1007/s40279-015-0303-1

Mason OJ, Holt R (2012) Mental health and physical activity interventions: a review of the qualitative literature. J Ment Health 21(3):274–284. https://doi.org/10.3109/09638237.2011.648344

McKinlay JB, McKinlay SM (1977) The Questionable Contribution of Medical Measures to the Decline of Mortality in the United States in the Twentieth Century. Milbank Mem Fund Q Health Soc 55(3):405. https://doi.org/10.2307/3349539

Mein G, Johal S, Grant RL, Seale C, Ashcroft R, Tinker A (2012) Predictors of two forms of attrition in a longitudinal health study involving ageing participants: an analysis based on the Whitehall II study. BMC Med Res Methodol 12:164. https://doi.org/10.1186/1471-2288-12-164

Mesquida C, Murphy J, Lakens D, Warne J (2022) Replication concerns in sports and exercise science: a narrative review of selected methodological issues in the field. R Soc Open Sci 9 (12):220946. https://doi.org/10.1098/rsos.220946

Morris JN, Heady JA, Raffle PA, Roberts CG, Parks JW (1953) Coronary heart-disease and physical activity of work. Lancet 262(6795):1053–1057. https://doi.org/10.1016/s0140-6736 (53)90665-5

Morrison A, Polisena J, Husereau D et al (2012) The effect of English-language restriction on systematic review-based meta-analyses: a systematic review of empirical studies. Int J Technol Assess Health Care 28(2):138–144. https://doi.org/10.1017/S0266462312000086

O'Donovan G, Lee IM, Hamer M, Stamatakis E (2017) Association of "Weekend Warrior" and Other Leisure Time Physical Activity Patterns With Risks for All-Cause, Cardiovascular Disease, and Cancer Mortality. JAMA Intern Med 177(3):335–342. https://doi.org/10.1001/jamainternmed.2016.8014

Omran AR (2005) The Epidemiologic Transition: a Theory of the Epidemiology of Population Change. Milbank Q 83(4):731–757. https://doi.org/10.1111/j.1468-0009.2005.00398.x

Palinkas LA (2014) Qualitative and mixed methods in mental health services and implementation research. J Clin Child Adolesc Psychol 43(6):851–861. https://doi.org/10.1080/15374416.2014.910791

Pate RR, Pratt M, Blair SN et al (1995) Physical Activity and Public Health: a Recommendation From the Centers for Disease Control and Prevention and the American College of Sports Medicine. JAMA 273(5):402–407. https://doi.org/10.1001/jama.1995.03520290054029

Physical Activity Guidelines Advisory Committee (2018) *2018 Physical Activity Guidelines Advisory Committee Scientific Report*. U.S. Department of Health and Human Services, Washington, DC

Riebe D (2018) *ACSM's Guidelines for Exercise Testing and Prescription*, 10th edn. Wolters Kluwer

Ross R, Blair SN, Arena R et al (2016) Importance of Assessing Cardiorespiratory Fitness in Clinical Practice: a Case for Fitness as a Clinical Vital Sign: a Scientific Statement From the American Heart Association. Circulation 134(24):e653–e699. https://doi.org/10.1161/CIR.0000000000000461

Ross R, Goodpaster BH, Koch LG et al (2019) Precision exercise medicine: understanding exercise response variability. Br J Sports Med 53(18):1141–1153. https://doi.org/10.1136/bjsports-2018-100328

Scherr J, Wolfarth B, Christle JW, Pressler A, Wagenpfeil S, Halle M (2013) Associations between Borg's rating of perceived exertion and physiological measures of exercise intensity. Eur J Appl Physiol 113(1):147–155. https://doi.org/10.1007/s00421-012-2421-x

Sjostrom M, Ainsworth B, Bauman A, Bull F, Hamilton-Craig C, Sallis J (2005) Guidelines for data processing analysis of the International Physical Activity Questionnaire (IPAQ) - Short and long forms. Accessed June 16, 2023. https://www.semanticscholar.org/paper/Guidelines-for-data-processing-analysis-of-the-and-Sjostrom-Ainsworth/efb9575f5c957b73c640f00950982e618e31a7be

Stubbs B, Vancampfort D, Rosenbaum S et al (2017) An examination of the anxiolytic effects of exercise for people with anxiety and stress-related disorders: a meta-analysis. Psychiatry Res 249:102–108. https://doi.org/10.1016/j.psychres.2016.12.020

Tipton CM (2014) The history of "Exercise Is Medicine" in ancient civilizations. Adv Physiol Educ 38(2):109. https://doi.org/10.1152/advan.00136.2013

Tremblay MS, Esliger DW, Tremblay A, Colley R (2007) Incidental movement, lifestyle-embedded activity and sleep: new frontiers in physical activity assessment. Can J Public Health 98 (Suppl 2):S208–S217

U.S. Department of Health and Human Services (2018) *Physical Activity Guidelines for Americans*, 2nd edn. U.S. Department of Health and Human Services, Washington, DC

Watson KB, Carlson SA, Carroll DD, Fulton JE (2014) Comparison of accelerometer cut points to estimate physical activity in US adults. J Sports Sci 32(7):660–669. https://doi.org/10.1080/02640414.2013.847278

Welk, G. Morrow, J. Saint-Maurice, P. (2017) Measures Registry User Guide: individual Physical Activity. National Collaborative on Childhood Obesity Research, Washington (DC). January 2017. http://nccor.org/tools-mruserguides/wp-content/uploads/2017/ NCCOR_MR_User_Guide_Individual_PA-FINAL.pdf

Werneck AO, Vancampfort D, Oyeyemi AL, Szwarcwald CL, Stubbs B, Silva DR (2020) Lifestyle behaviors among 4,343 Brazilian adults with severe mental illness and 55,859 general population controls: data from the Brazilian National Health Survey. Br J Psychiatry 42(3):245–249. https://doi.org/10.1590/1516-4446-2019-0621

Applied Physical Activity Epidemiology: Relationship of Physical Activity and Exercise Exposures with Health Outcomes

Jessica (Jess) S. Gorzelitz

Contents

Abstract There is consistent, strong evidence that both physical activity and exercise lowers risk of many diseases, yet the way physical activity and exercise are measured varies substantially. Physical activity epidemiologists use observational designs (such as cohort or case-control studies) to examine the relationship of physical activity exposures and outcomes including mortality, or disease-specific such as incident type 2 diabetes mellitus, cancer, cardiovascular disease, anxiety, or depression. These same observational designs can be used to examine the associations of prevalent disease in reducing symptom burden in those who are physically active compared to those who are not, specifically examining anxiety and

J. (J). S. Gorzelitz (✉)
Department of Health and Human Physiology, College of Liberal Arts and Sciences, University of Iowa, Iowa City, IA, USA

Department of Obstetrics and Gynecology, Carver College of Medicine, Holden Comprehensive Cancer Center, University of Iowa, Iowa City, IA, USA
e-mail: jessica-gorzelitz@uiowa.edu

Curr Topics Behav Neurosci (2024) 67: 23–36
https://doi.org/10.1007/7854_2024_489
Published Online: 31 July 2024

depression. There is great statistical power and often large sample sizes in physical activity epidemiologic studies with valid and reliable assessment tools but known limitations such as measurement error or social desirability bias. Overall, physical activity epidemiology can be a useful tool to understand the influence of exercise and physical activity on risk of disease. In this chapter, we will explore the strength, consistency, and sources of evidence primarily using a physical activity epidemiology lens.

Keywords Exercise · Exposure · Mortality · Outcome · Physical activity · Physical activity epidemiology · Prevention · Treatment

1 Introduction

Understanding the association of physical activity with health outcomes is critical to understanding population health, exercise prescription, and how to use exercise in a therapeutic capacity. The discipline of physical activity epidemiology provides us with the ability to measure these associations in different populations and with various outcomes. Physical activity epidemiology allows for the population-level assessment of physical activity (including exercise) to understand the distribution and determinants of the behavior, and to understand the relationships between physical activity participation and disease/health outcomes. The previous chapter has provided an overview of this field including exposure and outcome assessment definitions. We will build on that introduction in this chapter to examine the influence of exercise and physical activity on risk of disease.

2 Mortality Outcomes

Death, or mortality as an outcome, is one of the critical outcomes in epidemiology. Mortality outcomes are often a strong endpoint as individuals will only experience this outcome once, which is both necessary and sufficient. Prospective cohort designs are ideal for assessing the association between exposures and mortality outcomes. Many cohort studies capture baseline or serial assessments of demographic and health behaviors over a period and follow individuals up for years or decades to assess who developed what outcomes. Mortality data can be collected from the National Death Index (NDI), Social Security Administration's death information, death certificates, or other registries. Depending on the data source, some mortality data also includes causes of death. Having specific attributable causes of death can be used analytically to go beyond all-cause mortality. Frequent mortality outcomes include cardiovascular disease (CVD)-specific deaths and cancer-related specific deaths. Given the population level of CVD and cancer, these outcomes are quite common with the high population burden of disease. Depending on the research question and the type of association to be studied with a given mortality outcome, physical activity (PA) epidemiology researchers can use previously collected exposures of physical activity to measure the associations. Through this chapter, we will discuss the relationship between PA and mortality, PA and physical disease outcomes, and PA and mental health outcomes. The chapter will finally describe benefits and limitations focused on observational designs, and ways to use and understand the physical activity epidemiologic literature.

2.1 *Observational Studies with Physical Activity and Mortality Risk*

When measuring mortality as an outcome, physical activity should be measured as the exposure which has occurred prior to the outcome of death. Some major cohort studies such as the Aerobic Center Longitudinal Study (ACLS) or the Nurses' Health Study have measured different constructs related to physical activity. ACLS has measured fitness, which is a consequence of exercise training, as the exposure to understand the relationship between fitness and mortality which is favorable (Barry et al. 2014). As discussed in subsequent sections, this fitness association has similarly been observed for all-cause, CVD, and cancer-specific mortality.

Larger cohort studies will often use global assessments of physical activity (Global Physical Activity Questionnaire-GPAQ, International Physical Activity Questionnaire-IPAQ) for population-level assessment. These instruments capture self-reported intensity and duration of typically aerobic physical activity. These exposures can then be expressed in mean hours per week of activity, or frequently aggregated into Metabolic Equivalents of Task (MET) levels per week. Many cohort studies have found a similar positive association between physical activity measures and all mortality outcomes (Brown and Gilmore 2019; Gorzelitz et al. 2022a, b; Hupin et al. 2015; Paffenbarger et al. 1993; Watts et al. 2022; Nocon et al. 2008; Blond et al. 2020; Friedenreich et al. 2020).

The National Health and Nutrition Examination Survey (NHANES) uses accelerometers (which provide a population-based device measurement of activity precision) and has also found similar associations between different PA exposures and mortality, with researchers finding accelerometer-measured physical activity associated with reduced mortality risk (Fishman et al. 2016; Schmid et al. 2015) as well as self-reported aerobic and muscle-strengthening exercises similarly associated with reduced mortality (Zhao et al. 2014; Navaneethan et al. 2014).

2.2 *Modality and Type of Physical Activity on Mortality Risk*

Most of physical activity epidemiology has historically focused on overall physical activity or aerobic exercise as an exposure, with limited interest and evidence on muscle-strengthening activities (Bennie et al. 2020). Per the 2018 Physical Activity Guidelines Advisory Committee (PAGAC) reports, all adults (including older adults above the age of 65) should participate in muscle-strengthening activities on two or more days per week (Committee PAGA 2018). Muscle-strengthening activities (MSA), including resistance exercise, strength training, weightlifting, and body weight exercises, made it into the guidelines based on the amount and quality of the evidence and health-promoting benefits of the activity. Comparatively, there is far less evidence on the associations of MSA on mortality outcomes often due to limited (if any) epidemiologic or survey assessments of MSA. Recently more research has emphasized the mortality benefits of MSA, finding positive mortality associations for weightlifting and all MSA following a J-shaped curve, suggesting that moderate levels of MSA are associated with reductions in risk but high levels of MSA are associated with an increased risk (Shailendra et al. 2022; Saeidifard et al. 2019). There is even new evidence of the unique and additive benefits of weightlifting in older adults, with researchers finding mortality reductions for weightlifting alone or in combination with aerobic exercise (Gorzelitz et al. 2022c).

3 Disease Outcomes

Like mortality as an outcome, disease outcomes can also be measured in prospective cohort designs. Observational study designs can be used to study the role of physical activity to determine if physical activity is associated with prevention of certain diseases, in this section using examples of cancer, diabetes, and heart disease. Additionally, observational designs commonly used in physical activity epidemiology can be used to determine if physical activity is associated with improved treatment or outcomes (treatment efficacy, tertiary prevention). This chapter will primarily focus on the associations of physical activity in preventing disease as primary prevention, with a brief mention of how physical activity can be used as treatment in populations already diagnosed with disease.

3.1 *Type 2 Diabetes Mellitus*

Type 2 diabetes mellitus (T2DM) is characterized by hyperglycemia, or high blood sugar, which is the primary clinical manifestation of this disease. T2DM incidence and prevalence rates are continuing to rise due to a multitude of factors (increasing age, poor diet, high sedentary behavior, low physical activity, other metabolic dysfunctions). The American Diabetes Association has published their position statement on physical activity/exercise and diabetes with recommendations for people with type 2 diabetes, in addition to less commonly examined outcomes, including T1DM, gestational diabetes mellitus, and prediabetes (Colberg et al. 2016).

LaMonte and colleagues have summarized the evidence on the role of physical activity and fitness in primary prevention of diabetes. In this narrative review, the authors summarize many primary literature studies and describe that the evidence base is only approximately 30 years old. Researchers report consistent observational literature from large cohort studies including the Iowa Women's Health Study, Nurses' Health Study, Women's Health Study, and the Women's Health Initiative Observational Study. These different studies have all yielded similar conclusions about the dose-response relationship between self-reported physical activity levels and lowered incident risk of T2DM (LaMonte et al. 2005). Other observational studies like the ACLS longitudinal cohort similarly found a significantly lowered risk of new T2DM with higher levels of cardiorespiratory fitness (LaMonte et al. 2005).

3.2 *Cancer*

Cancer is often considered a disease of the old, and as our population continues to age, we are expected to have more cancer diagnoses than ever before. One of the challenges with studying physical activity and cancer prevention is the latent and long-term etiology of the disease. Given that older adults are at higher risk of developing cancer, is important to study individuals at an age where they are cancer-free if wanting to measure the association of PA and risk of cancer. Not all cancers have similar etiology, which further complicates the study. For example, testicular cancer is most common in young, college-aged males where multiple myeloma is most common in adults above age 70 (Siegel et al. 2023). Therefore, when studying the association of PA with cancer prevention, most researchers will use large prospective study designs that enroll adults with long follow-up periods.

The American College of Sports Medicine convened an International Multidisciplinary Roundtable on Exercise and Cancer in March 2018, with one of the subsquent publications led by Dr. Alpa Patel. This expert roundtable conducted a thematic review and meta-analysis of previous studies reporting on the association of physical activity and risk of cancer. In this report, researchers were able to examine multiple cancer outcomes and concluded that physical activity was indeed associated with lowered risks of breast, colon, endometrial, kidney, bladder, esophageal, and stomach cancers. The one exception in that report was melanoma, a malignant form of skin cancer, which was found to have a positive risk association with physical activity (Patel et al. 2019). Since those who are physically active are more likely to do outdoor activities in sunlight such as outdoor biking, running, walking, or other outdoor recreation, it is speculated that sun exposure during active time is responsible for this positive risk association. This seminal report provides strong evidence for the role of physical activity in preventing certain types of cancer, however this report is limited in the inferences able to be made. These studies provide evidence on the association and not the causal relationship between physical activity and cancer, especially knowing that there are highly heterogenous etiologies of different types of cancer. One of the criteria for causality is biological plausibility, and each of the seven identified sites has different biological pathways by which physical activity could influence the etiologic pathways of carcinogenesis. It is important to note that the goal of these types of studies is not to establish a causal relationship or to determine the biological pathway, but rather to summarize the evidence and inform public health and healthcare-related recommendations for individuals to lower their risk of cancer.

For adults with a cancer diagnosis, there is similar evidence on the benefits of physical activity and exercise in improving eight commonly reported symptoms by cancer survivors including low health-related quality of life, cancer-related fatigue, anxiety, depression, lymphedema, low bone quality, poor physical function, and sleep impairments (Campbell et al. 2019).

3.3 Cardiovascular Disease

Exercise and physical activity are both established preventive factors for atherosclerotic cardiovascular disease. A statement from the Council on Clinical Cardiology and the Council on Nutrition, Physical Activity, and Metabolism summarizes the evidence and provides a position statement firmly in favor of both exercise and physical activity behaviors (Thompson et al. 2003). Similar to the evidence bases for diabetes and cancer, many volumes of observational research have consistently shown an inverse protective relationship between physical activity and incident cardiovascular disease (Arnett et al. 2019; Archer and Blair 2011; Haennel and Lemire 2002; Alves et al. 2016; Wahid et al. 2016). While there has been strong mechanistic literature to understand the etiologic process of the disease in terms of exactly how exercise and physical activity impacts the vascular, observational research allows for comparisons of active vs inactive individuals to provide population-level inferences of the preventive benefit for physical activity (Lee et al. 1997; Blair and Jackson 2001). Similar research has been done for fitness, revealing the highest levels of cardiorespiratory fitness associated with lowered risk of cardiovascular disease (Al-Mallah et al. 2018; Ozemek et al. 2018). Recent Cochrane systematic reviews have revealed the benefits of cardiac rehabilitation, a form of structured exercise following a CVD incident or diagnosis, to be beneficial for assisting with treatment and lowered mortality in coronary heart disease ((Dibben et al. 2021) as well for adults with heart failure (Long et al. 2019).

4 Mental Health Outcomes: Anxiety and Depression

Mental health outcomes can similarly be studied using physical activity epidemiology approaches to examine the association of PA with prevention of mental health issues or with the efficacy of treatment for those who have a diagnosed condition. Approximately 40 million adults (~19%) in the United States experience anxiety disorders each year, with an estimated 31% of US adults experiencing an anxiety disorder at some point in their lives (Kessler et al. 2004). Women have been more likely to experience an anxiety disorder than men, however these estimates are imperfect and perhaps subject to bias for reporting (Beesdo et al. 2010). Estimates for major depressive disorder reveal that 6.7% of adults (8% of women and 5% of men) had at least one major depressive episode in the past year impacting an estimated 17 million US adults. Approximately 20% of all adults are likely to experience at least one bout of depression in their lifetime (Kessler et al. 2004). Estimates for both anxiety and depression are measured at the population level, which, like all other study tools, have known limitations in their measurement abilities. These limitations are described at the end of the chapter.

4.1 Physical Activity as Primary Prevention

Studies examining relationships of physical activity and exercise on disease outcomes will often assume a person is disease-free at exposure assessment and then assess later risk of developing an outcome. Studying those who are free from disease without any indications of clinical precursors on reducing risk of developing the outcome is primary prevention.

4.1.1 Anxiety

In prospective cohort studies, where participants are free from disease at baseline, physical activity has been associated with lower prevalence of major anxiety disorders. Data from the National Comorbidity Survey have revealed that those who report regular physical activity have lower adjusted odds of agoraphobia (36–48% lower), social phobia (35–42% lower), specific phobias (22–32% lower), generalized anxiety disorder (24–42% lower), and panic attacks (27–41% lower) (Goodwin 2003). Much work has been published addressing the potential mechanisms of this relationship elsewhere, and that is beyond the scope of this chapter. This association of lowered risk with physical activity and risk of incident anxiety symptoms/disorder is consistent (see Chapter “Traumatic Psychedelic Experiences”).

4.1.2 Depression

Just as with anxiety, so too can PA epidemiology designs help us to understand the relationship between physical activity and the prevention of depression. In a systematic review and meta-analysis of over 111 reports representing over three million adults, researchers found that overall, there was 25–37% lower odds of depressive symptoms among physically active individuals compared to physically inactive individuals. After adjustment for potential confounders, there was still 18–25% lower odds of depressive symptoms in active individuals, when compared to inactive individuals (Dishman et al. 2021). Of note, this meta-analysis included original studies of adolescents, adults and older adults form 11 countries in five continents, providing strong evidence for the role of physical activity in preventing depression.

In addition to the measures of self-reported PA levels as a measure of activity, researchers can also test the consequences of regular physical activity and exercise by using measures of fitness in association with mental health outcomes. Researchers were able to conduct a prospective study measuring cardiorespiratory fitness and depressive symptoms over time. Participants had their fitness measured at enrollment and incidence rates of new depressive symptoms were measured over time. Researchers found a risk reduction of 31% in men and 44% in women for those who had moderate levels of cardiorespiratory fitness. For those with the highest levels of cardiorespiratory fitness, the risk reduction was 51% in men and 54% in women. This prospective study combines some of the strongest and objective measures of fitness coupled with the prospective design allowing for the temporal association to see which came first – the inactivity or the depression (Sui et al. 2009). Overall, the evidence base is quite extensive illustrating the role of exercise in preventing depression (see Chapter "Management of 'Challenging Experiences' and 'Bad Trips' in Psychedelic-assisted Therapy and Recreational Use").

4.2 Physical Activity as Treatment

Oftentimes physical activity epidemiology designs allow for a simple exposure outcome relationship with one measure of exposure. There are examples of certain studies and designs that have used repeated measures or longitudinal designs to examine how changes in physical activity or exercise behaviors are associated with changes or reductions to disease status or progression.

4.2.1 Mortality

The Harvard Alumni Study began as a prospective cohort study that has since included multiple longitudinal assessments of participants, making it an ideal data source to examine changes in physical activity over time on disease risk.

Paffenbarger et al. in 1993 found that men who increased their moderate to vigorous intensity physical activity (MVPA) was associated with a 23% lowered risk of mortality, and for those men who maintained their already high levels of MVPA was also associated with a 41% lowered risk of CVD-specific mortality (Paffenbarger et al. 1993). When comparing these similar changes in physical activity in women, Gregg et al. found that in a sample of ~9,000 women aged 65 years or older, those who increased their physical activity over time had 48% lower risk of all-cause mortality, and those women who were active and remained active had a 36% lowered risk of all-cause mortality (Gregg et al. 2003).

Fitness, whether cardiovascular or muscular, is often the consequence of training or repeated, intentional exposures to exercise to improve physiological function. The Aerobic Center Longitudinal Study is one of the unique epidemiologic studies as researchers there will measure participant fitness as an exposure to examine outcomes over time. With these repeated exposures over time, researchers can quantify the effect of changing one's fitness on their risk of disease. One seminal contribution was Blair et al. (1995) publication examining over 9,000 men aged 20–82 at baseline with fitness tests and 5 years of follow-up to examine the role of changing fitness profiles on mortality. Men who were initially categorized as unfit at baseline, but then became fit at the five-year follow-up had a 44% reduction in mortality. Men who were fit at both assessments had a 67% lowered risk of mortality, and those men who were fit initially but still improved their fitness measures at follow-up had a 28% lower risk of mortality when compared to men who went from unfit to fit (Blair et al. 1995).

Early studies in physical activity epidemiology were almost exclusively focused on men, which is certainly not the whole picture nor is it generalizable to human health. Additional studies have replicated these findings in women, older adults, and individuals with racial and ethnic diversity (Gregg et al. 2003; Mok et al. 2019; Talbot et al. 2007; Imboden et al. 2019; Braun 1970).

4.2.2 Anxiety

Understanding the role of PA for treatment in those who have already been diagnosed with anxiety is similarly consistent. The 2018 Physical Activity Guidelines for Americans state that there is strong evidence that physical activity reduces anxiety symptoms in those with anxiety disorders (Committee PAGA 2018). Reviews have varied in conclusions specifically on whether exercise is either better than or no different from or not as effective as other treatments (see Chapter "Traumatic Psychedelic Experiences" for a comprehensive overview).

4.2.3 Depression

While physical activity is strongly associated with the prevention of depression, less is known about the role of exercise in treating those with clinical depression. A meta-analysis of 14 randomized controlled trials found that chronic exercise reduced depression symptoms by 1.1 standard deviations. This meta-analysis included studies with participants as having depression diagnosed via any method and also included any severity of depression, but notably most of the studies found in this review enrolled non-clinical participants which limits generalizability of findings (Lawlor and Hopker 2001). More recent reviews of reviews have similarly concluded that exercise can appropriately be recommended for depression in combination with other treatments (Daley 2008). Newer randomized studies have been conducted to examine the dose and viability of physical activity as a treatment for depression. The DOSE study randomly assigned participants aged 20–45 to one of four exercise treatments or to an exercise placebo control of stretching and flexibility. Exercise treatments varied by energy expenditure (low dose, public health dose) and frequency (3 vs. 5 days per week). Researchers found a 47% reduction in depressive symptoms following 12 weeks of exercise for those in the public health dose group (independent of weekly frequency) (Dunn et al. 2005). The SMILE study was a prospective, randomized controlled trial with 202 adults. Individuals were randomly assigned to one of four conditions including supervised group exercise, home-based exercise, anti-depressant medication, and placebo group. Researchers found that the aerobic exercise groups were as effective as drug treatments for those adults aged 40+ with diagnosed major depressive disorder (Blumenthal et al. 2021). Finally, the TREAD study found that exercise was a viable treatment for adults aged 18–75 with MDD that did not respond to drug treatment (Trivedi et al. 2011), leading to clinical guidelines for prescription and dosing of exercise/PA as treatment for adults with MDD (Rethorst and Trivedi 2013). These results from multiple, high quality randomized controlled trials suggest that exercise may be a viable alternative treatment for Major Depressive Disorder (see Chapter "Management of 'Challenging Experiences' and 'Bad Trips' in Psychedelic-assisted Therapy and Recreational Use" for a comprehensive overview).

5 Tradeoffs and Limitations of Observational Designs in Physical Activity Epidemiology

While there are many benefits to using large population-based research and PA epidemiology methods to assess the aforementioned relationships between PA and various health outcomes (e.g., mortality, cancer, CVD, T2D, depression, anxiety), there are also known limitations. Most large, population-based cohort studies will rely on self-report for these outcomes. Self-report is the most efficient and least invasive (with respect to privacy) for the participant; however, it is prone to measurement error including artifacts of social desirability bias and participant outcomes expectations (Brenner and DeLamater 2014; Motl et al. 2005).

Furthermore, when assessing mental health conditions, it's incredibly important to use validated, reliable tools for the given population. This is a key research design consideration that will pay dividends for many individuals. In addition to having a validated assessment tool, it is important to determine if the outcome is symptoms of a condition or a diagnosis of the condition. The former can be captured on tools like the Center for Epidemiologic Studies Depression Scale (CES-D); however, the latter will require confirmation on the part of the participant that indeed there is a diagnosis. Some researchers will ask about certain medications to confirm or assess the severity of illness in that it may require medication treatment. Furthermore, anxiety and depression are often recurring or chronic conditions, as compared to heart disease or diabetes, and thus the temporal sequence and relationship can be challenging. Oftentimes researchers will want a hard timeline of the relationship between these components, and that is simply very challenging to do given the nature of these conditions.

For research contexts, it is important to consider symptoms (vs. diagnoses); treated vs. untreated (medication/CBT/therapy/combination); assessment timing (relative to symptom burden); and finally, state vs. trait anxiety or depression. These considerations will be more fully explored in future chapters of this textbook (see Chapters "Classification and Differential Diagnoses of Psychedelic-related Crises and Disorders" and "Traumatic Psychedelic Experiences"), specifically exploring experimental designs.

Finally, a major limitation of this type of epidemiologic research is the population reach. Individuals who volunteer for and participate in this type of research are often times WEIRD: Western, Educated, Industrialized, Rich, and Democratic (WEIRD) (Schulz et al. 2018; Hendriks et al. 2019). While there is nothing inherently wrong with any of these characteristics, these attributes do not represent the full entirety of our populations (Muthukrishna et al. 2020). Furthermore, research studies are often prone to the healthy adherer effect and any given population we study will be inherently healthier than those who do not agree to participate in exercise studies (Shrank et al. 2011; Ladova et al. 2014). Any of our inferences for any study are only as good as those who are recruited, we cannot extrapolate beyond the sample characteristics.

6 Conclusion

There are known tradeoffs when using epidemiologic designs versus experimental studies. Observational designs generally allow for larger sample sizes, and often can include longer-term follow-up with longitudinal or prospective designs. These large sample sizes will often require certain types of measurement tools, such as self-report or standardized, validated questionnaires. These tools have limitations including measurement error, recall bias in certain contexts, and social desirability bias. These are important considerations depending on the disease or outcome studied. Historically there has been a stigma against certain conditions including certain mental health conditions, some of these stigmas may be persistent in certain populations or age groups. Subsequent chapters will address experimental designs and more clinical utility for assessing these conditions and more. Overall, physical activity epidemiology can be a useful tool to understand the influence of exercise and physical activity on risk of disease.

References

Al-Mallah MH, Sakr S, Al-Qunaibet A (2018) Cardiorespiratory fitness and cardiovascular disease prevention: an update. Curr Atheroscler Rep 20:1–9

Alves AJ, Viana JL, Cavalcante SL, Oliveira NL, Duarte JA, Mota J et al (2016) Physical activity in primary and secondary prevention of cardiovascular disease: overview updated. World J Cardiol 8(10):575

Archer E, Blair SN (2011) Physical activity and the prevention of cardiovascular disease: from evolution to epidemiology. Prog Cardiovasc Dis 53(6):387–396

Arnett DK, Blumenthal RS, Albert MA, Buroker AB, Goldberger ZD, Hahn EJ et al (2019) 2019 ACC/AHA guideline on the primary prevention of cardiovascular disease: executive summary: a report of the American College of Cardiology/American Heart Association Task Force on Clinical Practice Guidelines. Circulation 140(11):e563–ee95

Barry VW, Baruth M, Beets MW, Durstine JL, Liu J, Blair SN (2014) Fitness vs. fatness on all-cause mortality: a meta-analysis. Prog Cardiovasc Dis 56(4):382–390

Beesdo K, Pine DS, Lieb R, Wittchen H-U (2010) Incidence and risk patterns of anxiety and depressive disorders and categorization of generalized anxiety disorder. Arch Gen Psychiatry 67(1):47–57

Bennie JA, Shakespear-Druery J, De Cocker K (2020) Muscle-strengthening exercise epidemiology: a new frontier in chronic disease prevention. Sports Med-Open 6(1):1–8

Blair SN, Jackson AS (2001) Physical fitness and activity as separate heart disease risk factors: a meta-analysis. pp 762–764

Blair SN, Kohl HW, Barlow CE, Paffenbarger RS, Gibbons LW, Macera CA (1995) Changes in physical fitness and all-cause mortality. JAMA 273(14):1093–1098

Blond K, Brinkløv CF, Ried-Larsen M, Crippa A, Grøntved A (2020) Association of high amounts of physical activity with mortality risk: a systematic review and meta-analysis. Br J Sports Med 54(20):1195–1201

Blumenthal JA, Babyak MA, Craighead WE, Davidson J, Hinderliter A, Hoffman B et al (2021) The role of comorbid anxiety in exercise and depression trials: secondary analysis of the SMILE-II randomized clinical trial. Depress Anxiety 38(2):124–133

Braun L (1999) Cardiorespiratory fitness, body mass index and all-cause mortality in women, ACLS 1970-1994

Brenner PS, DeLamater JD (2014) Social desirability bias in self-reports of physical activity: is an exercise identity the culprit? Soc Indic Res 117:489–504

Brown JC, Gilmore LA (2019) Physical activity reduces the risk of recurrence and mortality in cancer patients. Exerc Sport Sci Rev

Campbell KL, Winters-Stone KM, Wiskemann J, May AM, Schwartz AL, Courneya KS et al (2019) Exercise guidelines for cancer survivors: consensus statement from international multidisciplinary roundtable. Med Sci Sports Exerc 51(11):2375–2390

Colberg SR, Sigal RJ, Yardley JE, Riddell MC, Dunstan DW, Dempsey PC et al (2016) Physical activity/exercise and diabetes: a position statement of the American Diabetes Association. Diabetes Care 39(11):2065–2079

Committee PAGA (2018) Physical activity guidelines advisory committee scientific report. US Department of Health and Human Services, Washington, pp F2–F33

Daley A (2008) Exercise and depression: a review of reviews. J Clin Psychol Med Settings 15:140–147

Dibben G, Faulkner J, Oldridge N, Rees K, Thompson DR, Zwisler A-D et al (2021) Exercise-based cardiac rehabilitation for coronary heart disease. Cochrane Database Syst Rev 11

Dishman RK, McDowell CP, Herring MP (2021) Customary physical activity and odds of depression: a systematic review and meta-analysis of 111 prospective cohort studies. Br J Sports Med 55(16):926–934

Dunn AL, Trivedi MH, Kampert JB, Clark CG, Chambliss HO (2005) Exercise treatment for depression: efficacy and dose response. Am J Prev Med 28(1):1–8

Fishman EI, Steeves JA, Zipunnikov V, Koster A, Berrigan D, Harris TA et al (2016) Association between objectively measured physical activity and mortality in NHANES. Med Sci Sports Exerc 48(7):1303

Friedenreich CM, Stone CR, Cheung WY, Hayes SC (2020) Physical activity and mortality in cancer survivors: a systematic review and meta-analysis. JNCI Cancer Spectr 4(1):pkz080

Goodwin RD (2003) Association between physical activity and mental disorders among adults in the United States. Prev Med 36(6):698–703

Gorzelitz J, Trabert B, Katki HA, Moore SC, Watts EL, Matthews CE (2022a) Independent and joint associations of weightlifting and aerobic activity with all-cause, cardiovascular disease and cancer mortality in the prostate, lung, colorectal and ovarian cancer screening trial. Br J Sports Med 56(22):1277–1283

Gorzelitz JS, Trentham Dietz A, Hampton JM, Spencer RJ, Costanzo E, Koltyn K et al (2022b) Mortality risk and physical activity across the lifespan in endometrial cancer survivors. Cancer Causes Control 33(3):455–461

Gorzelitz J, Trabert B, Katki HA, Moore SC, Watts EL, Matthews CE (2022c) Independent and joint associations of weightlifting and aerobic activity with all-cause, cardiovascular disease and cancer mortality in the prostate, lung, colorectal and ovarian cancer screening trial. Br J Sports Med:bjsports-2021-105315

Gregg EW, Cauley JA, Stone K, Thompson TJ, Bauer DC, Cummings SR et al (2003) Relationship of changes in physical activity and mortality among older women. JAMA 289(18):2379–2386

Haennel RG, Lemire F (2002) Physical activity to prevent cardiovascular disease. How much is enough? Can Fam Physician 48(1):65–71

Hendriks T, Warren MA, Schotanus-Dijkstra M, Hassankhan A, Graafsma T, Bohlmeijer E et al (2019) How WEIRD are positive psychology interventions? A bibliometric analysis of randomized controlled trials on the science of well-being. J Posit Psychol 14(4):489–501

Hupin D, Roche F, Gremeaux V, Chatard J-C, Oriol M, Gaspoz J-M et al (2015) Even a low-dose of moderate-to-vigorous physical activity reduces mortality by 22% in adults aged≥60 years: a systematic review and meta-analysis. Br J Sports Med:bjsports-2014

Imboden MT, Harber MP, Whaley MH, Finch WH, Bishop DL, Fleenor BS et al (2019) The association between the change in directly measured cardiorespiratory fitness across time and mortality risk. Prog Cardiovasc Dis 62(2):157–162

Kessler RC, Berglund P, Chiu WT, Demler O, Heeringa S, Hiripi E et al (2004) The US National Comorbidity Survey Replication (NCS-R): design and field procedures. Int J Methods Psychiatr Res 13(2):69–92

Ladova K, Vlcek J, Vytrisalova M, Maly J (2014) Healthy adherer effect – the pitfall in the interpretation of the effect of medication adherence on health outcomes. J Eval Clin Pract 20(2):111–116

LaMonte MJ, Blair SN, Church TS (2005) Physical activity and diabetes prevention. J Appl Physiol 99(3):1205–1213

Lawlor DA, Hopker SW (2001) The effectiveness of exercise as an intervention in the management of depression: systematic review and meta-regression analysis of randomised controlled trials. BMJ 322(7289):763

Lee I, Paffenbarger R, Hennekens C (1997) Physical activity, physical fitness and longevity. Aging Clin Exp Res 9:2–11

Long L, Mordi IR, Bridges C, Sagar VA, Davies EJ, Coats AJ et al (2019) Exercise-based cardiac rehabilitation for adults with heart failure. Cochrane Database Syst Rev 1

Mok A, Khaw K-T, Luben R, Wareham N, Brage S (2019) Physical activity trajectories and mortality: population based cohort study. BMJ:365

Motl RW, McAuley E, DiStefano C (2005) Is social desirability associated with self-reported physical activity? Prev Med 40(6):735–739

Muthukrishna M, Bell AV, Henrich J, Curtin CM, Gedranovich A, McInerney J et al (2020) Beyond western, educated, industrial, rich, and democratic (WEIRD) psychology: measuring and mapping scales of cultural and psychological distance. Psychol Sci 31(6):678–701

Navaneethan SD, Kirwan JP, Arrigain S, Schold JD (2014) Adiposity measures, lean body mass, physical activity and mortality: NHANES 1999–2004. BMC Nephrol 15(1):1–9
Nocon M, Hiemann T, Müller-Riemenschneider F, Thalau F, Roll S, Willich SN (2008) Association of physical activity with all-cause and cardiovascular mortality: a systematic review and meta-analysis. Eur J Prev Cardiol 15(3):239–246
Ozemek C, Laddu DR, Lavie CJ, Claeys H, Kaminsky LA, Ross R et al (2018) An update on the role of cardiorespiratory fitness, structured exercise and lifestyle physical activity in preventing cardiovascular disease and health risk. Prog Cardiovasc Dis 61(5–6):484–490
Paffenbarger RS Jr, Hyde RT, Wing AL, Lee IM, Jung DL, Kampert JB (1993) The association of changes in physical-activity level and other lifestyle characteristics with mortality among men. N Engl J Med 328(8):538–545
Patel AV, Friedenreich CM, Moore SC, Hayes SC, Silver JK, Campbell KL et al (2019) American College of Sports Medicine roundtable report on physical activity, sedentary behavior, and cancer prevention and control. Med Sci Sports Exerc 51(11):2391
Rethorst CD, Trivedi MH (2013) Evidence-based recommendations for the prescription of exercise for major depressive disorder. J Psychiatr Pract 19(3):204–212
Saeidifard F, Medina-Inojosa JR, West CP, Olson TP, Somers VK, Bonikowske AR et al (2019) The association of resistance training with mortality: a systematic review and meta-analysis. Eur J Prev Cardiol 26(15):1647–1665
Schmid D, Ricci C, Leitzmann MF (2015) Associations of objectively assessed physical activity and sedentary time with all-cause mortality in US adults: the NHANES study. PLoS One 10(3): e0119591
Schulz J, Bahrami-Rad D, Beauchamp J, Henrich J (2018) The origins of WEIRD psychology. Available at SSRN 3201031
Shailendra P, Baldock KL, Li LK, Bennie JA, Boyle T (2022) Resistance training and mortality risk: a systematic review and meta-analysis. Am J Prev Med
Shrank WH, Patrick AR, Alan BM (2011) Healthy user and related biases in observational studies of preventive interventions: a primer for physicians. J Gen Intern Med 26:546–550
Siegel RL, Miller KD, Wagle NS, Jemal A (2023) Cancer statistics, 2023. CA Cancer J Clin 73(1): 17–48
Sui X, Laditka JN, Church TS, Hardin JW, Chase N, Davis K et al (2009) Prospective study of cardiorespiratory fitness and depressive symptoms in women and men. J Psychiatr Res 43(5): 546–552
Talbot LA, Morrell CH, Fleg JL, Metter EJ (2007) Changes in leisure time physical activity and risk of all-cause mortality in men and women: the Baltimore Longitudinal Study of Aging. Prev Med 45(2–3):169–176
Thompson PD, Buchner D, Piña IL, Balady GJ, Williams MA, Marcus BH et al (2003) Exercise and physical activity in the prevention and treatment of atherosclerotic cardiovascular disease: a statement from the Council on Clinical Cardiology (Subcommittee on Exercise, Rehabilitation, and Prevention) and the Council on Nutrition, Physical Activity, and Metabolism (Subcommittee on Physical Activity). Circulation 107(24):3109–3116
Trivedi MH, Greer TL, Church TS, Carmody TJ, Grannemann BD, Galper DI et al (2011) Exercise as an augmentation treatment for nonremitted major depressive disorder: a randomized, parallel dose comparison. J Clin Psychiatry 72(5):6911
Wahid A, Manek N, Nichols M, Kelly P, Foster C, Webster P et al (2016) Quantifying the association between physical activity and cardiovascular disease and diabetes: a systematic review and meta-analysis. J Am Heart Assoc 5(9):e002495
Watts EL, Matthews CE, Freeman JR, Gorzelitz JS, Hong HG, Liao LM et al (2022) Association of leisure time physical activity types and risks of all-cause, cardiovascular, and cancer mortality among older adults. JAMA Netw Open 5(8):e2228510-e
Zhao G, Li C, Ford ES, Fulton JE, Carlson SA, Okoro CA et al (2014) Leisure-time aerobic physical activity, muscle-strengthening activity and mortality risks among US adults: the NHANES linked mortality study. Br J Sports Med 48(3):244–249

Multiple Sex- and Circuit-Specific Mechanisms Underlie Exercise-Induced Stress Resistance

Margaret K. Tanner, Simone M. Mellert, Isabella P. Fallon, Michael V. Baratta, and Benjamin N. Greenwood

Contents

M. K. Tanner and B. N. Greenwood (✉)
Department of Psychology, University of Colorado Denver, Denver, CO, USA
e-mail: Benjamin.greenwood@ucdenver.edu

S. M. Mellert
Department of Integrative Biology, University of Colorado Denver, Denver, CO, USA

I. P. Fallon
Department of Neurobiology, Duke University School of Medicine, Durham, NC, USA

M. V. Baratta
Department of Psychology and Neuroscience, University of Colorado Boulder, Boulder, CO, USA

Curr Topics Behav Neurosci (2024) 67: 37–60
https://doi.org/10.1007/7854_2024_490
Published Online: 31 July 2024

Abstract Prior physical activity reduces the risk of future stress-related mental health disorders including depression, anxiety, and post-traumatic stress disorder. Rodents allowed to engage in voluntary wheel running are similarly protected from behavioral consequences of stress. The present review summarizes current knowledge on mechanisms underlying exercise-induced stress resistance. A conceptual framework involving the development (during exercise) and expression (during stress) of stress resistance from exercise is proposed. During the development of stress resistance, adaptations involving multiple exercise signals and molecular mediators occur within neural circuits orchestrating various components of the stress response, which then respond differently to stress during the expression of stress resistance. Recent data indicate that the development and expression of stress resistance from exercise involve multiple independent mechanisms that depend on sex, stressor severity, and behavioral outcome. Recent insight into the role of the prefrontal cortex in exercise-induced stress resistance illustrates these multiple mechanisms. This knowledge has important implications for the design of future experiments aimed at identifying the mechanisms underlying exercise-induced stress resistance.

Keywords Anxiety · Physical activity · Serotonin · Sex differences · Wheel running

1 Introduction

Exposure to stress is a primary contributing factor to the development of mental health disorders such as depression, anxiety, and post-traumatic stress disorder. These stress-related mental health disorders are the most common mental health disorders in the world, accounting for 7% of the global burden of disease and 19% of all years lived with disability (Rehm and Shield 2019). In 2019, depression disorders were the second and anxiety disorders the eighth leading contributor to years lived with disability (Ferrari et al. 2022). Understanding factors that can facilitate the prevention (stress resistance) and recovery from (stress resilience) the effects of stress on mental health is thus of utmost importance.

Stress resistance and resilience factors include experiential factors over which we have some degree of control. One experiential factor that both enables protection against, and facilitates recovery from, stress-related disorders is physical activity. Indeed, sedentary behavior is recognized as a leading risk factor for development of future stress-related mental health disorders, whereas a history of physical activity reduces this risk (Chekroud et al. 2018; Laird et al. 2023; Harvey et al. 2018). Additionally, physical activity, alone or in combination with traditional behavioral or pharmacological treatments, can aid in the recovery from stress-related disorders (Heissel et al. 2023; Hegberg et al. 2019; Bjorkman and Ekblom 2022; Stonerock et al. 2015). A handful of pharmaceutical treatment options with varying degrees of efficacy are employed for the treatment of existing stress-related disorders (Cipriani et al. 2018). However, no drugs are currently available to reduce the risk of development of future stress-related mental health disorders. Therefore, the ability of physical activity to provide protection against the future onset of these disorders (here simply referred to as stress resistance) is of particular importance.

Understanding the features of, and mechanisms underlying, exercise-induced stress resistance could potentially inform strategies to maximize the stress-protective potential of physical activity and reveal circuit and/or molecular targets for novel prophylactic pharmaceuticals (*exercise mimetics*). With these objectives in mind, this review discusses current knowledge and issues pertaining to the mechanisms by which physical activity enables stress resistance. First, we discuss the understudied aspect of sex differences in exercise-induced stress resistance and the intriguing possibility that females are more responsive than males to stress resistance from exercise. Next, we present a conceptual framework for the study of exercise-induced stress resistance which includes the development (during exercise) and expression (during stress) of stress resistance. Applying this framework to our model system of acute inescapable stress clarifies the involvement of exercise signals, molecular mediators, and circuit adaptations to exercise-induced stress resistance. Finally, recent insights into sex- and circuit-determinants of exercise-induced stress resistance support the notion that exercise recruits multiple, independent, stress resistance mechanisms which act in parallel to enable protection from stress. These topics are presented with an eye toward limitations of current experimental approaches and ideas to guide future research.

2 Sex Differences in Exercise-Induced Stress Resistance

Women are more likely than men to suffer from stress-related mental health disorders (Pavlidi et al. 2023; Steel et al. 2014). Additionally, some stress resistance factors (Fallon et al. 2020) and pharmacological treatment strategies (Bigos et al. 2009) are reported to work better in males compared to females. Given the recent emphasis on prevention (Insel and Scolnick 2006; Simmons et al. 2021), it is particularly relevant that physical activity is a readily available and inexpensive stress prophylactic strategy for both sexes. Understanding potential sex differences in exercise-induced stress resistance could inform the implementation of physical activity programs designed to optimize benefits for a particular sex.

2.1 Limitations of Clinical Studies on Exercise-Induced Stress Resistance

Longitudinal studies including both female and male subjects indicate that exercise can reduce the risk for development of future stress-related disorders in both sexes.

For example, Pearce and colleagues recently estimated that achieving the recommended minimum amount of weekly physical activity reduces the risk of developing future depression in both sexes by an impressive 25% (Pearce et al. 2022). However, the duration, frequency, or intensity of physical activity required to enable stress resistance, or the magnitude of the protective effect of exercise against a stressor of a given severity, could differ between the sexes in subtle ways that are difficult to detect in longitudinal studies. Intervention studies that assign groups of people to exercise and stress conditions could, in theory, be designed to detect potential sex differences. Clinical intervention studies, however, are limited to the use of relatively mild stressors, since it is unethical to expose people to the types of stressors that lead to stress-related mental health disorders. Characteristics of resistance to the effects of mild stress could differ from those of severe stress, so it cannot be assumed that results of intervention studies using mild stressors resemble what would be found in response to severe stress. Intervention studies can more readily be designed to detect potential sex differences in therapeutic effects of exercise, by assigning groups of humans with stress-related mental health disorders to exercise groups. However, mechanisms underlying therapeutic exercise are not necessarily the same as those of prophylactic exercise. These are some of the reasons why our understanding of sex differences in exercise-induced stress resistance remains limited.

2.2 Considerations for the Use of Pre-Clinical Models to Study Sex Differences in Exercise-Induced Stress Resistance

Given the limitations of clinical studies, preclinical models could be particularly useful in the study of sex differences in stress resistance. Preclinical models allow groups of rodents to be assigned to sedentary or exercise conditions, followed by exposure to no stress or a laboratory stressor resulting in behaviors resembling symptoms of stress-related mental health disorders. Acute, rather than chronic, stress models are advantageous here, as the exercise intervention can be precisely timed to occur prior to stress administration. It is difficult to differentiate effects of physical activity on stress resistance vs. recovery in studies that allow animals to continue exercising during chronic stress exposure.

There are various ways, both voluntary and involuntary, to increase the physical activity status of rodents. Since removing the control over aspects of one's environment is aversive (Leotti et al. 2010), chronic stress produced by forced exercise could overshadow potential stress-protective effects. Voluntary physical activity paradigms are thus most appropriate. Voluntary physical activity paradigms are easily employed in the laboratory, as rodents demonstrate robust spontaneous wheel running behavior on wheels placed in their cages (voluntary wheel running; VWR). An additional benefit of preclinical studies is that they allow the measurement and manipulation of factors potentially involved in exercise-induced stress resistance, and so are useful for identification of underlying mechanisms.

An important, and often overlooked, consideration in the design of experiments aimed at determining sex differences in exercise-induced stress resistance is that the experiments need to be designed to statistically compare female and male subjects. Designs that simply include subjects of both sexes and determine effects of exercise in each sex separately are insufficient for determination of sex differences. Designs allowing interactions between sex and exercise to be compared statistically are necessary (Garcia-Sifuentes and Maney 2021). Additional factors such as housing conditions, breeding history, estrous phase, controlling for handling and potential stress effects of estrous phase monitoring, and hormonal status should also be considered (Becker et al. 2005).

2.3 Exercise-Induced Stress Resistance Is Sex Divergent

We have been using an acute, inescapable stress (IS) model, to study sex difference in exercise-induced stress resistance. The IS procedure involves restraining rats in tubes and delivering a series of 100 unpredictable, inescapable electric shocks to the tail. IS produces behaviors in rats that resemble symptoms of stress-related mental health disorders, including anxiety-like social avoidance, exaggerated fear-like potentiated shock-elicited freezing, and depression-like deficits in shuttle box escape (Christianson et al. 2008; Maier and Watkins 2005). Importantly, the behavioral consequences of IS are dependent on the uncontrollability of the stressor and do not occur if shocks of equal duration and intensity can be terminated by the subject (Christianson et al. 2008; Maier and Watkins 2005).

We and others have previously reported that long-term VWR (e.g., 4–6 weeks) prevents anxiety-, fear-, and depression-like behavioral consequences of IS in adult male (Dishman 1997; Dishman et al. 1997; Duman et al. 2008; Greenwood et al. 2003a, 2012a) and female (Tanner et al. 2019) rats. We have also observed a potential sex difference in the duration of VWR required to enable stress resistance; whereby 3 weeks of VWR prevents IS-induced social avoidance and exaggerated fear in females (Fallon et al. 2020), but fails to prevent IS-induced exaggerated fear or the shuttle box escape deficit in males (the effects of 3 weeks of VWR on IS-induced social avoidance in males was not assessed; (Greenwood et al. 2005a)). These time course studies, however, were performed in different laboratories, years apart, and did not allow statistical comparison of the effects of exercise between sexes.

To determine if sex differences in exercise-induced stress resistance exist, we recently allowed adult (age-matched), female and male Sprague-Dawley rats voluntary access to in-cage running wheels for either 3 or 6 weeks prior to exposure to a single session of IS. As expected, 6 weeks of VWR prevented IS-induced social avoidance and exaggerated fear in both sexes. Confirming prior work with females and males separately, we observed a sex difference in exercise-induced stress resistance, whereby 3 weeks of VWR prevented IS-induced exaggerated fear in females, but not in males. Surprisingly, however, 3 weeks of VWR did protect males from IS-induced social avoidance (Tanner et al. 2023).

This study reveals that long-term physical activity may be equally effective at enabling stress resistance in females and males, but the duration of physical activity required to enable stress resistance differs between the sexes in a stress outcome-dependent manner. A similar duration of physical activity seems to enable protection against anxiety-like consequences of stress between sexes, whereas exaggerated fear-like effects of stress seem to be more readily prevented by physical activity in females compared to males. Physical activity also seems to reduce depression-like behavior more readily in female mice than males. Both VWR (Elias et al. 2023) and treadmill training (Munive et al. 2016; Naghibi et al. 2021) have been reported to reduce immobility time in the tail suspension test in female, but not male, mice.

If these preclinical data apply to humans, they suggest that women at risk for mental health disorders should prioritize physical activity as a prophylactic strategy. Unfortunately, this sex difference will be difficult to confirm in humans, given the difficulty in performing clinical intervention studies on stress resistance. One longitudinal study following more than 100,000 adults over 3 years did find a sex difference like that observed in rodents. Kim et al. (2019) found that transitioning from a sedentary lifestyle to a physically active one reduced the incidence of new depression within a year in women, but not in men (Kim et al. 2019).

2.4 Sex Differences in Exercise-Induced Stress Resistance Resemble Sex Differences in Other Exercise Effects

The observation that females are more responsive than males to exercise-induced stress resistance does not seem to be an anomaly among sex differences in response to exercise. Physical activity can improve cognitive function more readily in female than in male rats and humans (Wang et al. 2015; Colcombe and Kramer 2003; Barha et al. 2017a; Barha and Liu-Ambrose 2018) and can reduce the risk for dementia to a greater extent in women than in men (Hogervorst et al. 2012; Laurin et al. 2001). A similar sex difference is found in studies examining the impact of physical activity on drug-seeking behaviors (Cosgrove et al. 2002; Lespine and Tirelli 2018; Smethells et al. 2020; Gallego et al. 2015), hypoalgesia (Rice et al. 2019), and mood (McDowell et al. 2016). Although conflicting reports exist (Peterson et al. 2014; Szuhany et al. 2015), growing evidence indicates that females are generally more responsive to the behavioral effects of physical activity than males. However, these effects may partly depend on exercise type (Barha et al. 2017a, b).

2.5 *Potential Mechanisms Underlying Sex Differences in Exercise-Induced Stress Resistance*

Relatively little work has explored whether mechanisms underlying exercise-induced stress resistance are sex divergent. Observed sex differences in the effects of physical activity on hippocampal microglial activation (Kohman et al. 2013), hippocampal neurotrophic factors (Naghibi et al. 2021; Uysal et al. 2015), and hippocampal volume (Varma et al. 2015) could all contribute to sex differences in exercise-induced stress resistance. Two factors that stand out as potentially driving sex differences in exercise effects, including stress resistance, are sex differences in exercise behavior and ovarian hormones.

Female rats run greater distances and faster speeds than males (Basso and Morrell 2017; Eikelboom and Mills 1988). Additionally, ovarian hormones during early life could alter the development of the female brain rendering it more susceptible to exercise effects (organizational effects). Alternatively, ovarian hormones during adulthood could interact with exercise, altering the way in which exercise impacts the brain (activational effects).

Sex differences in VWR behavior are dependent on activational effects of ovarian hormones (Gentry and Wade 1976; Slonaker 1924), particularly estrogen (Gorzek et al. 2007), thus the involvement of sex differences in VWR behavior and activational effects of ovarian hormones in mediating sex differences in exercise-induced stress resistance can be simultaneously tested by removing ovarian hormones. We removed ovaries from adult, female rats, and observed that female VWR behavior now resembled that of males (Tanner et al. 2023). Interestingly, despite the elimination of the female pattern of VWR and the absence of activational effects of ovarian hormones during exercise, 3 weeks of VWR still produced rapid protective effects against both IS-induced social avoidance and exaggerated fear (Tanner et al. 2023).

Whether ovarian hormones are similarly unnecessary for other sex differences in response to exercise is unknown. It does seem that ovarian hormones are not required for exercise to impact brain and behavior in females, in general. For example, VWR rapidly upregulates BDNF gene expression (Berchtold et al. 2001) and cell proliferation (Jin et al. 2008) in the hippocampus of ovariectomized female rats, and both VWR and treadmill training have been reported to protect ovariectomized rats from impairments in cognition produced by sleep deprivation (Saadati et al. 2015; Rajizadeh et al. 2020). Future work will be needed to determine mechanisms underlying sex differences in exercise-induced stress resistance. Since very little research on sex differences in underlying mechanisms has been completed, the following section on mechanisms focuses on work done in males.

3 A Conceptual Framework for Mechanisms Underlying Exercise-Induced Stress Resistance Includes the Development and Expression of Stress Resistance

A conceptual framework to aid in consideration of the variety of mechanisms underlying exercise-induced stress resistance is provided in Fig. 1. For physical activity to alter the brain and behavioral response to stress, physical activity must produce experience-dependent plasticity prior to stressor exposure which then allows the physically active organism to respond to the stressor differently than it would if it were sedentary. Mechanisms of exercise-induced stress resistance can, therefore, be thought of in terms of those mechanisms underlying the *development* of exercise-induced stress resistance (i.e., during exercise prior to stress) and the *expression* of exercise-induced stress resistance (i.e., during stress after a period of physical activity).

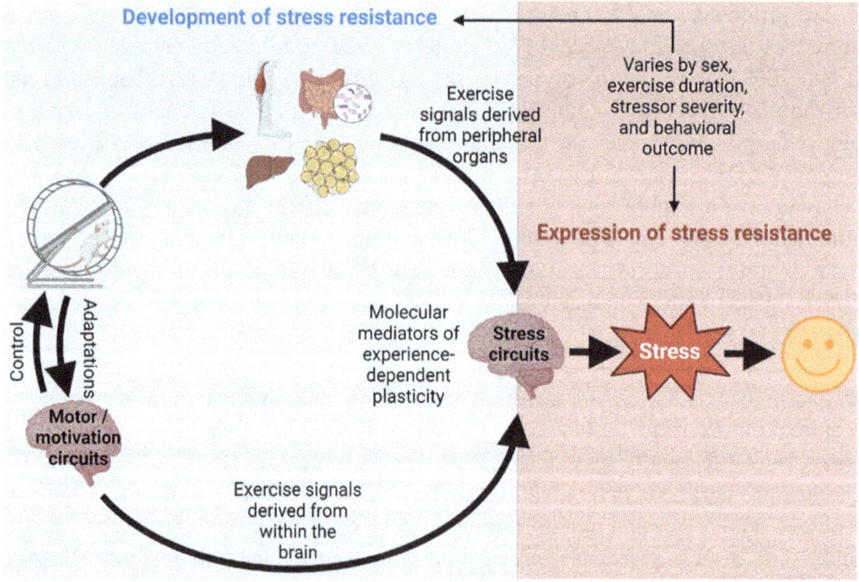

Fig. 1 A conceptual framework for exercise-induced stress resistance involves the development and expression of stress resistance. During the development of stress resistance (left, blue box), neural circuits governing physical activity motivation, planning, execution, and reinforcement determine the extent to which the organism engages in physical activity. Repeated physical activity triggers exercise signals, either derived from peripheral organs and/or motor/motivation circuits themselves, which communicate the experience of physical activity to brain stress-responsive circuits responsible for stress-induced behavioral outcomes. Exercise signals converge on molecular mediators which produce experience-dependent plasticity within these circuits. During the expression of stress resistance (right, red box), stress circuits altered by prior physical activity now respond to stressful events in a manner that is protective against stress-induced behavioral outcomes

4 The Development of Stress Resistance from Exercise Involves Molecular Mediators, Neural Circuits, and Exercise Signals

As we saw with the protective effects of VWR against the behavioral consequences of IS, stress resistance from exercise develops over time with repeated exercise. The development of stress resistance from exercise refers to exercise-induced physiological adaptations occurring during exercise which mediate the later expression of stress resistance. The molecular mediators of these adaptations, the circuits in which these adaptations occur, and the signals through which exercise initiates these adaptations are all critical to the development of stress resistance (Fig. 1).

Physical activity status of the organism modulates many aspects of physiology, including central gene expression (Loughridge et al. 2013; Tong et al. 2001), alterations in neurotransmitter systems (Dishman 1997; Greenwood 2019; Greenwood and Fleshner 2011; Petzinger et al. 2015), synaptic plasticity (Cotman et al. 2007; de Sousa Fernandes et al. 2020), proliferation and survival of adult-born neurons (Vivar et al. 2013; van Praag 2008), and brain structure (Wilckens et al. 2021; Chen et al. 2020). Despite these many exercise adaptations, identifying which, if any, of these are causally related to stress resistance is challenging. The types of experiments required to determine causation are especially difficult to perform with long-term exercise studies. Loss-of-function experiments require candidate mechanistic factors to be altered in physically active organisms for the entire duration of the exercise period (during the development of stress resistance), but not during exposure to stress (during the expression of stress resistance). Similarly, gain-of-function experiments require factors to be increased for a long period of time prior to, but not during, stress. Compounding the issue is the possibility that physical activity utilizes several different mechanisms acting in parallel to enable stress resistance. For example, several molecular mediators could drive the experience-dependent neural plasticity required for stress resistance. Likely candidates include brain-derived neurotrophic factor (Duman et al. 2008; Cotman et al. 2007), mammalian target of rapamycin (Lloyd et al. 2017; Moya et al. 2020), and delta-FosB (Mul et al. 2018; Werme et al. 2002). Additionally, several exercise signals derived from peripheral organs (e.g., muscle, fat, liver, gut) or the brain itself (e.g., from circuits directly governing voluntary exercise) could be responsible for communicating the experience of exercise to these molecular mediators. Irisin (Islam et al. 2021), IL-6 (Severinsen and Pedersen 2020), and lactate (Xue et al. 2022), for example, have been shown to be important for periphery-to-brain communication during exercise.

Since physical activity most obviously involves the movement of skeletal muscle, it is not surprising that research efforts aimed at identifying factors responsible for communicating the experience of exercise to the brain circuits underlying behavioral effects of exercise (referred to here as exercise signals) have been focused on signals released from muscle or other peripheral organs. We must keep in mind, however, that these signals could also come from within the brain itself. Potential sources of central exercise signals include neural circuits involved in the motivation for physical activity, and/or the planning, performance, and/or reinforcing effects of physical activity. Our understanding of these circuits has been growing in recent years (Tanner et al. 2022; Dohnalova et al. 2022). Interestingly, there is evidence that circuits governing voluntary exercise vary by sex (Tanner et al. 2022). It is therefore plausible that sex differences in central signals could drive sex differences in the development of exercise-induced stress resistance.

5 Multiple Mechanisms Underly the Expression of Exercise-Induced Stress Resistance

The expression of stress resistance from exercise occurs during the discrete timepoint of stressor exposure. The changes in the brain established during the development of stress resistance enable stress-reactive neural circuitry orchestrating neurochemical, behavioral, and physiological stress responses to react differently during stress than they would if the organism were sedentary (Fig. 1). Results of recent experiments reveal a common take-home message: there are multiple mechanisms mediating the expression of stress resistance from exercise. These mechanisms can vary depending on sex, stressor severity, duration of prior exercise, and behavioral outcome. The multiple stress resistance mechanisms also seem to act in parallel to enable the expression of stress resistance from exercise. This is perhaps not surprising, given the wide variety of physiological and neural systems impacted by exercise.

5.1 Non-specific Effects of Exercise

In addition to changes in specific stress-responsive systems, non-specific effects of exercise, such as altered pain sensitivity or locomotor activity, could contribute to the expression of stress resistance from exercise. VWR can reduce pain sensitivity, raising the possibility that VWR could reduce sensitivity to stressors such as electric shock. VWR, however, does not alter sensitivity to shock in mice (Duman et al. 2008) or rats (Tanner et al. 2023), nor does it globally blunt stress responses to electric shock stressors, as would be expected if VWR reduced shock sensitivity. For example, VWR has no effect on IS-induced increases in corticosterone (Tanner et al. 2023; Fleshner 2000; Speaker et al. 2014), plasma cytokines (Speaker et al. 2014), or cFos in several stress-responsive brain regions (Greenwood et al. 2003b, 2005a). Regarding altered pain sensitivity contributing to protection against the behavioral effects of IS specifically, IS itself produces massive analgesia (Maier et al. 1982) that begins early during IS treatment (Maier and Keith 1987). Importantly, the behavioral sequelae of IS occur whether or not these antinociceptive effects of IS are present, indicating that IS outcomes are independent of pain sensitivity (MacLennan et al. 1982; Hemingway and Reigle 1987).

Animal models of stress-related mental health disorders often depend on subjects engaging in locomotor activity, whereby general increases in movement could be mistaken for anxiolytic or antidepressant effects. This raises the possibility that potential effects of VWR on locomotor activity could contribute to the observed behavioral effects. IS reduces locomotor activity during exposure to a novel environment for at least 48 h following IS, but VWR has no impact on this stress-induced reduction in locomotor activity in either sex (Tanner et al. 2023). This observation suggests that the stress-protective effect of VWR against IS outcomes is not attributable to alterations in locomotor activity. Future research utilizing different animal models should consider how these factors and others (such as increased fitness) contribute to the behavioral effects of exercise in that model.

5.2 The Role of the Hypothalamic-Pituitary-Adrenal Axis

Physical activity blunts the response of the hypothalamic-pituitary-adrenal (HPA) axis to stress (see (Nowacka-Chmielewska et al. 2022) for a recent review). Given the link between HPA activity and stress-related mental health disorders (Sapolsky 2000; McEwen et al. 2015), it is easy to conceptualize how constraint over stress-induced activity of the HPA axis could contribute to stress resistance from exercise. However, severely stressful events are more likely to precipitate the formation of stress-related disorders than mild stressors, and the impact of VWR on the HPA axis seems to be limited to mild-moderate stressors (Campeau et al. 2010). VWR, for example, fails to attenuate the HPA response to IS in either sex (Tanner et al. 2023; Fleshner 2000; Speaker et al. 2014). Attenuation of HPA responses could, therefore, contribute to exercise-induced resistance to the effects of relatively mild stressors, but protective effects of exercise against behavioral consequences of more severe stressors likely involve mechanisms other than blunting of the HPA axis. These data demonstrate that mechanisms underlying the expression of stress resistance from exercise can vary depending on stressor severity.

5.3 Constraint over Stress-Induced Activity of Dorsal Raphe Nucleus Serotonergic Neurons Is One Mechanism Underlying the Expression of Stress Resistance from Exercise

Since VWR prevents the anxiety- and depression-like consequences of acute exposure to severe, inescapable stressors, such as IS, it is of particular interest to consider the mechanisms involved in this effect. One of the benefits of using the IS model is that the neural circuits mediating the behavioral consequences of IS have been well-defined. Briefly, brain regions involved in processing specific aspects of IS, including the locus coeruleus (LC), lateral habenula, and bed nucleus of the stria terminalis (BNST), have converging excitatory inputs to the dorsal raphe nucleus (DRN). Activity of these inputs during severe stressors drives the activity of DRN serotonergic (5-HT) neurons, measured both by cFos (Grahn et al. 1999) and extracellular 5-HT released into the DRN from DRN axon collaterals (Maswood et al. 1998). The excessive extracellular 5-HT in the DRN during IS desensitizes inhibitory somatodendritic 5-HT_{1A} autoreceptors, leaving DRN 5-HT neurons in a sensitized state (Rozeske et al. 2011). Sensitized DRN 5-HT neurons now respond to future mild stressors with excessive 5-HT release in DRN projection sites. Among these DRN projection sites are the basolateral amygdala (Amat et al. 1998), where sensitized 5-HT release produced by prior IS leads to anxiety-like (Christianson et al. 2010) and exaggerated fear-like (potentiated shock-elicited freezing) behavior (Strong et al. 2011), and the dorsal striatum, where sensitized 5-HT release produces depression-like behavior, such as the shuttle box escape deficit (Strong et al. 2011). In addition to being necessary and sufficient for the behavioral consequences of IS (Maier and Watkins 2005), increased 5-HT activity in the DRN is also involved in the anxiety- and depression-like consequences of social defeat (Hammack et al. 2012; Amat et al. 2010). Moreover, manipulations that prevent the behavioral outcomes of IS, including behavioral control over the stressor (Amat et al. 2006), prophylactic ketamine (Amat et al. 2016), dominance rank (Coleman et al. 2023), and 6 weeks of VWR (Greenwood and Fleshner 2011), all prevent IS-induced DRN hyperactivity and sensitization.

In adult, male rats, 6 weeks of VWR attenuates IS-induced expression of the neural activation marker cFos within DRN 5-HT neurons (Greenwood et al. 2003a), indicating that 6 weeks of VWR prevents the excessive activation of DRN 5-HT neurons that occurs during IS. Six weeks of VWR also prevents IS-induced sensitization of extracellular 5-HT in the DS in response to a few foot shocks, assessed with microdialysis (Clark et al. 2015). These data suggest that the expression of stress resistance from 6 weeks of VWR involves the prevention of stress-induced hyperactivity and subsequent sensitization of DRN 5-HT neurons, at least in males. An increase in central 5-HT is thought to contribute to central fatigue (Davis and Bailey 1997). Constraint over DRN 5-HT activity could thus be a training adaptation that evolved to help delay fatigue during exercise. Whether DRN constraint is involved in the expression of stress resistance from exercise in females is a critical unanswered question.

Understanding how exercise comes to constrain the DRN 5-HT response to severe stressors could reveal novel insights into mechanisms underlying the development and expression of exercise-induced stress resistance. Two possibilities seem most plausible. First, exercise could alter inputs to the DRN that regulate 5-HT activity during stress. Second, exercise could produce adaptations within DRN 5-HT neurons themselves which alter their response to stress. There is evidence that exercise does both, as outlined below. Additionally, exercise modifies the response of DRN efferent circuits mediating specific behavioral consequences of IS to 5-HT and/or stress. Multiple independent mechanisms involving the 5-HT system thus seem to contribute to the expression of stress resistance from exercise (Fig. 2).

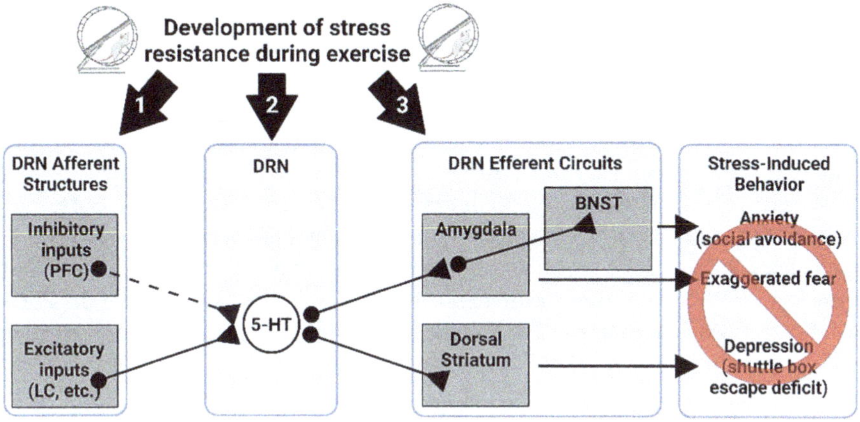

Fig. 2 Circuit-specific determinants of exercise-induced stress resistance. Repeated physical activity initiates the development of the stress-resistant brain by acting on at least 3 unique neural substrates, each of which could be sufficient for the expression of stress resistance: (1) Repeated physical activity potentiates activity of inhibitory inputs to the dorsal raphe nucleus (DRN; dotted line) and potentially reduces activity of excitatory DRN inputs (solid lines) during stress. (2) Repeated physical activity alters factors intrinsic to DRN serotonergic (5-HT) neurons that influence their response to stress. (3) Repeated physical activity produces adaptations within DRN efferent circuits that are the proximal mediators of behavioral-specific stress outcomes. These stress-protective mechanisms develop at different rates during exercise, depending on sex and circuit-specific behavioral outcome

5.4 The Role of Inputs to the Dorsal Raphe Nucleus

One means by which exercise could attenuate the DRN 5-HT response to stress is by modifying the activity during stress of brain structures with modulatory control over DRN 5-HT neurons. VWR could, for example, attenuate excitatory drive or increase inhibitory drive to DRN 5-HT neurons from afferent structures during IS. Numerous DRN afferent structures are sensitive to physical activity and are thus prime candidates for exercise-induced modulation of DRN activity, as discussed previously (Greenwood 2019; Greenwood and Fleshner 2008, 2011; Nicastro and Greenwood 2016). Exercise produces a hyperdopaminergic state within midbrain-striatal pathways which we have argued simultaneously contributes to both a more positive outlook on stress and DRN constraint in physically active compared to sedentary organisms (Greenwood 2019). The medial prefrontal cortex (PFC) also stands out due to its established role in stress resistance and resilience (Maier et al. 2006). Based on our prior data, it was thought that the PFC was not involved in stress resistance from exercise (Christianson and Greenwood 2014; Greenwood et al. 2013). However, here we present new data which provides an updated view on the role of the PFC in exercise-induced stress resistance.

The PFC has an inhibitory influence over DRN 5-HT activity, mediated through a glutamatergic projection from the medial PFC to the DRN, where PFC glutamatergic terminals synapse on GABAergic neurons which inhibit 5-HT neural activity (Amat et al. 2005). Physical activity induces neuroplasticity in the PFC and enhances PFC function in both humans (Erickson and Kramer 2009; Soshi et al. 2021) and animals (Dong et al. 2018; Brockett et al. 2015). Given the established sensitivity of the PFC to exercise and the ability of the PFC to inhibit the DRN during IS, we sought to determine whether the PFC is necessary for the expression of exercise-induced stress resistance. After 3 weeks of wheel access to establish habitual VWR, male rats received sham surgery or bilateral lesions of the medial PFC, where DRN-projecting neurons are located. Rats ran an additional 3 weeks following surgery, for a total of 6 weeks of VWR, prior to exposure to no stress or IS. The idea was that if the PFC-to-DRN pathway inhibits the DRN during IS in VWR rats, then removing this inhibitory influence over DRN activity would prevent the expression of stress resistance and would restore the behavioral consequences of IS in previously physically active rats. To our surprise, PFC lesions had absolutely no effect on exercise-induced stress resistance. Six weeks of VWR still protected male rats from the behavioral consequences of IS despite PFC lesions (Greenwood et al. 2013).

These data indicate that the PFC is not necessary for the expression of exercise-induced stress resistance. But does this negative experimental result mean that physical activity does not utilize the PFC or the PFC-to-DRN pathway for stress resistance at all? A limitation of lesion and inactivation studies is that if negative results are found, it cannot be assumed that the inhibited region or circuit is not utilized for stress resistance in an intact animal. Indeed, if multiple independent mechanisms act in parallel to enable stress resistance from exercise, then inhibiting one should have no impact on another and stress resistance should remain intact. If multiple neural circuits contribute to DRN constraint during stress in previously physically active animals, the only way to know whether a certain circuit contributes to DRN constraint is to measure its activity during stress.

We designed an experiment to determine whether prior exercise recruits the PFC-to-DRN circuit during stress (Fig. 3a). Red fluorescent retro-beads were injected into the DRN of male rats (Fig. 3c). These beads reveal DRN-projecting PFC neurons by traveling retrograde from axon terminals in the DRN to cell bodies of origin in DRN afferent structures. After 6 weeks of sedentary or VWR conditions, rats were exposed to IS and cFos was quantified in DRN-projecting PFC neurons with immunohistochemistry (Fig. 3b; $n = 6$/group). We found that 6 weeks of VWR increased the percentage of DRN-projecting PFC neurons expressing cFos following IS (*unpaired t-test* $p < 0.0001$; Fig. 3d).

These data provide an updated view on the role of the PFC in exercise-induced stress resistance. VWR does indeed recruit the PFC-to-DRN inhibitory circuit during stress, which likely contributes to DRN inhibition (Fig. 3e). However, since VWR still enables stress resistance even in the absence of the PFC, this activity is not necessary for stress resistance from exercise and exercise must utilize an additional mechanism that can enable stress resistance in the absence of the PFC (Fig. 3f).

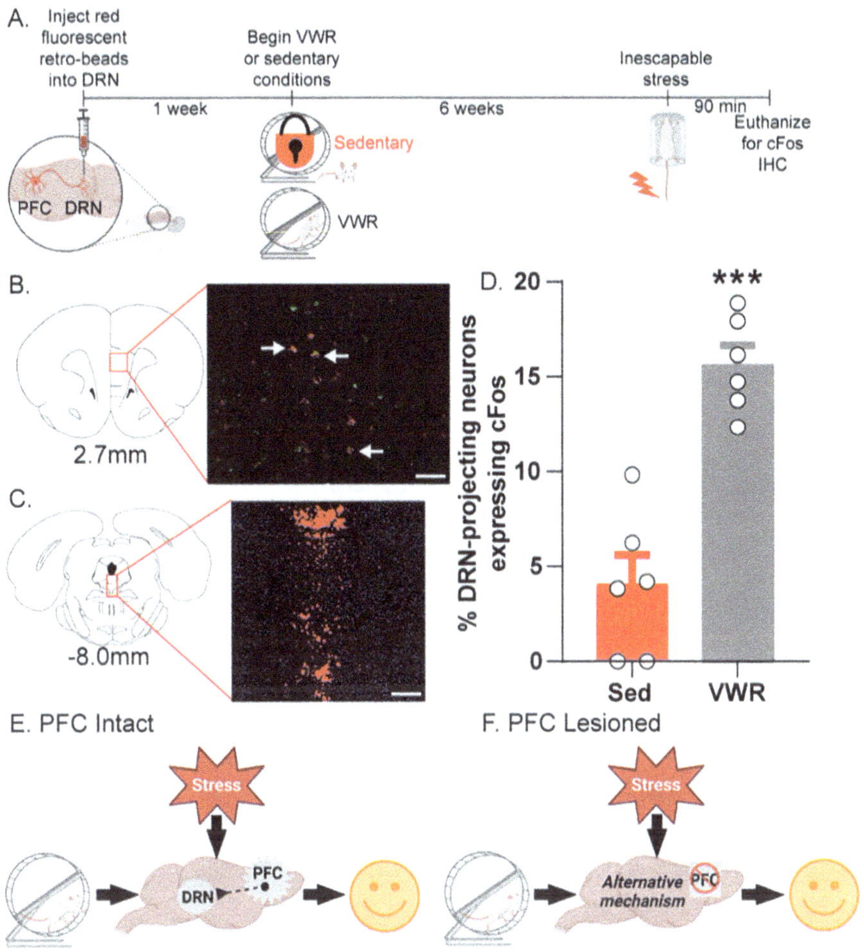

Fig. 3 Prior physical activity recruits a prefrontal cortex (PFC)-to-dorsal raphe nucleus (DRN) pathway during inescapable stress. (**a**) Experimental timeline. Rats received red retro-beads in the DRN prior to 6 weeks of sedentary (Sed) or voluntary wheel running (VWR) conditions. All rats were exposed to inescapable tail shock stress and then euthanized 90 min later. PFC was processed for cFos immunohistochemistry (IHC) and the % of DRN-projecting PFC neurons containing cFos was quantified. (**b**) cFos (green) expression within DRN-projecting PFC neurons (red). Arrows point to DRN-projecting neurons expressing cFos. Distance noted is relative to Bregma. Scale bar = 200 μm. (**c**) Injection site of red retro-beads in the DRN. Distance noted is relative to Bregma. Scale bar = 200 μm. (**d**) Relative to Sed rats, prior VWR increased cFos expression within DRN-projecting PFC neurons during inescapable stress. ***$p < 0.0001$. (**e**) Cartoon illustrating that VWR rats with intact PFC utilize the PFC-to-DRN pathway during stress. (**f**) In PFC-lesioned rats, VWR rats no longer utilize the PFC-to-DRN pathway but are still protected from stress through an alternative mechanism. This illustrates that multiple mechanisms are involved in the expression of exercise-induced stress resistance. Figure created in BioRender. Brain images adapted from the rat brain atlas of Paxinos and Watson (Paxinos and Watson 1998)

5.5 The Role of Exercise Adaptations Within Dorsal Raphe Nucleus 5-HT Neurons

Another possible way in which prior physical activity could attenuate activity of DRN 5-HT neurons during stress is by impacting DRN 5-HT neurons directly (during the development of stress resistance), altering factors intrinsic to 5-HT neurons which modulate their response to stress (during the expression of stress resistance; Fig. 2). Reported adaptations within DRN 5-HT neurons in response to VWR, such as increased 5-HT content in the DRN and DRN projection sites (Dishman 1997; Dishman et al. 1997) and altered gene expression within DRN 5-HT neurons (Loughridge et al. 2013), are consistent with this possibility. For example, VWR increases mRNA coding for the 5-HT$_{1A}$ inhibitory autoreceptor in the DRN in male rats (Greenwood et al. 2003a). An increase in 5-HT$_{1A}$ autoinhibition could help constrain the DRN 5-HT response to stress. This mechanism may only contribute to stress resistance after long-term VWR in males, since 6 weeks, but not 3 weeks, of VWR increases 5-HT$_{1A}$ mRNA (Greenwood et al. 2005b). Other DRN adaptations occur more rapidly (Greenwood et al. 2005b). Future work should determine if similar intrinsic changes in DRN 5-HT neurons occur in females and clarify whether these adaptations contribute to DRN constraint during stress in both sexes.

5.6 The Role of Exercise Adaptations Within Dorsal Raphe Nucleus Efferent Circuits

Constraint over stress-induced activity of DRN 5-HT neurons, either via modulation of the activity of DRN inputs or adaptations within DRN 5-HT neurons, is one potential mechanism underlying the expression of stress resistance from long-term exercise. However, it doesn't appear to be the only one. In male rats, 3 weeks of VWR does not attenuate the activity of DRN 5-HT neurons during IS, as 6 weeks does (Greenwood et al. 2005a). How then, does 3 weeks of VWR prevent IS-induced social avoidance in males?

IS-induced social avoidance depends on IS-induced sensitization of 5-HT release in the amygdala (Christianson et al. 2010). Since 3 weeks of VWR fails to constrain the DRN response to IS, 3 weeks of VWR could instead be preventing IS-induced social avoidance by acting on DRN efferent circuits that mediate the anxiety-like consequences of sensitized 5-HT release. The BNST is implicated in anxiety (Walker et al. 2003), and stress-induced social avoidance involves a circuit from the amygdala to the BNST (Vantrease et al. 2022). Three weeks of VWR does not alter cFos in response to IS in the amygdala (Greenwood et al. 2005a), but it does alter gene expression (Fox et al. 2023) and the cFos response to IS (Greenwood et al. 2005a) in the BNST. Anxiogenic effects in a fear-potentiated startle paradigm produced by 2 weeks of VWR were recently shown to be mediated by a reduction in 5-HT_{2C} receptor expression in the BNST (Fox et al. 2023).

Together, these data are consistent with the possibility that rapid exercise-induced adaptations in the BNST could lead to protection from social avoidance and other anxiety-like behaviors, despite a failure of 3 weeks of VWR to prevent IS-induced sensitization of the DRN. Of course, longer-duration physical activity could also produce adaptations within DRN efferent circuits that underly fear- and depression-like effects of stress. Indeed, 6 weeks of VWR reduces the expression and sensitivity of 5-HT_{2C} receptors in the amygdala and dorsal striatum (Greenwood et al. 2012b). Since the behavioral consequences of IS are dependent on 5-HT_{2C} receptor signaling in these regions (Christianson et al. 2010; Strong et al. 2009, 2011), the expression of stress resistance induced by a long duration of prior physical activity could involve both DRN constraint and resistance to the post-synaptic effects of 5-HT in DRN projection sites mediating specific IS-induced behaviors. This prior work was done in males. Whether physical activity produces neural adaptations in DRN projection sites in females is currently unknown and should be addressed in future studies.

6 Conclusions

Physical activity enables robust protection against behavioral effects of stress, including stress-related mental health disorders in humans and anxiety-, fear-, and depression-like behaviors in animals. Mechanisms underlying stress resistance from exercise can be thought of in terms of those driving the development of stress resistance (during exercise) and those driving its expression (during stress). The development of stress resistance involves multiple exercise signals and molecular mediators and varies depending on sex and behavioral outcome. The expression of stress resistance from exercise similarly involves multiple independent mechanisms acting in parallel, as is demonstrated by the role of the PFC in exercise-induced stress resistance. This knowledge has important implications for the design and interpretation of causal experiments aimed at identifying underlying mechanisms. The inescapable stress model in rodents reveals a key role of DRN constraint in mediating exercise-induced stress resistance, but the specific involvement of exercise adaptations within DRN inputs, within the DRN itself, and within DRN projection sites require further clarification. Much of our knowledge of the mechanisms underlying exercise-induced stress resistance comes from work using male subjects. The recent realization that female rodents are more responsive than males to exercise-induced stress resistance highlights the need to expand our mechanistic knowledge in females. Whether this sex difference extends to humans is a critical unanswered question, but regardless of the answer, identifying the mechanisms driving this sex difference could reveal novel targets that could be harnessed by both sexes to bolster the stress-protective impact of exercise.

References

Amat J, Matus-Amat P, Watkins LR, Maier SF (1998) Escapable and inescapable stress differentially alter extracellular levels of 5-HT in the basolateral amygdala of the rat. Brain Res 812(1–2):113–120

Amat J, Baratta MV, Paul E, Bland ST, Watkins LR, Maier SF (2005) Medial prefrontal cortex determines how stressor controllability affects behavior and dorsal raphe nucleus. Nat Neurosci 8(3):365–371

Amat J, Paul E, Zarza C, Watkins LR, Maier SF (2006) Previous experience with behavioral control over stress blocks the behavioral and dorsal raphe nucleus activating effects of later uncontrollable stress: role of the ventral medial prefrontal cortex. J Neurosci 26(51):13264–13272

Amat J, Aleksejev RM, Paul E, Watkins LR, Maier SF (2010) Behavioral control over shock blocks behavioral and neurochemical effects of later social defeat. Neuroscience 165(4):1031–1038

Amat J, Dolzani SD, Tilden S, Christianson JP, Kubala KH, Bartholomay K et al (2016) Previous ketamine produces an enduring blockade of neurochemical and behavioral effects of uncontrollable stress. J Neurosci 36(1):153–161

Barha CK, Liu-Ambrose T (2018) Exercise and the aging brain: considerations for sex differences. Brain Plast 4(1):53–63

Barha CK, Davis JC, Falck RS, Nagamatsu LS, Liu-Ambrose T (2017a) Sex differences in exercise efficacy to improve cognition: a systematic review and meta-analysis of randomized controlled trials in older humans. Front Neuroendocrinol 46:71–85

Barha CK, Falck RS, Davis JC, Nagamatsu LS, Liu-Ambrose T (2017b) Sex differences in aerobic exercise efficacy to improve cognition: a systematic review and meta-analysis of studies in older rodents. Front Neuroendocrinol 46:86–105

Basso JC, Morrell JI (2017) Using wheel availability to shape running behavior of the rat towards improved behavioral and neurobiological outcomes. J Neurosci Methods 290:13–23

Becker JB, Arnold AP, Berkley KJ, Blaustein JD, Eckel LA, Hampson E et al (2005) Strategies and methods for research on sex differences in brain and behavior. Endocrinology 146(4): 1650–1673

Berchtold NC, Kesslak JP, Pike CJ, Adlard PA, Cotman CW (2001) Estrogen and exercise interact to regulate brain-derived neurotrophic factor mRNA and protein expression in the hippocampus. Eur J Neurosci 14(12):1992–2002

Bigos KL, Pollock BG, Stankevich BA, Bies RR (2009) Sex differences in the pharmacokinetics and pharmacodynamics of antidepressants: an updated review. Gend Med 6(4):522–543

Bjorkman F, Ekblom O (2022) Physical exercise as treatment for PTSD: a systematic review and meta-analysis. Mil Med 187(9–10):e1103–e1e13

Brockett AT, LaMarca EA, Gould E (2015) Physical exercise enhances cognitive flexibility as well as astrocytic and synaptic markers in the medial prefrontal cortex. PLoS One 10(5):e0124859

Campeau S, Nyhuis TJ, Sasse SK, Kryskow EM, Herlihy L, Masini CV et al (2010) Hypothalamic pituitary adrenal axis responses to low-intensity stressors are reduced after voluntary wheel running in rats. J Neuroendocrinol 22(8):872–888

Chekroud SR, Gueorguieva R, Zheutlin AB, Paulus M, Krumholz HM, Krystal JH et al (2018) Association between physical exercise and mental health in 1.2 million individuals in the USA between 2011 and 2015: a cross-sectional study. Lancet Psychiatry 5(9):739–746

Chen FT, Hopman RJ, Huang CJ, Chu CH, Hillman CH, Hung TM et al (2020) The effect of exercise training on brain structure and function in older adults: a systematic review based on evidence from randomized control trials. J Clin Med 9(4)

Christianson JP, Greenwood BN (2014) Stress-protective neural circuits: not all roads lead through the prefrontal cortex. Stress 17(1):1–12

Christianson JP, Paul ED, Irani M, Thompson BM, Kubala KH, Yirmiya R et al (2008) The role of prior stressor controllability and the dorsal raphe nucleus in sucrose preference and social exploration. Behav Brain Res 193(1):87–93

Christianson JP, Ragole T, Amat J, Greenwood BN, Strong PV, Paul ED et al (2010) 5-hydroxytryptamine 2C receptors in the basolateral amygdala are involved in the expression of anxiety after uncontrollable traumatic stress. Biol Psychiatry 67(4):339–345

Cipriani A, Furukawa TA, Salanti G, Chaimani A, Atkinson LZ, Ogawa Y et al (2018) Comparative efficacy and acceptability of 21 antidepressant drugs for the acute treatment of adults with major depressive disorder: a systematic review and network meta-analysis. Lancet 391(10128): 1357–1366

Clark PJ, Amat J, McConnell SO, Ghasem PR, Greenwood BN, Maier SF et al (2015) Running reduces uncontrollable stress-evoked serotonin and potentiates stress-evoked dopamine concentrations in the rat dorsal striatum. PLoS One 10(11):e0141898

Colcombe S, Kramer AF (2003) Fitness effects on the cognitive function of older adults: a meta-analytic study. Psychol Sci 14(2):125–130

Coleman PT, Costanza-Chavez GW, Martin HN, Amat J, Frank MG, Sanchez RJ et al (2023) Prior experience with behavioral control over stress facilitates social dominance. bioRxiv

Cosgrove KP, Hunter RG, Carroll ME (2002) Wheel-running attenuates intravenous cocaine self-administration in rats: sex differences. Pharmacol Biochem Behav 73(3):663–671

Cotman CW, Berchtold NC, Christie LA (2007) Exercise builds brain health: key roles of growth factor cascades and inflammation. Trends Neurosci 30(9):464–472

Davis JM, Bailey SP (1997) Possible mechanisms of central nervous system fatigue during exercise. Med Sci Sports Exerc 29(1):45–57

de Sousa Fernandes MS, Ordonio TF, Santos GCJ, Santos LER, Calazans CT, Gomes DA et al (2020) Effects of physical exercise on neuroplasticity and brain function: a systematic review in human and animal studies. Neural Plast 2020:8856621

Dishman RK (1997) Brain monoamines, exercise, and behavioral stress: animal models. Med Sci Sports Exerc 29(1):63–74

Dishman RK, Renner KJ, Youngstedt SD, Reigle TG, Bunnell BN, Burke KA et al (1997) Activity wheel running reduces escape latency and alters brain monoamine levels after footshock. Brain Res Bull 42(5):399–406

Dohnalova L, Lundgren P, Carty JRE, Goldstein N, Wenski SL, Nanudorn P et al (2022) A microbiome-dependent gut-brain pathway regulates motivation for exercise. Nature 612(7941):739–747

Dong J, Zhao J, Lin Y, Liang H, He X, Zheng X et al (2018) Exercise improves recognition memory and synaptic plasticity in the prefrontal cortex for rats modelling vascular dementia. Neurol Res 40(1):68–77

Duman CH, Schlesinger L, Russell DS, Duman RS (2008) Voluntary exercise produces antidepressant and anxiolytic behavioral effects in mice. Brain Res 1199:148–158

Eikelboom R, Mills R (1988) A microanalysis of wheel running in male and female rats. Physiol Behav 43(5):625–630

Elias E, Zhang AY, White AG, Pyle MJ, Manners MT (2023) Voluntary wheel running promotes resilience to the behavioral effects of unpredictable chronic mild stress in male and female mice. Stress 26(1):2203769

Erickson KI, Kramer AF (2009) Aerobic exercise effects on cognitive and neural plasticity in older adults. Br J Sports Med 43(1):22–24

Fallon IP, Tanner MK, Greenwood BN, Baratta MV (2020) Sex differences in resilience: experiential factors and their mechanisms. Eur J Neurosci 52(1):2530–2547

Ferrari AJ, Santomauro DF, Mantilla Herrera AM, Shadid J, Ashbaugh C, Erskine HE et al (2022) Global, regional, and national burden of 12 mental disorders in 204 countries and territories, 1990–2019: a systematic analysis for the global burden of disease study 2019. Lancet Psychiatry 9:137–150

Fleshner M (2000) Exercise and neuroendocrine regulation of antibody production: protective effect of physical activity on stress-induced suppression of the specific antibody response. Int J Sports Med 21(S1):S14–SS9

Fox JH, Boucher MN, Abedrabbo KS, Hare BD, Grimmig BA, Falls WA et al (2023) Exercise reduces the anxiogenic effects of meta-chlorophenylpiperazine: the role of 5-HT2C receptors in the bed nucleus of the stria terminalis. Front Synaptic Neurosci 14:1067420

Gallego X, Cox RJ, Funk E, Foster RA, Ehringer MA (2015) Voluntary exercise decreases ethanol preference and consumption in C57BL/6 adolescent mice: sex differences and hippocampal BDNF expression. Physiol Behav 138:28–36

Garcia-Sifuentes Y, Maney DL (2021) Reporting and misreporting of sex differences in the biological sciences. eLife:10

Gentry RT, Wade GN (1976) Sex differences in sensitivity of food intake, body weight, and running-wheel activity to ovarian steroids in rats. J Comp Physiol Psychol 90(8):747–754

Gorzek JF, Hendrickson KC, Forstner JP, Rixen JL, Moran AL, Lowe DA (2007) Estradiol and tamoxifen reverse ovariectomy-induced physical inactivity in mice. Med Sci Sports Exerc 39(2):248–256

Grahn RE, Will MJ, Hammack SE, Maswood S, McQueen MB, Watkins LR, Maier SF (1999) Activation of serotonin-immunoreactive cells in the dorsal raphe nucleus in rats exposed to an uncontrollable stressor. Brain Res 826(1):35–43

Greenwood BN (2019) The role of dopamine in overcoming aversion with exercise. Brain Res 1713:102–108

Greenwood BN, Fleshner M (2008) Exercise, learned helplessness, and the stress-resistant brain. Neuromolecular Med 10(2):81–98

Greenwood BN, Fleshner M (2011) Exercise, stress resistance, and central serotonergic systems. Exerc Sport Sci Rev 39(3):140–149

Greenwood BN, Foley TE, Day HE, Campisi J, Hammack SH, Campeau S et al (2003a) Freewheel running prevents learned helplessness/behavioral depression: role of dorsal raphe serotonergic neurons. J Neurosci 23(7):2889–2898

Greenwood BN, Kennedy S, Smith TP, Campeau S, Day HE, Fleshner M (2003b) Voluntary freewheel running selectively modulates catecholamine content in peripheral tissue and c-Fos expression in the central sympathetic circuit following exposure to uncontrollable stress in rats. Neuroscience 120(1):269–281

Greenwood BN, Foley TE, Burhans D, Maier SF, Fleshner M (2005a) The consequences of uncontrollable stress are sensitive to duration of prior wheel running. Brain Res 1033(2): 164–178

Greenwood BN, Foley TE, Day HE, Burhans D, Brooks L, Campeau S et al (2005b) Wheel running alters serotonin (5-HT) transporter, 5-HT(1A), 5-HT(1B), and alpha(1b)-adrenergic receptor mRNA in the rat raphe nuclei. Biol Psychiatry 57(5):559–568

Greenwood BN, Loughridge AB, Sadaoui N, Christianson JP, Fleshner M (2012a) The protective effects of voluntary exercise against the behavioral consequences of uncontrollable stress persist despite an increase in anxiety following forced cessation of exercise. Behav Brain Res 233(2): 314–321

Greenwood BN, Strong PV, Loughridge AB, Day HE, Clark PJ, Mika A et al (2012b) 5-HT(2C) receptors in the basolateral amygdala and dorsal striatum are a novel target for the anxiolytic and antidepressant effects of exercise. PLoS One 7(9):e46118

Greenwood BN, Spence KG, Crevling DM, Clark PJ, Craig WC, Fleshner M (2013) Exercise-induced stress resistance is independent of exercise controllability and the medial prefrontal cortex. Eur J Neurosci 37(3):469–478

Hammack SE, Cooper MA, Lezak KR (2012) Overlapping neurobiology of learned helplessness and conditioned defeat: implications for PTSD and mood disorders. Neuropharmacology 62(2): 565–575

Harvey SB, Overland S, Hatch SL, Wessely S, Mykletun A, Hotopf M (2018) Exercise and the prevention of depression: results of the HUNT cohort study. Am J Psychiatry 175(1):28–36

Hegberg NJ, Hayes JP, Hayes SM (2019) Exercise intervention in PTSD: a narrative review and rationale for implementation. Front Psych 10:133

Heissel A, Heinen D, Brokmeier LL, Skarabis N, Kangas M, Vancampfort D et al (2023) Exercise as medicine for depressive symptoms? A systematic review and meta-analysis with meta-regression. Br J Sports Med

Hemingway RB 3rd, Reigle TG (1987) The involvement of endogenous opiate systems in learned helplessness and stress-induced analgesia. Psychopharmacology (Berl) 93(3):353–357

Hogervorst E, Clifford A, Stock J, Xin X, Bandelow S (2012) Exercise to prevnet cognitive decline and alzheimer's disease: for whom, when, what, and (most importantly) how much? J Alzheimers Dis Parkinsonism 2(2)

Insel TR, Scolnick EM (2006) Cure therapeutics and strategic prevention: raising the bar for mental health research. Mol Psychiatry 11(1):11–17

Islam MR, Valaris S, Young MF, Haley EB, Luo R, Bond SF et al (2021) Exercise hormone irisin is a critical regulator of cognitive function. Nat Metab 3(8):1058–1070

Jin J, Jing H, Choi G, Oh MS, Ryu JH, Jeong JW et al (2008) Voluntary exercise increases the new cell formation in the hippocampus of ovariectomized mice. Neurosci Lett 439(3):260–263

Kim SY, Park JH, Lee MY, Oh KS, Shin DW, Shin YC (2019) Physical activity and the prevention of depression: a cohort study. Gen Hosp Psychiatry 60:90–97

Kohman RA, Bhattacharya TK, Wojcik E, Rhodes JS (2013) Exercise reduces activation of microglia isolated from hippocampus and brain of aged mice. J Neuroinflammation 10:114

Laird E, Rasmussen CL, Kenny RA, Herring MP (2023) Physical activity dose and depression in a cohort of older adults in the Irish longitudinal study on ageing. JAMA Netw Open 6(7): e2322489

Laurin D, Verreault R, Lindsay J, MacPherson K, Rockwood K (2001) Physical activity and risk of cognitive impairment and dementia in elderly persons. Arch Neurol 58(3):498–504

Leotti LA, Iyengar SS, Ochsner KN (2010) Born to choose: the origins and value of the need for control. Trends Cogn Sci 14(10):457–463

Lespine LF, Tirelli E (2018) Evidence for a long-term protection of wheel-running exercise against cocaine psychomotor sensitization in adolescent but not in adult mice. Behav Brain Res 349:63–72

Lloyd BA, Hake HS, Ishiwata T, Farmer CE, Loetz EC, Fleshner M et al (2017) Exercise increases mTOR signaling in brain regions involved in cognition and emotional behavior. Behav Brain Res 323:56–67

Loughridge AB, Greenwood BN, Day HE, McQueen MB, Fleshner M (2013) Microarray analyses reveal novel targets of exercise-induced stress resistance in the dorsal raphe nucleus. Front Behav Neurosci 7:37

MacLennan AJ, Drugan RC, Hyson RL, Maier SF, Jt M, Barchas JD (1982) Dissociation of long-term analgesia and the shuttle box escape deficit caused by inescapable shock. J Comp Physiol Psychol 96(6):904–912

Maier SF, Keith JR (1987) Shock signals and the development of stress-induced analgesia. J Exp Psychol Anim Behav Process 13(3):226–238

Maier SF, Watkins LR (2005) Stressor controllability and learned helplessness: the roles of the dorsal raphe nucleus, serotonin, and corticotropin-releasing factor. Neurosci Biobehav Rev 29(4–5):829–841

Maier SF, Drugan RC, Grau JW (1982) Controllability, coping behavior, and stress-induced analgesia in the rat. Pain 12(1):47–56

Maier SF, Amat J, Baratta MV, Paul E, Watkins LR (2006) Behavioral control, the medial prefrontal cortex, and resilience. Dialogues Clin Neurosci 8(4):397–406

Maswood S, Barter JE, Watkins LR, Maier SF (1998) Exposure to inescapable but not escapable shock increases extracellular levels of 5-HT in the dorsal raphe nucleus of the rat. Brain Res 783(1):115–120

McDowell CP, Campbell MJ, Herring MP (2016) Sex-related differences in mood responses to acute aerobic exercise. Med Sci Sports Exerc 48(9):1798–1802

McEwen BS, Bowles NP, Gray JD, Hill MN, Hunter RG, Karatsoreos IN et al (2015) Mechanisms of stress in the brain. Nat Neurosci 18(10):1353–1363

Moya NA, Tanner MK, Smith AM, Balolia A, Davis JKP, Bonar K et al (2020) Acute exercise enhances fear extinction through a mechanism involving central mTOR signaling. Neurobiol Learn Mem 176:107328

Mul JD, Soto M, Cahill ME, Ryan RE, Takahashi H, So K et al (2018) Voluntary wheel running promotes resilience to chronic social defeat stress in mice: a role for nucleus accumbens DeltaFosB. Neuropsychopharmacology 43(9):1934–1942

Munive V, Santi A, Torres-Aleman I (2016) A concerted action of estradiol and insulin like growth factor I underlies sex differences in mood regulation by exercise. Sci Rep 6:25969

Naghibi S, Shariatzadeh Joneydi M, Barzegari A, Davoodabadi A, Ebrahimi A, Eghdami E et al (2021) Treadmill exercise sex-dependently alters susceptibility to depression-like behaviour, cytokines and BDNF in the hippocampus and prefrontal cortex of rats with sporadic Alzheimer-like disease. Physiol Behav 241:113595

Nicastro TM, Greenwood BN (2016) Central monoaminergic systems are a site of convergence of signals conveying the experience of exercise to brain circuits involved in cognition and emotional behavior. Curr Zool 62(3):293–306

Nowacka-Chmielewska M, Grabowska K, Grabowski M, Meybohm P, Burek M, Malecki A (2022) Running from stress: neurobiological mechanisms of exercise-induced stress resilience. Int J Mol Sci 23(21)

Pavlidi P, Kokras N, Dalla C (2023) Sex differences in depression and anxiety. Curr Top Behav Neurosci 62:103–132

Paxinos G, Watson C (1998) The rat brain in stereotaxic coordinates. Academic Press, New York

Pearce M, Garcia L, Abbas A, Strain T, Schuch FB, Golubic R et al (2022) Association between physical activity and risk of depression: a systematic review and meta-analysis. JAMA Psychiatry 79(6):550–559

Peterson AB, Hivick DP, Lynch WJ (2014) Dose-dependent effectiveness of wheel running to attenuate cocaine-seeking: impact of sex and estrous cycle in rats. Psychopharmacology (Berl) 231(13):2661–2670

Petzinger GM, Holschneider DP, Fisher BE, McEwen S, Kintz N, Halliday M et al (2015) The effects of exercise on dopamine neurotransmission in Parkinson's disease: targeting neuroplasticity to modulate basal ganglia circuitry. Brain Plast 1(1):29–39

Rajizadeh MA, Esmaeilpour K, Motamedy S, Mohtashami Borzadaranb F, Sheibani V (2020) Cognitive impairments of sleep-deprived ovariectomized (OVX) female rats by voluntary exercise. Basic Clin Neurosci 11(5):573–586

Rehm J, Shield KD (2019) Global burden of disease and the impact of mental and addictive disorders. Curr Psychiatry Rep 21(2):10

Rice D, Nijs J, Kosek E, Wideman T, Hasenbring MI, Koltyn K et al (2019) Exercise-induced hypoalgesia in pain-free and chronic pain populations: state of the art and future directions. J Pain 20(11):1249–1266

Rozeske RR, Evans AK, Frank MG, Watkins LR, Lowry CA, Maier SF (2011) Uncontrollable, but not controllable, stress desensitizes 5-HT1A receptors in the dorsal raphe nucleus. J Neurosci 31(40):14107–14115

Saadati H, Esmaeili-Mahani S, Esmaeilpour K, Nazeri M, Mazhari S, Sheibani V (2015) Exercise improves learning and memory impairments in sleep deprived female rats. Physiol Behav 138: 285–291

Sapolsky RM (2000) Glucocorticoids and hippocampal atrophy in neuropsychiatric disorders. Arch Gen Psychiatry 57(10):925–935

Severinsen MCK, Pedersen BK (2020) Muscle-organ crosstalk: the emerging roles of Myokines. Endocr Rev 41(4):594–609

Simmons JM, Winsky L, Zehr JL, Gordon JA (2021) Priorities in stress research: a view from the U.S. National Institute of Mental Health. Stress 24(2):123–129

Slonaker JR (1924) The effect of pubescence, oestruation and menopause on the voluntary activity in the albino rat. Am J Physiol

Smethells JR, Greer A, Dougen B, Carroll ME (2020) Effects of voluntary exercise and sex on multiply-triggered heroin reinstatement in male and female rats. Psychopharmacology (Berl) 237(2):453–463

Soshi T, Andersson M, Kawagoe T, Nishiguchi S, Yamada M, Otsuka Y et al (2021) Prefrontal plasticity after a 3-month exercise intervention in older adults relates to enhanced cognitive performance. Cereb Cortex 31(10):4501–4517

Speaker KJ, Cox SS, Paton MM, Serebrakian A, Maslanik T, Greenwood BN et al (2014) Six weeks of voluntary wheel running modulates inflammatory protein (MCP-1, IL-6, and IL-10) and DAMP (Hsp72) responses to acute stress in white adipose tissue of lean rats. Brain Behav Immun 39:87–98

Steel Z, Marnane C, Iranpour C, Chey T, Jackson JW, Patel V et al (2014) The global prevalence of common mental disorders: a systematic review and meta-analysis 1980-2013. Int J Epidemiol 43(2):476–493

Stonerock GL, Hoffman BM, Smith PJ, Blumenthal JA (2015) Exercise as treatment for anxiety: systematic review and analysis. Ann Behav Med 49(4):542–556

Strong PV, Greenwood BN, Fleshner M (2009) The involvement of 5-HT2C receptors in the dorsal striatum in learned helplessness behaviors. Society for Neuroscience 2009 Abstract Viewer/ Itinerary Planner

Strong PV, Christianson JP, Loughridge AB, Amat J, Maier SF, Fleshner M et al (2011) 5-hydroxytryptamine 2C receptors in the dorsal striatum mediate stress-induced interference with negatively reinforced instrumental escape behavior. Neuroscience 197:132–144

Szuhany KL, Bugatti M, Otto MW (2015) A meta-analytic review of the effects of exercise on brain-derived neurotrophic factor. J Psychiatr Res 60:56–64

Tanner MK, Fallon IP, Baratta MV, Greenwood BN (2019) Voluntary exercise enables stress resistance in females. Behav Brain Res 369:111923

Tanner MK, Davis JKP, Jaime J, Moya NA, Hohorst AA, Bonar K et al (2022) Duration- and sex-dependent neural circuit control of voluntary physical activity. Psychopharmacology (Berl) 239(11):3697–3709

Tanner MK, Hohorst AA, Mellert SM, Loetz EC, Baratta MV, Greenwood BN (2023) Female rats are more responsive than are males to the protective effects of voluntary physical activity against the behavioral consequences of inescapable stress. Stress:1–14

Tong L, Shen H, Perreau VM, Balazs R, Cotman CW (2001) Effects of exercise on gene-expression profile in the rat hippocampus. Neurobiol Dis 8(6):1046–1056

Uysal N, Kiray M, Sisman AR, Camsari UM, Gencoglu C, Baykara B et al (2015) Effects of voluntary and involuntary exercise on cognitive functions, and VEGF and BDNF levels in adolescent rats. Biotech Histochem 90(1):55–68

van Praag H (2008) Neurogenesis and exercise: past and future directions. Neuromolecular Med 10(2):128–140

Vantrease JE, Avonts B, Padival M, DeJoseph MR, Urban JH, Rosenkranz JA (2022) Sex differences in the activity of basolateral Amygdalar neurons that project to the bed nucleus of the Stria terminalis and their role in anticipatory anxiety. J Neurosci 42(22):4488–4504

Varma VR, Chuang YF, Harris GC, Tan EJ, Carlson MC (2015) Low-intensity daily walking activity is associated with hippocampal volume in older adults. Hippocampus 25(5):605–615

Vivar C, Potter MC, van Praag H (2013) All about running: synaptic plasticity, growth factors and adult hippocampal neurogenesis. Curr Top Behav Neurosci 15:189–210

Walker DL, Toufexis DJ, Davis M (2003) Role of the bed nucleus of the stria terminalis versus the amygdala in fear, stress, and anxiety. Eur J Pharmacol 463(1–3):199–216

Wang S, Chen L, Zhang L, Huang C, Xiu Y, Wang F et al (2015) Effects of long-term exercise on spatial learning, memory ability, and cortical capillaries in aged rats. Med Sci Monit 21:945–954

Werme M, Messer C, Olson L, Gilden L, Thoren P, Nestler EJ et al (2002) Delta FosB Regulates Wheel Running. J Neurosci 22(18):8133–8138

Wilckens KA, Stillman CM, Waiwood AM, Kang C, Leckie RL, Peven JC et al (2021) Exercise interventions preserve hippocampal volume: a meta-analysis. Hippocampus 31(3):335–347
Xue X, Liu B, Hu J, Bian X, Lou S (2022) The potential mechanisms of lactate in mediating exercise-enhanced cognitive function: a dual role as an energy supply substrate and a signaling molecule. Nutr Metab (Lond) 19(1):52

Exercise and Neuroendocrinology

Anthony C. Hackney

Contents

Abstract The intent of this chapter is to provide a basic overview and explanation of the neuroendocrine system responses to an exercise session in healthy adults. Specifically, the physiologic mechanisms for inducing the hormonal responses of this system and the scope of such responses are addressed. Additionally, factors that augment or attenuate exercise hormonal responses are presented as well as issues related to sex differences, and exercise-related hormonal dysfunction.

Keywords Endocrines · Homeostasis · Hormones · Physical Activity · Stress

---

A. C. Hackney (✉)
University of North Carolina, Chapel Hill, NC, USA,
e-mail: ach@email.unc.edu

Curr Topics Behav Neurosci (2024) 67: 61–76
https://doi.org/10.1007/7854_2024_491
Published Online: 23 July 2024

1 Introduction

Public health and medical research support the benefit of regular physical activity to aid the overall health of well-being of people (Surgeon General Report 1996). Recommendations for the quantity and quality of such physical activity have been developed by leading health agencies such as the World Health Organization (WHO) and the American College of Sports Medicine (Bull et al. 2020). To this end, exercise scientists continue to pursue and understand the physiological changes associated with the multitude of physical activities that the human body can perform. This is critical as the adaptive processes (physiological mechanisms) and time course for such responses need to be fully elucidated in order to optimize the health and performance benefits of women and men.

Along these lines, many scientists interested in the physiology of exercise recognize the critical roles of the hormonal responses by the neuroendocrine system to the physical stress of exercise. This is a logical pursuit due to the critical roles hormones play in bringing about homeostatic adjustments in many bodily systems to allow coping with the demands of any stressful activity, such as an exercise session (Johnson et al. 1992). The hormonal reactions to exercise occur in order to bring about and aid in inducing appropriate cardiovascular-respiratory adjustments (e.g., the hormones, norepinephrine, epinephrine), accelerating energy metabolism (e.g., growth hormone, cortisol), and maintaining proper thermoregulation accommodations (e.g., aldosterone). In addition, some physiological changes occur which can be classified as stress-reactive responses (e.g., immunological changes) which may not physiologically aid directly in the exercise session but are a consequence of the activity (Hackney 2006).

The endocrine system's release of hormones works via a stimulus-response model. Specifically, some form of humoral, neural, or hormonal stimuli (primarily within the blood) provokes an endocrine gland to produce and secrete its hormone (s) in the circulatory system. In turn, the circulating hormones invoke changes in target tissues to allow for accommodations to the original stimulus (Nussey and Whitehead 2001). The regulation of circulating hormones revolves around feedback loops, which are chiefly of a negative feedback type (Nussey and Whitehead 2001). A classic illustration of this would be when a person eats, this raises blood glucose, which provokes pancreatic insulin release, which then in turn promotes target tissue (e.g., skeletal muscle) update of glucose from the blood.

This chapter intends to present an overview of the sequence of events that bring about the multi-faceted changes in circulating hormones in reaction to an exercise session. It is recognized that entire volumes have been written concerning this topic but herein space limitations dictate a truncated discussion of the topic. For that reason, this dialogue is delimited to hormonal responses found in healthy, mature adults who are not overweight/obese and prior to experiencing either menopause or andropause. Furthermore, in the context of this discussion, relative to sex the focus herein is on cis-gender men and women. Therefore, the terms *female* and *male* are used as adjectives, and *man/men* and *woman/women* as nouns.

2 Hormonal Roles: Physiology of Exercise

Figure 1 presents a schematic illustration of the physiological consequences of the hormonal responses to an exercise session. Noticeably, there are wide-ranging effects of hormones, and without their appropriate response (i.e., neuroendocrine activity) the ability of an exerciser, whether recreational or a competitive athlete, to perform physical exercise would be severely compromised and sub-optimal.

Furthermore, in examining Fig. 1, it is important to recognize that there are several critical factors that determine the magnitude of the change in the circulating hormone levels during a specific exercise session (i.e., not all exercise is the same in effect on the neuroendocrine system). These factors are: (1) the intensity at which the exercise is performed, (2) the duration for which the exercise is performed, (3) the modality involved in doing the exercise (e.g., running vs. lifting-weights), (4) the ambient environment that the exercise is performed in (hot vs. cold weather), and (5) the physical fitness level of the person performing the exercise, in the sense of, are they a novice or experienced and considered "trained" (Borer 2003; Galbo 1983; McMurray and Hackney 2000; Viru and Viru 2001).

Relative to the first, in nearly a universal fashion, the greater the intensity of an exercise session the larger the magnitude of the hormonal change in the circulation. And, in the case of some hormones, for example, cortisol and growth hormone, until a critical threshold of exercise intensity is reached (~50–60% of maximal intensity) there are only slight changes in circulating levels. Secondly, prolonging the duration

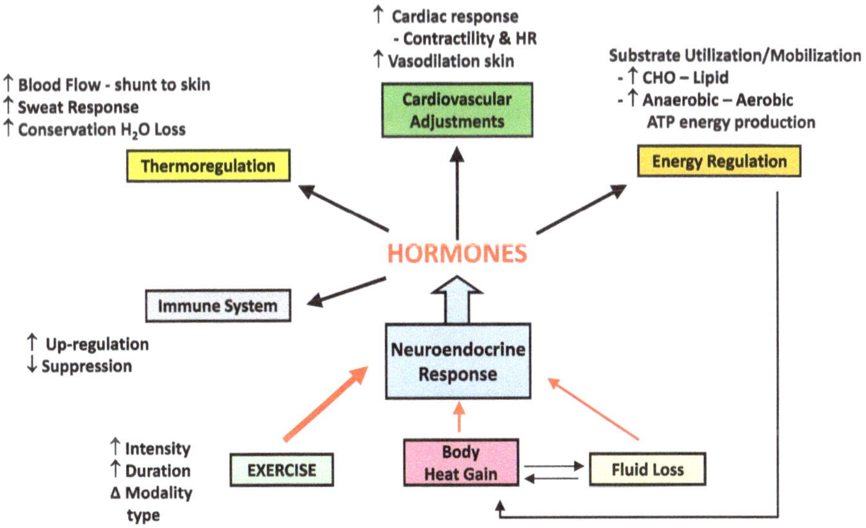

Fig. 1 The roles that the neuroendocrine system plays via hormones to bring about the physiological adjustments to allow exercise in humans. Abbreviations: ATP = adenosine triphosphate (i.e., the energy currency of the body), CHO = carbohydrate, H_2O = water, HR = heart rate. Red arrows indicate factors that invoke and amplify neuroendocrine hormonal changes/responses

of an exercise session typically has a similar effect to augment the magnitude of the hormonal change. The mechanism for these gradual increases seems to be a combination of events influencing the hormone levels, such as blood hemoconcentration, increased hormonal production and reduced metabolic clearance of hormones from the blood (Hackney 2006). However, in some circumstances, excessive duration can actually have the opposite effect, and hormonal levels have been reported to be reduced (Galbo 1983; Galbo et al. 1977) (*N.B.*, what defines "excessive duration" is relative to the level of training an individual has undergone). Examples of hormones that can be reduced in such scenarios are testosterone and thyroid-stimulating hormone (Galbo et al. 1977; Opstad and Aakvaag 1981). Relative to this last point, the specific level of duration to invoke reductions is an issue of some debate as this phenomenon is not well studied. As to exercise modality, this can be a complicated factor to characterize, as issues such as muscle mass involvement (e.g., lifting weights with only upper body muscularity versus only the lower body), the dynamic versus static movement pattern aspects of the muscular contractile process, and the body position (sitting down, upright, supine) can all play modifying roles to affect hormones (Athanasiou et al. 2023). Similarly, the ambient environment can drastically modify responses. For example, exercising in hot, humid weather can exacerbate the thermoregulatory responses of individuals (i.e., greater body core temperature increases), which leads to a substantially greater magnitude of hormonal changes (Cross et al. 1996). Interestingly, becoming dehydrated while exercising mimics the physiological responses of a hot environment, even if the ambient temperature is comfortable and thermoneutral (Costill et al. 1981). Along these same lines, exercising in water can impact thermoregulation influences as water immersion results in an enhanced conductive heat transfer in or out of the body depending upon the water temperature. Furthermore, with water immersion, the influence of the external hydrostatic forces of the water changes fluid distributions and dynamics in the body and as such hormonal responses. Finally, the greater the level of physical fitness an individual exhibits, typically the less the magnitude of the hormonal response to an exercise session. With increases in an individual's physical fitness level, there is a greater degree of sensitivity within the neuroendocrine system and target tissue receptor responsiveness due to increased receptor expression (Goodman 2002; Hackney and Lane 2015); hence, less hormone is necessary to invoke needed physiological accommodations.

3 Hormonal Exercise Response Model

As already noted, there is a multitude of hormone changes in a single exercise session, primarily increased circulating blood levels (*N.B.,* although typically insulin becomes decreased during an exercise session). Many of these hormonal responses are not independent of one another but are highly interrelated and interdependent. To illustrate this point, what is termed the *Hormonal Exercise Response Model* (HERM) is presented herein. Essentially, this model describes how the hormonal responses to most exercise sessions can be organized into a series of three interactive phases which invoke either differing responses or amplification of ongoing responses (Hackney and Lane 2015).

3.1 Phase I

The first phase of this model deals with the hormonal response immediately at the onset of exercise, with these responses taking just seconds to occur. These responses revolve around the increased sympathetic nervous system activation that occurs with the onset of bodily motion. This increased sympathetic nervous system activity can also be a result of anticipation of the ensuing exercise—which is most certainly the case in sports competition scenarios. This increased sympathetic nervous system activity results in catecholamines (norepinephrine and epinephrine) released at target tissues directly, as well as elevations in circulating catecholamines (specifically norepinephrine) from so-called sympathetic "spillover" effects (Luger et al. 1988).

This effect is further amplified by the sympathetic connection to the adrenal medullary gland which in turn adds to the circulating catecholamine (epinephrine release via medulla is greater than norepinephrine) response. Simultaneous with these sympathetic-adrenal medullary actions, pancreatic insulin secretion begins to be inhibited, while glucagon secretion becomes stimulated, the latter resulting mobilization of glucose for liver glycogen stores and fatty acid from adipocytes. This entire process involves a feed-forward mechanism of the central nervous system to drive these initial responses, although the events are also modified by peripheral afferent neural input from sensory receptors, in particular, those of skeletal muscle once movement commences (Luger et al. 1988; Hackney et al. 2011; Sutton et al. 1968).

3.2 Phase II

The intermediate or secondary phase takes slightly longer to develop but is still typically very fast beginning usually in much less than a minute from the onset of exercise. In this stage, the hypothalamus begins the process of releasing hormones such as thyrotropin-releasing factor, corticotrophin-releasing factor (i.e., CRF; see Table 1 for standard neuroendocrine abbreviations), and growth hormone-releasing factor in an attempt to provoke changes at the anterior pituitary gland to stimulate the release of specific hormones from this endocrine gland. As the pituitary begins to respond to the hypothalamic stimulus there is the release of the various "trophic hormones" from it into circulation. These hormones in turn begin to affect their specific peripheral target endocrine glands to stimulate additional hormonal release. One of the most rapidly acting elements in this cascade of events is the hypothalamic-pituitary-adrenal cortical interaction where CRF brings about

Table 1 Hormone (hormone-like agents) abbreviations utilized in neuroendocrinology. Note that some hormones have more than one abbreviation, as well as more than one proper name. *N.B.*, hormone abbreviations are not standardized within the scientific literature, as such, what is presented is the most commonly cited term(s) used for specific hormones. This list is not exhaustive but focuses on the hormones most prevalent in exercise-related research

Abbreviation	Hormone
ACTH	Adrenocorticotropic hormone
ALD	Aldosterone
ADH	Antidiuretic hormone Arginine vasopressin
AMH	Anti-Müllerian hormone
CCKCCK-PZ	Cholecystokinin
CRH CRF	Corticotropin-releasing hormone Corticotropin-releasing factor
C	Cortisol
CT	Calcitonin
DHEA	Dehydroepiandrosterone
DHEA-S	Dehydroepiandrosterone-sulfate
E Epi	Epinephrine Noradrenaline
EPO	Erythropoietin
E_2	Estradiol-β-17
FSH	Follicle-stimulating hormone
GHRL	Ghrelin
GH	Growth hormone
GIP	Glucose-dependent insulinotropic polypeptide
GHIH	Growth hormone-inhibiting hormone
GLP-1	Glucagon-like peptide 1
GnRH	Gonadotropin-releasing hormone
GHRH	Growth hormone-releasing hormone
hCG	Human chorionic gonadotropin
IGF-1	Insulin-like growth factor-1
IL-6	Interleukin-6
LH	Luteinizing hormone
MSH	Melanocyte-stimulating hormone
NE NEpi	Norepinephrine Noradrenaline
OT	Oxytocin
P_4	Progesterone
17-OHP	17-Hydroxyprogesterone
PIH	Prolactin-inhibiting hormone
PRH	Prolactin-releasing hormone
PRL	Prolactin
PTH	Parathyroid hormone Parathormone
PYY	Peptide YY

(continued)

Table 1 (continued)

Abbreviation	Hormone
SS SST SOM	Somatostatin
T Testo	Testosterone
T_3	Triiodothyronine
T_4	Thyroxine Tetraiodothyronine
TRH TRF	Thyrotropin-releasing hormone Thyrotropin-releasing factor
TSH	Thyroid-stimulating hormone

adrenocorticotrophic hormone release and that in turn ultimately brings about cortisol release from the adrenal cortex. The duration of this phase is highly variable, being dependent on the intensity of the exercise session and the overall physical fitness level of the individual who is exercising, the phase length being prolonged the greater the exercise intensity and, or the lower the fitness level of the individual (Borer 2003; Galbo 1983; Luger et al. 1987; Deuster et al. 1989).

3.3 *Phase III*

As an exercise session is continued (i.e., extending the duration) there is a transition beyond the intermediate phase into the third phase of response which is a more prolonged state of responsiveness. In this third phase, the responses of the sympathetic-adrenal axis are augmented by other hormones from the anterior and posterior pituitary (e.g., growth hormone, prolactin, antidiuretic hormone) and the peripheral endocrine glands subordinated to pituitary regulation (such as testosterone, thyroxine, triiodothyronine, insulin-like growth factor-1 [i.e., hepatic release of insulin-like growth factor-1 (IGF-1) into blood; muscle release of mechano-growth factor, a local IGF-1 variant]). As fluids shift from the vascular space and total body water stores are compromised due to sweating for heat dissipation and inter-compartmental fluid movement, the renin-angiotensin-aldosterone system (RAAS) is activated (inducing vasoconstrictive actions and water resorption action at the kidney). Additionally, during this phase, the skeletal muscle begins to release select cytokines (e.g., interleukin-6 [IL-6]), hormonal-like agents, into the circulation which affect other hormones to be released (e.g., cortisol) which can have actions to signal energy substrate mobilization and well as immune response (Borer 2003).

Phases I and II of the model propose that neural factors are the primary stimuli regulating the hormonal responses to exercise; however, in the third phase of the model, there is an ever-increasing influence of the humoral and hormonal factors that regulate the overall responses due to the changes in the *internal milieu*. This shifting of primary regulatory factors allows an increasing reliance upon feedback rather than a feed-forward control mechanism to determine the magnitude of the hormonal response. The influence of humoral and hormonal stimuli in modulating hormonal levels is magnified as the exercise duration is extended and energy substrate availability issues cause shifts in energy fuel usage (i.e., decreased carbohydrate increased lipid), or hydration issues affecting the blood volume (i.e., hemoconcentration and, or dehydration) begin compromising the thermoregulatory ability and leading to greater heat storage within the body affecting hormonal responses (e.g., increased heat storage increased norepinephrine, and epinephrine release). Interestingly, the core temperature changes with exercise (and also via ambient environment amplification) result in an exceedingly greater hormonal response for several key endocrine agents (Borer 2003; Galbo 1983; Galbo et al. 1977; Luger et al. 1987; Daly et al. 2005).

3.4 HERM Synopsis

In summary, the proposed HERM model consists of: Phase I – an immediate phase primarily driven by neural mechanism, Phase II – the intermediate phase involves neural and pituitary mechanisms controlling responses, and Phase III – the humoral-based adjustment phase involving principally more slow-acting pituitary mechanisms. The model, while not perfectly inclusive of all hormonal responses, does provide an organized framework for the endocrine responses to exercise activities and helps to illustrate the highly interactive complexity of these responses. Furthermore, the HERM model gives the context for the sequence of changes and the direction of change. However, it is important for scientists studying exercise hormonal responses to recognize that the intensity and duration of the exercise sessions are critical factors in determining the magnitude of the hormonal change response observed. The objective of this model is to provide a conceptual framework to aid scientists in understanding the endocrine responses and reactivity to the stress of an exercise session.

4 Recovery from Exercise

Findings from many research studies suggest the neuroendocrine response to an acute exercise session appears very transient in nature and rebounds during the recovery from exercise. That is, the hormonal levels return to baseline or slightly below basal values relatively rapidly within 1 to 2 h (i.e., the latter being a biphasic response event within the negative feedback regulatory system) (Goodman 2002). The exception to these findings is when the exercise is highly demanding such as a competitive event or exercising to exhaustion (Galbo 1983; Galbo et al. 1977). In such scenarios the recovery of the hormonal levels to normal, resting points can take several days (Anderson et al. 2016).

Interestingly, it has been reported that if athletes performed two intense training sessions (>75% maximal aerobic capacity [VO_{2max}]) in a single day, their circulating cortisol levels were significantly depressed at night compared to daytime basal levels. Furthermore, the more intense these day-time exercise sessions (i.e., higher intensity of workload), the greater the magnitude of the night-time suppression observed (Viru et al. 2007). While not typical in exercise programs, such exercise training practices are something that does occur in the training regimes of elite athletes (McKay et al. 2022).

5 Mechanisms of Hormonal Responses

The physiological mechanisms by which hormone levels (i.e., concentrations) in the blood change during exercise are due to three factors: *1*) alterations in circulatory vascular fluid content, *2*) rate of hormonal metabolic clearance, and *3*) glandular production/secretion responses. It is important to recognize that blood concentration changes (increases) with exercise are not always due to the increased endocrine gland secretion (*N.B.*, a common belief; (Viru 1985)) (Borer 2003; McMurray and Hackney 2000; Viru and Viru 2001).

Circulatory vascular fluid content alterations are a result of plasma fluid moving into or out of the vascular spaces. Throughout an exercise session, it is common for plasma volume "shifts" to occur, specifically a loss of plasma fluid, resulting in hemoconcentration (increased levels) of any substance trapped within the blood vessel. This would apply to large molecular weight hormones or hormones bound to large molecular weight carrier proteins (e.g., >20,000 Da, but capillaries [glomerular vs. systemic] vary on this point) (Kaufman and Lamster 2002; Renkin 1986). For example, in a 30-to 90-min exercise session of moderate intensity (~50–60% of maximum) a 10–15% hemoconcentration effect of the blood could be expected (Van Beaumont 1973; Dill and Costill 1974). This plasma volume reduction is exacerbated if the ambient environment condition creates a greater thermal load invoking a heavy sweating response and, or an individual does not consume adequate fluids during exercise in an attempt to maintain hydration (Costill et al. 1981). Conversely, during the recovery from exercise (when ample fluids are consumed) there tends to be a rebound effect and fluid moves into the vascular space to cause a hemodilution effect on substances such as hormones (Borer 2003; McMurray and Hackney 2000; Viru and Viru 2001; Hackney and Viru 2008).

The rate of metabolic clearance of a hormone is a physiologic concept concerned with hormone removal from the blood. Hormones are "taken up" by receptor mechanisms at their target tissues which can implement physiological actions at the tissue, what are termed *downstream* events (Goodman 2002). Additionally, though, some clearance is through the degradation–deactivation removal process at non-target tissues such as hepatocytes, or via renal filtration and clearance.

Increases and/or decreases in the rate of metabolic clearance can consequentially impact blood hormonal concentration dramatically (Cadoux-Hudson et al. 1985).

Glandular secretion involves the endocrine gland producing the hormone in response to a stimulus and the gland secreting more of the hormone in the circulatory system. This usually results in increases in blood concentration levels, but it is possible for the rate of secretory increases and the rate of metabolic clearance removal to match one another (Goodman 2002; Widmaier 1992). In such a case as this, the concentration of the hormone in the blood does not seem to increase but the turnover has, and the target tissues are being presented with greater amounts of the hormone (Goodman 2002).

Regrettably, it is a prevalent belief that if hormonal levels increase in the blood, then the physiological actions of that hormone are going to be activated (e.g., increased [↑] cortisol → ↑ catabolism). This is a gross oversimplification of endocrinological events concerning hormones. Indeed, hormonal concentration in the blood is a key fundamental determinant of activation of physiological actions, but not the only one; that is, in order for a hormone to initiate and activate a physiological process in target tissue there must be:

(a) Adequate numbers of target tissue cells expressing functional receptors for the hormone.
(b) Adequate numbers of functional receptors on the cells of target tissue.
(c) A high-affinity level of the receptors on the cells for the hormone.
(d) Sufficient post-receptor amplification mechanisms within the cells (Goodman 2002; Widmaier 1992).

This last point is extremely critical and can be influenced by the behavior and actions of the exerciser. For example, if anabolism is to be initiated in the target tissue, not only does an anabolic hormone need to be present, but adequate amounts of amino acids must be consumed in the diet and be available for cellular mechanisms (biochemical pathways, enzymes) to utilize in specific protein synthesis processes. These downstream amplification mechanisms are essential for a positive adaptation to occur in skeletal muscle, arguably the most critical tissue–organ affecting exercise capacity (Brooks et al. 2004; Hackney and Constantini 2020).

Hormonal assessment has traditionally been collected on blood specimens (plasma or serum). Contemporary technology has allowed saliva to be used as part of the evaluation of endocrine function with exercise. The factors that affect drug availability in saliva are generally true also for salivary hormones. Most hormones enter saliva by passive diffusion across the acinar cells. Most of these hormones are lipid-soluble (i.e., steroids). Small polar molecules do not readily diffuse across cells and instead enter saliva through the tight junctions between cells (i.e., ultrafiltration; Quissell 1993). The molecular weight cut-off for ultrafiltration is 100–200 Da. This relatively small molecular size prevents many hormones from entering saliva from serum by means of ultrafiltration. In addition, active transport does not appear to facilitate hormone transfer into saliva (Vining and McGinley 1986).

6 Sex and Exercise Endocrinology

6.1 *Menstrual Cycle and Pregnancy*

Adult men and women share a multitude of physiological similarities in how their bodies function. Yet there are specific neuroendocrine differences that result in vast hormonal differences between the sexes. Most obviously a majority of these hormonal differences revolve around the reproductive systems in men compared to women. Women of reproductive age have a highly dynamic system for ensuring menstrual cycle hormonal changes to allow uterine and follicle development. The menstrual cycle can be divided into various phases based on the reproductive hormonal change. This consists of a repeating pattern of menses to start the follicular phase, followed by ovulation which leads into the luteal phase (Elliott-Sale et al. 2021). Evidence supports that the phase of the menstrual cycle influences to some degree the magnitude of hormonal responses to an exercise session. For example, during the luteal phase of the cycle, growth hormone and cortisol responses can be elevated over that of the follicular (Hackney 2021; Bunt 1990), while the exercise-induced suppression of ghrelin is blunted (Moniz et al. 2023). However, these menstrual cycle phase effects on the hormonal responses to exercise are not universally reported (Kanaley et al. 1992a, b), which may relate in some instances to the quality of the science being conducted on the menstrual cycle and exercise in women (Elliott-Sale et al. 2021).

An in-depth discussion of the impact of pregnancy on the neuroendocrine system is beyond the scope of this chapter. Nonetheless, it is important to mention, even if in brief. As the mother goes through gestation, reproductive hormones such as estradiol-β-17 and progesterone gradually increase to super-physiological levels as do some non-reproductive hormones (Kumar and Magon 2012). While there is growing research on the safety and efficacy of exercise during pregnancy, there is limited work on exercise hormonal responses in pregnant women (ACOG 2015). Interestingly it was reported from a study that cortisol and prolactin (two highly reactive stress hormones) did not significantly change with maximal exercise to volitional fatigue at 35 weeks of pregnancy. However, in this study, the resting levels of these hormones were many-fold higher than non-pregnancy levels (Hackney et al. 1993). The authors speculated this might have represented some type of "physiological ceiling effect" (i.e., resting levels of the hormones were at near maximal glandular secretion and hence when stressed with exercise there was a limited ability for further production).

6.2 *Women Versus Men*

Additionally, when comparing the sexes directly it is notable that women typically have higher resting levels of growth hormone, estradiol-β-17, progesterone, luteinizing hormone follicle stimulating hormone, androstenedione, IGF-1, prolactin, and leptin than men (Rifai 2018; Clinical Laboratory Reference Values 2019; Bunt et al. 1986; Pincus et al. 1996). Conversely, women have lower testosterone levels than men. However, it should be noted these differences can be influenced by age, menstrual cycle phase-status, and, or if hormonal contraceptive is being used by a woman (Lucis and Lucis 1972; Gardner and Shoback 2018).

These initial hormone differences (just noted) in a resting state can result in exercise response changes differing if men and women are directly compared (Rifai 2018; Clinical Laboratory Reference Values 2019; Vislocky et al. 2008). Interestingly, there are scenarios where no pre-exercise resting differences existed between the sexes; however, during and at post-exercise hormone response differences develop (e.g., norepinephrine and epinephrine) (Horton et al. 1998). The causality of this occurrence is unclear but may related to muscle mass size differences between the sexes, or muscle fiber type variations, and/or cardiovascular system disparities (e.g., in general women have smaller hearts, less blood volume, and cardiac output than men) (Huxley 2007).

Currently, the quantity and extent of exercise neuroendocrinology research on women is lacking. This is regrettable as women represent 50% of the population of the world and their engagement in exercise and sport is growing exponentially and shows no signs of abatement, and as such there deserves to be more robust research on female exercise physiology (Cowley et al. 2021; Laxdal 2023).

7 Neuroendocrine Dysfunction with Exercise

It has been well established that excessive exercise training, or even appropriate training without adequate nutritional caloric intake results in neuroendocrine dysfunction (Stellingwerff et al. 2021). Most prominent in these effects are disruptions to reproductive hormones resulting in lower resting levels of testosterone, estradiol-β-17, progesterone, luteinizing hormone, and follicle-stimulating hormone in both sexes (Borer 2003; Hackney 2017). These resting changes are not limited to just reproductive hormones as triiodothyronine has been found to be reduced as has leptin, while cortisol, prolactin, and growth hormone are elevated. The changes also manifest as altered exercise responses for some hormones (Meeusen et al. 2010), which can disrupt energy metabolism regulation as well as the adaptive response to exercise training (Hackney 2006; Stellingwerff et al. 2021). This entire area of research regrettably is understudied and is limited in application and practicality as published papers tend to be underpowered or are case studies.

8 Conclusions

The neuroendocrine system is highly responsive to stress, and the hormonal changes it evokes are means to accommodate and adjust (*deal with*) the disruption of homeostasis brought on by stress exposure. Physical exercise is a stressor to the human body (Hackney 2006). In many connotations, the term stress is viewed or interpreted as negative, however, exposure to exercise, in a repeated fashion has been shown to be beneficial and lead to many positive health outcomes (Surgeon General Report 1996; Brooks et al. 2004). Hence, not all stress exposure is bad and within realms that the body can cope with and adapt to.

The hormonal changes induced by exercise are multi-faceted and bring about diverse changes principally in energy metabolism, cardiovascular-respiratory parameters, and thermoregulatory temperature control as homeostatic adjustment responses. The physiological changes in these systems allow humans to perform a variety of exercise forms (modalities) across a spectrum of intensity efforts, and in varying environments.

The neuroendocrine system shows a great deal of plasticity and is able to adapt to the stress of repetitive exercise exposure. These adaptations allow for a more sensitive response to physical exercise sessions, and more tightly regulated accommodations in the physiological systems being controlled. The result of which is an improved exercise capacity and tolerance. The system, however, is not perfect and can be compromised in functionality and subsequent physiological effects by poor exercise approaches (e.g., too much) or dietary nutrition inadequacies (Mountjoy et al. 2023). This said, the neuroendocrine system is remarkable in its ability to allow us to change our physiological systems so we as humans can perform exercise.

References

American College of Obstetricians and Gynecologists (2015) ACOG committee opinion No. 650: physical activity and exercise during pregnancy and the postpartum period. Obstet Gynecol 126: 135–142

Anderson T, Lane AR, Hackney AC (2016) Cortisol and testosterone dynamics following exhaustive endurance exercise. Eur J Appl Physiol 116(8):1503–1509. https://doi.org/10.1007/s00421-016-3406-y

Athanasiou N, Bogdanis GC, Mastorakos G (2023) Endocrine responses of the stress system to different types of exercise. Rev Endocr Metab Disord 24:251–266

Borer KT (2003) Exercise endocrinology. Human Kinetics Publishers, Champaign

Brooks GA, Fahey TD, Baldwin KM (2004) Exercise physiology: human bioenergetics and its applications. McGraw Hill, New York

Bull FC, Al-Ansari SS, Biddle S, Borodulin K, Buman MP, Cardon G, Carty C, Chaput JP, Chastin S, Chou R, Dempsey PC, DiPietro L, Ekelund U, Firth J, Friedenreich CM, Garcia L, Gichu M, Jago R, Katzmarzyk PT, Lambert E, Leitzmann M, Milton K, Ortega FB, Ranasinghe C, Stamatakis E, Tiedemann A, Troiano RP, van der Ploeg HP, Wari V, Willumsen JF (2020) World Health Organization 2020 guidelines on physical activity and sedentary behaviour. Br J Sports Med 54(24):1451–1462. https://doi.org/10.1136/bjsports-2020-102955

Bunt JC (1990) Metabolic actions of estradiol: significance for acute and chronic exercise responses. Med Sci Sports Exerc 22(3):286–290

Bunt JC, Boileau RA, Bahr JM, Nelson RA (1986) Sex and training differences in human growth hormone levels during prolonged exercise. J Appl Physiol 61(5):1796–1801. https://doi.org/10.1152/jappl.1986.61.5.1796

Cadoux-Hudson TA, Few JD, Imms FJ (1985) The effect of exercise on the production and clearance of testosterone in well trained young men. Eur J Appl Physiol Occup Physiol 54(3): 321–325. https://doi.org/10.1007/BF00426153

Clinical Laboratory Reference Values (2019) In: Laposata M (ed) Laposata's laboratory medicine: diagnosis of disease in the clinical laboratory, 3rd edn. McGraw Hill, New York

Costill DL, Coté R, Fink WJ, Van Handel P (1981) Muscle water and electrolyte distribution during prolonged exercise. Int J Sports Med 2(3):130–134. https://doi.org/10.1055/s-2008-1034598

Cowley ES, Olenick AA, McNulty KL, Ross EZ (2021) "Invisible sportswomen": the sex data gap in sport and exercise science research. Women Sport Phys Activity J 29(2):146–151

Cross MC, Radomski MW, VanHelder WP, Rhind SG, Shephard RJ (1996) Endurance exercise with and without a thermal clamp: effects on leukocytes and leukocyte subsets. J Appl Physiol 81(2):822–829. https://doi.org/10.1152/jappl.1996.81.2.822

Daly W, Seegers CA, Rubin DA, Dobridge JD, Hackney AC (2005) Relationship between stress hormones and testosterone with prolonged endurance exercise. Eur J Appl Physiol 93(4): 375–380. https://doi.org/10.1007/s00421-004-1223-1

Deuster PA, Chrousos GP, Luger A, DeBolt JE, Bernier LL, Trostmann UH, Kyle SB, Montgomery LC, Loriaux DL (1989) Hormonal and metabolic responses of untrained, moderately trained, and highly trained men to three exercise intensities. Metabolism 38(2):141–148. https://doi.org/10.1016/0026-0495(89)90253-9

Dill DB, Costill DL (1974) Calculation of percentage changes in volumes of blood, plasma, and red cells in dehydration. J Appl Physiol 37(2):247–248. https://doi.org/10.1152/jappl.1974.37.2.247

Elliott-Sale KJ, Minahan CL, de Jonge XAKJ, Ackerman KE, Sipilä S, Constantini NW, Lebrun CM, Hackney AC (2021) Methodological considerations for studies in sport and exercise science with women as participants: a working guide for standards of practice for research on women. Sports Med 51(5):843–861. https://doi.org/10.1007/s40279-021-01435-8

Galbo H (1983) Hormonal and metabolic adaptation to exercise. Georg Thieme Verlag, Stuttgart

Galbo H, Hummer L, Peterson IB, Christensen NJ, Bie N (1977) Thyroid and testicular hormone responses to graded and prolonged exercise in man. Eur J Appl Physiol Occup Physiol 36(2): 101–106. https://doi.org/10.1007/BF00423117

Gardner DG, Shoback D (eds) (2018) Normal hormone reference ranges. Chapter appendix in Greenspan's basic & clinical endocrinology, 10th edn. McGraw Hill, New York

Goodman HM (2002) Endocrinology concepts for medical students. Adv Physiol Educ 25:213–224

Hackney AC (2006) Stress and the neuroendocrine system: the role of exercise as a stressor and modifier of stress. Expert Rev Endocrinol Metab 1(6):783–792. https://doi.org/10.1586/17446651.1.6.783

Hackney AC (2017) Doping, performance-enhancing drugs, and hormones in sport mechanisms of action and methods of detection. Elsevier, New York

Hackney AC (2021) Menstrual cycle hormonal changes and energy substrate metabolism in exercising women: a perspective. Int J Environ Res Public Health 18(19):10024. https://doi.org/10.3390/ijerph181910024

Hackney AC, Constantini NW (2020) Endocrinology of physical activity and sport. Springer, New York

Hackney AC, Lane AR (2015) Exercise and the regulation of endocrine hormones. Prog Mol Biol Transl Sci 135:293–311. https://doi.org/10.1016/bs.pmbts.2015.07.001

Hackney AC, Viru A (2008) Research methodology: endocrinologic measurements in exercise science and sports medicine. J Athl Train 43(6):631–639. https://doi.org/10.4085/1062-6050-43.6.631

Hackney AC, Katz VL, Watson WJ, Gall MM, McMurray RG (1993) Alpha-fetoprotein, prolactin and cortisol responses to maximal exercise during pregnancy. Scand J Med Sci Sports 3(1): 41–45. https://doi.org/10.1111/j.1600-0838.1993.tb00359.x

Hackney AC, Viru M, VanBruggen M, Janson T, Karelson K, Viru A (2011) Comparison of the hormonal responses to exhaustive incremental exercise in adolescent and young adult males.

Arq Bras Endocrinol Metabol 55(3):213–218. https://doi.org/10.1590/s0004-27302011000300006

Horton TJ, Pagliassotti MJ, Hobbs K, Hill JO (1998) Fuel metabolism in men and women during and after long-duration exercise. J Appl Physiol 85(5):1823–1832. https://doi.org/10.1152/jappl.1998.85.5.1823

Huxley VH (2007) Sex and the cardiovascular system: the intriguing tale of how women and men regulate cardiovascular function differently. Adv Physiol Educ 31(1):17–22. https://doi.org/10.1152/advan.00099.2006

Johnson EO, Kamilaris TC, Chrousos GP, Gold PW (1992) Mechanisms of stress: a dynamic overview of hormonal and behavioral homeostasis. Neurosci Biobehav Rev 16(2):115–130. https://doi.org/10.1016/s0149-7634(05)80175-7

Kanaley JA, Boileau RA, Bahr JM, Misner JE, Nelson RA (1992a) Cortisol levels during prolonged exercise: the influence of menstrual phase and menstrual status. Int J Sports Med 13(4):332–336. https://doi.org/10.1055/s-2007-1021276

Kanaley JA, Boileau RA, Bahr JA, Misner JE, Nelson RA (1992b) Substrate oxidation and GH responses to exercise are independent of menstrual phase and status. Med Sci Sports Exerc 24(8):873–880

Kaufman E, Lamster IB (2002) The diagnostic applications of saliva – a review. Crit Rev Oral Biol Med 13(2):197–212. https://doi.org/10.1177/154411130201300209

Kumar P, Magon N (2012) Hormones in pregnancy. Niger Med J 53(4):179–183. https://doi.org/10.4103/0300-1652.107549

Laxdal A (2023) The sex gap in sports and exercise medicine research: who does research on females? Scientometrics 128:1987–1994

Lucis OJ, Lucis R (1972) Oral contraceptives and endocrine changes. Bull World Health Organ 46(4):443–450

Luger A, Deuster PA, Kyle SB, Gallucci WT, Montgomery LC, Gold PW, Loriaux DL, Chrousos GP (1987) Acute hypothalamic-pituitary-adrenal responses to the stress of treadmill exercise. Physiologic adaptations to physical training. N Engl J Med 316(21):1309–1315. https://doi.org/10.1056/NEJM198705213162105

Luger A, Deuster PA, Gold PW, Loriaux DL, Chrousos GP (1988) Hormonal responses to the stress of exercise. Adv Exp Med Biol 245:273–280. https://doi.org/10.1007/978-1-4899-2064-5_22

McKay AKA, Stellingwerff T, Smith ES, Martin DT, Mujika I, Goosey-Tolfrey VL, Sheppard J, Burke LM (2022) Defining training and performance caliber: a participant classification framework. Int J Sports Physiol Perform 17(2):317–331. https://doi.org/10.1123/ijspp.2021-0451

McMurray RG, Hackney AC (2000) The endocrine system and exercise. In: Garrett W, Kirkendall D (eds) Exercise and sports science. Williams & Wilkins Publisher, Philadelphia

Meeusen R, Nederhof E, Buyse L, Roelands B, de Schutter G, Piacentini MF (2010) Diagnosing overtraining in athletes using the two-bout exercise protocol. Br J Sports Med 44(9):642–648. https://doi.org/10.1136/bjsm.2008.049981

Moniz SC, McCarthy SF, Broad AA, Medeiros PJ, Hazell TJ (2023) The exercise-induced suppression of acylated ghrelin is blunted in the luteal phase of the menstrual cycle compared to the follicular phase following vigorous-intensity exercise. Appetite 182:106425. https://doi.org/10.1016/j.appet.2022.106425

Mountjoy M, Ackerman KE, Bailey DM, Burke LM, Constantini N, Hackney AC, Heikura IA, Melin A, Pensgaard AM, Stellingwerff T, Sundgot-Borgen JK, Torstveit MK, Jacobsen AU, Verhagen E, Budgett R, Engebretsen L, Erdener U (2023) 2023 International Olympic Committee's (IOC) consensus statement on relative energy deficiency in sport (REDs). Br J Sports Med 57(17):1073 1097. https://doi.org/10.1136/bjsports-2023-106994

Nussey S, Whitehead S (2001) Endocrinology: an integrated approach. BIOS Scientific Publishers, Oxford

Opstad PK, Aakvaag A (1981) The effect of a high calory diet on hormonal changes in young men during prolonged physical strain and sleep deprivation. Eur J Appl Physiol Occup Physiol 46(1):31–39. https://doi.org/10.1007/BF00422172

Pincus SM, Gevers EF, Robinson IC, van den Berg G, Roelfsema F, Hartman ML, Veldhuis JD (1996) Females secrete growth hormone with more process irregularity than males in both humans and rats. Am J Physiol 270(1 Pt 1):E107–E115. https://doi.org/10.1152/ajpendo.1996.270.1.E107

Quissell DO (1993) Steroid hormone analysis in human saliva. Ann N Y Acad Sci 20(694): 143–145. https://doi.org/10.1111/j.1749-6632.1993.tb18348.x

Renkin EM (1986) Some consequences of capillary permeability to macromolecules: Starling's hypothesis reconsidered. Am J Physiol 250(5 Pt 2):H706–H710. https://doi.org/10.1152/ajpheart.1986.250.5.H706

Rifai N (2018) Tietz fundamentals of clinical chemistry and molecular diagnostics, 8th edn. Elsevier, New York

Stellingwerff T, Heikura IA, Meeusen R, Bermon S, Seiler S, Mountjoy ML, Burke LM (2021) Overtraining syndrome (OTS) and relative energy deficiency in sport (RED-S): shared pathways, symptoms and complexities. Sports Med 51:2251–2280

Surgeon General Report (1996) Surgeon general's report on physical activity and health. From the Centers for Disease Control and Prevention. JAMA 276(7):522

Sutton J, Young JD, Lazarus L, Hickie JB, Maksvytis J (1968) Hormonal changes during exercise. Lancet 2(7581):1304–1305. https://doi.org/10.1016/s0140-6736(68)91806-0

Van Beaumont W (1973) Red cell volume with changes in plasma osmolarity during maximal exercise. J Appl Physiol 35(1):47–50. https://doi.org/10.1152/jappl.1973.35.1.47

Vining RF, McGinley RA (1986) Hormones in saliva. Crit Rev Clin Lab Sci 23(2):95–146. https://doi.org/10.3109/10408368609165797

Viru A (1985) Hormonal ensemble in exercise – hormones in muscular activity, vol 1. CRC Pres, Boca Raton

Viru A, Viru M (2001) Biochemical monitoring of sport training. Human Kinetics, Champaign

Viru A, Viru M, Karelson K, Janson T, Siim K, Fischer K, Hackney AC (2007) Adrenergic effects on adrenocortical cortisol response to incremental exercise to exhaustion. Eur J Appl Physiol 100(2):241–245. https://doi.org/10.1007/s00421-007-0416-9

Vislocky LM, Gaine PC, Pikosky MA, Martin WF, Rodriguez NR (2008) Gender impacts the post-exercise substrate and endocrine response in trained runners. J Int Soc Sports Nutr 5:7. https://doi.org/10.1186/1550-2783-5-7

Widmaier EP (1992) Metabolic feedback in mammalian endocrine systems. Horm Metab Res 24: 147–153

Part II
Exercise and Laboratory Studies of Mood and Cognition

Exercise and Acute Cognitive Enhancement

Jennifer L. Etnier, Samuel W. Kibildis, and Samantha L. DuBois

Contents

Abstract Substantial research has explored the effects of a single session of exercise on cognitive performance. Meta-analytic reviews provide summary statistics relative to these effects when the cognitive task is performed while exercising and when performed after exercise. When performed concomitantly the effects of exercise are typically small or negative with the exception of moderate effects for speed of cognitive performance during moderate intensity exercise. When cognitive performance is assessed following the cessation of exercise, the effects are typically positive. When considering the effects following exercise, most studies have focused on executive function and episodic memory as the cognitive outcomes of interest with meta-analyses supporting small-to-moderate benefits. When potential moderators related to exercise parameters or participants characteristics are examined, results suggest that 20 minutes of moderate to vigorous intensity exercise consistently benefits performance with larger effects evident for executive function tasks with higher cognitive demands, for long-term memory tasks when exercise is performed prior to information encoding, and when considering individuals with lower initial cognitive performance levels. These observed benefits are likely due to the impacts acute exercise can have on physiological arousal, brain function, and brain neurotrophic factors. Advancements in our understanding of the potential of a

J. L. Etnier (✉), S. W. Kibildis, and S. L. DuBois
Department of Kinesiology, University of North Carolina Greensboro, Greensboro, NC, USA
e-mail: jletnier@uncg.edu

Curr Topics Behav Neurosci (2024) 67: 79–102
https://doi.org/10.1007/7854_2024_492
Published Online: 31 August 2024

single session of exercise to benefit cognitive performance are important for exercise prescription and to inform chronic exercise programs designed to benefit cognition.

Keywords Cognitive Performance · Executive Function · Memory · Physical Activity

1 Introduction and Chapter Overview

As discussed in the previous chapters, exercise has been shown to impact neurotransmitters, neuroendocrinology, and neurocircuitry. These impacts in the central nervous system have the potential to influence cognitive performance in both the short-term (acute) and the long-term (chronic). Cognitive performance is defined as the performance of tasks that require mental action or brain activity and that demonstrate proficiency in a variety of mental processes such as attention, executive function, information processing, memory, and perception. In this chapter, we focus on the effects of exercise on acute cognitive performance and look at cognitive domains that have been consistently explored in the literature.

In the published literature on acute exercise and cognitive performance, the cognitive domain that has received the most empirical attention is executive function. Executive function is defined as a higher order cognitive process that regulates more basic cognitive functions and emotions toward goal-directed behavior. Several cognitive abilities commonly included under the umbrella of executive function include inhibition, planning, set shifting, and working memory. Inhibition is the ability to restrain impulses and behaviors. Planning is defined as the ability to choose the necessary actions and perform those actions in the correct order to achieve a goal. Set shifting (sometimes referred to as cognitive flexibility) refers to the ability to switch mental focus to adapt to changing situations. And lastly, working memory is the ability to temporarily hold and manipulate a limited amount of information in mind. This ability to actively manipulate information to meet task goals is what characterizes working memory as an executive function ability and distinguishes it from the other forms of memory we will review in this chapter. The research interest in the effects of acute exercise on executive function probably reflects the perception that exercise-induced effects on these higher order cognitive abilities are likely to have the greatest impact on overall cognitive performance.

In addition to executive function, memory is another cognitive domain that has recently received substantial attention in the acute exercise and cognition literature. Memory is a broad cognitive construct that encompasses our ability to encode (initially learn), consolidate (stabilize learned information into long-term memory), and retrieve (access at a later time) information. Given the wide variety of information we encounter on a daily basis, memory is often classified into different subtypes according to the nature of the information involved. For example, episodic memory is the ability to remember details of specific events and experiences (e.g., what you ate for dinner last night). In contrast, semantic memory refers to the ability to remember general knowledge, concepts, and facts (e.g., your date of birth). And finally, procedural memory is the ability to unconsciously remember information necessary to perform different actions and skills (e.g., how to ride a bike). In addition to classification according to the type of information to be remembered, memory is also frequently classified according to the length of time the information needs to be retained. Short-term memory typically refers to the retention of information for a few seconds or 1–2 min after information encoding. In contrast, long-term memory is characterized by the retention of information for longer periods of time, usually several minutes, hours, or even days. The effect of acute exercise on memory performance is particularly interesting because exercise-induced benefits may be observed following an extended period of time after exercise cessation. In other words, although most research suggests that the effects of acute exercise on overall cognitive performance tend to be relatively transient (Chang et al. 2012a), evidence also suggests that the effects on memory specifically may be more enduring with larger effects reported for long-term memory than for short-term memory (Roig et al. 2013).

Due to the attention the cognitive domains of executive function and memory have received in the acute exercise literature, this chapter will focus on the effects of acute exercise on these cognitive abilities following the cessation of exercise. We will set the stage with a brief historical overview of research investigating the effects of acute exercise on overall cognitive performance. Then, we will transition to a discussion of the effects of acute exercise on executive function and memory specifically. We will examine meta-analytic reviews (quantitative reviews that allow for the assessment of standardized measures of an effect which are called effect sizes) assessing the effects of acute exercise on these cognitive abilities and provide exemplars to demonstrate the types of empirical studies conducted in these areas. We will also discuss the exercise parameters that may influence the overall effect of acute exercise on executive function and memory, as well as the mechanisms that may underlie the relationship between acute exercise and these cognitive abilities. Finally, we will conclude the chapter with a consideration of the practical implications of acute exercise-induced benefits to executive function and memory and why continuing to study the effects of acute exercise is important for achieving the long-term goal of developing precise chronic exercise recommendations that maximize the benefits of exercise to cognitive performance.

2 Historical Overview

From a historical perspective, there has been interest in the potential effects of acute exercise on cognitive performance for almost a century. Early studies were focused on understanding the relationship between muscle tension and mental performance efficiency with the earliest work looking at how performance of a mental task influenced performance of a muscular tension task. Subsequent research then flipped the question by exploring how performance of a muscular tension task influenced mental task performance. For example, Bills (1927) published a study examining the effects of a handgrip dynamometer task on addition, associative memory (i.e., remembering the relationship between unrelated items), and perception performance when these cognitive tasks were performed during the handgrip exercise. A similar protocol was used in many of the early exercise and cognition studies (Bills 1927; Bills and Stauffacher 1937; Courts 1939; Stauffacher 1937) in that they focused on cognitive performance during very brief, high-intensity anaerobic exercises. Scientists during this time were interested in identifying the optimal intensity of the exercise, understanding the role of the demands of the cognitive task, and exploring the influence of the fatigue level of participants. Since that early research, scientists have begun to explore how other types of acute exercise with a wide range of durations and intensities affect cognitive performance. They have continued to explore how exercise affects cognitive performance when exercise and cognitive tasks are performed simultaneously and have expanded the question to assess cognitive performance following the cessation of exercise.

Meta-analytic reviews of this literature generally show that the effects of acute exercise on cognitive performance during exercise are relatively small (Chang et al. 2012a) or negative (Lambourne and Tomporowski 2010; Loprinzi et al. 2019; McMorris and Hale 2012). However, when studies are limited to those that measure speed of cognitive performance during moderate intensity exercise, McMorris and Hale do report a moderate benefit (effect size, ES = 0.48). These overall findings may be explained by the reticular-activating hypofrontality model (Dietrich and Audiffren 2011) which posits that during exercise, resources in the central nervous system are prioritized to support the physical demands of the exercise itself thus limiting the resources available for cognitive processing. According to this model, performance on frontal-lobe dependent tasks (also referred to as explicit tasks which include measures of executive function) is expected to be hindered when simultaneously completed with exercise. By contrast, performance on tasks that are not frontal-lobe dependent (also referred to as implicit tasks which include speed of performance measures for simple tasks) is expected to benefit from exercise due to an exercise-induced release of catecholamines. Further discussion of the effects of acute exercise on cognitive performance assessed during exercise is beyond the scope of this chapter. Instead, we will focus on cognitive performance assessed following the cessation of an acute bout of exercise.

Meta-analytic reviews focused on the effects of acute exercise on cognitive performance following the cessation of the exercise bout are relatively consistent in reporting small benefits (Chang et al. 2012a; Roig et al. 2013; Lambourne and Tomporowski 2010; Loprinzi et al. 2019; McMorris and Hale 2012; Oberste et al. 2019), with the average effect size ranging from 0.10 to 0.52. Given that numerous meta-analytic reviews have provided summary statistics relative to the effects of a single bout of exercise on cognitive performance, Chang et al. (in press) conducted an umbrella review of this literature (a meta-review of systematic reviews and meta-analyses) and reported an overall effect size of 0.31. Examination of moderators indicated that the largest effects were evident when cognitive performance was assessed following the cessation of exercise (ES = 0.30) as compared to during exercise (ES = 0.03) and when either executive function (ES = 0.34) or attention (ES = 0.34) was assessed. Effects for memory were also significant and positive with an average effect size of 0.22.

3 Acute Exercise and Executive Function

The effect of acute exercise on executive function is a topic of significant interest in the exercise and cognition literature. Evidence suggests that executive function follows a common developmental pattern, as it emerges within the first few years of life, strengthens throughout childhood and adolescence (Best and Miller 2010), peaks in adulthood, and then declines with advancing age due to normal cognitive aging or disease-related cognitive deficits, such as mild cognitive impairment or Alzheimer's disease (Buckner 2004; Kirova et al. 2015) (see Fig. 1).

The various cognitive processes involved in executive function can be assessed with a wide range of cognitive tasks. Performance on many of these tasks can be measured with both time-dependent (e.g., response time) and accuracy (e.g., number of correct responses) scores to provide multiple estimates of executive functioning. Interestingly, in some cases researchers will assess accuracy simply to ensure that participants aren't sacrificing speed for accuracy. Once they establish that accuracy is maintained at a high level, they then present speed of performance data as the primary outcome. In the acute exercise and cognition literature, researchers have largely used measures of working memory and inhibition (Ishihara et al. 2021; Moreau and Chou 2019) (e.g., Stroop Test) but have also been interested in assessing set shifting (e.g., Wisconsin Card Sorting Test) and planning (e.g., Tower of London Task).

Chang and colleagues conducted a study that serves as a useful exemplar of the methodology commonly used in the acute exercise and executive function literature.

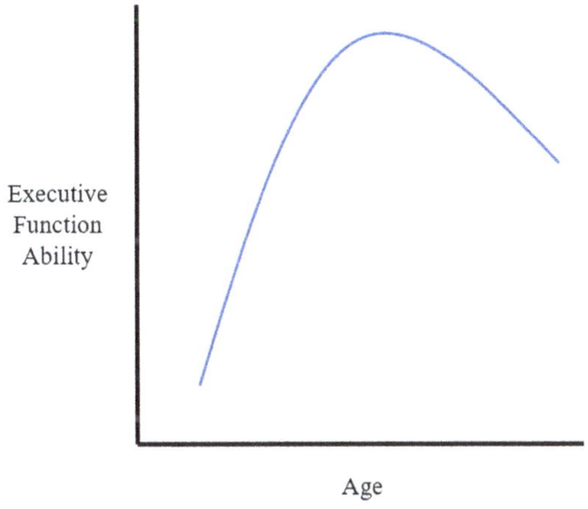

Fig. 1 Theoretical depiction of the relationship between age and executive function ability

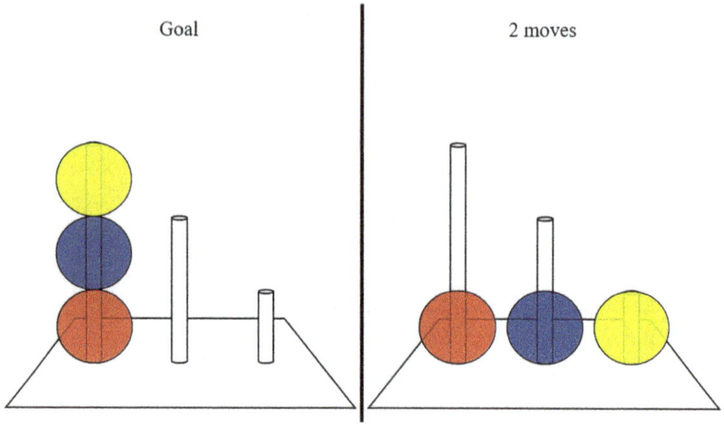

Fig. 2 The Tower of London Task requires the participant to move 3 beads on pegs from a starting state (right) to a goal state (left) in a specified number of moves (2 moves in this example). The tallest peg can hold all 3 beads, the middle peg can hold 2 beads, and the shortest peg can hold 1 bead. Only one bead can be moved at a time, and only the top bead on a peg can be moved

In this study, 42 young adults performed the Tower of London Task before and after either cycling for 30 min at moderate intensity or reading exercise materials for the same amount of time. During the Tower of London Task, participants are asked to move beads on pegs from a starting state to a goal state in a prescribed number of moves. The tallest peg can hold 3 beads, the middle peg can hold 2 beads, and the shortest peg can only hold 1 bead. Only the top bead on a peg can be moved and beads must be moved one at a time (see Fig. 2). Results showed that the acute exercise group improved their performance on the Tower of London test from pre- to post-exercise, while the reading group exhibited no change in performance.

The Stroop Test (Stroop 1935) and Wisconsin Card Sorting Test (WCST; Heaton et al. (1993)) are widely used measures of executive function in the acute exercise and cognition literature (Etnier and Chang 2009). When performing the Stroop Test, participants are instructed to name colors as quickly as possible in three different conditions. The Stroop Word condition displays color names written in black ink and the participant must read the color name. The Stroop Color condition displays colors in rectangles and the participant must identify the color. The Stroop Color-Word condition displays color names printed in a different color of ink and the participant must identify the color of ink. The Stroop Color-Word condition is considered a measure of inhibition because the participant has to inhibit the prepotent response of reading the color name.

The WCST is primarily a measure of set shifting but is also thought to reflect inhibition, working memory, and selective attention (Miyake et al. 2000). During the WCST, participants are required to sort 128 response cards based on one of three characteristics (i.e., color, form, or number) present on 4 stimulus cards (see Fig. 3). The experimenter provides feedback to the participants to let them know if their response is correct or incorrect as they try to determine the characteristic of the card

Fig. 3 The WCST includes 4 stimulus cards and 128 response cards to be sorted based on three characteristics (i.e., color, form, or number)

being used for sorting (the sorting characteristic). The sorting characteristic changes after every 10 responses.

In an acute exercise and executive function study conducted by Chang et al. (2012b), researchers administered both the Stroop Test and the WCST to 40 children with attention-deficit hyperactivity disorder (ADHD), a common neurobehavioral disorder that is associated with deficiencies in executive function (Semrud-Clikeman et al. 2008). Participants completed the Stroop and WCST both before and after either running for 30 min at moderate intensity or sitting to watch a video about exercise for the same amount of time. Results showed that the acute exercise group exhibited greater improvements in pre- to post-condition performance on the Stroop Color-Word condition and WCST compared to the seated rest group.

Interest in the effects of acute exercise on executive function and the large number of studies conducted in this area have made it a topic worthy of meta-analytic review. In an early meta-analysis examining the effects of acute exercise on cognitive performance, Chang et al. (2012a) found a positive effect of acute exercise on executive function during exercise (ES = 0.26), immediately following exercise (ES = 0.19), and after a delay following exercise (ES = 0.17). It is worth noting that this positive effect of acute exercise on executive function during exercise is different from the effect of acute exercise on overall cognitive performance during exercise which suggests that executive function is positively impacted regardless of the timing of the exercise bout. Since this meta-analysis, three additional meta-analyses have examined the effects of acute exercise on executive function specifically (Ishihara et al. 2021; Moreau and Chou 2019; Ludyga et al. 2016).

In 2016, Ludyga et al. (2016) conducted a meta-analytic review of the literature investigating the effects of moderate intensity acute aerobic exercise on executive function, including both time-dependent and accuracy measures of set shifting, inhibition, and working memory. Since executive function has been shown to change across the lifespan, the moderating role of age in the acute exercise and executive function relationship was examined. Age was categorized into four groups: preadolescent children (6–12 years old), adolescents (13–17 years old), young adults (18–35 years old), and older adults (50+ years old). The age range of 36–49 years old was not included as there were no empirical studies investigating the effects of acute exercise on executive function for this age group. The moderating role of fitness was also examined, as cross-sectional evidence has consistently shown that higher levels of fitness are associated with better performance on executive function tasks (Hillman et al. 2005; Huang et al. 2015). Fitness was categorized according to maximal oxygen uptake (VO_2max) into three groups: low-, average-, and high-fit. The researchers included 40 studies and found that acute exercise had a positive effect on both time-dependent (ES = 0.35) and accuracy (ES = 0.22) measures of executive function. The specific executive function task used and the fitness level of participants were not found to be significant moderators. However, age was found to be a significant moderator but only for time-dependent measures of executive function, such that older adults (ES = 0.67) and preadolescent children (ES = 0.54) demonstrated the largest benefits from acute exercise. The effect was small in young adults (ES = 0.20) and nonsignificant in adolescents.

While the meta-analysis conducted by Ludyga et al. (2016) specifically examined moderate intensity acute exercise, Moreau and Chou (2019) conducted a meta-analytic review of the effects of high-intensity acute exercise on executive function. The authors chose to focus on high-intensity exercise because they argued that shorter, high-intensity exercise may have the potential to help address the inactivity crisis as it is more efficient and engaging than longer duration, lower-intensity exercise while still conferring health benefits comparable to traditional longer duration exercise. In this meta-analytic review, the researchers included over 1,100 participants from 28 studies comparing the effects of high-intensity exercise to low-to-moderate intensity exercise and seated rest. A significant positive effect was found for high-intensity acute exercise compared to seated rest (ES = 0.34), but not compared to low-to-moderate intensity exercise (ES = 0.07). This lack of a difference between the effects for high-intensity and low-to-moderate intensity exercise suggests that acute exercise may benefit executive function regardless of exercise intensity. The potential moderating role of exercise intensity will be discussed in greater detail later in this chapter.

In a more recent meta-analysis, Ishihara et al. (2021) adopted a slightly different approach and conducted a meta-analysis of individual participant data regarding the effects of acute aerobic exercise on executive function. By using individual participant data, as compared to a traditional meta-analysis based on comparisons among studies, they were able to investigate how differences in baseline cognitive performance may impact the effects of acute exercise. Previous evidence demonstrates that acute exercise may be more beneficial for individuals with worse performance at baseline (Drollette et al. 2014). Furthermore, they assessed how executive function task demands influenced the relationship between acute exercise and executive function by grouping tasks as either high- or low-demand. They hypothesized that acute exercise would confer greater benefits to higher-demand executive function tasks. Like Ludyga et al. (2016), both time-dependent and accuracy measures were considered in this review. Results demonstrated that acute exercise improved response time, with greater improvements in response time observed for tasks that placed higher demands on executive function and for individuals with slower baseline response times. Additionally, while the main effect of acute exercise on accuracy was not significant, the interaction between acute exercise and baseline accuracy was significant, with greater improvements in accuracy after acute exercise observed for participants with lower accuracy at baseline. Thus, performance at baseline was a significant moderator of the effect of acute exercise on both time-dependent and accuracy measures of executive function. This finding suggests that acute exercise may serve to improve executive function in those most in need of that improvement.

Summarily, researchers have used a wide variety of tasks to assess the effects of acute exercise on executive function in varying populations. Overall, the evidence indicates that acute aerobic exercise can benefit various aspects of executive function. This acute exercise-induced benefit may be greatest for performance on tasks that place higher demands on executive function (Ishihara et al. 2021) and for individuals with lower levels of baseline executive function (Ishihara et al. 2021; Ludyga et al. 2016). As such, future investigations may seek to explore the effects of exercise in populations with weaker executive function abilities (e.g., children, older adults, ADHD) and continue to investigate variables related to exercise and cognition that could moderate the effect of acute exercise on executive function.

4 Acute Exercise and Memory

As previously discussed, memory has received substantial attention in the acute exercise and cognition literature in recent years. Specifically, research has primarily focused on how a single session of exercise affects short- and long-term episodic memory. As a reminder, episodic memory is the ability to remember details of specific events or experiences, including contextual, temporal, spatial, and emotional information about a past event/experience. For example, the ability to remember the details of your 10th birthday party (where it was, what you did, who was there, a gift you received, etc.) is an example of long-term episodic memory. In this chapter, we consider episodic memory by describing meta-analytic reviews of this literature and providing exemplars of studies exploring how the timing of the bout of exercise relative to encoding and consolidation may influence the observed effects.

Additional information about exercise and memory can be found in Chapter "Exercise and Memory" which considers memory more broadly.

A wide range of cognitive tasks have been used in the literature to assess the effects of acute exercise on episodic memory. These tasks have employed a variety of stimuli, including visual (e.g., objects and faces), auditory (e.g., words and numbers), and spatial (e.g., locations and routes) information. One of the most common episodic memory tasks used in the acute exercise literature is a word-list learning paradigm, like the Rey Auditory Verbal Learning Test (de Sousa et al. 2012) or the California Verbal Learning Test (Delis et al. 1987). During these word-list learning tasks, participants are exposed to a list of unrelated words (usually 15–20 words) and then immediately asked to recall as many words as they can remember to assess short-term episodic memory. The list of words is then repeated and recalled multiple times to measure learning with repeated exposure. Sometimes participants are then exposed to a second list of unrelated words to assess proactive (i.e., memory for previous information impairing learning of new information) and retroactive (i.e., new information impairing memory for previously learned information) interference. Finally, to measure long-term episodic memory, participants can also be asked to recall the original list of words after a delay period of 15 min, 30 min, 24 h, or longer. In addition to word-list learning paradigms like this, other common assessments of episodic memory found in the acute exercise literature include visual object recognition tasks, number sequence tasks, and short story learning tasks.

The results of early studies suggested that acute exercise had null or minimal effects on memory performance when performance was assessed following a single session of exercise (Etnier et al. 1997; Tomporowski 2003). Similarly, the meta-analytic review conducted by Chang et al. (2012a) also found nonsignificant effects of acute exercise on memory performance both immediately following exercise (ES = 0.05) and after a delay following exercise (ES = −0.03). However, when performance was classified according to the specific type of memory measure used, acute exercise was found to have small to moderate positive effects on free recall (ES = 0.49) and visual short-term memory (ES = 0.23) following exercise. This meta-analysis therefore provided evidence that the effects of acute exercise on memory are not domain general but vary based on the specific type of memory ability being assessed.

In a more recent meta-analytic review, Roig et al. (2013) examined data from 22 studies exclusively focused on memory (episodic and semantic). Acute aerobic exercise was shown to have an overall small positive effect on the memory of healthy adults (ES = 0.22). When broken down by retention interval, acute exercise was found to have a small positive effect on short-term memory (ES = 0.26) and a moderate positive effect on long-term memory (ES = 0.52). This finding of a more robust effect of acute exercise on long-term memory is particularly interesting because it suggests that the benefits of acute aerobic exercise to memory are not transient but can be relatively long-lasting. Unlike other cognitive abilities that see the benefits of acute exercise dissipate after a few minutes (e.g., attention, executive

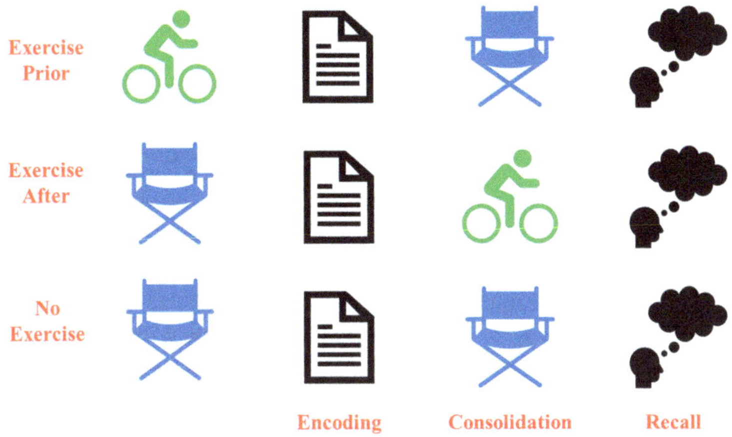

Fig. 4 Schematic of the protocol typically used in studies examining the timing of acute exercise relative to episodic memory encoding, consolidation, and recall. These studies frequently include three conditions: acute exercise prior to information encoding, acute exercise after information encoding during memory consolidation, and no exercise. The parameters of the acute exercise bout can be manipulated to examine how different modalities, durations, and intensities of exercise impact episodic memory performance. The delay between information encoding and recall can also be varied to explore the durability of acute exercise effects on episodic memory

function, and processing speed), the benefits of acute exercise to memory may persist even after an extended period of time has passed since exercise cessation.

It has been hypothesized that the durable effects of acute exercise on memory exist because acute exercise confers benefits to both the initial encoding of information and the ongoing memory consolidation process that continues after encoding has finished. A substantial body of research has been conducted to test this hypothesis by manipulating the timing of acute exercise relative to information encoding, memory consolidation, and information recall. For example, in a study conducted by Labban and Etnier (2011), young adults were randomly assigned to one of three conditions: acute exercise prior to encoding a short story, acute exercise after encoding a short story, or no exercise. Figure 4 depicts the protocol for these conditions and the timing of exercise relative to encoding, consolidation, and recall. Results showed that the young adults in the exercise prior to encoding group had significantly better story recall 35 min after exposure to the story compared to the young adults in the no-exercise group. No differences in story recall were found between the exercise prior to encoding group and exercise after encoding group as well as between the exercise after encoding group and the no-exercise group. This finding therefore provides preliminary evidence suggesting that acute exercise facilitates the greatest improvement in episodic memory when performed prior to encoding and consolidation rather than during the consolidation process.

Continuing this line of research investigating the timing of acute exercise relative to information encoding and memory consolidation, Labban and Etnier (2018) next examined how acute exercise prior to encoding, acute exercise after encoding, and no exercise impacted the word-list recall of younger adults 60-min and 24 h after exposure to the list of words. In this study, the researchers implemented a within-subjects design, whereby every participant completed each of the three exercise conditions. Additionally, the implementation of a 60-min delay following exposure to the to-be-remembered information provided adequate time for exercise completion and recovery prior to word recall so the effects of acute exercise on memory consolidation could be better isolated from the potential confound of acute exercise effects on information retrieval. Similar to the results of the previously discussed study, the acute exercise prior to encoding condition had significantly better word recall at the 60-min and 24 h delay intervals compared to the no-exercise condition. No significant differences were found between the acute exercise prior to encoding and acute exercise after encoding conditions, as well as between the acute exercise after encoding and no-exercise conditions. This study therefore provided further evidence that acute exercise benefits to episodic memory are greatest when exercise is performed prior to information encoding. Furthermore, it also showed that acute exercise benefits to long-term episodic memory can be observed for up to 24 h following information exposure.

With numerous studies having been conducted to investigate how the timing of acute exercise relative to encoding and consolidation affects episodic memory performance, a meta-analytic review of the literature was conducted by Loprinzi et al. (2019). Loprinzi et al. reported that acute exercise prior to information encoding (ES = 0.11), during early consolidation (ES = 0.47), and during late consolidation (ED = 1.05) enhanced episodic memory performance. In contrast, acute exercise performed during information encoding had a negative effect on episodic memory performance (ES = −0.12). While these summary statistics are inconsistent with the results of the two previously discussed empirical studies that found acute exercise prior to information encoding conferred the greatest benefit to episodic memory, it is worth noting that this meta-analysis identified several variables that may moderate the temporal effects of acute exercise on episodic memory, including age, exercise intensity, and retention interval (short-term or long-term memory). For example, a significant positive effect for acute exercise prior to information encoding was found for long-term memory (ES = 0.19) but not for short-term memory (ES = −0.01). The temporal relationship between acute exercise and episodic memory therefore appears to be influenced by complex interactions between participant characteristics, exercise parameters, and features of the specific memory task used.

To date, the vast majority of acute exercise and episodic memory studies have been conducted with younger adults, with far fewer studies focusing on children and older adults. However, children and older adults are potentially important target populations for acute exercise studies due to the developing and declining nature of their episodic memory abilities, respectively. In children, whose episodic memory abilities are still developing as they mature, acute exercise may confer greater benefits. Preliminary evidence is promising, as acute aerobic exercise has been shown to benefit the word-list recall of preadolescent children (Etnier et al. 2014; Pesce et al. 2009). In the study conducted by Pesce et al. (2009), children aged

11–12 years completed a word-list recall task following three experimental conditions: a physical education class involving aerobic circuit training, a physical education class involving team games, and a no-exercise control condition. Results showed that word recall immediately after exposure was higher following the team games compared to the control condition, but word recall after a 12-min delay was higher following both team games and aerobic circuit training compared to the control condition. These findings suggest that acute exercise performed during physical education classes may facilitate better episodic memory performance in preadolescent students. Furthermore, the cognitive and social demands of the acute exercise bout (i.e., team games versus aerobic circuit training) may also influence acute exercise-induced benefits for children's episodic memory performance.

In contrast to children whose episodic memory abilities are still developing, older adults experience a decline in episodic memory as part of the normal cognitive aging process. Acute exercise may therefore confer greater benefits to older adults by reducing the severity of this age-related decline to preserve or even improve their episodic memory abilities. Thus far, the preliminary evidence is positive as acute aerobic exercise has been shown to benefit the word-list recall (Etnier et al. 2021) and visual-image recall (Segal et al. 2012) of older adults. Using a similar experimental design to the previously discussed Labban and Etnier study conducted with younger adults, Etnier et al. (2021) examined how acute exercise prior to encoding, acute exercise after encoding, and no exercise influenced older adults' performance on a word-list learning paradigm where word recall occurred immediately after exposure and 24 h later. Results showed that in the acute exercise prior to encoding condition, older adults had significantly better immediate recall compared to the acute exercise after encoding and no-exercise conditions. For 24-h recall, word recall was significantly better in the acute exercise prior to encoding condition than the after encoding condition, which in turn was significantly better than the no-exercise condition. The authors concluded that the benefits observed for acute exercise performed after encoding were significant in this study, but not in their previous studies, because of the deleterious effect of aging on episodic memory performance.

Collectively, the existing body of research demonstrates that acute aerobic exercise can benefit certain aspects of episodic memory, especially when performed prior to information encoding and when recall occurs after a delay period. However, more research across the lifespan is needed so we can better understand how acute exercise affects the episodic memory of specific age groups, especially children and older adults. In addition to age, specific exercise parameters (e.g., modality, duration, and intensity) have also been shown to influence the effect acute exercise has on episodic memory and other cognitive abilities, including executive function. We will therefore now turn our attention to the research examining the moderating effects of these exercise parameters on the acute exercise and cognitive performance relationship.

5 The Influence of Acute Exercise Parameters

Although the previously discussed empirical studies and meta-analytic reviews provide some insights about relevant moderators of the acute exercise and cognitive performance relationship, we will focus on moderating variables that are relevant to the exercise experience itself. When considering the effects of acute exercise on cognitive performance, important parameters that should be considered include the exercise modality (i.e., type of exercise), intensity (e.g., light, moderate, vigorous), and duration (e.g., 10 min, 30 min).

Exercise modality is a relatively complex construct because of the various ways in which exercise type can be categorized. For example, meta-analytic reviewers have coded exercise modality to allow for comparisons between aerobic, anaerobic, muscular resistance, combination exercise, and accelerometer-based reports of physical activity (Chang et al. 2012a), aerobic versus cognitively-engaging exercise (de Greeff et al. 2018), and aerobic versus resistance exercise (Oberste et al. 2021). Results from these reviews must be interpreted cautiously because of the overlap between exercise types and other relevant parameters like exercise duration and intensity. Furthermore, since the vast majority of the research has focused on aerobic exercise (e.g., 92% of effects in Chang et al.; 83% of effects in de Greeff et al.; 77% of effects in Oberste et al. 2021), it is likely that the results for all other types of exercise are unreliable due to the small number of studies. Overall, when limited to studies focused only on aerobic exercise, findings are consistent with the overall conclusions drawn from meta-analytic reviews regarding small to moderate benefits of acute exercise on cognitive performance (Chang et al. 2012a; de Greeff et al. 2018; Oberste et al. 2021; Chueh et al. 2022).

Taking a somewhat unique approach to the definition of exercise type, Lambourne and Tomporowski (2010) made comparisons between the effects of ergometer cycling and treadmill running on cognitive performance following exercise. Results showed that while both modes of exercise yielded small benefits, effects were larger for cycling than for treadmill running. The authors attributed this difference to the potentially larger attentional demands required to sustain a prescribed pace and maintain balance while running on a treadmill compared to the lesser attentional demands required to cycle on a stationary ergometer. Subsequently, when examining the effects of acute exercise on episodic memory specifically, Loprinzi et al. (2019) also compared effect sizes between ergometer cycling and treadmill-based walking or running. When exercise was performed prior to information encoding, memory benefits were evident for studies that implemented cycling (ES = 0.46) but not for studies that used treadmill walking or running. Similarly, when exercise was performed in the early stages of memory consolidation, the authors found that benefits were only evident for studies using cycling (ES = 1.17) but not for studies examining treadmill walking or running. When exercise was performed in the late stages of memory consolidation, there were no effects for cycling, but treadmill walking or running resulted in a large benefit (ES = 1.31). To the best of our knowledge, there is currently no published empirical study that has directly compared the effects of treadmill exercise to cycling exercise on acute cognitive performance. In fact, to expand upon this point, there are relatively few studies that have directly compared the cognitive benefits of acute exercise relative to exercise modality (Hung et al. 2020). This lack of direct comparisons between different modes of exercise is important because of the challenge of basing conclusions on summary statistics. That is, findings across meta-analytic reviews are somewhat limited because they can only be examined at the levels identified as relevant by the reviewers and because they cannot account for third order variables (e.g., exercise duration, exercise intensity, cognitive domain) that might offer alternative explanations of the findings. Additional theory-driven research in this area would therefore be of benefit as we continue to work toward the goal of being able to prescribe exercise as a means of improving cognitive performance.

Like exercise mode, relatively little attention has been paid to understanding the potential dose-response relationship between exercise duration and cognitive performance. Meta-analytic reviews can again provide us with summary information. Chang et al. (2012a) categorized acute exercise duration as 0–10, 11–20, and >20 min and found positive effects only for exercise of 11–20 min or >20 min. Specific to studies looking at memory, Loprinzi and colleagues (2019) compared effects for short (<20 min), medium (20–40 min), and long (>40 min) acute exercise durations and reported that effects were only significant and positive for long duration exercise performed during the early memory consolidation period. By contrast, Roig and colleagues (2013) reported that effects for short-term memory were largest when exercise duration was <20 min while the effects for long-term memory were best when exercise was <20 min or 20–40 min. Specific to executive function tasks, Moreau and Chou (2019) categorized duration as 0–5, 6–10, 11–20, and >20 min and found no significant differences as a function of acute exercise duration. These contrasting findings are perplexing and again emphasize the challenge of using summary statistics to answer questions about moderating variables. When summarized across meta-analytic reviews, the umbrella review conducted by Chang et al. (in press) reported that benefits were evident for exercise performed for more than 20 min (ES = 0.28) and exercise performed for less than 20 min (ES = 0.20). This summary finding suggests that benefits are generally evident for acute exercise regardless of exercise duration.

Given the aforementioned limitations of using summary data to inform our understanding of the dose-response relationship between exercise duration and cognitive outcomes, empirical studies are important to allow for a direct comparison across levels within a single study. Chang et al. (2015) conducted a series of studies comparing the effects of no exercise with exercising at moderate intensity for 10, 20, or 45 min. In a sample of young men (20–22 years), Chang et al. showed that both speed of performance and accuracy on the congruent and incongruent trials of the Stroop Test were significantly better after 20 min of exercise compared to all other durations of exercise. In a sample of late middle-aged adults (55–65 years), Chang et al. (2019) found that response times on the congruent, neutral, and incongruent trials of the Stroop Test were best after 20 min of exercise. Chen et al. (2018)

conducted a similar study with late middle-aged adults and reported that 20 min of exercise resulted in the best performance on a different measure of executive function (task switching). Overall, these findings are consistent in suggesting that when considering 10, 20, or 45 min of moderate intensity exercise, beneficial effects for executive function are greatest after 20 min of exercise.

The final exercise parameter we will discuss is exercise intensity. Exercise intensity has been extensively studied with numerous empirical studies comparing performance across intensity levels and many meta-analytic reviews considering intensity as a moderator. Results from meta-analytic reviews suggest that intensity moderates the observed effects of acute exercise on cognitive performance. Across all cognitive outcomes, Chang et al. (2012a) found that only very light, light, and moderate intensity exercise resulted in small positive benefits for cognitive performance performed immediately after exercise. However, when cognitive performance was assessed after a delay following exercise cessation, benefits were observed for light, moderate, hard, and very hard intensity exercise. These findings are intriguing because they may indicate that when cognitive performance is performed immediately after exercise, a lower-intensity exercise is recommended, but that for more enduring effects, higher intensity exercise is better. Focusing on studies examining memory performance, Roig et al. (2013) reported that effects for short-term memory were best at low intensity. They also found that intensity did not impact long-term memory. By contrast, in their meta-analytic review that was also focused on memory, Loprinzi et al. (2019) found that vigorous intensity exercise resulted in the largest memory benefits for exercise performed prior to encoding (ES $= 0.54$) and during early consolidation (ES $= 1.09$). It is difficult to compare these findings to those of the meta-analysis conducted by Roig and colleagues because the analyses weren't conducted separately for short- and long-term memory. However, if most of the studies included in the Loprinzi et al. review were looking at long-term memory, then these results would be consistent. In a meta-analysis looking at the executive function performance of children following acute exercise, Moreau and Chou (2019) found that effects for both lower-intensity and high-intensity exercise were beneficial.

Results from the Chang et al. (in press) umbrella review suggest that there is no significant difference in effect sizes as a function of intensity. However, despite the lack of significance of this finding, an examination of the effect sizes suggests that there might be something of an inverted-U relationship between intensity and performance with the largest effects evident at an intensity described as moderate to vigorous (ES $= 0.57$). On either side of this intensity level, effects were smaller: ES $= 0.28$ for moderate intensity, ES $= 0.24$ for vigorous to maximal intensity, and ES $= 0.10$ for very light to light intensity exercise. Again, these summary findings across meta-analytic reviews are somewhat limited because they can only be examined at the levels identified as relevant by the reviewers and because they cannot account for third order variables (e.g., exercise duration, exercise intensity, cognitive domain) that might offer alternative explanations of the findings. That is, for example, if most of the studies performed at moderate to vigorous intensity also measured executive function while the studies at other intensity levels measured information processing, the findings that we are attributing to intensity could actually be reflective of the cognitive domain assessed rather than intensity per se.

To address the limitations of using summary data from meta-analytic reviews, it is important to also consider empirical studies in which intensity level was manipulated within a single study so that the inferences can be directly linked to intensity level per se and not to other variables. Chang and Etnier (2009) conducted a study that explored the effects of exercise intensity of resistance exercise on cognitive performance on two tasks that assess different cognitive domains. They randomly assigned young adults to either no exercise or 40%, 70%, or 100% of 10-repetition maximal effort on six resistance exercises. Following the treatment condition, participants performed the Stroop Test and the Paced Auditory Serial Addition Task (PASAT). The PASAT is a task that requires participants to calculate the sum of every pair of single digit numbers as they are read in a single list. So, for example, as a string of numbers are read, they would say aloud the sum of the first and second number, they would then hear the third number and have to say the sum of the second and third number, they would then hear the fourth number and have to say the sum of the third and fourth number, and so on. This is a very challenging task that assesses attention and working memory. Results from this study showed that there was a significant linear relationship between exercise intensity and performance on the Stroop neutral condition and a significant curvilinear relationship for the Stroop incongruent condition. For the PASAT, the relationship was curvilinear and explained increasingly more of the variance as the task difficulty increased. These findings are fascinating because they suggest that the relationship between intensity and performance may be task-specific with more simple measures of information processing benefiting from higher intensity levels of exercise while more complex measures that require executive function benefit most from moderate intensity exercise.

It is actually very challenging to study exercise intensity or exercise duration in isolation because the two so naturally go together in real-world exercise activities. In other words, if people are exercising at a high intensity, they typically only exercise for a short period of time. By contrast, if exercising at a low intensity, they are able to exercise for a longer period of time. For this reason, in an important extension to this literature, Chang et al. (2023) conducted a study in which they held exercise volume constant to test the effects of bouts of acute exercise that differed in terms of both intensity and duration. Older adults carrying a genetic risk factor for Alzheimer's disease (the epsilon 4 allele of apolipoprotein E) were asked to perform the Tower of London task (a measure of executive function) following four treatment conditions. The treatment conditions were rest for 30 min, lower-intensity exercise for 40 min, moderate intensity exercise for 30 min, or higher intensity exercise for 16 min. Results showed that performance was better for all exercise conditions relative to the control condition and there was no difference in performance between the exercise conditions. This study is intriguing because the focus was on exercise volume rather than looking at intensity with duration held constant or duration with intensity held constant.

6 Mechanisms

While empirical studies and meta-analyses have provided consistent evidence of acute exercise-induced benefits to cognitive performance, less is known about the mechanisms underlying this relationship. To date, several potential mechanisms have been proposed, including both psychological and biological changes following a single session of exercise. From a psychological perspective, acute exercise may enhance arousal to facilitate better cognitive performance (Audiffren 2009; McGaugh 2006). According to the Inverted-U Hypothesis initially applied to the relationship between exercise and cognitive performance by Davey (1973), exercise can induce small to large changes in arousal that can then positively or negatively impact cognitive performance. Exercise at low intensities may not increase arousal enough to enhance cognition. However, high-intensity exercise may stimulate too much arousal and be detrimental to cognitive performance. Moderate intensity exercise is therefore predicted to stimulate the optimal amount of arousal to enhance cognitive functioning. As was previously discussed, this hypothesis is somewhat supported by evidence from an umbrella review (Chang et al. (in press)).

Regarding executive function, the effect of acute exercise on neural activation has been extensively examined using electroencephalogram (EEG). EEG is a neuroimaging technique that measures electrical signals produced by the brain on the scalp in real time. Given the distance between the sites producing the signals and where the signals are measured on the scalp, EEG may be limited with regard to spatial resolution, but the technique has strong temporal resolution. Specifically, most of this research has used event-related potentials (ERPs), which represent the electrical activity of large groups of neurons that can be broken down into time-locked components. ERPs are used to assess the electrical activity of the brain in preparation for, or in response to, a stimulus or response (Coles et al. 1990). When exploring the relationship between acute exercise and executive function, the most well-studied ERP component is P3 (also known as P300). This is a positive-going deflection that occurs approximately 300 ms after the presentation of a stimulus with a topographical maximum over the parietal region (Polich and Kok 1995). The P3 is thought to represent the allocation of attentional resources during stimulus engagement (Polich 2007). Meta-analytic reviews show that acute exercise leads to decreased P3 latency and increased P3 amplitude, with this increase in P3 amplitude associated with better executive function (Kao et al. 2022; Martínez et al. 2023). The nature of the P3 component and the changes observed following acute exercise suggest that acute exercise may improve the allocation of attentional resources which subsequently leads to improvements on executive function tasks. Other ERP components that have been assessed in the acute exercise and executive function literature include N2 (also known as N200, which is thought to relate to inhibition), contingent negative variation (CNV), and N450. Results for these other components are less conclusive, though there is some evidence of acute exercise effects on N2 and N450 (Chang et al. 2017). Ultimately, ERP evidence supports an acute exercise effect on neural activation that relates to improvements in executive function, and researchers continue to use ERPs to explore the nuances of the acute exercise and cognition relationship.

In the acute exercise and episodic memory literature, acute exercise has been proposed to stimulate several physiological changes that benefit the underlying neural processes that give rise to our episodic memory abilities. For example, previous research has shown that acute aerobic exercise can increase levels of brain-derived neurotrophic factor (BDNF), a protein that plays an important role in regulating and maintaining several different neural processes that enable learning and memory (Dinoff et al. 2017; Szuhany et al. 2015). Additionally, a neuroimaging technique known as functional magnetic resonance imaging (fMRI) has been used to investigate how acute aerobic exercise influences neural activation during episodic memory tasks. fMRI is a noninvasive technique that indirectly measures neural activity by detecting changes in blood flow to different brain regions in response to changes in neural activity. This change in blood flow at the site of neural activity produces a signal (i.e., the blood oxygen level-dependent signal) that can be measured with fMRI to provide insight into different aspects of brain function during cognitive tasks. In a study conducted by Slutsky-Ganesh et al. (2020) using fMRI, acute aerobic exercise performed both before and after encoding was shown to improve the word-list recall of younger adults and reduce neural activation in brain regions that support memory processing, including the lingual gyrus which plays a role in the encoding of visual memories as well as the superior temporal gyrus and planum temporale which are involved in auditory processing and language. This reduction in neural activation coupled with improved memory performance suggests that acute aerobic exercise may enhance younger adults' neural efficiency during episodic memory tasks. Acute aerobic exercise may therefore induce rapid physiological adaptations that enhance the neural processes underlying episodic memory.

7 Why Study Exercise and Acute Cognitive Enhancement

Thus far, we believe this review has demonstrated that engaging in a single session of exercise can benefit cognitive performance following exercise cessation. Meta-analytic reviews and empirical studies have shown that this acute exercise-induced benefit extends to specific aspects of both executive function and episodic memory performance. Research has also identified several moderating variables that can influence the effect acute exercise has on cognitive performance, including the exercise parameters of modality, duration, and intensity. Furthermore, we have briefly discussed some of the potential mechanisms through which acute exercise may enhance cognitive performance. However, at this point in the chapter you may be asking yourself, "Why is understanding the effects of acute exercise on cognitive performance important?". This is an excellent question!

First, understanding the effects of acute exercise on cognitive performance has important practical implications for improving our performance on everyday tasks that require cognitive processing. For example, young adults in college are frequently required to take exams to assess their learning. If we consider the results of the previously discussed Labban and Etnier (2018) study that examined the effects of acute exercise on young adults' long-term episodic memory, then it might be beneficial for young adults to exercise prior to studying for an exam in order to enhance their memory for the studied information 24 h later. Similarly, the findings of the study conducted by Pesce et al. (2009) with elementary school children suggest that incorporating exercise sessions into the school day (e.g., physical education classes, recess time, movement breaks) can enhance children's learning and memory. Research examining the effects of afterschool exercise programs has also shown that acute exercise can improve children's performance on executive function tasks and mathematics assessments, as well as benefit certain markers of brain function (Davis et al. 2011; Hillman et al. 2014). Acute exercise-induced benefits to episodic memory and executive function may therefore have positive implications for the academic performance of children and young adults.

The practical applications of acute exercise enhancements to cognitive performance do not just include students in academic settings but can also be extended to the daily functioning of the rapidly expanding older adult population. As will be discussed in more detail in a later chapter focused on exercise and age-related cognitive declines (see Chapter "Exercise and Protection from Age-Related Cognitive Decline"), older adults are susceptible to declines in cognitive performance as part of the normal aging process. However, if we consider the results of the study conducted by Etnier et al. (2021), acute exercise may be a useful strategy for mitigating some of the normal age-related declines in episodic memory experienced by older adults. Engaging in acute exercise may therefore help older adults better perform everyday tasks that require episodic memory (e.g., remembering items on a grocery list, remembering the location of their parked car). Recent investigations have also found a positive effect of both acute aerobic and resistance exercise on the executive function performance of older adults with mild cognitive impairment (Tsai et al. 2018). Acute exercise-induced enhancements to older adults' cognitive performance therefore have important implications for the maintenance of daily functioning and overall quality of life.

In addition to these everyday practical applications, examining the effects of acute exercise on cognitive performance is also critical to achieving the long-term goal of being able to prescribe chronic exercise as a strategy for improving cognitive performance. The physiological changes that accompany a single session of exercise may be the building blocks for the chronic exercise-induced brain changes that facilitate long-term benefits to cognitive performance (El-Sayes et al. 2019). Studying the effects of acute exercise can therefore further our understanding of the mechanisms through which chronic exercise confers enduring benefits to cognitive performance. Furthermore, acute exercise studies allow us to examine which exercise parameters are optimal for enhancing specific aspects of cognitive performance. Subsequently, we can begin to develop precise exercise recommendations that maximize the benefits of chronic exercise for cognitive performance. For example, meta-analytic reviews (Chang et al. 2012a; Roig et al. 2013) and empirical studies (Etnier et al. 2021; Callow et al. 2022) suggest that 20 min of moderate intensity aerobic exercise may be optimal for eliciting acute exercise-induced benefits to the episodic memory of older adults. Chronic aerobic exercise interventions aimed at improving the everyday memory function of older adults may therefore want to design exercise programs that encourage older adults to engage in 20 min bouts of moderate intensity aerobic exercise several days per week. However, before we can achieve this long-term goal of prescribing precise chronic exercise programs to enhance cognitive performance, more research examining moderating variables and underlying mechanisms of the acute exercise-induced cognitive enhancements is needed.

References

Audiffren M (2009) Acute exercise and psychological functions: a cognitive-energetic approach. In: Exercise and cognitive function. Wiley Blackwell, Hoboken, pp 3–39

Best JR, Miller PH (2010) A developmental perspective on executive function. Child Dev 81(6): 1641–1660

Bills AG (1927) The influence of muscular tension on the efficiency of mental work. Am J Psychol 38:227–251

Bills AG, Stauffacher JC (1937) The influence of voluntarily induced tension on rational problem solving. J Psychol 4(2):261–271

Buckner RL (2004) Memory and executive function in aging and AD: multiple factors that cause decline and reserve factors that compensate. Neuron 44(1):195–208

Callow DD, Pena GS, Stark CEL, Smith JC (2022) Effects of acute aerobic exercise on mnemonic discrimination performance in older adults. J Int Neuropsychol Soc 15:1–10

Chang YK, Etnier JL (2009) Exploring the dose-response relationship between resistance exercise intensity and cognitive function. J Sport Exerc Psychol 31(5):640–656

Chang YK, Labban JD, Gapin JI, Etnier JL (2012a) The effects of acute exercise on cognitive performance: a meta-analysis. Brain Res 1453:87–101

Chang YK, Liu S, Yu HH, Lee YH (2012b) Effect of acute exercise on executive function in children with attention deficit hyperactivity disorder. Arch Clin Neuropsychol 27(2):225–237

Chang YK, Chu CH, Wang CC, Wang YC, Song TF, Tsai CL et al (2015) Dose-response relation between exercise duration and cognition. Med Sci Sports Exerc 47(1):159–165

Chang Y, Alderman BL, Chu C, Wang C, Song T, Chen F (2017) Acute exercise has a general facilitative effect on cognitive function: a combined ERP temporal dynamics and BDNF study. Psychophysiology 54(2):289–300

Chang YK, Chen FT, Kuan G, Wei GX, Chu CH, Yan J et al (2019) Effects of acute exercise duration on the inhibition aspect of executive function in late middle-aged adults. Front Aging Neurosci 11:227

Chang YK, Karageorghis CI, Wang CC, Li RH, Chen FT, Fang RY et al (2023) Effects of exercise intensity and duration at a predetermined exercise volume on executive function among Apolipoprotein E (APOE)-ε4 carriers. Curr Psychol [Internet]. 11 Jun 2022 [cited 14 May 2023]; Available from https://doi.org/10.1007/s12144-022-03257-x

Chen FT, Etnier JL, Wu CH, Cho YM, Hung TM, Chang YK (2018) Dose-response relationship between exercise duration and executive function in older adults. J Clin Med 7(9):279

Chueh TY, Hsieh SS, Tsai YJ, Yu CL, Hung CL, Benzing V et al (2022) Effects of a single bout of moderate-to-vigorous physical activity on executive functions in children with attention-deficit/hyperactivity disorder: a systematic review and meta-analysis. Psychol Sport Exerc 58:102097
Coles MG, Gratton G, Fabiani M. Event-related brain potentials. 1990;
Courts FA (1939) Relations between experimentally induced muscular tension and memorization. J Exp Psychol 25:235–256
Davey C (1973) Physical exertion and mental performance. Ergonomics 16(5):595–599
Davis CL, Tomporowski PD, McDowell JE, Austin BP, Miller PH, Yanasak NE et al (2011) Exercise improves executive function and achievement and alters brain activation in overweight children: a randomized, controlled trial. Health Psychol 30(1):91
de Greeff JW, Bosker RJ, Oosterlaan J, Visscher C, Hartman E (2018) Effects of physical activity on executive functions, attention and academic performance in preadolescent children: a meta-analysis. J Sci Med Sport 21(5):501–507
de Sousa MS, Malloy-Diniz LF, Hamdan AC (2012) Validity convergent and reliability test-retest of the Rey auditory verbal learning test. Clin Neuropsychiatry 9(3):129–137
Delis DC, Kramer JH, Kaplan E, Ober BA. California verbal learning test – second edition. 1987;
Dietrich A, Audiffren M (2011) The reticular-activating hypofrontality (RAH) model of acute exercise. Neurosci Biobehav Rev 35(6):1305–1325
Dinoff A, Herrmann N, Swardfager W, Lanctot KL (2017) The effect of acute exercise on blood concentrations of brain-derived neurotrophic factor in healthy adults: a meta-analysis. Eur J Neurosci 46(1):1635–1646
Drollette ES, Scudder MR, Raine LB, Moore RD, Saliba BJ, Pontifex MB et al (2014) Acute exercise facilitates brain function and cognition in children who need it most: an ERP study of individual differences in inhibitory control capacity. Dev Cogn Neurosci 7:53–64
El-Sayes J, Harasym D, Turco CV, Locke MB, Nelson AJ (2019) Exercise-induced neuroplasticity: a mechanistic model and prospects for promoting plasticity. Neuroscientist 25(1):65–85
Etnier JL, Chang YK (2009) The effect of physical activity on executive function: a brief commentary on definitions, measurement issues, and the current state of the literature. J Sport Exerc Psychol 31(4):469–483
Etnier JL, Salazar W, Landers DM, Petruzzello SJ, Han M, Nowell P (1997) The influence of physical fitness and exercise upon cognitive functioning: a meta-analysis. J Sport Exerc Psychol 19(3):249–277
Etnier J, Labban JD, Piepmeier A, Davis ME, Henning DA (2014) Effects of an acute bout of exercise on memory in 6th grade children. Pediatr Exerc Sci 26(3):250–258
Etnier JL, Vance JC, Ueno A (2021) Effects of acute exercise on memory performance in middle-aged and older adults. J Aging Phys Act 1(aop):1–8
Heaton RK . . . et al (1993) Wisconsin Card Sorting Test (WCST) [Internet]. Revised and expanded. Odessa, FL. (P.O. Box 998, Odessa 33556): Psychological Assessment Resources, [1993] ©1993. Available from: https://search.library.wisc.edu/catalog/9910126849002121
Hillman CH, Castelli DM, Buck SM (2005) Aerobic fitness and neurocognitive function in healthy preadolescent children. Med Sci Sports Exerc 37(11):1967
Hillman CH, Pontifex MB, Castelli DM, Khan NA, Raine LB, Scudder MR et al (2014) Effects of the FITKids randomized controlled trial on executive control and brain function. Pediatrics 134(4):e1063–e1071
Huang T, Tarp J, Domazet SL, Thorsen AK, Froberg K, Andersen LB et al (2015) Associations of adiposity and aerobic fitness with executive function and math performance in Danish adolescents. J Pediatr 167(4):810–815
Hung CL, Tseng JW, Chao HH, Hung TM, Wang HS (2020) Effect of acute exercise mode on serum brain-derived neurotrophic factor (BDNF) and task switching performance. J Clin Med. 24 Sep 2018 [cited 29 Nov 2020];7(10). Available from: https://www.ncbi.nlm.nih.gov/pmc/articles/PMC6209934/

Ishihara T, Drollette ES, Ludyga S, Hillman CH, Kamijo K (2021) The effects of acute aerobic exercise on executive function: a systematic review and meta-analysis of individual participant data. Neurosci Biobehav Rev 128:258–269
Kao SC, Chen FT, Moreau D, Drollette ES, Amireault S, Chu CH et al (2022) Acute effects of exercise engagement on neurocognitive function: a systematic review and meta-analysis on P3 amplitude and latency. Int Rev Sport Exerc Psychol:1–43
Kirova AM, Bays RB, Lagalwar S (2015) Working memory and executive function decline across normal aging, mild cognitive impairment, and Alzheimer's disease. Biomed Res Int:2015
Labban JD, Etnier JL (2011) Effects of acute exercise on long-term memory. Res Q Exerc Sport 82(4):712–721
Labban JD, Etnier JL (2018) The effect of acute exercise on encoding and consolidation of long-term memory. J Sport Exerc Psychol 40(6):336–342
Lambourne K, Tomporowski P (2010) The effect of exercise-induced arousal on cognitive task performance: a meta-regression analysis. Brain Res 1341:12–24
Loprinzi PD, Blough J, Crawford L, Ryu S, Zou L, Li H (2019) The temporal effects of acute exercise on episodic memory function: systematic review with meta-analysis. Brain Sci 9(4):87
Ludyga S, Gerber M, Brand S, Holsboer-Trachsler E, Pühse U (2016) Acute effects of moderate aerobic exercise on specific aspects of executive function in different age and fitness groups: a meta-analysis. Psychophysiology 53(11):1611–1626
Martínez RM, Chen C, Fan YT, Wu HH, Du PS, Chen GY et al (2023) Meta-analysis of electroencephalographic correlates and cognitive performance for acute exercise-induced modulation. Neuropsychobiology:1–19
McGaugh JL (2006) Make mild moments memorable: add a little arousal. Trends Cogn Sci 10(8): 345–347
McMorris T, Hale BJ (2012) Differential effects of differing intensities of acute exercise on speed and accuracy of cognition: a meta-analytical investigation. Brain Cogn 80(3):338–351
Miyake A, Friedman NP, Emerson MJ, Witzki AH, Howerter A, Wager TD (2000) The unity and diversity of executive functions and their contributions to complex "frontal lobe" tasks: a latent variable analysis. Cogn Psychol 41(1):49–100
Moreau D, Chou E (2019) The acute effect of high-intensity exercise on executive function: a meta-analysis. Perspect Psychol Sci 14(5):734–764
Oberste M, Javelle F, Sharma S, Joisten N, Walzik D, Bloch W et al (2019) Effects and moderators of acute aerobic exercise on subsequent interference control: a systematic review and meta-analysis. Front Psychol 10:2616
Oberste M, Sharma S, Bloch W, Zimmer P (2021) Acute exercise-induced set shifting benefits in healthy adults and its moderators: a systematic review and meta-analysis. Front Psychol 12: 528352
Pesce C, Crova C, Cereatti L, Casella R, Bellucci M (2009) Physical activity and mental performance in preadolescents: effects of acute exercise on free-recall memory. Ment Health Phys Act 2(1):16–22
Polich J (2007) Updating P300: an integrative theory of P3a and P3b. Clin Neurophysiol 118(10): 2128–2148
Polich J, Kok A (1995) Cognitive and biological determinants of P300: an integrative review. Biol Psychol 41(2):103–146
Roig M, Nordbrandt S, Geertsen SS, Nielsen JB (2013) The effects of cardiovascular exercise on human memory: a review with meta-analysis. Neurosci Biobehav Rev 37(8):1645–1666
Segal SK, Cotman CW, Cahill LF (2012) Exercise-induced noradrenergic activation enhances memory consolidation in both normal aging and patients with amnestic mild cognitive impairment. J Alzheimers Dis 32(4):1011–1018
Semrud-Clikeman M, Pliszka S, Liotti M (2008) Executive functioning in children with attention-deficit/hyperactivity disorder: combined type with and without a stimulant medication history. Neuropsychology 22(3):329

Slutsky-Ganesh AB, Etnier JL, Labban JD (2020) Acute exercise, memory, and neural activation in young adults. Int J Psychophysiol 158:299–309

Stauffacher JC (1937) The effect of induced muscular tension upon various phases of the learning process. J Exp Psychol 21:26–46

Stroop JR (1935) Studies of interference in serial verbal reactions. J Exp Psychol 18(6):643

Szuhany KL, Bugatti M, Otto MW (2015) A meta-analytic review of the effects of exercise on brain-derived neurotrophic factor. J Psychiatr Res 60:56–64

Tomporowski PD (2003) Effects of acute bouts of exercise on cognition. Acta Psychol (Amst) 112(3):297–324

Tsai CL, Ukropec J, Ukropcová B, Pai MC (2018) An acute bout of aerobic or strength exercise specifically modifies circulating exerkine levels and neurocognitive functions in elderly individuals with mild cognitive impairment. NeuroImage Clin 17:272–284

Exercise and Acute Affect

Steven J. Petruzzello and Jonathan R. North

Contents

Abstract The study of the affective impact of physical activity and exercise has a long and fraught history. In this chapter we outline some of the longstanding issues, in the hopes that research moving forward will be more mindful of these. Included are: (a) what we are actually trying to assess; (b) measurement issues; (c) the timing of affective measurement; (d) research that highlights what we know and don't know; and (e) suggestions for moving forward.

Keywords Ecological Momentary Assessment · Genetics · Measurement · Temporal Dynamics

If you are in a bad mood, go for a walk. If you are still in a bad mood, go for another walk.
Attributed to Hippocrates

S. J. Petruzzello (✉) and J. R. North
Department of Kinesiology and Community Health, University of Illinois at Urbana-Champaign, Urbana, IL, USA
e-mail: petruzze@illinois.edu

Curr Topics Behav Neurosci (2024) 67: 103–124
https://doi.org/10.1007/7854_2024_493
Published Online: 31 July 2024

According to Ekkekakis and Brand (2019), the question "How does exercise make people feel?" led to the development of the field of study known as exercise psychology in the 1960s. The vast majority of that early research focused on examining whether exercise led to reductions in negative emotions, namely anxiety and depression. While there was nothing inherently wrong with this line of inquiry, particularly considering the prevalence of such mental-health problems, this approach was also rather limiting as there is potentially much more than people might feel as a result of exercise or, more broadly, physical activity. For example, one of the most consistently reported effects from exercise, even at lower intensities, is an increased feeling of energy. Robert Thayer referred to this phenomenon as the "energizing effect" of exercise (Thayer 2001), paradoxical in that expending energy could leave one feeling more energetic. What follows is an overview of some of the major issues in studying the impact of physical activity and exercise on *affect* along with a brief overview of the literature and recommendations for future research in this domain.

1 Issues of Conceptualization: Affect, Emotions, or Moods?

The literature examining feeling states associated with physical activity and exercise has been messy and confusing, in large part because of the equating of the terms *affect, emotion,* and *mood.* An uninformed reading of studies would lead one to believe that these terms all refer to the same thing (see 3, 4), which of course they do not. While affect, emotion, and mood are all feeling states and are related to one another, each has important differences from the others. These terms, and the constructs they represent, are distinct and should not be used as if they were synonyms. Indeed, an entire volume has been devoted to this very issue because of its importance and should be required reading for those working in this area (Ekkekakis 2013).

Emotions (e.g., fear, guilt, pride) are brief states of feeling that are not always present. Importantly, an emotion is elicited following an appraisal where something (e.g., object, person, or event) is judged to either directly impact or have the potential to impact, an individual's well-being (Lazarus 1991). As such, an emotion is an immediate, often fairly intense but short-lived, response to some stimulus (Scherer 2005) (e.g., an exam elicits anxiety, winning the lottery elicits joy, completing a challenging workout elicits a sense of accomplishment) that requires some level of cognitive input. Simply put, emotions occur following a cognitive evaluation ascribing a level of meaning to an occurrence.

Moods (e.g., irritation, cheerfulness) are also feeling states and, like emotions, have a cognitive basis, and can augment or impede goal-directed behavior. While moods are present most of the time, they are different from emotions in at least three important ways:

1. While emotions are usually brief, moods typically occur over a longer time frame, lasting for hours or days (Nowlis and Nowlis 1956).
2. Whereas the cause of an emotion is often identifiable, the cause of a mood is sometimes unable to be determined (Morris 1992).
3. The intensity and variability of emotions are greater than moods (Nowlis and Nowlis 1956).

Affect is the most basic element of all feeling states and is always present (Russell and Barrett 1999). Distinguished from emotions and moods, affect is a general "valenced" response (i.e., good–bad, positive–negative, pleasure–displeasure). As such, affect subsumes all emotions and moods. Affect is considered more basic than emotion in that it is evolutionarily more primitive and does not require any cognitive input (Russell 2003). The primitive nature of affect is adaptive, much like a reflex. Consider that an immediate response to pain – that it is unpleasant – does not require any cognitive input to feel unpleasant. A valenced response is an affect, but only becomes an emotion or a mood once some level of cognitive elaboration takes place (Ekkekakis and Petruzzello 2000).

It could be argued that prior to 2000, the vast majority of what was examined regarding the impact of exercise on feeling states was exclusively about emotions and moods (e.g., anxiety, depression). The implicit assumption was that exercise could potentially reduce anxiety or depression but wouldn't have an impact on any other affective states (e.g., exercise would not improve perceptions of energy, increase feelings of relaxation, or minimize a sense of fatigue). Very little work had taken a broader perspective to first determine if such a focus on specific feeling states was even warranted. That began to change when the focus of research went from discrete states (i.e., a categorical approach) to a perspective that could theoretically capture a broader range of affective states (i.e., a dimensional approach). The dimensional approach assumes that affective states are interrelated and can be accurately captured by a small number of dimensions (as few as two), thereby providing a more encompassing picture of the affective dynamics occurring as a result of a bout of exercise. With as few as two dimensions (e.g., valence, activation), a bout of exercise might be shown to result in increased energy during and immediately upon completion of, and increased relaxation during the recovery from, the activity.

2 Categorical Versus Dimensional Approaches

We, and others, argued at the time that dimensional models provide an opportunity to study affective phenomena in a more desirable manner (see Biddle and Mutrie (2001); Ekkekakis and Petruzzello (2002)). Even earlier, Gauvin and Brawley (1993) noted the utility of the dimensional approach to examining affect in the exercise domain:

> [The dimensional] approach seems better suited to the understanding of exercise and affect because the models stemming from it are intended to be broad, encompassing conceptualizations of affective experience. Because the affective experience that accompanies exercise has not been thoroughly described, a model of affect that has a wider breadth is more likely to capture the essence of exercise induced affect than a model that, at the outset, limits the focus of investigation to specific emotions. (p. 152)

The model adopted and still primarily being used is the 2-dimensional circumplex model of affect (Ekkekakis and Petruzzello 1999, 2002; Russell 1980, 1997). Essentially, the model posits that the orthogonal dimensions of affective valence (pleasure–displeasure) and activation (low–high) can be used to capture affect through the various combinations of the dimensions. What results are 4 quadrants: pleasant–activated (e.g., excited, enthusiastic, energetic); pleasant–unactivated (e.g., relaxed, calm, tranquil); unpleasant–unactivated (e.g., tired, fatigued, bored); and unpleasant–activated (e.g., anxious, tense, nervous). As such, the circumplex model provides a more general representation of affect and is able to capture both positive and negative affective states. Furthermore, the dimensions of valence and activation can each be assessed with a single-item measure, thus allowing for more frequent assessment as well as assessments before, during, and after exercise.

3 Issues of Measurement: Generalized or Specific?

There is a plethora of measures available for assessing affective states, including more generalized measures and supposedly "exercise-specific" measures.

3.1 *Generalized Measures*

Measures of mood, emotion, or affect that are not specific to any particular context are considered "general" measures. The most used measures in the exercise literature include the Profile of Mood States (POMS), the Positive and Negative Affect Schedule (PANAS), and the Activation–Deactivation Adjective Check List (AD ACL). Of course, this does not include the State-Trait Anxiety Inventory or the Beck Depression Inventory, but these are not included here because they are specific to anxiety and depression, not affect more generally.

Profile of Mood States The POMS (McNair et al. 1971) is one of the most frequently used questionnaires in the exercise (and sport for that matter) psychology literature. It was originally developed for use in psychiatric populations, particularly military veterans with anxiety or post-traumatic stress disorder (and formerly known as the Psychiatric Outpatient Mood Scale). Despite being designed to capture feeling states relevant in assessing the effects of psychotropic drugs, the POMS has been routinely used to examine mood changes in response to various exercise manipulations for nearly 40 years. The POMS, comprised of a 65-item adjective rating scale, yields scores on six different mood subscales. Five of the subscales assess negative moods (tension–anxiety, depression–dejection, anger–hostility, fatigue–inertia, and confusion–bewilderment) with only one assessing positive mood (vigor–activity).

Each of the 65 adjectives (e.g., exhausted, alert, terrified, ready to go, miserable, tense, confused, uneasy, fatigued) is rated on a 5-point scale ranging from 0 (not at all) to 4 (extremely), according to how the individual feels. The instructions given for making the assessments (i.e., response set) can be varied so that the respondent reflects on how they feel "right now," "today," "over the past week," "over the past month," "over the past year," or even "over a lifetime." While it is not designed for a specific context, a problem with the POMS is that it only assesses six distinct mood states. Its use in an exercise study thereby implicitly assumes that one or more of the six mood states measured are influenced by an exercise manipulation. Of course, these six states likely do not reflect the entire domain of mood. Furthermore, the POMS is problematic because of its length. Thus, assessing mood multiple times in a single exercise session creates participant burden. Even though shorter and alternative versions are available, they remain limited in the range of moods that they assess (see Ekkekakis (2013) for a detailed critique).

Positive and Negative Affect Schedule The PANAS (Watson et al. 1988) is another relatively popular measure of feeling states in the exercise domain. Much shorter than the POMS, this 20-item questionnaire was developed to capture the two-dimensional (positive affect, negative affect) model proposed by Watson and Tellegen (1985). It therefore assesses two orthogonal (i.e., conceptually independent) dimensions: positive affect (PA; enthusiastic, alert) and negative affect (NA; anger, disgust). Respondents, according to how they feel, rate each of the 20 adjectives (e.g., interested, distressed, excited, jittery, ashamed) along a 5-point scale, ranging from 1 (very slightly or not at all) to 5 (extremely). Response sets range from "you feel this way right now, that is, at the present moment" to "you generally feel this way, that is, how you feel on average." Although the PANAS provides a broader capture of potential feeling states than the POMS, it only assesses the high-activation ends of PA (e.g., excitement, enthusiasm) and NA (e.g., nervousness, fright), but not the low-activation ends (e.g., PA: relaxation, calmness; NA: sadness, tiredness). Watson and Clark created a modified version, the PANAS-X (Watson and Clark 1994) to capture each dimension more fully, but the revised measure has been infrequently used in the exercise literature.

Activation–Deactivation Adjective Check List (AD ACL) Thayer (1986, 1989) created the AD ACL, using a theoretical framework incorporating two bipolar dimensions of activation and arousal (see Fig. 1). The dimension of **Energetic Arousal (EA)** is designed to capture feelings ranging from energy, vigor, and liveliness to feelings of fatigue and tiredness. An important distinction for the exercise domain is that Thayer proposed EA to reflect changes in gross physical activity. The second dimension referred to as **Tense Arousal (TA)** is designed to capture feelings ranging from subjective tension to placidity, quietness, and stillness.

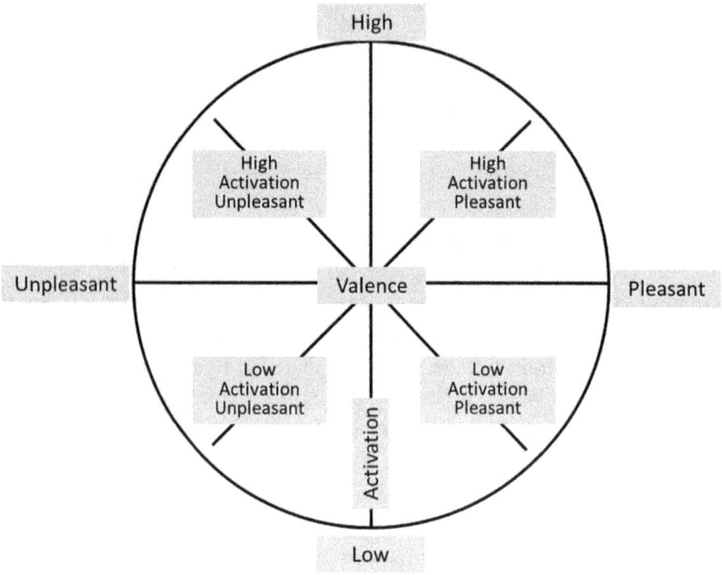

Fig. 1 The circumplex model of affect. (*Adapted* from (Ekkekakis and Petruzzello 2002))

The two dimensions were thought to vary as a function of energy expenditure. Specifically, Thayer proposed they would be positively related (i.e., EA and TA both either increase or decrease) from low to moderate levels and negatively related from moderate to high levels of energy expenditure (i.e., EA increases, TA decreases). The AD ACL is the 20-item measure developed to capture the various feeling states characterized in the 2-dimensional model. Respondents rate the way they are feeling using a 4-point scale ranging from "definitely feel" to "definitely do not feel." The resultant scoring provides values for both EA and TA, along with the component subscales of Energy and Tiredness (i.e., EA dimension) and Tension and Calmness (i.e., TA dimension). Given its more complete capture of possible feeling states, it is puzzling why this measure has not been used more frequently by exercise researchers.

Feeling Scale Finally, the FS (Hardy and Rejeski 1989) is a single-item measure of valence or hedonic tone (good–bad/pleasant–unpleasant). In this regard, it is much like a visual analog scale. While it is often thought to be an "exercise-specific" measure, it is hardly context-specific. Responses are made along an 11-point continuum ranging from −5 (very bad) to +5 (very good), with 0 as a neutral midpoint. It was designed for use as an in-task (e.g., during exercise) measure of affective valence and has been used extensively in studies examining affect during acute exercise bouts.

3.2 Exercise-Specific Measures

There was a movement in the early 1990s that led to the development of several "exercise-specific" self-report measures of feeling states. This was driven by two major concerns with existing general self-report instruments: (1) measures like the POMS and the PANAS were not sensitive enough to exercise stimuli, containing irrelevant items or items that were often unchanged as a result of exercise manipulations; and (2) exercise has unique and distinct properties that available measures failed to detect. In other words, the argument was made that feeling states (i.e., affect) are changed in the exercise context in ways that are distinct from other contexts.[1] Out of this dissatisfaction, three multi-item scales – the Exercise-induced Feeling Inventory (EFI), Subjective Exercise Experiences Scale (SEES), and Physical Activity Affect Scale (PAAS) – and one single-item scale – Feeling Scale (FS) – were developed.

Exercise-Induced Feeling Inventory The EFI (Gauvin and Rejeski 1993) was developed as a 12-item measure of exercise-induced feeling states, composed of four subscales (positive engagement, revitalization, physical exhaustion, and tranquility). Each of these subscales was thought to represent the affective experience associated with exercise that was not adequately captured with existing instruments. These subscales are supposedly analogous to enthusiasm, energy, fatigue, and calmness (Gauvin and Spence 1998).

Subjective Exercise Experiences Scale The SEES (McAuley and Courneya 1994) was also developed as a 12-item measure of the subjective experiences unique to exercise (with supposed attention to the positive dimensions of affect resulting from exercise), comprised of three subscales (positive well-being, psychological distress, and fatigue). Psychological well-being is analogous to positive affect, whereas psychological distress is analogous to negative affect. An extensive discussion of the numerous issues inherent in both the EFI and the SEES is available in separate articles by Ekkekakis and Petruzzello (2001a, b).

Physical Activity Affect Scale Lox et al. (2000), attempting to exploit the "unique" aspects of the EFI and the SEES while simultaneously reducing the overlap between them, developed the PAAS, another 12-item exercise-specific measure consisting of the four EFI subscales and the psychological distress subscale of the SEES. The PAAS has not been used extensively in the literature to this point.

A shortcoming of these exercise-specific scales is the fact that each was developed in the absence of any theoretical framework (see Ekkekakis and Petruzzello (2000, 2001a, b) for extensive critiques). Interestingly enough, neither the EFI, SEES, nor PAAS has actually addressed, in any substantive manner, the two

[1] It is worth noting that no evidence was (or has been) shown that exercise has such distinct, unique properties. As such, the already existing affect measures *should* work well in assessing any affective changes occurring as a result of exercise.

concerns (see above) that led to their development. What these instruments have done, more than anything else, is create additional confusion and messiness in an already messy area of research. Ekkekakis has gone to great lengths to highlight the importance in deciding on the most appropriate affect measure to use in their research (see (Ekkekakis 2013)).

4 Temporal Dynamics of Affective Responses

Timing when affective responses are measured is particularly important, not only for understanding the temporal dynamics of affective responses to exercise but also for examining whether any consistent dose–response effects may occur. As Ekkekakis and Brand noted (Ekkekakis and Brand 2019), timing of affective assessments in much of the research was restricted to before and after an acute bout of exercise due to the cumbersome nature of the oft-used multi-item measures. Systematic investigation of affect is now more routinely done before, *during*, and after a bout of exercise.

As has now been shown repeatedly, the affect experienced during an exercise bout may be rather different from what is seen if only before- and after-exercise assessments are made. As shown in Fig. 2, if only before- and after-exercise responses were assessed (Path A), the assumption would be that the exercise resulted in an increasingly more pleasant affective state. However, Path B and Path C show that the same exercise, for 2 different people (or potentially even the same person on different days), yields much different affective responses during the activity, even

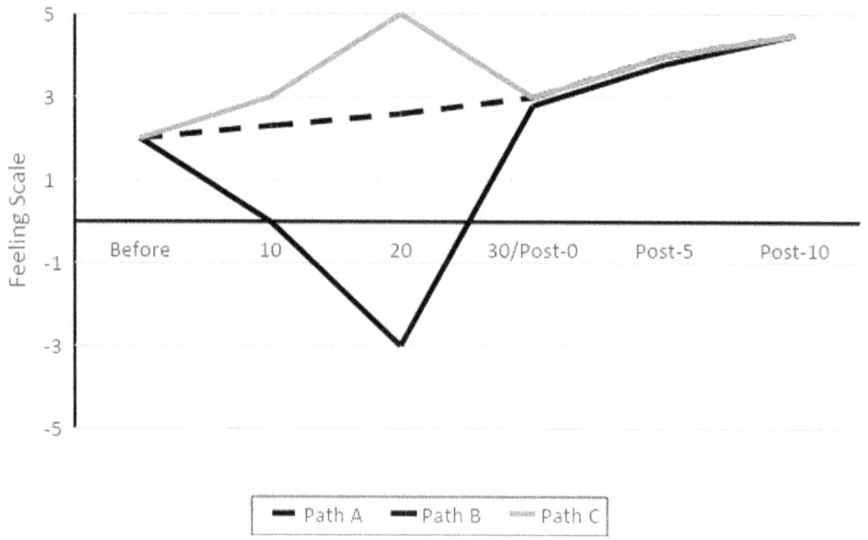

Fig. 2 Potentially divergent patterns of affective dynamics to the same exercise bout

though the affective state at the end and during recovery is the same. This clearly has important implications for exercise prescription. An individual with an affective pattern revealed in Path B (i.e., decreasing pleasantness/increasing unpleasantness), even with the rebound seen afterward, might be less likely to do such an activity again. As Emmons and Diener (1986) showed, time spent engaging in various activities is predicted by how people feel when doing them. And of course, Williams et al. (2008) showed that even a small increase in pleasantness during treadmill walking predicted increased time spent being active 6 and 12 months later. The post-activity pleasant feelings might not be enough to "override" the unpleasant feelings experienced during the activity.

All told, the "methodological platform" for guiding the study of affective responses to exercise, originally outlined by Ekkekakis and Petruzzello (1999) and reiterated numerous times since (e.g., (Ekkekakis et al. 2011)), has been nicely encapsulated by Ekkekakis and Brand (2019). In sum, they note that such investigations should: (1) utilize a dimensional approach (e.g., affect circumplex); (2) make repeated affective assessments (i.e., before, during, and after); (3) examine patterns of variation as well as group level changes; and (4) standardize the intensity of the exercise itself with more individualized metabolic landmarks (e.g., ventilatory threshold).

5 Research on Affective Responses to Exercise

5.1 "Real World," Free-Living Responses

As noted at the outset, physical activity and acute bouts of exercise have long been believed to create a "feel better" effect. However, there has been a growing recognition of the fact that physical activity and exercise do not always result in pleasant feelings (Ekkekakis and Brand 2019; Dunton et al. 2023). Indeed, for some people (e.g., sedentary, highly deconditioned individuals) and for some types of physical activity (e.g., high intensity), physical activity and exercise can result in rather unpleasant feelings (e.g., (Ekkekakis et al. 2011; Ekkekakis and Lind 2006)). With increasing access to technology that is both mobile and wearable, investigators have leveraged such technological advances to study affective responses to physical activity "in the wild," that is, in response to everyday physical activity experiences people have outside of controlled laboratory environments (Dunton 2017; Liao et al. 2015). This methodology, called experience sampling or ecological momentary assessment (EMA), often pairs accelerometers with smartphones to capture an individual's experiences (e.g., affective states, context, perceptions) at various times during their normal daily routines, including before, during, and after physical activity. Many of these studies use a variation of random prompting schemes, what Dunton et al. (2016) describe as *interval-contingent* sampling [i.e., prompts given at preset times (e.g., daily at 8:00 a.m., 4:00 p.m.), *signal-contingent* sampling (i.e., prompts randomly throughout the day), or *event-contingent* sampling [i.e., individual is tasked with completing survey during or following a predetermined behavior (e.g., bout of physical activity)]. A more recent strategy, *context-sensitive* EMA (CS-EMA) automatically prompts the individual to respond based on information from a smartphone or other devices or sensors (e.g., accelerometer), allowing collection of information related to specific behaviors (e.g., exercise, sedentariness) within specific contexts (Dunton et al. 2016). For example, if the accelerometer indicated a certain level of movement, a prompt would be sent to the smartphone requesting a response from the participant. As noted by Dunton et al. (2023), such "emerging technologies pose incredible opportunities for real-time, real-world interventions to enhance how one feels while engaging in physical activity or to boost the effects of those affective states on future behavior."

An initial review (Liao et al. 2015) analyzed the early research utilizing EMA to study the relationships between affect and physical activity within a relatively short time frame from the physical activity stimulus itself. The review included studies where (a) affective states and subsequent physical activity were examined and (b) affective states following physical activity were studied. Liao et al. concluded that when positive affective states were reported, there was a greater likelihood of engaging in physical activity within the following several hours; negative affective states appeared unrelated to subsequent physical activity. Additional studies from this review examined affective states following physical activity, with Liao et al. concluding that higher positive affect (e.g., greater feelings of energy) generally occurred for the subsequent several hours. Reductions in negative affect following physical activity were not consistently seen in these studies.

Perhaps more than anyone else, Genevieve Dunton and her team have pioneered EMA research examining physical activity and affect. Building from the 2015 review, Liao et al. (2017a) conducted an EMA study with a relatively large sample ($N = 110$) of non-active adults (~40 years old; engaged in <150 min·week^{-1} of physical activity). Both light (LPA) and moderate-to-vigorous (MVPA) physical activity were captured via accelerometry, with random smartphone prompts (8 prompts within a 15.5 h time period for 4 consecutive days) to assess affective states [e.g., positive affect, negative affect, energy, and fatigue; "how __________ (happy/stressed/sad/cheerful/angry/tired/calm/energetic/tense) were you feeling?"] experienced just prior to the prompt. More MVPA took place within the 15 min following responses indicating greater positive affect coupled with lower negative affect and fatigue. Greater amounts of LPA and MVPA were related to higher feelings of energy for up to 30 min following the activity. In a separate study, Liao et al. (2017b) assessed current physical activity level and affect (e.g., happy, stress, energetic, tired) as well as MVPA 6 and 12 months later. Using random prompting across 4 days, they were able to examine reported affect (via EMA prompts) at times when participants reported performing physical activity (e.g., exercise, sports, walking), as well as the extent to which these affective responses during activity were associated with MVPA (via accelerometry) measured 6 and 12 months later. They showed that those who reported feeling energetic during physical activity were performing more daily MVPA 6 and 12 months later. In addition, less negative affect (assessed as anxious, stressed, angry, depressed) during physical activity was associated with more daily MVPA minutes 12 months later. They interpreted this as reflecting an increase of 91 min per week of MVPA 12 months later for every decrease of 1 standard deviation in negative affect during physical activity. Dunton et al. (2023), in discussing this work, proposed that a reduction of ~1/3 standard deviation in negative affect (as assessed in the Liao study (Liao et al. 2017b)) would result in 30 more min/week of MVPA, the equivalent of ~200 min/week in overall physical activity (including light intensity). This is again more evidence of the importance of the affective response during physical activity and its impact on future physical activity behavior. As Liao et al. (2015) stated "Knowledge gained from this topic could offer valuable insights for promoting daily physical activity adoption and maintenance in non-clinical populations." (p. 1).

Reichert et al. (2017) sought to differentiate physical activity in the form of exercise from non-exercise physical activity (NEA; walking to the bus stop) and whether there was any differential relationship with affect (which they incorrectly refer to as "mood"). A sample of adults ($N = 93$, 23.4 years, 62% female) wore an accelerometer and were prompted over a 7-day recording period to report their affective states (well–unwell; content–discontent; full energy–without energy; awake–tired; relaxed–tense; calm–agitated) using a visual analog scale format (score of 0–100) and a sophisticated sampling scheme. Prompts were triggered based on location (using GPS information), at fixed times (8:00 a.m., 10:20 p.m.), and every 40–100 min throughout the day. Reichert et al. reported that exercise activity resulted in increased valence (i.e., greater pleasantness) and calmness, whereas NEA increased perceptions of energy and wakefulness but decreased calmness. They noted that it may be worthwhile to distinguish between exercise activity and NEA as they seemingly have different impacts on affect.

As Dunton noted in her 2017 paper, a limitation of EMA is that it is a time-intensive methodology (Dunton 2017). It is also an invasive methodology. However, it has the potential to not only reveal more within-person variability than more traditional laboratory-based approaches, but to also better account for dynamic change. All of this could assist in better explaining and predicting physical activity and its effects, something existing theories have not done particularly well to date.

Next, we turn to the more traditional approach to studying affective responses to exercise. The emphasis is on research that examined varying levels of exercise intensity, duration, or their combination on affective responses. Separate sections focus on research examining pre- and post-exercise affective responses and more recent studies that also examine affective responses during exercise.

5.2 *Before- and After-Exercise Affective Responses*

Research, primarily in non-clinical samples, has generally shown that exercise results in an increase in positively valenced affective states (e.g., energy; see Reed and Buck (2009); Reed and Ones (2006) and Chapter "Psychedelic-associated Affective and Anxiety Disorders") and a reduction in negatively valenced affective states (e.g., tension). This is particularly the case when the exercise intensity is light to moderate in nature (e.g., up to ~60–65% of maximal capacity) or is based on a relative workload (e.g., %VO_{2peak}). As intensities increase to higher levels, feelings of fatigue are often increased. However, as Ekkekakis and Petruzzello (1999) pointed out, fitness mediates this effect at the high intensity end of the continuum when absolute intensities are used (i.e., less fit individuals experience significant increases in fatigue); fitness has no effect on higher intensities that are relativized. Another important factor in all of this is when the post-exercise measure is taken. These more negatively valenced feeling states are usually only seen immediately post-exercise, are often gone within 5 min after completion of the exercise bout, and completely reversed thereafter (i.e., increased energy).

As alluded to earlier, an interesting effect of exercise on affect is the paradoxical energizing effect. Perhaps the first study to demonstrate this effect showed that 10 min of brisk walking at a self-selected pace significantly elevated feelings of energy for up to 120 min following the walk (Thayer 1987). As noted by Thayer (2001), moderate amounts (both in duration and intensity) of aerobic exercise usually energize the exerciser. Saklofske et al. (1992) also showed increased energy, along with decreased tension (both assessed via the AD ACL), following walks of 4 and 10 min in duration. We have shown (Ekkekakis et al. 2000) in a series of studies including more naturalistic as well as controlled laboratory settings that 10 to 15 min walking consistently resulted in increased activation and more positive affective valence, essentially reflecting increased energy and decreased tension (using the AD ACL, FS, FAS). The typical finding is that positive affect tends to increase from pre- to post-exercise following exercise intensities that are more moderate in nature (Reed and Ones 2006; Reed 2013). Stated another way, positively valenced affective states (e.g., energy, vigor, exhilaration) are increased, whereas negatively valenced affective states (e.g., fatigue, tension) are either reduced or unchanged, following moderate-intensity aerobic exercise (see also Reed and Ones (2006); Reed (2013)). Immediately following high-intensity exercise, negative affective states are often increased while positive affective states are decreased (e.g., (Acevedo et al. 1994)), especially in individuals who are less fit. It isn't unusual to see, in more fit individuals, that positive affect may be improved following high-intensity exercise (see (Ekkekakis and Petruzzello 1999)). Backhouse et al. (2007) suggested that this might just as likely reflect the relief that the exercise bout is over.

Of course, there are numerous studies that have assessed feeling states before and after exercise using measures that have already been described as problematic for various reasons (e.g., POMS, PANAS). This is a potentially defensible position when the sample being examined has a psychiatric diagnosis (e.g., (Brellenthin et al. 2021; Crombie et al. 2018, 2019, 2021; Meyer et al. 2016)). However, given the explicit focus of the current chapter, these studies are cited for the interested reader, but are not discussed in the context of affective responses to exercise.

5.3 *In-Task Exercise Responses*

One of the greatest conundrums for the field is the fact that exercise generally makes people feel better once they have finished. This should result in more people exercising, but that is not the case (see, for example, Backhouse et al. (2007)). As Thayer (2001) put it, "The extensive evidence about the value of exercise should have the gyms packed, the running tracks crowded, and the sidewalks filled with throngs of people walking. But they aren't" (p. 32). We now know that it may very well be that even though people feel better *after* exercise, how they feel *during* exercise may be part of the problem. The work of Emmons and Diener (1986) nicely pointed out that the affect experienced during an activity is a good predictor of future engagement in that activity. As postulated by hedonic theory (Ekkekakis 2009; Kahneman 1999; Williams 2008), a person is not likely to continue an activity that is not fun to do or isn't enjoyable. Despite the abundance of research examining pre- to post-exercise changes in affect, studies examining "in-task" affective responses have been rare, although that has been changing (see Ekkekakis et al. (2011)).

One reason for the change has been the development and use of single-item or very brief multi-item self-report scales for measuring affect. This has made the assessment of affect during exercise more viable. The Feeling Scale has been extensively used in such investigations, often being coupled with a measure of arousal/activation (e.g., Felt Arousal Scale; (Svebak and Murgatroyd 1985)) to allow a dimensional assessment of affect. The replicable finding, routinely shown in a variety of samples and across various types of exercise (e.g., (Ekkekakis et al. 2011; Benjamin et al. 2012; Sheppard and Parfitt 2008a, b); see Ekkekakis (2013)), from this research is that, as exercise intensity increases, affect becomes progressively less positive/more negative (as measured by the FS) and more activated (as assessed by the FAS). It is also worth noting that, while use of single-item scales has been criticized, there is a growing body of literature showing them to be valid and reliable measures (Allen et al. 2022).

Another important issue in the study of affective responses has been the "prescription" of exercise intensity. Traditionally, the intensity of an exercise bout has been determined as a percentage of some maximal value (e.g., heart rate or maximal/peak aerobic capacity). Assigning an exercise intensity as a percentage of this maximal capacity has been thought to relativize the workload so that two people with different maximal capacities end up exercising at similar metabolic levels. This assumption has been demonstrated to be inaccurate, with individuals exercising at the same relative workload having very dissimilar metabolic responses (Ekkekakis and Petruzzello 1999; Katch et al. 1978). For example, exercise at a "moderate" intensity (e.g., 60% VO_{2peak}) leaves some individuals working almost exclusively aerobically, while others may need a significant degree of anaerobic metabolism to complete the same workload. It has been demonstrated that as the degree of reliance on anaerobic metabolism increases (e.g., as shown in increasing levels of blood lactate), affective valence gets progressively less positive/more negative (Ekkekakis et al. 2011).

A much more useful metabolic landmark in the study of exercise intensity effects is the lactate threshold (Ekkekakis 2003), that point where blood lactate concentrations increase faster than the rate it can be removed from the blood. As a result of this excess lactate, anaerobic energy systems are increasingly utilized during exercise. The **ventilatory threshold (VT)**, a less invasive approximation of the lactate threshold, thus also reflects the shift to increasingly anaerobic metabolism as exercise intensity progresses during exercise. Research using the lactate or ventilatory threshold as a basis for prescribing exercise intensity has yielded much more consistent and coherent results. For example, Parfitt, Rose, and Burgess (Parfitt et al. 2006) showed that exercising for 20 min at an intensity above the lactate threshold resulted in significantly greater negative affect than exercising for 20 min either below the lactate threshold or at a self-selected intensity. Interestingly, participant's self-selected intensity was very similar to the intensity that corresponded to the lactate threshold.

A final issue in examining affective responses is the potential that individual differences can have in such responses to exercise. The "nomothetic" approach (i.e., responses averaged across individuals) has been the predominant approach but could very easily obscure critical differences in inter-individual affective responses between (Ekkekakis 2005). An "ideographic" approach (i.e., responses examined at the individual level for differences) can reveal distinctions in affective responses that are missed with a nomothetic approach. For example, Van Landuyt and colleagues (Van Landuyt et al. 2000) demonstrated that a 30-min bout of moderate-intensity (60% VO_{2peak}) cycling resulted in essentially no change in FS ratings. However, the lack of change was due to nearly half (48%) of the participants reporting an increase in affective valence (greater pleasantness), 35% reporting a decrease in affective valence (less pleasantness), with the remainder (~17%) reporting no change. The lack of change at the group level masked the variability in affective response. This is a useful demonstration that even at exercise intensities considered moderate in nature, affective responses are not necessarily positive, likely because a third of the individuals were exercising above their VT. This is despite the fact that the pre- to post-exercise changes are generally positive (see also (Ekkekakis et al. 2000)).

An interesting example of research examining the intensity-affect link was done by Zenko et al. (2016). In this work, they assigned participants to an exercise session where intensity was gradually: (a) increased over the 15-min bout, ending with the intensity being 120% of the workload corresponding to the ventilatory threshold, or (b) gradually decreased, starting at 120% of the ventilatory threshold and decreasing from there. As expected, when intensity was "ramped up," the vast majority of participants (77%) reported decreased pleasantness/increased unpleasantness whereas when the intensity was "ramped down," most participants (83%) reported increased pleasantness/decreased unpleasantness. Zenko et al. thus demonstrated that a positive slope of pleasure (i.e., from the ramped down protocol) resulted in more pleasantness and enjoyment post-exercise as well as more positive remembered pleasure and forecasted pleasure for future exercise as much as a week later.

Given the previous discussion, it is worth noting what has been delineated regarding the spectrum of exercise intensity and affective responses. Ekkekakis (2005) has conceptualized this idea very nicely within the Dual-Mode Model. Essentially, when exercise is done at intensities below the lactate/ventilatory threshold, affective responses tend to be fairly homogeneous and pleasant (i.e., nearly everyone feels good at such intensities). When the intensity is increased to a level proximal to the lactate/ventilatory threshold, affective responses may be pleasant or unpleasant and are quite variable (i.e., some feel pleasant, some unpleasant, some indifferent). Finally, as exercise intensity exceeds the lactate/ventilatory threshold, homogeneity of affective responding is the norm again, except now it is primarily unpleasant (i.e., nearly everyone feels less pleasant/more unpleasant).

While less studied than aerobic forms of exercise, resistance exercise has received some attention regarding its affective impact. Like the research discussed above, as the intensity of the resistance exercise increases, affective valence becomes increasingly less positive/more negative. For example, Focht et al. (2015) showed that intensities based on 40% of an individual's 1-repetition maximum (1-RM) and self-selected intensity (which approximated 57% 1-RM) yielded primarily increased affective valence, whereas 70% 1-RM showed a progressive decline in valence. Greene and Petruzzello (2015) showed a similar finding. Whereas the same protocol performed at 100% 10-RM (keeping in mind that a 10-RM approximates ~70–75% of a person's 1-RM, thus 100% 10-RM is about 70% 1-RM) resulted in a decreasing valence response, 70% of an individual's 10-RM (a more moderate-intensity stimulus) yielded stable and somewhat increasing affective valence across a bout consisting of 3 sets of 10 repetitions of 7 different resistance exercises. While only a partial picture, resistance exercise seems to result in a similar pattern of affective responses as aerobic forms of exercise.

To summarize, affect becomes increasingly less positive/more negative as the intensity of exercise increases. Moderate-intensity exercise, when accounting for the lactate or ventilatory threshold, tends to yield more positive affective changes. However, it is critically important to account for individual difference factors (Ekkekakis 2013; Ekkekakis et al. 2011). To date, exercise duration has received relatively little systematic attention, but it seems that as duration progresses, affective valence seems to follow an intensity-dependent pattern, often tending to decrease (i.e., become less pleasant; (Ekkekakis and Petruzzello 1999)). It is worth noting that when it comes to exercise duration, it doesn't seem to have much of an impact on subsequent adherence (Rhodes et al. 2009).

6 Making a Place for Genetic Contributions to Affective Responding

Since at least 2007, a case has been made for the potential influence of genetics on affective responses to exercise (Bryan et al. 2007; de Geus and Moor 2008). In their paper, de Geus and De Moor argued that genetics may contribute significantly to the variability in acute affective responses, specifically that there might be individual differences in genetic sensitivity. A major point of their argument is that genetics influences the pleasant/unpleasant impact that exercise has. Specifically, their hypothesis is that for the regular exerciser, the pleasant (i.e., appetitive) effects are amplified whereas for the non-exerciser the unpleasant (i.e., aversive) effects hold sway. While there have not been many investigations of these propositions, there has been some promising work. In a study involving primarily twin pairs, Schutte et al. (2017) examined affective responses to submaximal and maximal exercise tests on both a cycle ergometer and treadmill, using the FS to assess affect during the exercise and the AD ACL to examine affect post-exercise.

The basic premise of the study by Schutte et al. (2017) was to examine the extent of heritability in affective responses during and following a bout of exercise. The submaximal exercise sessions involved 4 incremental increases in the workload, with each stage lasting 5 min (20 min total). Not surprisingly, FS responses decreased with increasing workloads (i.e., intensity), then rebounded during the 5-min post-exercise recovery period. For the AD ACL, Energy increased while Calmness and Tiredness decreased during the recovery period. To address the main hypothesis, the heritability estimate (i.e., the proportion of the affective response accounted for by the genetic contribution) for FS was 15%. For Energy and Calmness, heritability estimates ranged from 19 to 37%. Those who engaged in more regular exercise also had more positive/less negative affective responses. Huppertz et al. (2014), from a sample of 5,095 18–50 year old twins and their siblings, showed a genetic link between enjoyment (or lack thereof) of physical activity and weekly time spent engaging in voluntary exercise, with heritability estimates of 44–47%. de Geus (2023), drawing from a variety of brain research in animals and humans, notes that there appears to be a strong convergence on brain circuitry involved in the rewards:punishment of engaging in physical activity and actually engaging in exercise. One example comes from the work of Flack et al. (2019), where they demonstrated a genetic contribution to the ability to tolerate exercise intensity, the relative reinforcement value of exercise, and the amount of moderate-to-vigorous physical activity individuals engaged in.

7 What About High-Intensity Interval Exercise/Training (HIIE/HIIT)?

With what has been presented so far, it becomes somewhat difficult to understand why HIIT has become such a popular exercise modality. It has been clearly established that as intensity increases, especially at levels beyond the lactate/ventilatory threshold, affective valence declines and becomes increasingly less pleasant/more unpleasant. Yet, HIIT has been a yearly top 10 fitness trend since 2014 (Thompson 2023). Such workouts involve repeated intervals of all-out/nearly all-out effort (i.e., high-intensity), interspersed with periods of rest or recovery (very low effort). These work:recovery scenarios vary greatly, from the "classic" Tabata-style ([20 s:10 s] × 8), to those more commonly used in research settings ([60 s:60 s] × 10), and any other variation imaginable. While often promoted as a "time-efficient strategy" for physical activity, time-use surveys have shown large increases in discretionary time in Western countries (e.g., +4–5 h per day in the USA, 1965–2003), most of which has not been allocated to exercise or physical activity (Sturm and Cohen 2019). Research examining affect and enjoyment to HIIT protocols usually compares HIIE to moderate-intensity continuous exercise (MICE), often reporting no difference following or that HIIE results in more negative valence during the activity. Enjoyment, reported following the exercise has ended, is often similar between HIIE and MICE.

So, if HIIT (or HIIE) is a supposedly enjoyable form of exercise, do people stick with it? Stork et al. (2018) compared affective (via FS) and enjoyment responses in a group of inactive adults (~21 years old) to 3 different cycling protocols: (a) moderate-intensity continuous exercise (MICT), entailing 45 min of cycling at ~70–75% HR_{max} (50 min total including warm-up and cool-down); (b) HIIT, comprised of 10 bouts of 60-s at ~85–90% interspersed with 60-s of rest, both done a total of 10 times (24 min total); and (c) sprint interval training (SIT), which involved 320-s "all-out" sprints with 2-min of rest between each (10 min total). Predictably, affect (via FS) declined slightly across the MICT session, and much more so during HIIT and SIT. Interestingly, enjoyment was rated similarly across all 3 sessions. The key piece of information was, when activity behavior tracked over the next month, MICT was chosen almost twice as frequently (6.1 times) as HIIT (3.5 times) and far more often than SIT (1.4 times). This highlights, again, that how people feel during an exercise activity is potentially an important piece of the adherence puzzle (Williams et al. 2008, 2012). There seems to be plenty of unanswered questions surrounding the affective impact that HIIT has on the individual exerciser. The interested reader can learn much more about the "extraordinary claims" that have been made regarding HIIT and where these claims almost always come up short (Ekkekakis and Biddle 2023; Ekkekakis and Tiller 2022; Ekkekakis et al. 2023).

8 Where To From Here?

It should be obvious that physical activity and exercise is a pleasant experience for some and an unpleasant experience for others. These affective experiences during the activity have been shown to predict future exercise behavior (e.g., (Williams 2008; Williams et al. 2012; Dunton et al. 2020)). An interesting avenue that research is beginning to pursue is to better identify, via assessment, and then modify affective responses in an effort to determine if this might be a viable intervention strategy (Dunton et al. 2023). One such approach that seems to have promise involves creating associations between physical activity and pleasure through the use of smart phones (Conroy and Kim 2021). This particular study showed that when users were exposed to ~2 min per day of these conditioned associations, this was further associated with increased amounts of physical activity over the course of 8 weeks.

As alluded to throughout this chapter, there remains a need to systematically examine plausible mechanisms to explain both the positive and negative affective responses to physical activity and exercise as well as accounting for individual differences. This could include the examination of cognitive constructs drawn from relevant theory (e.g., perceived autonomy), exploration of biological mediators (e.g., endocannabinoids, brain activation, temperature), how somatosensory and cognitive factors interact at various exercise intensities, and continued investigation of genetic contributions to affective responses.

There is also a need to more carefully and systematically examine activity modalities beyond what are typically aerobic in nature. More attention is needed regarding resistance exercise and high-intensity interval exercise, especially with respect to the timing of affective measurement. It also bears repeating that choice of affective measure is crucially important, with measures that allow assessment of valence and activation seemingly providing the best capture of the possible affective space.

The road ahead appears to hold a great deal of promise for better understanding the impact that exercise has on affect and the impact such affective responses have on physical activity behavior. Borrowing once again from Schutte et al. (2017), some people associate exercise and physical activity with pleasantness and experience it as such. Unfortunately, others associate exercise with unpleasantness. Coupled with the recommendations from Dunton et al. (2023), individualizing the exercise "prescription" to highlight the positive aspects and reduce the aversive aspects to the extent possible seems a fruitful intervention approach.

References

Acevedo EO, Rinehardt KF, Kraemer RR (1994) Perceived exertion and affect at varying intensities of running. Res Q Exerc Sport 65:372–376

Allen MS, Iliescu D, Greiff S (2022) Single item measures in psychological science. Eur J Psychol Assess 38:1–5

Backhouse SH, Ekkekakis P, Biddle SJH, Foskett A, Williams C (2007) Exercise makes people feel better but people are inactive: paradox or artifact? J Sport Exerc Psychol 29:498–517
Benjamin CC, Rowlands A, Parfitt G (2012) Patterning of affective responses during a graded exercise test in children and adolescents. Pediatr Exerc Sci 24:275–288
Biddle SJH, Mutrie N (2001) Psychology of physical activity: determinants, well-being and interventions. Routledge, New York
Brellenthin AG, Crombie KM, Hillard CJ, Brown RT, Koltyn KF (2021) Psychological and endocannabinoid responses to aerobic exercise in substance use disorder patients. Subst Abus 42:272–283
Bryan A, Hutchison KE, Seals DR, Allen DL (2007) A transdisciplinary model integrating genetic, physiological, and psychological correlates of voluntary exercise. Health Psychol 26:30–39
Conroy DE, Kim I (2021) HeartPhone: mobile evaluative conditioning to enhance affective processes and promote physical activity. Health Psychol 40:988–997
Crombie KM, Brellenthin AG, Hillard CJ, Koltyn KF (2018) Psychobiological responses to aerobic exercise in individuals with posttraumatic stress disorder. J Trauma Stress 31:134145
Crombie KM, Leitzelar BN, Brellenthin AG, Hillard CJ, Koltyn KF (2019) Loss of exercise- and stress-induced increases in circulating 2-arachidonoylglycerol concentrations in adults with chronic PTSD. Biol Psychol 145:1–7
Crombie KM, Cisler JM, Hillard CJ, Koltyn KF (2021) Aerobic exercise reduces anxiety and fear ratings to threat and increases circulating endocannabinoids in women with and without PTSD. Ment Health Phys Act 20:100366
de Geus EJC (2023) Genetic pathways underlying individual differences in regular physical activity. Exerc Sport Sci Rev 51:2–18
de Geus EJC, Moor MHM (2008) A genetic perspective on the association between exercise and mental health. Ment Health Phys Act 1:53–61
Dunton GF (2017) Ecological momentary assessment in physical activity research. Exerc Sport Sci Rev 45:48–54
Dunton GF, Dzubur E, Intille S (2016) Feasibility and performance test of a real-time sensor-informed context-sensitive ecological momentary assessment to capture physical activity. J Med Internet Res 18:e106
Dunton GF, Kaplan JT, Monterosso J, Pang RD, Mason TB, Kirkpatrick MG et al (2020) Conceptualizing health behaviors as acute mood-altering agents: implications for cancer control. Cancer Prev Res 13:343–350
Dunton GF, Crosley-Lyons R, Rhodes RE (2023) Affective response during real-world physical activity as an intervention mediator. Exerc Sport Sci Rev 51:140–149
Ekkekakis P (2003) Pleasure and displeasure from the body: perspectives from exercise. Cognit Emot 17:213–239
Ekkekakis P (2005) Exercise and affect – the study of affective responses to acute exercise: the dual-mode model. In: Stelter R, Roessler KK (eds) New approaches to sport and exercise psychology. Meyer and Meyer Sport, Oxford, pp 119–146
Ekkekakis P (2009) Let them roam free? Physiological and psychological evidence for the potential of self-selected exercise intensity in public health. Sports Med 39:857–888
Ekkekakis P (2013) The measurement of affect, mood, and emotion: a guide for health-behavioral research. Cambridge, New York
Ekkekakis P, Biddle SJH (2023) Extraordinary claims in the literature on high-intensity interval training (HIIT): IV. Is HIIT associated with higher long-term exercise adherence? Psychol Sport Exerc 64:102295
Ekkekakis P, Brand R (2019) Affective responses to and automatic affective valuations of physical activity: fifty years of progress on the seminal question in exercise psychology. Psychol Sport Exerc 42:130–137
Ekkekakis P, Lind E (2006) Exercise does not feel the same when you are overweight: the impact of self-selected and imposed intensity on affect and exertion. Int J Obes (Lond) 30:652–660

Ekkekakis P, Petruzzello SJ (1999) Acute aerobic exercise and affect: current status, problems and prospects regarding dose-response. Sports Med 28:337–374
Ekkekakis P, Petruzzello SJ (2000) Analysis of the affect measurement conundrum in exercise psychology: I. Fundamental issues. Psychol Sport Exerc 1:71–88
Ekkekakis P, Petruzzello SJ (2001a) Analysis of the affect measurement conundrum in exercise psychology: II. A conceptual and methodological critique of the exercise-induced feeling inventory. Psychol Sport Exerc 2(1):1–26
Ekkekakis P, Petruzzello SJ (2001b Oct) Analysis of the affect measurement conundrum in exercise psychology. III. A conceptual and methodological critique of the subjective exercise experiences scale. Psychol Sport Exerc 2(4):205–232
Ekkekakis P, Petruzzello SJ (2002) Analysis of the affect measurement conundrum in exercise psychology: IV. A conceptual case for the affect circumplex. Psychol Sport Exerc 3(1):35–63
Ekkekakis P, Tiller NB (2022) Extraordinary claims in the literature on high-intensity interval training: II. Are the extraordinary claims supported by extraordinary evidence? Kinesiol Rev 12: 144–157
Ekkekakis P, Hall EE, VanLanduyt LM, Petruzzello SJ (2000) Walking in (affective) circles: can short walks enhance affect? J Behav Med 23(3):245–275
Ekkekakis P, Parfitt G, Petruzzello SJ (2011) The pleasure and displeasure people feel when they exercise at different intensities: decennial update and progress towards a tripartite rationale for exercise intensity prescription. Sports Med 41:641–671
Ekkekakis P, Vallance J, Wilson PM, Ewing GC (2023) Extraordinary claims in the literature on high-intensity interval training (HIIT): III. Critical analysis of four foundational arguments from an interdisciplinary lens. Psychol Sport Exerc 66:102399
Emmons RA, Diener E (1986) A goal-affect analysis of everyday situational choices. J Res Pers 20: 309–326
Flack K, Pankey C, Ufholz K, Johnson L, Roemmich JN (2019) Genetic variations in the dopamine reward system influence exercise reinforcement and tolerance for exercise intensity. Behav Brain Res 375:112148
Focht BC, Garver MJ, Cotter JA, Devor ST, Lucas AR, Fairman CM (2015) Affective responses to acute resistance exercise performed at self-selected and imposed loads in trained women. J Strength Cond Res 29:3067–3074
Gauvin L, Brawley LR (1993) Alternative psychological models and methodologies for the study of exercise and effect. In: Seraganian P (ed) Exercise psychology: the influence of physical exercise on psychological processes. Wiley, New York, pp 146–171
Gauvin L, Rejeski WJ (1993) The exercise-induced feeling inventory: development and initial validation. J Sport Exerc Psychol 15:403–423
Gauvin L, Spence JC (1998) Measurement of exercise-induced changes in feeling states, affect, mood, and emotions. In: Duda JL (ed) Advances in sport and exercise psychology measurement. Fitness Information Technology, Morgantown, pp 325–336
Greene DR, Petruzzello SJ (2015) More isn't necessarily better: examining the intensity-affect-enjoyment relationship in the context of resistance exercise. Sport Exerc Perform Psychol 4:75–87
Hardy CJ, Rejeski WJ (1989) Not what, but how one feels: the measurement of affect during exercise. J Sport Exerc Psychol 11:304–317
Huppertz C, Bartels M, Jansen IE, Boomsma DI, Willemsen G, Moor MHM et al (2014) A twin-sibling study on the relationship between exercise attitudes and exercise behavior. Behav Genet 44:45–55
Kahneman D (1999) In: Kahneman D, Diener E, Schwarz N (eds) Objective happiness. Russell Sage Foundation, New York, pp 3–25
Katch V, Weltman A, Sady S, Freedson P (1978) Validity of the relative percent concept for equating training intensity. Eur J Appl Physiol Occup Physiol 39:219–227
Lazarus RS (1991) Cognition and motivation in emotion. Am Psychol 46:352–367

Liao Y, Shonkoff ET, Dunton GF (2015) The acute relationships between affect, physical feeling states, and physical activity in daily life: a review of current evidence. Front Psychol 6:1975
Liao Y, Chou CP, Huh J, Leventhal A, Dunton G (2017a) Examining acute bi-directional relationships between affect, physical feeling states, and physical activity in free-living situations using electronic ecological momentary assessment. J Behav Med 40:445–457
Liao Y, Chou CP, Huh J, Leventhal A, Dunton G (2017b) Associations of affective responses during free-living physical activity and future physical activity levels: an ecological momentary assessment study. Int J Behav Med 24:513–519
Lox CL, Jackson S, Tuholski SW, Wasley D, Treasure DC (2000) Revisiting the measurement of exercise-induced feeling states: the physical activity affect scale (PAAS). Measure Physic Educ Exerc Sci 4:79–95
McAuley E, Courneya KS (1994) The subjective exercise experiences scale (SEES): development and preliminary validation. J Sport Exerc Psychol 16:163–177
McNair DM, Lorr M, Droppleman LF (1971) Manual for the profile of mood states. Educational and Industrial Testing Service, San Diego
Meyer JD, Koltyn KF, Stegner AJ, Kim JS, Cook DB (2016) Influence of exercise intensity for improving depressed mood in depression: a dose-response study. Behav Ther 47:527–537
Morris WN (1992) A functional analysis of the role of mood in affective systems. In: Clark MS (ed) Review of personality and social psychology, vol 13, pp 256–293
Nowlis V, Nowlis HH (1956) The description and analysis of mood. Ann N Y Acad Sci 65:345–355
Parfitt G, Rose EA, Burgess WM (2006) The psychological and physiological responses of sedentary individuals to prescribed and preferred intensity exercise. Br J Health Psychol 11: 39–53
Reed J (2013) Effect of acute and regular aerobic physical activity on positive activated affect. In: Ekkekakis P (ed) Routledge handbook of physical activity and mental health. Routledge, New York, pp 413–421
Reed J, Buck S (2009) The effect of regular aerobic exercise on positive-activated affect: a meta-analysis. Psychol Sport Exerc 10:581–594
Reed J, Ones DS (2006) The effect of acute aerobic exercise on positive activated affect: a meta-analysis. Psychol Sport Exerc 7:477–514
Reichert M, Tost H, Reinhard I, Schlotz W, Zipf A, Salize HJ et al (2017) Exercise versus nonexercise activity: E-diaries unravel distinct effects on mood. Med Sci Sports Exerc 49: 763–773
Rhodes RE, Warburton DE, Murray H (2009) Characteristics of physical activity guidelines and their effect on adherence: a review of randomized trials. Sports Med 39:355–375
Russell JA (1980) A circumplex model of affect. J Pers Soc Psychol 39:1161–1178
Russell JA (1997) How shall an emotion be called? In: Plutchik R, Conte HR (eds) Circumplex models of personality and emotions. American Psychological Association, Washington, pp 205–222
Russell JA (2003) Core affect and the psychological construction of emotion. Psychol Rev 110: 145–172
Russell JA, Barrett LF (1999) Core affect, prototypical emotional episodes, and other things called emotion: dissecting the elephant. J Pers Soc Psychol 76(5):805–819
Saklofske DH, Blomme GC, Kelly IW (1992) The effect of exercise and relaxation on energetic and tense arousal. Pers Individ Differ 13:623–625
Scherer KR (2005) What are emotions? And how can they be measured? Soc Sci Info 44(4): 695–729
Schutte NM, Nederend I, Hudziak JJ, Bartels M, Geus EJC (2017) Heritability of the affective response to exercise and its correlation to exercise behavior. Psychol Sport Exerc 31:139–148
Sheppard KE, Parfitt G (2008a) Acute affective responses to prescribed and self-selected exercise intensities in young adolescent boys and girls. Pediatr Exerc Sci 20:129–141

Sheppard KE, Parfitt G (2008b) Patterning of physiological and affective responses during a graded exercise test in sedentary men and boys. J Exerc Sci Fit 6:121–129
Stork MJ, Gibala MJ, Martin Ginis KA (2018) Psychological and behavioral responses to interval and continuous exercise. Med Sci Sports Exerc 50:2110–2121
Sturm R, Cohen DA (2019) Free time and physical activity among Americans 15 years or older: cross-sectional analysis of the American Time Use Survey. Prev Chronic Dis 16:E133
Svebak S, Murgatroyd S (1985) Metamotivational dominance: a multimethod validation of reversal theory constructs. J Pers Soc Psychol 48:107–116
Thayer RE (1986) Activation–deactivation adjective check list: current overview and structural analysis. Psychol Rep 58:607–614
Thayer RE (1987) Energy, tiredness, and tension effects of a sugar snack versus moderate exercise. J Pers Soc Psychol 52:119–125
Thayer RE (1989) The biopsychology of mood and arousal. Oxford University Press, New York
Thayer RE (2001) Calm energy: how people regulate mood with food and exercise. Oxford University Press, New York
Thompson WR (2023) Worldwide survey of fitness trends for 2023. ACSMs Health Fit J 27:9–18
Van Landuyt L, Ekkekakis P, Hall EE, Petruzzello SJ (2000) Throwing the mountains into the lakes: on the perils of nomothetic conceptions of the exercise–affect relationship. J Sport Exerc Psychol 22:208–234
Watson D, Clark LA (1994) The PANAS-X: manual for the positive and negative affect schedule–expanded form. University of Iowa, Iowa City
Watson D, Tellegen A (1985) Toward a consensual structure of mood. Psychol Bull 98:219–235
Watson D, Clark LA, Tellegen A (1988) Development and validation of brief measures of positive and negative affect: the PANAS scales. J Pers Soc Psychol 54:1063–1070
Williams DM (2008) Exercise, affect, and adherence: an integrated model and a case for self-paced exercise. J Sport Exerc Psychol 30:471–496
Williams DM, Dunsiger S, Ciccolo J, Lewis B, Albrecht A, Marcus B (2008) Acute affective response to a moderate-intensity exercise stimulus predicts physical activity participation 6 and 12 months later. Psychol Sport Exerc 9:231–245
Williams DM, Dunsiger S, Jennings EG, Marcus BH (2012) Does affective valence during and immediately following a 10-min walk predict concurrent and future physical activity? Ann Behav Med 44(1):43–51
Zenko Z, Ekkekakis P, Ariely D (2016) Can you have your vigorous exercise and enjoy it too? Ramping intensity down increases postexercise, remembered, and forecasted pleasure. J Sport Exerc Psychol 38:149–159

Exercise and Fear and Safety Learning

Ameera Azar, Troy Hubert, Thomas G. Adams, Josh M. Cisler, and Kevin M. Crombie

Contents

A. Azar
Department of Psychiatry and Behavioral Sciences, The University of Texas at Austin Dell Medical School, Austin, TX, USA

Department of Neuroscience, The University of Texas at Austin, Austin, TX, USA

T. Hubert
Department of Psychology, University of Kentucky, Lexington, KY, UK

T. G. Adams
Department of Psychology, University of Kentucky, Lexington, KY, UK

Department of Psychiatry, Yale School of Medicine, New Haven, CT, USA

J. M. Cisler
Department of Psychiatry and Behavioral Sciences, The University of Texas at Austin Dell Medical School, Austin, TX, USA

Institute for Early Life Adversity Research, The University of Texas at Austin Dell Medical School, Austin, TX, USA

K. M. Crombie (✉)
Department of Psychiatry and Behavioral Sciences, The University of Texas at Austin Dell Medical School, Austin, TX, USA

Department of Kinesiology, The University of Alabama, Tuscaloosa, AL, USA
e-mail: kmcrombie@ua.edu

Curr Topics Behav Neurosci (2024) 67: 125–140
https://doi.org/10.1007/7854_2024_494
Published Online: 23 July 2024

Abstract Fear conditioning paradigms have been studied for over 100 years and are of great interest to the behavioral and clinical sciences given that several safety learning processes (e.g., extinction learning and recall) are thought to be fundamental to the success of exposure-based therapies for anxiety and related disorders. This chapter provides an overview of preclinical and clinical investigations that examined the effects of exercise on initial fear acquisition, fear extinction learning and consolidation, and return of fear outcomes. This chapter highlights the collective body of evidence suggesting that exercise administered after extinction learning enhances the consolidation and subsequent recall of extinction memories to a greater extent than exercise administered prior to extinction learning. This suggests that the addition of exercise after exposure therapy sessions may improve treatment outcomes for people with anxiety and related disorders. Potential mechanisms are discussed in addition to suggestions for future research to improve our understanding of the effects of exercise on fear conditioning and extinction outcomes.

Keywords Behavioral tagging · Exposure therapy · Extinction recall · Fear conditioning · Fear extinction · Memory encoding · Memory retrieval · Posttraumatic stress disorder

1 Clinical Relevance for Examining the Effects of Exercise on Fear Conditioning and Extinction Outcomes

Fear learning and memory processes are commonly studied with laboratory-based paradigms that assess fear acquisition, extinction, and recall (see Fig. 1). Fear learning and memory have long been a topic of interest among behavioral scientists, neuroscientists, and mental health professionals. In addition to improving our fundamental knowledge of these processes, there has also been a surge of interest in incorporating our understanding of fear learning and memory processes to enhance treatments for mental health disorders (Fullana et al. 2020). Exposure-based psychotherapies are some of the most common and efficacious treatment options for individuals with anxiety disorders, obsessive-compulsive disorder (OCD), or posttraumatic stress disorder (PTSD). These therapies are believed to rely heavily on the principles of fear extinction and extinction recall (Rothbaum and Davis 2003) (see Fig. 1). For example, trauma exposure can be considered a conditioning event whereby the traumatic event (e.g., car accident) serves as an unconditioned stimulus (US) and associated details (e.g., sound of screeching tires) serve as conditioned stimuli (CS+). Re-exposure to trauma-related details (CS+) then provokes conditioned threat responses (e.g., increased anxiety and re-traumatization expectancies). As part of therapy, the patient is gradually reintroduced and repeatedly exposed to components of the fear memory (i.e., trauma-related stimuli) in a safe context. This process mimics fear extinction processes (extinction learning, consolidation, and recall) and is thought to weaken the

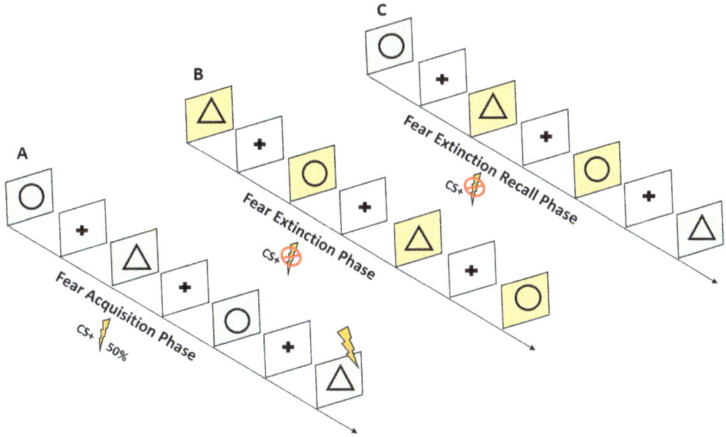

Fig. 1 Overview of a typical human lab-based fear conditioning paradigm assessing fear acquisition (**a**), extinction (**b**), and extinction recall (**c**). During fear acquisition (**a**), participants learn to associate the presentation of a given stimulus (CS+; depicted here as a triangle) with the occurrence of an aversive outcome (US; depicted here as an electric shock), presented in an acquisition context (depicted here as a green background). During fear extinction (**b**), the conditioned stimuli (CS+ and CS-) are presented in the extinction context (depicted here as an orange background), and aversive outcomes no longer follow presentation of the CS+. During the fear extinction recall phase (**c**) participants are presented both conditioned stimuli (CS+ and CS-) in both the acquisition and extinction contexts, without any aversive outcomes administered. The extinction recall phase often involves two runs, with an experimental procedure (e.g., single electric shock) administered in between runs in order to promote reinstatement. During the fear extinction recall phase, researcher can assess fear responding (often via physiological and cognitive indices) to assess several relevant return of fear outcomes (e.g., spontaneous recovery, context renewal, and reinstatement). This figure was adapted from a figure included in a recent review by Crombie et al. (2023a)

predictive value of the CS+ to elicit distress and a conditioned threat response. Essentially, learned inhibition of the fear memory is accomplished following the acquisition and consolidation of new "safety" learning (Fullana et al. 2020; Rothbaum and Davis 2003; Cisler et al. 2014; Craske et al. 2008a; Crombie et al. 2021a).

Unfortunately, even when successful fear extinction learning occurs, original fears are likely to return (Quirk and Mueller 2008). For example, increased fear responding following adequate extinction learning can occur following the passage of time (i.e., spontaneous recovery), after the presentation of context cues of the original trauma (i.e., context renewal), or upon re-exposure to the unconditioned stimulus that directly contributed to the initial fear acquisition (i.e., reinstatement) (Crombie et al. 2021b) (Fig. 2). It has been demonstrated that the return of fear can be a critical barrier to the successful treatment of anxiety and related disorders (e.g., PTSD and OCD) with therapeutic exposure (Craske et al. 2008a, 2014). It has also been demonstrated that greater reductions in distress between subsequent exposure therapy sessions and decreased return of fear between sessions are predictive of greater overall treatment response (Craske et al. 2008b). These findings suggest that

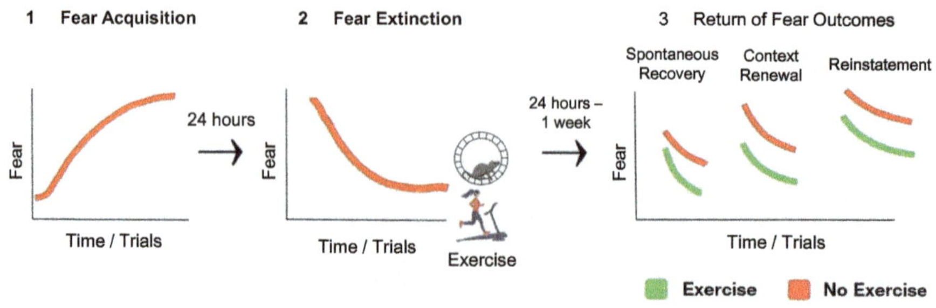

Fig. 2 General depiction of animal and human fear conditioning paradigm involving administration of exercise following extinction learning and subsequent effects on return of fear outcomes

enhanced extinction learning consolidation and greater retrieval of inhibitory safety learning are important therapeutic targets for enhancing the efficacy of exposure-based therapy. Given this, it makes sense for researchers to use fear conditioning paradigms to develop means for enhancing several of these fear and safety learning and memory processes to improve the efficacy of exposure-based therapies.

Fear conditioning paradigms have been well-established and studied for over 100 years (Fullana et al. 2020; Bach et al. 2023; Lonsdorf et al. 2017; Vervliet and Boddez 2020). Recently, researchers have investigated manipulating different aspects of extinction learning to improve understanding of potential adjunct treatments for exposure-based therapies (e.g., enhanced consolidation of extinction learning and extinction recall). The vast majority of such investigations have been pharmacological (e.g., D-cycloserine, methylene blue, yohimbine) (Ebrahimi et al. 2020; Inslicht et al. 2021; Smits et al. 2013, 2014; Zoellner et al. 2017). However, there has also been a recent interest in examining the potential role of exercise, given that it generally results in a small but significant cognitive enhancing effect and engages multiple neural systems thought to play an important role in fear and safety learning processes (Tanner et al. 2018; Crombie et al. 2023a). The proceeding sections of this chapter will provide a brief overview of the preclinical (animal) and clinical (human) studies that have shaped our current understanding, followed by a synopsis of potential mechanisms responsible for exercise-induced enhancement of extinction learning and recall, and finally, suggestions for future research.

2 Animal Studies

Most early investigations examining the effects of exercise on fear conditioning outcomes were conducted in mice and rats. Although slightly different paradigms and protocols have been used, the collective body of evidence has primarily examined: 1) the effect of chronic exercise (i.e., wheel running) before fear conditioning on subsequent fear acquisition, fear extinction learning, extinction recall, and return of fear outcomes (e.g., spontaneous recovery, context renewal, reinstatement), and 2) the effect of acute exercise (i.e., wheel running) before, during, or after fear extinction on subsequent fear extinction learning, extinction recall, and return of fear outcomes. The following section provides a summary of some of the seminal findings in each of these domains. For a comprehensive overview, see Tanner et al.'s review out of Benjamin Greenwood's lab at the University of Colorado at Denver (Tanner et al. 2018).

2.1 Chronic Exercise Prior to Fear Conditioning

Studies conducted to date suggest that chronic wheel running of a sufficient duration prior to contextual fear conditioning enhances the initial acquisition of the fear memory (Burghardt et al. 2006; Falls et al. 2010; Greenwood et al. 2009; Van Hoomissen et al. 2004). The effect of chronic wheel running enhancing the initial fear acquisition has been demonstrated in male mice and rats with wheel running access ranging from 2 to 8 weeks (Burghardt et al. 2006; Falls et al. 2010; Greenwood et al. 2009; Van Hoomissen et al. 2004), with 1 week having null effects (Greenwood et al. 2009). There is also evidence to suggest that chronic wheel running in male rats before fear conditioning contributes to delayed extinction learning and impaired extinction recall during subsequent extinction recall phases (Greenwood et al. 2009). However, Tanner et al. speculated that the delayed within-session extinction observed following chronic exercise is likely due to exercise's effect on strengthening the initial encoding of the fear memory as opposed to impairing extinction mechanisms (Tanner et al. 2018). This hypothesis is supported by findings that suggest that chronic wheel running by male rats does not impair initial fear extinction learning in protocols that assess fear extinction learning immediately after a fear acquisition protocol (Greenwood et al. 2003, 2005), which suggests that chronic exercise delays extinction learning only when there is an opportunity for exercise to boost consolidation of the original fear memory (Tanner et al. 2018). In contrast to the number of investigations conducted examining the effects of chronic wheel running prior to fear acquisition, there has only been one investigation conducted that examined the influence of chronic wheel running after fear acquisition on subsequent fear extinction learning (Greenwood et al. 2009). In this study, Greenwood et al. provided male rats with 1 or 6 weeks of wheel running access and found that neither condition influenced fear extinction learning (Greenwood et al. 2009).

2.2 Acute Exercise Before, During, or After Fear Extinction

Several studies examined the effect of acute exercise before, during, or after fear extinction on initial fear extinction learning, extinction recall, and fear relapse. An early investigation in this area reported that 3 h of wheel running access prior to or immediately following fear extinction learning contributed to improved extinction recall (24 h later) among male rats, compared to male rats with locked running wheels or with wheel access delayed until 6 h after extinction learning sessions (Siette et al. 2014). Additionally, Jacquart et al. reported that 30-min of wheel access 1 or 2 h prior to fear extinction, or 3 h of wheel access immediately prior to fear extinction, did not influence initial fear extinction learning or return of fear (extinction consolidation) in male rats (Jacquart et al. 2017).

Male rats provided with running wheel access during a cued fear extinction training session exhibited lower fear responses during an extinction recall test (1 week later) compared to male rats that were provided with running wheel access prior to fear extinction (Mika et al. 2015). In line with this finding, acute access to running wheels after a contextual fear extinction session and an extinction retention test (24 h later; i.e., a proximal memory test that also served as an additional extinction training session) contributed to enhanced extinction recall in male rats when tested 10 days later (Bouchet et al. 2017). Although wheel running did not influence the consolidation of fear extinction learning or return of fear outcomes in female rats, both male and female rats that ran more during the fear extinction learning consolidation phase exhibited stronger extinction memories during the extinction recall and return of fear tests (Bouchet et al. 2017). Similarly, Crombie et al. demonstrated in male wild-type mice (but not cannabinoid type-1 receptor knockout mice) that voluntary running wheel access in between extinction training sessions (daily training for 4 days) resulted in significantly less (moderate to large effects) freezing during tests of extinction retention compared to mice that were housed with locked running wheels. Additionally, this study revealed that there were moderate to strong relationships between the number of wheel running revolutions and freezing time during extinction training sessions, indicating that mice that ran more in between extinction training sessions exhibited less freezing during subsequent sessions (Crombie et al. 2018). However, a limitation of this study was that mice were provided with unlimited access from immediately after a given extinction training session ended until immediately before the next extinction training session (24h later).

Collectively, these findings led researchers from several recent reviews to conclude that the timing of exercise in relation to fear extinction is highly relevant, such that exercise during or shortly after extinction training limits the return of fear to a greater extent than if administered prior to extinction training (Tanner et al. 2018; Crombie et al. 2023a). The conclusions from these reviews were also supported via meta-analytic evidence that demonstrated that the timing of exercise administration was a significant moderator in determining whether exercise enhances extinction processes among rodents (Roquet and Monfils 2018). Additional preclinical research examining sex differences is warranted. The collective body of evidence suggests that the pro-extinction effect of exercise may be blunted, but not abated, in female rats. Additional research is needed, but it may be the case that female rats need to engage in a greater volume of exercise in order to achieve comparable pro-extinction effects as male rats (Bouchet et al. 2017).

3 Human Studies

More recently, investigations into the effects of aerobic exercise on fear conditioning outcomes have been conducted in humans with and without psychiatric disorders. In contrast to the preclinical body of evidence, early exercise and fear conditioning investigations in human populations have solely examined the effects of exercise administered after fear extinction learning on subsequent extinction recall. This is in large part because the preclinical investigations preceded the clinical investigations, and therefore human studies were primarily designed based on the collective body of knowledge from the animal studies.

Keyan and Bryant conducted the first human trial among a sample of adult men and women (mean age = 20.64 ± 3.18) without a psychiatric disorder and reported that 20-min of stationary cycling (cycling at maximal resistance while still maintaining a cadence of 60–70 rpm [RPMs]) following an extinction training session contributed to significantly lower fear-potentiated startle to CS+ stimuli (24 h later) compared to participants from a light-intensity aerobic exercise control group (20-min of cycling with no resistance while maintaining a cadence of 60–70 RPMs) (Keyan and Bryant 2019). This was also the first human study to examine potential neurobiological mechanisms – as the authors reported that the beneficial effect of exercise was moderated by the presence of the BDNF Val66Met polymorphism (Keyan and Bryant 2019). Although this was a seminal study, one limitation of this investigation was that the researchers implemented a 2-day design, during which fear acquisition and fear extinction occurred on day 1, with extinction recall tests occurring on day 2. As a result, it is difficult to determine if the aerobic exercise group exhibited reduced fear responding during extinction recall tests due to the effects of exercise on the consolidation of fear conditioning or extinction (Keyan and Bryant 2019).

More recently, a study among women (mean age = 30.88 ± 7.15) with PTSD stemming from interpersonal violence (i.e., physical and/or sexual assault) reported that individuals that engaged in 30-min of moderate-intensity aerobic exercise (70–75% age-adjusted maximum heart rate [MHR] on a treadmill) after extinction learning exhibited significantly lower threat expectancy ratings to conditioned stimuli during a delayed (24 h later) reinstatement test compared to individuals that were administered 30-min of a light-intensity aerobic exercise control condition (Crombie et al. 2021b). Importantly, exercise did not appear to affect spontaneous recovery or context renewal and did not affect skin conductance responses during return of fear testing. A follow-up report indicated that the effect of aerobic exercise on threat expectancy ratings during reinstatement was statistically mediated by exercise-induced increases in circulating brain-derived neurotrophic factor (BDNF) and N-arachidonoylethanolamine (i.e., Anandamide, AEA), but not homovanillic acid (HVA) or 2-arachidonoylglycerol (2-AG) (Crombie et al. 2021c). Importantly, this study implemented a three-day design during which fear acquisition, fear extinction, and fear extinction recall phases occurred on separate days, with all study visits occurring at the same time of day (1800 CST) for all participants (Crombie et al. 2021b). Additionally, researchers examined the effects of aerobic exercise on fear extinction recall using a 3-day category-based protocol (i.e., images of tools and animals used as conditioned stimuli). Results revealed that women without a clinical diagnosis (mean age = 33.00 ± 9.82) that were administered 30-min of moderate-intensity aerobic exercise (i.e., 70–75% MHR on a stationary cycle while maintaining a cadence of 60–70 RPMs) after extinction learning exhibited reduced threat expectancy ratings during delayed (24 h after extinction learning) tests of fear recall compared to individuals that were administered 30-min of light-intensity aerobic exercise (i.e., less than 50% MHR on a stationary cycle) (Crombie et al. 2023b).

Most recently, a 3-day study was conducted among 60 healthy, young men (mean age = 24.30 ± 4.40) to determine whether a single bout of vigorous-intensity aerobic exercise (20-min of running at 85% heart rate reserve) administered after an extinction training session enhanced physiological indices (skin conductance and pupillometry) of extinction memory retrieval during delayed (24 h after extinction session) spontaneous recovery and reinstatement tests (Jentsch et al. 2023). This study also tested if vigorous-intensity exercise influenced the generalization of safety learning to two novel, but perceptually similar, stimuli. Contrary to their hypothesis, the researchers reported that the exercise group exhibited impaired extinction memory retrieval during spontaneous recovery, as evident via stronger differential skin conductance and pupillary responses to the original conditioned stimuli and the novel stimuli, compared to the control group (Jentsch et al. 2023). Importantly, the authors note that although participants exhibited a slight decrease in differential conditioned fear responses throughout the extinction training session, there was still evidence of inadequate extinction at the end of the extinction session (i.e., higher ratings to CS+ compared to CS-). This led the researchers to suggest that exercise administered after unsuccessful or incomplete fear extinction may have boosted the consolidation of the original fear memory trace, which was evident via greater physiological responses during the return of fear tests (Jentsch et al. 2023). These results are in contrast to the aforementioned trials which administered moderate-intensity aerobic exercise (i.e., different exercise stimulus) in women (i.e., different sex examined) and reported either an exercise-enhancing effect on limiting physiological indices of return of fear outcomes (Keyan and Bryant 2019) or no difference in physiological indices between exercise and control groups (Crombie et al. 2021b, 2023b). Unfortunately, the Jentsch et al. trial did not assess cognitive indices of return of fear outcomes, which was the outcome that yielded significant effects in the Crombie et al. studies (Crombie et al. 2021b, 2023b). Ultimately, this study suggests that adequate and complete extinction memory formation may be a prerequisite for vigorous-intensity aerobic exercise to enhance extinction memory retrieval in regularly healthy men (Jentsch et al. 2023). This hypothesis will need to be tested in future studies.

Collectively, the human studies are promising and warrant further consideration. Although the findings from the majority of investigations suggest an exercise-enhancing effect (Crombie et al. 2021b, 2023b; Keyan and Bryant 2019), the findings are nuanced, which may be due to different population characteristics (i.e., PTSD vs control), paradigm characteristics (i.e., standard fear conditioning vs category-based fear conditioning protocol), or exercise intensity parameters (i.e., moderate-intensity vs vigorous-intensity aerobic exercise). For instance, the study in women with PTSD revealed an exercise-enhancing effect during a reinstatement test (Crombie et al. 2021b), which was in contrast to the study in women without a clinical diagnosis (Crombie et al. 2023b), which found an exercise-enhancing effect during both extinction recall tests. Given that these studies used slightly different fear conditioning paradigms, the researchers from these studies speculated that it may be the case that aerobic exercise only influences threat expectancy ratings in women with PTSD after a reactivation event occurs, or perhaps women with PTSD may also exhibit reduced threat expectancy ratings prior to a reactivation event (similar to women without PTSD) if category-based fear conditioning protocols are administered (Crombie et al. 2023b). Exercise intensity may also play an important role as the studies (Crombie et al. 2021b, 2023b; Keyan and Bryant 2019) that reported a beneficial effect of exercise on return of fear outcomes administered moderate-intensity aerobic exercise, whereas the one study that did not report a beneficial effect administered vigorous-intensity aerobic exercise (Jentsch et al. 2023).

In addition to fear conditioning protocols, there have also been several recent attempts in humans to examine the effects of exercise on exposure therapy or therapy analogue outcomes (Jacquart et al. 2017; Bryant et al. 2023; Voorendonk et al. 2022; Weisman and Rodebaugh 2020; Young-McCaughan et al. 2022; Bischoff et al. 2018). As alluded to earlier, this makes intuitive sense because much of the research examining the relationship between exercise and fear conditioning is being conducted to improve our understanding of the potential utility of exercise as an adjunct/augmentative strategy to enhance the efficacy of psychotherapies (e.g., exposure-based therapy) for clinical anxiety and PTSD populations. Given that this chapter focuses exclusively on fear conditioning outcomes, we refer the reader to a recent comprehensive review that discusses the results of these investigations (Crombie et al. 2023a). In brief, the findings from the few studies in this domain have been mixed, which highlights the importance for future studies to incorporate experimental designs and methods based on the prior lab-based fear conditioning and extinction protocols (Crombie et al. 2023a).

4 Potential Mechanisms for Exercise-Induced Enhancement of Extinction Learning and Recall

Given that enhanced consolidation of extinction learning and subsequent retrieval of said learning is thought to be an important therapeutic target for exposure-based therapies, it is important to briefly consider some plausible mechanisms that may be at least partially responsible for the exercise-induced enhancement of extinction learning and recall. Although a comprehensive overview of plausible mechanisms is beyond the scope of this chapter (see (Crombie et al. 2023a)), it is important to briefly describe a few interwoven, synergistic mechanisms (see Fig. 3).

One potential mechanism involves the behavioral tagging hypothesis (Crombie et al. 2023a; Dunsmoor et al. 2022; de Carvalho et al. 2013, 2014), which is an extension of the synaptic tagging and capture process (Frey and Morris 1997). The behavioral tagging hypothesis postulates that some behavioral experiences result in the synthesis of plasticity-related products that can get captured by a "synaptic tag" set by some recently completed "weak" learning experience, thereby contributing to enhanced long-term memory formation; said otherwise, plasticity-related proteins/ neuromodulatory factors released in close proximity to initial memory encoding can lead to greater long-term memory consolidation (Crombie et al. 2021c, 2023a; Dunsmoor et al. 2022; de Carvalho et al. 2013, 2014; Ritchey et al. 2016). Based on this notion, it may be the case that aerobic exercise performed in conjunction with extinction learning serves as a behavioral event that increases plasticity-related products, which then get captured by the initial LTP tag set during extinction learning, thereby contributing to greater enhancement of extinction learning (Crombie et al. 2021c, 2023a). Although more research in humans is needed, recent research from a clinical study among women with PTSD offered support for this hypothesis as greater exercise-induced increases in AEA and BDNF were found to mediate the relationship between aerobic exercise and enhanced cognitive indices of extinction recall (Crombie et al. 2021c).

Another potential mechanism involves exercise-induced enhancement of pattern separation, and more broadly, episodic memory (Crombie et al. 2023a). Engaging

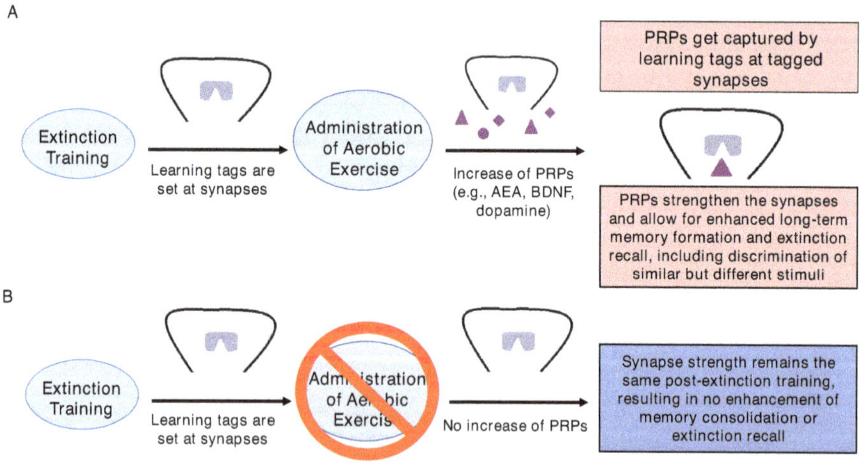

Fig. 3 Overview of hypothesized synergistic mechanisms involved in exercise-induced enhanced consolidation of extinction learning and enhanced recall. This figure was adapted from a figure included in a recent review by Crombie and colleagues (Crombie et al. 2023a). When extinction learning occurs, a tag is set at hippocampal synapses. If aerobic exercise is administered during the post-consolidation window, exercise-induced increases in plasticity-related products (PRPs) get captured by the tag, which then strengthens the tagged synapses, resulting in enhanced long-term memory consolidation (see panel **a**). If aerobic exercise is not administered in the post-extinction consolidation window, enhanced long-term memory consolidation is less likely to occur (see panel **b**)

with every detail in the environment is ultimately impossible for humans, and therefore details are analyzed based on importance and relevance. Humans' ability to retain and retrieve specific details about important events is advantageous for future decision-making (i.e., catalog information for future retrieval). For example, someone with PTSD may want to recall particular details about the event (or about information covered in therapy) to adequately differentiate between safe and threatening contexts in similar and dissimilar contexts. This process of distinguishing similar experiences to keep stored memories separate from one another is otherwise known as pattern separation. Individuals with PTSD often have considerable difficulty with pattern separation, which is why a significant portion of exposure-based psychotherapy focuses on forming new safety memories to inhibit the initial trauma memory. This method also helps individuals with PTSD learn to discern a safe contextual environment from a threatening one and to further limit avoidance behaviors and reduce symptomology (Crombie et al. 2023a). In other words, greater pattern separation is thought to be beneficial for adults with PTSD.

Although this is a relatively new area of investigation, initial human evidence suggests that aerobic exercise is capable of improving pattern separation and episodic memory in adults (Crombie et al. 2023a; Déry et al. 2013; Suwabe et al. 2017, 2018; Marin Bosch et al. 2020; van Dongen et al. 2016). For instance, participants administered an acute bout of aerobic exercise (35-min of submaximal interval training on stationary cycling) after a declarative memory task, exhibited greater retention of picture-location associations and increased hippocampal pattern similarity for correct responses (48 h later) compared to a no-exercise control condition (van Dongen et al. 2016). Additionally, an acute bout of aerobic exercise and greater baseline aerobic fitness are associated with enhanced pattern separation during a mnemonic discrimination task (Déry et al. 2013; Suwabe et al. 2017, 2018). Several animal studies also revealed that wheel running enhanced discrimination between distinct, but similar, stimuli and contexts (Greenwood et al. 2009; Bolz et al. 2015; Creer et al. 2010; Wu et al. 2015; Islam et al. 2021). Collectively, these results provide initial support to suggest that perhaps exercise enhances extinction learning via its effects on pattern separation. In other words, aerobic exercise may be effective because it enhances an individual's ability to correctly discriminate similar but importantly different stimuli (e.g., CS+ in acquisition context vs CS+ in the extinction context). In addition to offering an explanation for the aforementioned lab studies, it is plausible that this may prove to be a mechanism by which incorporating exercise into exposure-based therapy sessions may help individuals distinguish between distinct safety and threat cues as they navigate their environment outside of the clinical setting (Crombie et al. 2023a).

5 Future Directions for Clinical Trials

There is much work to be done to improve our understanding of the effect of exercise on fear conditioning outcomes, and ultimately to improve our understanding of the potential for incorporating exercise with exposure-based therapies to improve clinical outcomes for individuals with anxiety disorders and PTSD. For example, future clinical trials need to continue to systematically examine the precise exercise parameters needed to enhance relevant outcomes of interest (e.g., enhanced extinction recall). For instance, 3 of the 4 human trials that reported a beneficial effect of exercise on return of fear outcomes administered moderate-intensity aerobic exercise (Crombie et al. 2021b, 2023b; Keyan and Bryant 2019), whereas the one investigation that did not report a beneficial effect administered vigorous-intensity exercise (Jentsch et al. 2023). Exercise intensity was not the only major difference between these investigations, but nonetheless, the differential findings highlight that further attention be devoted to teasing out the potential relevance of exercise intensity. Relatedly, future investigations should continue to assess and report pertinent information about the exercise stimulus administered as part of the study (e.g., duration of exercise bout, exercise intensity, and methods for assessing intensity), in addition to the timing of the exercise bout in relation to the fear conditioning protocol (e.g., prior to or following extinction learning) (Crombie et al. 2023a). Future research should also continue to examine potential sex differences. Much of our understanding from preclinical studies is derived from studies involving male rodents (Tanner et al. 2018). Among human studies conducted to date, one study has examined men and women (Keyan and Bryant 2019), two have assessed only women (Crombie et al. 2021b, 2023b), and one has assessed only men (Jentsch et al. 2023). Additionally, future clinical trials should examine individuals with psychiatric disorders, for which we may expect to see the biggest effect, and who are arguably the most important populations of interest from a clinical perspective (Crombie et al. 2023a). To date, there has only been one trial among individuals with PTSD (Crombie et al. 2021b).

It is also important for future trials to improve our understanding of potential mechanisms that contribute to an exercise-enhancing effect on extinction learning in some cases. For instance, the collection of peripheral biomarkers prior to and following exercise may help determine relevant systems involved (or not involved) in the cognitive enhancing effects of exercise (Crombie et al. 2023a). Relatedly, when possible, researchers should expand our understanding beyond peripheral biomarkers by having participants complete fear conditioning protocols while in the MRI environment. Incorporating modern advances in functional neuroimaging should significantly improve our understanding of potential neural mechanisms responsible for enhanced extinction memory consolidation (Crombie et al. 2023a). Additionally, beyond extinction learning specific mechanisms, future research should also consider including other relevant outcomes that may mediate or moderate effects, such as participant's physical activity and exercise history, motivation for physical activity/exercise, and potential expectancy/placebo effects (Lindheimer et al. 2020) (see Chap. 19: The Placebo Effect in Exercise and Mental Health Research).

Lastly, future investigations should consider examining: 1) other exercise modalities beyond aerobic exercise (e.g., resistance training), 2) paradigms that use generalized stimuli (perhaps trauma-related) as conditioned stimuli, and 3) the effect of exercise on the return of fear outcomes beyond 24 h (Crombie et al. 2023a).

Additionally, although it is more demanding from a data collection perspective, future research investigations should utilize 3-day fear conditioning paradigms where fear acquisition, fear extinction, and extinction recall/return of fear outcomes are assessed on separate days. As alluded to earlier, doing so allows researchers to more clearly define the impact of exercise on the consolidation of extinction learning compared to 2-day designs where acquisition and extinction phases occur on the same day (Crombie et al. 2023a).

6 Conclusion

Overall, this chapter provided an overview of animal and human studies that administered exercise (or provided wheel running access in the case of animal studies) at different phases of fear conditioning and extinction paradigms to examine subsequent effects of exercise on initial fear acquisition, fear extinction learning, extinction recall, and return of fear outcomes (e.g., spontaneous recovery, context renewal, reinstatement). As discussed, enhanced extinction learning consolidation and greater retrieval of inhibitory safety learning are important therapeutic targets for enhancing the efficacy of exposure-based therapies among adults with anxiety and related disorders. Importantly, the collective body of evidence to date suggests that exercise administered after extinction learning enhances the consolidation and subsequent recall of extinction memories, but these effects may be moderated by biological sex and exercise parameters (e.g., intensity). This chapter also highlighted that there is much work to be done to improve our understanding of the relationship between exercise and enhanced consolidation of extinction learning; research is needed to characterize mechanisms, sex differences, and to optimize exercise parameters. Despite the need for additional research, initial investigations discussed in this chapter suggest that there may be potential for precisely timed exercise to play an important augmentative role in enhancing exposure-based therapies for anxiety and related disorders.

References

Bach DR, Sporrer J, Abend R, Beckers T, Dunsmoor JE, Fullana MA et al (2023) Consensus design of a calibration experiment for human fear conditioning. Neurosci Biobehav Rev 148:105146

Bischoff S, Wieder G, Einsle F, Petzold MB, Janßen C, Mumm JLM et al (2018) Running for extinction? Aerobic exercise as an augmentation of exposure therapy in panic disorder with agoraphobia. J Psychiatr Res 101:34–41

Bolz L, Heigele S, Bischofberger J (2015) Running improves pattern separation during novel object recognition. Brain Plast 1(1):129–141

Bouchet CA, Lloyd BA, Loetz EC, Farmer CE, Ostrovskyy M, Haddad N et al (2017) Acute exercise enhances the consolidation of fear extinction memory and reduces conditioned fear relapse in a sex-dependent manner. Learn Mem 24(8):358–368

Bryant RA, Dawson KS, Azevedo S, Yadav S, Cahill C, Kenny L et al (2023) Augmenting trauma-focused psychotherapy for post-traumatic stress disorder with brief aerobic exercise in Australia: a randomised clinical trial. Lancet Psychiatry 10(1):21–29

Burghardt PR, Pasumarthi RK, Wilson MA, Fadel J (2006) Alterations in fear conditioning and amygdalar activation following chronic wheel running in rats. Pharmacol Biochem Behav 84(2):306–312

Cisler JM, Steele JS, Lenow JK, Smitherman S, Everett B, Messias E et al (2014) Functional reorganization of neural networks during repeated exposure to the traumatic memory in posttraumatic stress disorder: an exploratory fMRI study. J Psychiatr Res 48(1):47–55

Craske MG, Kircanski K, Zelikowsky M, Mystkowski J, Chowdhury N, Baker A (2008a) Optimizing inhibitory learning during exposure therapy. Behav Res Ther 46(1):5–27

Craske MG, Waters AM, Lindsey Bergman R, Naliboff B, Lipp OV, Negoro H et al (2008b) Is aversive learning a marker of risk for anxiety disorders in children? Behav Res Ther 46(8): 954–967

Craske MG, Treanor M, Conway CC, Zbozinek T, Vervliet B (2014) Maximizing exposure therapy: an inhibitory learning approach. Behav Res Ther 58:10–23

Creer DJ, Romberg C, Saksida LM, van Praag H, Bussey TJ (2010) Running enhances spatial pattern separation in mice. Proc Natl Acad Sci U S A 107(5):2367–2372

Crombie KM, Brellenthin AG, Hillard CJ, Koltyn KF (2018) Endocannabinoid system involvement in exercise-induced enhancement of fear extinction in mice. Med Sci Sports Exerc 50(5S):96

Crombie KM, Privratsky AA, Schomaker CM, Heilicher M, Ross MC, Sartin-Tarm A et al (2021a) The influence of FAAH genetic variation on physiological, cognitive, and neural signatures of fear acquisition and extinction learning in women with PTSD. Neuroimage Clin 33:102922

Crombie KM, Sartin-Tarm A, Sellnow K, Ahrenholtz R, Lee S, Matalamaki M et al (2021b) Aerobic exercise and consolidation of fear extinction learning among women with posttraumatic stress disorder. Behav Res Ther 142:103867

Crombie KM, Sartin-Tarm A, Sellnow K, Ahrenholtz R, Lee S, Matalamaki M et al (2021c) Exercise-induced increases in anandamide and BDNF during extinction consolidation contribute to reduced threat following reinstatement: preliminary evidence from a randomized controlled trial. Psychoneuroendocrinology 132:105355

Crombie KM, Adams TG, Dunsmoor JE, Greenwood BN, Smits JA, Nemeroff CB et al (2023a) Aerobic exercise in the treatment of PTSD: an examination of preclinical and clinical laboratory findings, potential mechanisms, clinical implications, and future directions. J Anxiety Disord 94:102680

Crombie KM, Azar A, Botsford C, Heilicher M, Moughrabi N, Gruichich TS et al (2023b) Aerobic exercise after extinction learning reduces return of fear and enhances memory of items encoded during extinction learning. Ment Health Phys Act 24:100510

de Carvalho MJ, Benetti F, Izquierdo I (2013) Behavioral tagging of extinction learning. Proc Natl Acad Sci U S A 110(3):1071–1076

de Carvalho MJ, Furini CRG, Benetti F, Izquierdo I (2014) Hippocampal molecular mechanisms involved in the enhancement of fear extinction caused by exposure to novelty. Proc Natl Acad Sci U S A 111(12):4572–4577

Déry N, Pilgrim M, Gibala M, Gillen J, Wojtowicz JM, Macqueen G et al (2013) Adult hippocampal neurogenesis reduces memory interference in humans: opposing effects of aerobic exercise and depression. Front Neurosci 7:66

Dunsmoor JE, Murty VP, Clewett D, Phelps EA, Davachi L (2022) Tag and capture: how salient experiences target and rescue nearby events in memory. Trends Cogn Sci 26(9):782–795

Ebrahimi C, Gechter J, Lueken U, Schlagenhauf F, Wittchen HU, Hamm AO et al (2020) Augmenting extinction learning with D-cycloserine reduces return of fear: a randomized, placebo-controlled fMRI study. Neuropsychopharmacology 45(3):499–506

Falls WA, Fox JH, MacAulay CM (2010) Voluntary exercise improves both learning and consolidation of cued conditioned fear in C57 mice. Behav Brain Res 207(2):321–331

Frey U, Morris RG (1997) Synaptic tagging and long-term potentiation. Nature 385(6616):533–536

Fullana MA, Dunsmoor JE, Schruers KRJ, Savage HS, Bach DR, Harrison BJ (2020) Human fear conditioning: from neuroscience to the clinic. Behav Res Ther 124:103528

Greenwood BN, Foley TE, Day HEW, Campisi J, Hammack SH, Campeau S et al (2003) Freewheel running prevents learned helplessness/behavioral depression: role of dorsal raphe serotonergic neurons. J Neurosci 23(7):2889–2898

Greenwood BN, Foley TE, Burhans D, Maier SF, Fleshner M (2005) The consequences of uncontrollable stress are sensitive to duration of prior wheel running. Brain Res 1033(2): 164–178

Greenwood BN, Strong PV, Foley TE, Fleshner M (2009) A behavioral analysis of the impact of voluntary physical activity on hippocampus-dependent contextual conditioning. Hippocampus 19(10):988–1001

Inslicht SS, Niles AN, Metzler TJ, Lipshitz SL, Otte C, Milad MR et al (2021) Randomized controlled experimental study of hydrocortisone and D-cycloserine effects on fear extinction in PTSD. Neuropsychopharmacology

Islam MR, Valaris S, Young MF, Haley EB, Luo R, Bond SF et al (2021) Exercise hormone irisin is a critical regulator of cognitive function. Nat Metab 3(8):1058–1070

Jacquart J, Roquet RF, Papini S, Powers MB, Rosenfield D, Smits JAJ et al (2017) Effects of acute exercise on fear extinction in rats and exposure therapy in humans: null findings from five experiments. J Anxiety Disord 50:76–86

Jentsch VL, Wolf OT, Otto T, Merz CJ (2023) The impact of physical exercise on the consolidation of fear extinction memories. Psychophysiology:e14373

Keyan D, Bryant RA (2019) Acute exercise-induced enhancement of fear inhibition is moderated by BDNF Val66Met polymorphism. Transl Psychiatry 9(1):131

Lindheimer JB, Szabo A, Raglin JS, Beedie C (2020) Advancing the understanding of placebo effects in psychological outcomes of exercise: lessons learned and future directions. Eur J Sport Sci 20(3):326–337

Lonsdorf TB, Menz MM, Andreatta M, Fullana MA, Golkar A, Haaker J et al (2017) Don't fear "fear conditioning": methodological considerations for the design and analysis of studies on human fear acquisition, extinction, and return of fear. Neurosci Biobehav Rev 77:247–285

Marin Bosch B, Bringard A, Logrieco MG, Lauer E, Imobersteg N, Thomas A et al (2020) Effect of acute physical exercise on motor sequence memory. Sci Rep 10(1):15322

Mika A, Bouchet CA, Bunker P, Hellwinkel JE, Spence KG, Day HEW et al (2015) Voluntary exercise during extinction of auditory fear conditioning reduces the relapse of fear associated with potentiated activity of striatal direct pathway neurons. Neurobiol Learn Mem 125:224–235

Quirk GJ, Mueller D (2008) Neural mechanisms of extinction learning and retrieval. Neuropsychopharmacology 33(1):56–72

Ritchey M, Murty VP, Dunsmoor JE (2016) Adaptive memory systems for remembering the salient and the seemingly mundane. Behav Brain Sci 39:e221

Roquet RF, Monfils MH (2018) Does exercise augment operant and Pavlovian extinction: a meta-analysis. J Psychiatr Res 96:73–93

Rothbaum BO, Davis M (2003) Applying learning principles to the treatment of post-trauma reactions. Ann N Y Acad Sci 1008:112–121

Siette J, Reichelt AC, Westbrook RF (2014) A bout of voluntary running enhances context conditioned fear, its extinction, and its reconsolidation. Learn Mem 21(2):73–81

Smits JAJ, Rosenfield D, Otto MW, Marques L, Davis ML, Meuret AE et al (2013) D-cycloserine enhancement of exposure therapy for social anxiety disorder depends on the success of exposure sessions. J Psychiatr Res 47(10):1455–1461

Smits JAJ, Rosenfield D, Davis ML, Julian K, Handelsman PR, Otto MW et al (2014) Yohimbine enhancement of exposure therapy for social anxiety disorder: a randomized controlled trial. Biol Psychiatry 75(11):840–846

Suwabe K, Hyodo K, Byun K, Ochi G, Yassa MA, Soya H (2017) Acute moderate exercise improves mnemonic discrimination in young adults. Hippocampus 27(3):229–234

Suwabe K, Byun K, Hyodo K, Reagh ZM, Roberts JM, Matsushita A et al (2018) Rapid stimulation of human dentate gyrus function with acute mild exercise. Proc Natl Acad Sci U S A 115(41): 10487–10492

Tanner MK, Hake HS, Bouchet CA, Greenwood BN (2018) Running from fear: exercise modulation of fear extinction. Neurobiol Learn Mem 151:28–34

van Dongen EV, Kersten IHP, Wagner IC, Morris RGM, Fernández G (2016) Physical exercise performed four hours after learning improves memory retention and increases hippocampal pattern similarity during retrieval. Curr Biol 26(13):1722–1727

Van Hoomissen JD, Holmes PV, Zellner AS, Poudevigne A, Dishman RK (2004) Effects of beta-adrenoreceptor blockade during chronic exercise on contextual fear conditioning and mRNA for galanin and brain-derived neurotrophic factor. Behav Neurosci 118(6):1378–1390

Vervliet B, Boddez Y (2020) Memories of 100 years of human fear conditioning research and expectations for its future. Behav Res Ther 135:103732

Voorendonk EM, Sanches SA, Mojet M, De Jongh A, Van Minnen A (2022) Combining a single session of prolonged exposure with physical activity in patients with PTSD: the effect of sequence. Ment Health Phys Act 21:100417

Weisman JS, Rodebaugh TL (2020) Testing the efficacy of a brief exercise intervention for enhancing exposure therapy outcomes. J Anxiety Disord 74:102266

Wu MV, Luna VM, Hen R (2015) Running rescues a fear-based contextual discrimination deficit in aged mice. Front Syst Neurosci 9:114

Young-McCaughan S, Peterson AL, Mintz J, Hale WJ, Dondanville KA, Borah EV et al (2022) Testing the role of aerobic exercise in the treatment of posttraumatic stress disorder (PTSD) symptoms in U.S. active duty military personnel: a pilot study. Cogn Behav Ther 10:1–16

Zoellner LA, Telch M, Foa EB, Farach FJ, McLean CP, Gallop R et al (2017) Enhancing extinction learning in posttraumatic stress disorder with brief daily imaginal exposure and methylene blue: a randomized controlled trial. J Clin Psychiatry 78(7):e782–e789

Exercise and Memory

Michele Schmitter and Janna Nonja Vrijsen

Contents

Abstract Our memory is crucial to our daily functioning, our identity, and how we interact with the world around us. Maintaining memory functioning is therefore important in many ways. Exercise positively affects our memory. Even a single exercise session can directly boost different memory types, including working, procedural, and episodic memory. Due to these benefits, exercise has great potential as an intervention in different settings, such as schools or rehabilitation centers, and

M. Schmitter (✉)
Behavioural Science Institute, Radboud University Nijmegen and Depression Expertise Center, Pro Persona Mental Health Care, Nijmegen, The Netherlands
e-mail: m.schmitter@propersona.nl

J. N. Vrijsen
Department of Psychiatry, Donders Institute for Brain, Cognition and Behaviour, Radboud University Medical Center, Nijmegen, The Netherlands

Depression Expertise Center, Pro Persona Mental Health Care, Nijmegen, The Netherlands
e-mail: janna.vrijsen@radboudumc.nl

Curr Topics Behav Neurosci (2024) 67: 141–154
https://doi.org/10.1007/7854_2024_495
Published Online: 31 July 2024

as a prevention strategy. In this chapter, we first give an overview of what memory is and why it is so relevant to our mental health. Next, we briefly discuss how, for whom, and under which circumstances exercise improves memory functioning. Then, we will summarize the literature of experimental studies investigating the direct effect of exercise on different memory types, while acknowledging current limitations and potential future directions. Finally, we briefly highlight the potential of exercise as intervention to preserve and improve memory.

Keywords Exercise · Memory · Mental health

1 Introduction

I often go on a run to clear my head. When I am brooding over something and notice that negative memories keep popping up, a run always helps. After running, I feel good, I can focus so much better, the negative memories become less potent, and I seem to glide through my tasks with more ease. Why is that? In addition to the direct effect on mood and affect (see Chapters "Exercise and Acute Affect and Mood States"), exercise also affects the functioning of our memory. This is very interesting both from a theoretical and mental health perspective. Understanding how and under which conditions exercise enhances memory – including memory processes related to mental health problems – gives insight in the cognitive and behavioral mechanisms of exercise. This in turn allows us to translate the findings from experimental research into prevention strategies and mental health care solutions. In this chapter, we start by briefly introducing relevant memory processes. This is followed by presenting experimental work on the effect of exercise on both "cold" (i.e., non-emotional) and "warm" (i.e., emotional) memory processes. Here we will briefly touch upon the clinical potential of the memory benefits for exercise, with a focus on mental health solutions. The evidence for the effect of exercise on memory is based mainly on aerobic or endurance exercise, although the clinical value and mechanisms of other types of exercise (i.e., strength, balance, flexibility) are gaining interest.

2 What Is Memory?

Take care of all your memories. For you cannot relive them. – Bob Dylan

Thinking back to our vacation or the precious moments we spent with our friends and families can enlighten us in difficult times and help maintain our mental well-being. They shape us as a person and how we perceive the world around us. Memories are therefore crucial to cherish and preserve. The formation of a memory can be divided into different phases, as well as different types of memory.

When new information enters the human system, it does so via working memory (WM). It is an essential component of our daily life as we rely on it constantly to perform many tasks throughout the day. WM is the ability to temporarily store and manipulate information and in turn comprises multiple processes including maintenance, updating, attending to, inhibiting, and encoding of information. The latter is what we could approach as the first phase of memory: most information is removed from WM, but a limited portion flows on to be encoded and consolidated (or: stored) in long-term memory. From long-term memory, information can be retrieved (i.e., brought to conscious memory) and re-consolidated again (or: re-worked based on relevant new/contextual information).

Long-term memory again comprises different subtypes of memory. In relation to exercise research, procedural memory (PM), semantic memory (SM) and episodic memory (EM) are important. PM refers to memory of how to complete tasks, for example how to ride a bike or make a cup of coffee. It enables us to perform a variety of everyday activities. SM is where information gets its meaning. This can include factual information (e.g., what a car is) and the emotional meaning of a memory.

Within EM, a distinction should be made between recall of nonpersonal information, as, for example, words one can learn during a computer task and later on recall, and autobiographical information. Autobiographical memory refers to recollections of our personal history and is unique to a person, e.g., where we parked our car in the morning, or that great road trip we took last summer. A mechanism necessary for unique episodic memories is called *pattern separation.* Pattern separation is the neurocomputational process that allows the brain to distinguish between similar experiences. An example of pattern separation is when we meet new people and try to remember their names and faces to avoid confusion. Overall, memory processes are essential to our daily functioning (without it we cannot do anything) and our conception of ourselves and the world around us. What affects these central memory processes and phases – including exercise – is hence incredibly valuable to understand.

3 Exercise and Memory

3.1 Biological Effects of Exercise Underlying Memory

Exercise has been found to have numerous biological effects on the brain (see Chapters "Exercise, Neurocircuitry, and Neurotransmitter Systems", "Exercise and Stress Resilience" and "Exercise and Neuroendocrinology and Immunology"), which enhance neuroplasticity in the prefrontal cortex (PFC) and the hippocampus; areas that subserve memory functioning. A single exercise session triggers several neuroplasticity-enhancing mechanisms (e.g., long-term potentiation [LTP], release of brain-derived neurotrophic factor [BDNF]), and on a regular basis these effects accumulate and induce structural and functional changes in the brain. Consistent with the improved functioning of the hippocampus and PFC, recent reviews conclude that acute exercise bouts (see Chapter "Exercise and Acute Cognitive Enhancement") and regular exercise benefit memory (Loprinzi et al. 2018; Roig et al. 2013).

3.2 Exercise Protocols as Moderator

The relations between exercise and memory are complex, and various studies also report null or adverse effects (Roig et al. 2013). For instance, it has been suggested that the effect of an acute exercise bout on the outcome of a memory task depends on the timing of exercise as different phases of memory formation might be affected differently by exercise (e.g., Roig et al. 2016; Loprinzi et al. 2019). Mechanistically, exercise prior to encoding may improve memory through the direct facilitation of LTP, while exercise during consolidation may enhance the stabilization of the memory trace through the exercise-induced increase in neurotrophins. Several studies showed that an exercise bout before encoding and during early and late consolidation can enhance memory functioning (Loprinzi et al. 2019).

The benefit of exercise depends on the intensity of the exercise and the type of memory assessed. For example, low-intensity exercise prior to encoding is beneficial to WM, whereas high-intensity exercise before encoding enhances EM (Loprinzi 2018). Similarly, the duration of the exercise bout can affect memory functioning: Shorter bouts (<40 min) seem to be more favorable than longer durations, possibly due to exercise-induced fatigue (Roig et al. 2013; Crush and Loprinzi 2017). In summary, the variation in exercise protocols complicates comparisons between studies and impedes the determination of the right amount of exercise to improve memory functioning. Standardization of exercise paradigms would facilitate a more effective leverage of each exercise study (Basso and Suzuki 2017).

3.3 Different Populations as Moderator

The effect of exercise on memory may also depend on the population under study. For example, several studies suggest that while exercise has the potential to enhance memory function, the extent of these gains appears to depend on baseline fitness levels. Specifically, individuals with lower fitness levels tend to experience more memory improvements after exercise (e.g., Roig et al. 2013; Pena et al. 2022). Likewise, sex assigned at birth (i.e., differential results for males and females for different aspects of executive functioning, (Barha et al. 2019)) and age may moderate the effect of exercise (i.e., younger adults show greater memory improvements after an exercise session (Roig et al. 2013)). While this effect regarding age is of interest from a developmental perspective, this result does not imply that exercise is not effective for older adults. Stute et al. (2020), for example, showed that a moderate-intensity exercise bout (15 min; cycling at 50% VO_{2max}) directly improves

WM performance in older adults (<64 years). However, more studies are needed that focus specifically on the direct effect of exercise on older adults. Regular exercise, on the other hand, has been intensively studied as a tool to postpone cognitive decline in the aging population (e.g., Blondell et al. 2014; Sofi et al. 2011; see Chapter "Exercise and Cognitive / Age-Related Decline"). Similarly, chronic exercise gained interest as an intervention for specific populations with memory problems such as dementia (e.g., Huang et al. 2022), post-stroke (e.g., Yang and Wang 2021), Parkinson's disease (PD) (e.g., Murray et al. 2014), depression (e.g., Morres et al. 2019; see Chapter "Exercise and Depression"), anxiety (e.g., Bartley et al. 2013; see Chapter "Exercise and Anxiety"), and posttraumatic stress disorder (e.g., Crombie et al. 2023a; see Chapter "Exercise and PTSD"). However, it should be noted that although most of these patients are troubled with varying types of specific memory problems, exercise can affect the broad spectrum of memory types and, thus, work differently for different (combinations of) mental and somatic conditions. In the following, we will therefore summarize the effects of exercise on different memory types and focus on experimental studies using acute exercise bouts to improve memory functioning in healthy and patient populations.

4 Exercise on Different Memory Types

4.1 Working Memory

WM is vital for our daily functioning and WM impairment affects our mental health. Hence, it is of functional and clinical interest to improve WM functioning among those with WM impairment. While a recent meta-analysis of randomized controlled trials showed that chronic exercise has a small, positive effect on WM in cognitively and physically healthy participants, results for single exercise sessions are less consistent (Rathore and Lom 2017). According to this meta-analysis, acute exercise does not directly improve WM in healthy participants (Rathore and Lom 2017). However, the studies included in the analysis of acute exercise were very heteroge neous, which limits the interpretation of the results. A crucial factor that affects the impact of exercise is the test used to assess WM. Different WM tasks (e.g., n-back task, digit span test, and Sternberg task) tap into different components of WM, which can lead to differential results. To illustrate, Gothe et al. (2013) did not find superiority for moderate-intensity exercise (20 min; treadmill at 60–70% of estimated maximal heart rate; MHR) compared to yoga or resting in improving WM assessed with the *n*-back task. In comparison, Pontifex et al. (2009) showed that acute treadmill exercise (30 min; 60–70% VO_{2max}) compared to resistance training (three sets of 8–12 repetitions at 80% of their one-repetition maximum) improved reaction time during the Sternberg test. Similar results are also found in clinical populations. Moriya et al. (2016), for example, investigated the effect of moderate-intensity exercise (15 min; cycling at 40% VO_{2max}) on WM in post-stroke patients with the Sternberg task and found positive results. In contrast, Meyer et al. (2022)

did not find superiority for cycling (30 min; perceived exertion (RPE) of 13) compared to rest in improving WM measured with the 2-back task in depressed patients.

A WM-based process involved in the onset and maintenance of depression and anxiety disorders is rumination. Rumination is defined as the repetitive replay of negative information (Nolen-Hoeksema et al. 2008) and is related to deficits in inhibition, shifting, and updating of WM. Although previous research showed that chronic exercise appears to diminish rumination (Craft 2005) and that an acute exercise bout (30 min; cycling at 60–70% MHR) can attenuate the effect of rumination after a stressor (Bernstein and McNally 2017, 2018), no direct decrease in rumination after moderate-intensity exercise (30 min; cycling at 70% MHR) has been observed (Schmitter et al. 2023). In conclusion, the results on the direct effect of exercise on WM are mixed and depend on the test used to assess WM. More research is warranted to identify which processes within the WM domain are affected by exercise.

4.2 Procedural Memory

Studies investigating the effect of exercise on PM usually use motor skill learning as an outcome. For example, Roig et al. (2012) showed that for healthy individuals, 15 min of intervallic cycling at high intensity (2 min of warm-up at 75 W, followed by 3 × 3 min blocks of high intensity, interspersed with 3 × 2 min blocks of low intensity at 50 W in between; where high-intensity workload was adjusted to each subject's VO_{2peak}), prior to and after practicing a visuomotor accuracy-tracking task, improved retention of the motor skill more than resting. Especially high-intensity exercise seems to benefit PM (Thomas et al. 2016), although this kind of intensity may not be feasible in rehabilitation centers where exercise could be implemented to improve motor skill learning. However, initial evidence suggests that moderate-intensity cycling before or after a motor skill task is sufficient to improve retention in patients with PD ((Steib et al. 2018); where exercise intensity was based on 60–70% VO_{2max}; (Wanner et al. 2021); where exercise intensity was based on ≥60 rpm [RPM] and W_{max}) and presymptomatic and early manifest Huntington's disease (HD) ((Andrews et al. 2023); where exercise intensity was based on 50–55% of heart rate reserve; HRR). Overall, clinical trials demonstrated so far that exercise is an effective strategy for symptom management in patients with neurodegenerative diseases (e.g., Mak et al. 2017). Although the effect of exercise on PM specifically has not been extensively investigated, the results so far are promising as well. Future studies would benefit from larger samples and extensive randomized controlled trials (RCT) that include motor skill learning as an outcome.

4.3 Semantic Memory

Though under-investigated (Loprinzi 2019), from a biological viewpoint it is plausible that acute exercise improves SM. For example, Won et al. (2019) investigated brain activation during a semantic memory task after exercise (30 min; cycling at RPE 15; vs rest), with aid of functional magnetic resonance imaging (fMRI), in a counter-balanced within-subject design, including healthy older adults. The authors found significantly greater brain activation during the task after exercise (vs rest), in the brain regions implicated in semantic memory (i.e., middle frontal, inferior temporal, middle temporal, fusiform gyri, and bilateral hippocampus) (Won et al. 2019). To the best of our knowledge, only one experimental study (Day and Loprinzi 2019) has specifically investigated the effect of acute exercise (15 min; treadmill at 70% MHR; vs rest) on general SM in healthy participants. In this study, participants were instructed to recall words from two specific categories (i.e., animals and vegetables). Exercise, however, did not result in more correct recall. Yet, chronic exercise (12 weeks, spinning three times per week, gradually increased in intensity and duration) was found to improve SM in older adults (<65 years) more than the control condition (Nocera et al. 2015). More studies are needed to draw conclusions about the (lack of) acute effects of exercise on SM.

4.4 Episodic Memory: Neutral and Emotional Recall

A meta-analysis (Loprinzi et al. 2019) showed that acute exercise before encoding or during consolidation can enhance EM. Most of the studies included in this meta-analysis used word-recall paradigms, indicating that an exercise bout can directly improve neutral, episodic memories. Less is known about the impact of exercise on emotional memories that evoke a strong emotional reaction. To illustrate, Keyan and Bryant (2017) invited healthy participants to the laboratory where they first watched a car accident film to trigger an emotional state, and then exercised at high intensity (stepping exercise based on the Karvonen formula: 50–85% (MHR – Rest HR) + Rest HR where MHR was estimated using an age calculated formula (220 beats per minute – age)) or walked for 10 min. The authors found that participants in the exercise condition reported more intrusive memories 2 days later (Keyan and Bryant 2017). However, these results have not been replicated so far. In a series of laboratory studies, Loprinzi et al. (2022) showed that acute exercise of different intensities did not affect emotional memory performance in healthy participants, using the same car accident film as Keyan and Bryan (2017). Crombie et al. (2023b) examined the effect of exercise on fear-related memory after extinction learning in healthy females. Participants first completed a fear acquisition training and then on the next day a fear extinction training followed by moderate-intensity exercise (30 min; cycling at 70–75% MHR) or low-intensity exercise (40–50% MHR). The next day, the participants returned to the lab for a fear recall test. The authors found that moderate-intensity exercise had a beneficial effect on memory for items learned during extinction training. Regarding the inconsistency of study results, more studies are needed to better understand how exercise affects the encoding and (re)-consolidation of emotional memories.

4.5 Episodic Memory: Autobiographical Memory

Another part of EM is autobiographical memory. Day and Loprinzi (2019) investigated the effect of acute exercise (15 min; treadmill at 70% MHR) on autobiographical recall, in which participants had to recall as many personal memories as possible from their childhood and adolescence. No differences were found between exercise and rest conditions. In MDD, autobiographical memories tend to be overly general, lacking details about place and time (Gotlib and Joormann 2010). Schmitter et al. (2023) assessed whether an acute exercise bout (30 min; cycling at 70% MHR) after a negative mood induction can improve the amount of detail in autobiographical memories in healthy participants but did not detect any differences between exercise and rest. In summary, exercise can directly improve the recall of neutral material, but the effect of exercise on emotional and autobiographical memories is less understood and needs further investigation.

4.6 Episodic Memory: Pattern Separation

Pattern separation is hypothesized to strongly depend upon hippocampal function, a canonically important brain region where different types of information are integrated to form a unique long-term memory. If pattern separation is impaired, mnemonic discrimination is poor. Importantly, here the term "discrimination" refers to the behavioral measure of the neural process. Mnemonic discrimination is often assessed with object recognition tasks (e.g., Mnemonic Similarity Task; MST). For example, participants are shown everyday objects during encoding and then asked to identify each object as "old," "new," or "similar" during recognition. Other discrimination tasks use variations in spatial information of previously seen objects or in the temporal lag between events. Discrimination difficulties arise as part of aging, but are also found in psychological disorders such as anxiety or depression (Leal and Yassa 2018). In anxiety disorders, poor pattern separation may manifest as fear generalization (Bernstein et al. 2021; Lange et al. 2017), where patients feel anxious in reaction to stimuli (e.g., physical sensations in panic disorder or social situations in social anxiety disorder) that resemble a past, threatening experience. Better discrimination between past events and similar, but different, new events may help reduce anxious responses to the similar new events, which is also the goal of exposure-based therapy (see Chapter "Exercise and Fear Learning and Extinction"). In depression, pattern separation is diminished for neutral information, but enhanced for negative information (Yassa and Stark 2011). Impaired pattern separation may result in a blurring of negative and neutral information, leading to a greater likelihood of interpreting ambiguous information in a negative way (Yassa and Stark 2011), in line with a depressotypic negative memory bias (e.g., Everaert et al. 2022).

Because exercise improves hippocampal functioning (e.g., Kandola et al. 2016), it is plausible that exercise exerts an effect on pattern separation. Initial evidence derived from studies with healthy, young adults (Déry et al. 2013; Suwabe et al. 2017, 2018) supports this assumption. For example, Suwabe et al. (2017) examined the effect of acute moderate-intensity exercise (10 min; cycling at 50% VO_{2max}; vs rest) on pattern separation assessed with the mnemonic discrimination task and found that the exercise condition showed better separation between highly similar items. Nauer et al. (2020) investigated whether regular exercise can also attenuate the age-related decline in mnemonic discrimination and found positive results. Comparably, Callow et al. (2022) showed in a paradigm that moderate-intensity cycling (13–15 RPE) directly elicited better mnemonic discrimination in older adults. Similarly, in a follow-up study, the authors found evidence for better discrimination after acute moderate exercise (30 min; RPE 15), using a counterbalanced design. Mnemonic discrimination after acute exercise (vs rest) was better among those participants with lower baseline fitness (Pena et al. 2022). To the best of our knowledge, no study directly investigated the effect of exercise on mnemonic discrimination in patients with anxiety or depression. Yet, compelling evidence shows that exercise is an effective treatment for both disorders (e.g., Morres et al. 2019; Stubbs et al. 2017). For anxiety disorders, the exercise-induced discrimination improvements may contribute to improved fear extinction learning (see Chapter "Exercise and Fear Learning and Extinction") and therefore, impact the anxiolytic effect of exercise. In depression, improving pattern separation through exercise may help reduce the negative memory bias, which is also a goal of cognitive-behavioral therapy for depression. Schmitter et al. (2023) examined this idea in an experimental setting. Interestingly, the authors found that aerobic exercise (30 min; cycling at 70% MHR) resulted in more negative memory bias than rest, among healthy participants with higher levels of rumination. From this study, it seems that a single exercise session is not sufficient to improve negative memory bias or possibly mnemonic discrimination. In conclusion, exercise seems to benefit pattern separation in the hippocampus which enables us to discriminate between similar experiences and might be reduced in psychological disorders and the aging population. Although the initial results with healthy participants are promising, future laboratory studies are needed to test the effect of exercise directly on mnemonic discrimination in clinical populations. Collectively, we have shown so far that exercise may directly benefit different types of memory in different populations, although results are mixed and future research warranted. In the following, we will briefly discuss the relevance and promise of implementing exercise in daily life and clinical settings.

5 Relevance and Promise of Exercise to Improve Memory Functioning

Our memory enables us to perform simple activities every day, to learn and understand, and to interact with the world around us. Therefore, maintaining and improving memory functioning is crucial in many ways. Because regular exercise positively affects our memory, it offers several opportunities as intervention and prevention program. For example, exercise in the school setting may directly boost the learning ability of children. Similarly, it can delay the onset of memory loss in populations at risk for dementia and prevent developing emotional disorders such as depression or anxiety.

Furthermore, acute exercise has the potential to directly improve treatment outcomes in clinical settings. Cognitive and physical therapies depend on the patient's learning and memory capacity. Because a single exercise session directly improves memory of cognitive and motor skills, strategically timing exercise in close proximity to therapy sessions could augment clinical outcomes of therapy. This concept has also been termed "exercise priming" (Moriarty et al. 2019). Exercise priming in practice involves acute exercise stimulation prior to or directly after treatment. For example, exercise before cognitive behavioral therapy for patients with anxiety and depression or before physical therapy for patients with HD or PD. We believe that exercise priming holds great promise as a simple and effective way to enhance the outcomes of cognitive-behavioral and physical therapies, partly due to exercise's impact on memory processes. By capitalizing on the immediate benefits of acute exercise, patients may experience faster recovery. With ongoing research in this area, exercise priming may become a valuable addition to a range of therapeutic approaches for a variety of conditions.

6 Conclusion

Exercise is a simple and safe intervention that may protect our memory function and promote our mental well-being. Therefore, it is crucial to prioritize the implementation of exercise in both our daily life and clinical settings. However, despite the recognized benefits, there is much to discover about the exact memory-based effects and mechanisms of change of exercise. Open question are: The exact domains of memory impacted by exercise, the precise mechanisms by which exercise influences memory, the most effective exercise protocols, and whether there exists a performance threshold beyond which exercise ceases to yield additional gains. Of course, all these factors are interactive (e.g., the ideal exercise protocols and performance thresholds likely vary for specific memory domains). Through thoughtful research programs and translational endeavors, we can move further toward in-depth understanding of the exercise-memory link, in turn contributing to mental health promotion.

References

Andrews SC, Kämpf L, Curtin D, Hinder M, Wenderoth N, Stout JC et al (2023) A single bout of moderate-intensity aerobic exercise improves motor learning in premanifest and early Huntington's disease. Front Psychol 14:971. https://doi.org/10.3389/fpsyg.2023.1089333

Barha CK, Hsu C-L, ten Brinke L, Liu-Ambrose T (2019) Biological sex: a potential moderator of physical activity efficacy on brain health. Front Aging Neurosci 11:329. https://doi.org/10.3389/fnagi.2019.00329

Bartley CA, Hay M, Bloch MH (2013) Meta-analysis: aerobic exercise for the treatment of anxiety disorders. Prog Neuropsychopharmacol Biol Psychiatry 45:34–39. https://doi.org/10.1016/j.pnpbp.2013.04.016

Basso JC, Suzuki WA (2017) The effects of acute exercise on mood, cognition, neurophysiology, and neurochemical pathways: a review. Brain Plast 2(2):127–152. https://doi.org/10.3233/BPL-160040

Bernstein EE, McNally RJ (2017) Acute aerobic exercise hastens emotional recovery from a subsequent stressor. Health Psychol 36(6):560–567. https://doi.org/10.1037/hea0000482

Bernstein EE, McNally RJ (2018) Exercise as a buffer against difficulties with emotion regulation: a pathway to emotional wellbeing. Behav Res Ther 109:29–36. https://doi.org/10.1016/j.brat.2018.07.010

Bernstein EE, van der Does F, Orr SP, McNally RJ (2021) Poor mnemonic discrimination predicts overgeneralization of fear. J Psychopathol Behav Assess 43:152–161. https://doi.org/10.1007/s10862-020-09846-z

Blondell SJ, Hammersley-Mather R, Veerman JL (2014) Does physical activity prevent cognitive decline and dementia?: a systematic review and meta-analysis of longitudinal studies. BMC Public Health 14(1):510. https://doi.org/10.1186/1471-2458-14-510

Callow DD, Pena GS, Stark CEL, Smith JC (2022) Effects of acute aerobic exercise on mnemonic discrimination performance in older adults. J Int Neuropsychol Soc:1–10. https://doi.org/10.1017/S1355617722000492

Craft LL (2005) Exercise and clinical depression: examining two psychological mechanisms. Psychol Sport Exerc 6(2):151–171. https://doi.org/10.1016/j.psychsport.2003.11.003

Crombie KM, Adams TG, Dunsmoor JE, Greenwood BN, Smits JA, Nemeroff CB et al (2023a) Aerobic exercise in the treatment of PTSD: an examination of preclinical and clinical laboratory findings, potential mechanisms, clinical implications, and future directions. J Anxiety Disord 94:102680. https://doi.org/10.1016/j.janxdis.2023.102680

Crombie KM, Azar A, Botsford C, Heilicher M, Moughrabi N, Gruichich TS et al (2023b) Aerobic exercise after extinction learning reduces return of fear and enhances memory of items encoded during extinction learning. Ment Health Phys Act 24:100510. https://doi.org/10.1016/j.mhpa.2023.100510

Crush EA, Loprinzi PD (2017) Dose-response effects of exercise duration and recovery on cognitive functioning. Percept Mot Skills 124(6):1164–1193. https://doi.org/10.1177/0031512517726920

Day S, Loprinzi PD (2019) Effect of acute moderate-intensity exercise on autobiographical episodic memory and semantic memory. OBM Integr Complement Med 4(3):1–18. https://doi.org/10.21926/obm.icm.1903042

Déry N, Pilgrim M, Gibala M, Gillen J, Wojtowicz JM, MacQueen G et al (2013) Adult hippocampal neurogenesis reduces memory interference in humans: opposing effects of aerobic exercise and depression. Front Neurosci:7. https://doi.org/10.3389/fnins.2013.00066

Everaert J, Vrijsen JN, Martin-Willett R, van de Kraats L, Joormann J (2022) A meta-analytic review of the relationship between explicit memory bias and depression: depression features an explicit memory bias that persists beyond a depressive episode. Psychol Bull 148(5–6):435–463. https://doi.org/10.1037/bul0000367

Gothe N, Pontifex MB, Hillman C, McAuley E (2013) The acute effects of yoga on executive function. J Phys Act Health 10(4):488–495. https://doi.org/10.1123/jpah.10.4.488

Gotlib IH, Joormann J (2010) Cognition and depression: current status and future directions. Annu Rev Clin Psychol 6:285–312. https://doi.org/10.1146/annurev.clinpsy.121208.131305
Huang X, Zhao X, Li B, Cai Y, Zhang S, Wan Q et al (2022) Comparative efficacy of various exercise interventions on cognitive function in patients with mild cognitive impairment or dementia: a systematic review and network meta-analysis. J Sport Health Sci 11(2):212–223. https://doi.org/10.1016/j.jshs.2021.05.003
Kandola A, Hendrikse J, Lucassen PJ, Yücel M (2016) Aerobic exercise as a tool to improve hippocampal plasticity and function in humans: practical implications for mental health treatment. Front Hum Neurosci 10:373. https://doi.org/10.3389/fnhum.2016.00373
Keyan D, Bryant RA (2017) Brief exercise enhances intrusive memories of traumatic stimuli. Neurobiol Learn Mem 141:9–13. https://doi.org/10.1016/j.nlm.2017.03.012
Lange I, Goossens L, Michielse S, Bakker J, Lissek S, Papalini S et al (2017) Behavioral pattern separation and its link to the neural mechanisms of fear generalization. Soc Cogn Affect Neurosci 12(11):1720–1729. https://doi.org/10.1038/s41593-017-0065-1
Leal SL, Yassa MA (2018) Integrating new findings and examining clinical applications of pattern separation. Nat Neurosci 21(2):163–173. https://doi.org/10.1038/s41593-017-0065-1
Loprinzi PD (2018) Intensity-specific effects of acute exercise on human memory function: considerations for the timing of exercise and the type of memory. Health Promot Perspect 8(4):255–262. https://doi.org/10.15171/hpp.2018.36
Loprinzi PD (2019) An integrated model of acute exercise on memory function. Med Hypotheses 126:51–59. https://doi.org/10.1016/j.mehy.2019.03.010
Loprinzi PD, Frith E, Edwards MK, Sng E, Ashpole N (2018) The effects of exercise on memory function among young to middle-aged adults: systematic review and recommendations for future research. Am J Health Promot 32(3):691–704. https://doi.org/10.1177/0890117117737409
Loprinzi P, Blough J, Crawford L, Ryu S, Zou L, Li H (2019) The temporal effects of acute exercise on episodic memory function: systematic review with meta-analysis. Brain Sci 9(4):87. https://doi.org/10.3390/brainsci9040087
Loprinzi P, Olafson D, Scavuzzo C, Lovorn A, Mather M, Frith E et al (2022) Effects of acute exercise on emotional memory. Cognit Emot 36(4):660–689. https://doi.org/10.1080/02699931.2022.2050890
Mak MK, Wong-Yu IS, Shen X, Chung CL (2017) Long-term effects of exercise and physical therapy in people with Parkinson disease. Nat Rev Neurol 13(11):689–703. https://doi.org/10.1038/nrneurol.2017.128
Meyer JD, Murray TA, Brower CS, Cruz-Maldonado GA, Perez ML, Ellingson LD et al (2022) Magnitude, timing and duration of mood state and cognitive effects of acute moderate exercise in major depressive disorder. Psychol Sport Exerc 61:102172. https://doi.org/10.1016/j.psychsport.2022.102172
Moriarty TA, Mermier C, Kravitz L, Gibson A, Beltz N, Zuhl M (2019) Acute aerobic exercise based cognitive and motor priming: practical applications and mechanisms. Front Psychol:10. https://doi.org/10.3389/fpsyg.2019.02790
Moriya M, Aoki C, Sakatani K (2016) Effects of physical exercise on working memory and prefrontal cortex function in post-stroke patients. Adv Exp Med Biol:203–208. https://doi.org/10.1007/978-3-319-38810-6_27
Morres ID, Hatzigeorgiadis A, Stathi A, Comoutos N, Arpin-Cribbie C, Krommidas C et al (2019) Aerobic exercise for adult patients with major depressive disorder in mental health services: a systematic review and meta-analysis. Depress Anxiety 36(1):39–53. https://doi.org/10.1002/da.22842
Murray DK, Sacheli MA, Eng JJ, Stoessl AJ (2014) The effects of exercise on cognition in Parkinson's disease: a systematic review. Transl Neurodegener 3(1):5. https://doi.org/10.1186/2047-9158-3-5
Nauer RK, Schon K, Stern CE (2020) Cardiorespiratory fitness and mnemonic discrimination across the adult lifespan. Learn Mem 27(3):91–103. https://doi.org/10.1101/lm.049197.118

Nocera JR, McGregor KM, Hass CJ, Crosson B (2015) Spin exercise improves semantic fluency in previously sedentary older adults. J Aging Phys Act 23(1):90–94. https://doi.org/10.1123/JAPA.2013-0107

Nolen-Hoeksema S, Wisco BE, Lyubomirsky S (2008) Rethinking rumination. Perspect Psychol Sci 3:5. https://doi.org/10.1111/j.1745-6924.2008.00088.x

Pena GS, Callow DD, Evans WS, Prior SJ, Smith JC (2022) Associations between cardiorespiratory fitness, monocyte polarization, and exercise-related changes in mnemonic discrimination performance in older adults. Exp Gerontol 169:111973. https://doi.org/10.1016/j.exger.2022.111973

Pontifex MB, Hillman CH, Fernhall BO, Thompson KM, Valentini TA (2009) The effect of acute aerobic and resistance exercise on working memory. Med Sci Sports Exerc 41(4):927–934. https://doi.org/10.1249/MSS.0b013e3181907d69

Rathore A, Lom B (2017) The effects of chronic and acute physical activity on working memory performance in healthy participants: a systematic review with meta-analysis of randomized controlled trials. Syst Rev 6(1):124. https://doi.org/10.1186/s13643-017-0514-7

Roig M, Skriver K, Lundbye-Jensen J, Kiens B, Nielsen JB (2012) A single bout of exercise improves motor memory. PLoS One 7(9):e44594. https://doi.org/10.1371/journal.pone.0044594

Roig M, Nordbrandt S, Geertsen SS, Nielsen JB (2013) The effects of cardiovascular exercise on human memory: a review with meta-analysis. Neurosci Biobehav Rev 37(8):1645–1666. https://doi.org/10.1016/j.neubiorev.2013.06.012

Roig M, Thomas R, Mang CS, Snow NJ, Ostadan F, Boyd LA et al (2016) Time-dependent effects of cardiovascular exercise on memory. Exerc Sport Sci Rev 44(2):81–88. https://doi.org/10.1249/JES.0000000000000078

Schmitter M, Vanderhasselt M-A, Spijker J, Smits JAJ, Vrijsen JN (2023) Working it out: can an acute exercise bout alleviate memory bias, rumination and negative mood? Cogn Behav Ther 52(3):232–245. https://doi.org/10.1080/16506073.2022.2164349

Sofi F, Valecchi D, Bacci D, Abbate R, Gensini GF, Casini A et al (2011) Physical activity and risk of cognitive decline: a meta-analysis of prospective studies. J Intern Med 269(1):107–117. https://doi.org/10.1111/j.1365-2796.2010.02281.x

Steib S, Wanner P, Adler W, Winkler J, Klucken J, Pfeifer K (2018) A single bout of aerobic exercise improves motor skill consolidation in Parkinson's disease. Front Aging Neurosci 10:328. https://doi.org/10.3389/fnagi.2018.00328

Stubbs B, Vancampfort D, Rosenbaum S, Firth J, Cosco T, Veronese N et al (2017) An examination of the anxiolytic effects of exercise for people with anxiety and stress-related disorders: a meta-analysis. Psychiatry Res 249:102–108. https://doi.org/10.1016/j.psychres.2016.12.020

Stute K, Hudl N, Stojan R, Voelcker-Rehage C (2020) Shedding light on the effects of moderate acute exercise on working memory performance in healthy older adults: an fNIRS study. Brain Sci 10(11):813. https://doi.org/10.3390/brainsci10110813

Suwabe K, Hyodo K, Byun K, Ochi G, Yassa MA, Soya H (2017) Acute moderate exercise improves mnemonic discrimination in young adults. Hippocampus 27(3):229–234. https://doi.org/10.1002/hipo.22695

Suwabe K, Byun K, Hyodo K, Reagh ZM, Roberts JM, Matsushita A et al (2018) Rapid stimulation of human dentate gyrus function with acute mild exercise. Proc Natl Acad Sci 115(41):10487–10492. https://doi.org/10.1073/pnas.1805668115

Thomas R, Johnsen LK, Geertsen SS, Christiansen L, Ritz C, Roig M et al (2016) Acute exercise and motor memory consolidation: the role of exercise intensity. PLoS One 11(7):e0159589. https://doi.org/10.1371/journal.pone.0159589

Wanner P, Winterholler M, Gaßner H, Winkler J, Klucken J, Pfeifer K et al (2021) Acute exercise following skill practice promotes motor memory consolidation in Parkinson's disease. Neurobiol Learn Mem 178:107366. https://doi.org/10.1016/j.nlm.2020.107366

Won J, Alfini AJ, Weiss LR, Michelson CS, Callow DD, Ranadive SM et al (2019) Semantic memory activation after acute exercise in healthy older adults. J Int Neuropsychol Soc 25(6): 557–568. https://doi.org/10.1017/S1355617719000171

Yang B, Wang S (2021) Meta-analysis on cognitive benefit of exercise after stroke. Complexity 2021:5569346. https://doi.org/10.1155/2021/5569346

Yassa MA, Stark CEL (2011) Pattern separation in the hippocampus. Trends Neurosci 34(10): 515–525. https://doi.org/10.1016/j.tins.2011.06.006

Part III
Exercise and Clinical Symptoms and Treatment

Exercise for the Prevention and Treatment of Depression

Brandon L. Alderman, Hannah M. Perdue, and Amber H. Sarwani

Contents

Abstract Depression is among the world's leading causes of disability and accounts for a significant loss of life. Despite large investments in research for antidepressants and psychotherapies, non-response, partial response, and small effects remain significant problems. Exercise and physical activity are two lifestyle behaviors that have been studied for well over half a century for the prevention and treatment of depression. The aim of this chapter is to summarize the current evidence base supporting the efficacy of exercise and physical activity in the prevention and treatment of depression, including evidence supporting exercise as a monotherapy and adjunct to antidepressant medication and psychotherapies. We conclude the chapter by outlining challenges to prescribing exercise for depression and general recommendations for encouraging behavioral adoption for individuals suffering from depression.

Keywords Behavioral Intervention · Depression · Exercise · Lifestyle Physical Activity · Mental Health

B. L. Alderman (✉), H. M. Perdue, and A. H. Sarwani
Department of Kinesiology and Health, Rutgers University, New Brunswick, NJ, USA
e-mail: alderman@kines.rutgers.edu

Curr Topics Behav Neurosci (2024) 67: 157–176
https://doi.org/10.1007/7854_2024_496
Published Online: 24 July 2024

1 Introduction

Depression is a common, debilitating, and often recurrent disease characterized by persistent feelings of sadness, worthlessness, and a lack of interest or pleasure in previously enjoyable or rewarding activities. Clinical depression, or major depressive disorder (MDD), goes well beyond the normal vicissitudes of life and can significantly impair a person's social, occupational, and overall well-being. Globally, more than 300 million people suffer from depression, and it is recognized as a leading cause of avoidable suffering worldwide (World Health Organization 2017). It is estimated that over 20 million adults in the United States will have at least one major depressive episode during their lifetime, a figure that represents over 8% of all US adults (Brody et al. 2018). The prevalence rate of major depression is higher among females compared to males and is highest among individuals aged 18–25 years and for those who report being two or more races (13.9%) (Substance Abuse and Mental Health Services Administration 2021). Although there are different types of depressive disorders, clinical depression, or MDD is often just referred to as "depression."

Depression is formally diagnosed using the Diagnostic and Statistical Manual of Mental Disorders, Fifth Edition (American Psychiatric Association 2013). Diagnostic criteria include the presence of at least one of the two cardinal symptoms of a depressed mood and diminished interest or pleasure in previously enjoyable activities (i.e., anhedonia). Additional symptoms of depression include a significant change in weight or appetite, insomnia or hypersomnia, psychomotor agitation or retardation, fatigue or loss of energy, feelings of worthlessness or excessive/inappropriate guilt, difficulty thinking and concentrating, and thoughts of death or suicide. To be diagnosed with depression, an individual must experience five or more of these symptoms for more than 2 weeks, which cause significant emotional distress or impair normal day-to-day functioning (World Health Organization 2017; American Psychiatric Association 2013). Depression is characterized by pronounced heterogeneity. Estimates indicate that at least 227 and possibly up to 681 different combinations of symptoms may result in a formal diagnosis of depression (Fried and Nesse 2015). In addition, people may suffer from more minor or subsyndromal depression or dysthymic disorder, which can significantly impair one's quality of life (Judd et al. 1994; Baune et al. 2009). At the extremes, depression is associated with an increased risk for numerous chronic health conditions (Moussavi et al. 2007) and death by suicide, particularly among adolescents and young adults (Orsolini et al. 2020).

The exact causes of depression remain unknown; it likely results from a combination of genetic, biological, and psychological vulnerability factors. Therefore, depression is a common but poorly understood health condition with imprecise causal origins. Depression is also highly comorbid with other mental health conditions. For instance, nearly three-quarters of all people who meet criteria for depression during their lifetime will also meet criteria for another psychiatric disorder, including anxiety and substance use disorders (DeRubeis et al. 2008). In the United

States, the economic burden of depression increased by 21.5% from 2000 to 2010 (from $173.2 billion to $210.5 billion, inflation-adjusted dollars), with approximately 45% attributable to direct costs, 5% to suicide-related costs, and 50% to workplace costs (Greenberg et al. 2015). Over this same time period, the proportion of adults who received treatment for depression remained stable, suggesting a critical need for effective approaches to prevent and treat depression.

2 Conventional Treatments for Depression

Conventional treatments for depression typically consist of antidepressant medications or psychotherapy. The most commonly prescribed antidepressants include selective serotonin reuptake inhibitors (e.g., citalopram, escitalopram, fluoxetine, paroxetine, and sertraline), serotonin and norepinephrine reuptake inhibitors (e.g., venlafaxine and desvenlafaxine), and norepinephrine and dopamine reuptake inhibitors (e.g., bupropion) (Luo et al. 2020; Milea et al. 2010). The primary mechanism of action of these medications involves inhibiting the transporters responsible for the reuptake of these brain monoamine neurotransmitters into presynaptic neurons, increasing their availability and thereby enhancing neurotransmission (Feighner 1999). This evidence led, in part, to the notion that these neurotransmitters play a significant role in the etiopathogenesis of depression (i.e., that depression is *caused* by low levels of serotonin, norepinephrine, or dopamine in the central nervous system) (Artigas et al. 2002). Although the monoamine hypothesis of depression provides a plausible neurobiological explanation for the actions of antidepressants, this mechanism does not provide a complete explanation for the etiology of depression (Jesulola et al. 2018; Hirschfeld 2000). Rather, it is likely that a combination of biological, psychological, and social hypotheses (e.g., genetic vulnerability, childhood experiences, hormonal imbalances, major life events, and low socioeconomic status) will help to explain this heterogeneous and etiologically complex disorder.

Although antidepressants are effective for many people, they are not consistently beneficial. A review of four meta-analyses of efficacy trials submitted to the US Food and Drug Administration and findings from the Sequenced Treatment Alternatives to Relieve Depression (STAR*D) trial, the largest open trial of antidepressant effectiveness ever conducted, concluded that antidepressants are only marginally efficacious compared to placebos (Pigott et al. 2010). In addition to their questionable efficacy, antidepressant medications can be costly (Arnaud et al. 2021) and are often associated with unpleasant or intolerable side effects (e.g., nausea, constipation, weight gain, upset stomach, sexual dysfunction, headache, drowsiness, and difficulty sleeping, among other side effects) (Cascade et al. 2009; Wang et al. 2018). The stigma and fear of dependency associated with taking antidepressant medications also preclude many individuals from seeking or continuing treatment (Mechanic 2002; Castaldelli-Maia et al. 2011). Although thousands of clinical trials have been conducted to examine the efficacy of antidepressant medications, many of these trials have been limited to 8–12 weeks. This short duration of clinical trials limits our understanding of the potential consequences of long-term antidepressant use. In a mixed methods study, patients reported that antidepressants improved their depression episodes and symptoms, yet many reported concerns about withdrawal and adverse emotional effects, such as feeling emotionally numb and addicted (Cartwright et al. 2016). In addition, some patients expressed a need for more information about the potential long-term risks of antidepressants and information and support to discontinue their use. Some evidence suggests that using psychotherapy while tapering from antidepressants may be a successful approach to prevent relapse, with an effect comparable to continuing medication use (Breedvelt et al. 2021).

Like antidepressant medications, hundreds of randomized controlled trials (RCTs) have examined the efficacy of psychological treatments for depression. Psychological therapies, including cognitive behavioral therapy (CBT), work by helping individuals learn to recognize and change negative thought patterns and behaviors to better cope with stressful life events and challenging situations. Other types of psychological therapies for depression include interpersonal psychotherapy, problem-solving therapy, and behavioral activation. These treatments have demonstrated comparable efficacy to antidepressant medications for reducing depression, and some studies suggest the benefits may be longer-lasting (Cuijpers and Gentili 2017). However, like antidepressants, response rates to psychotherapies vary widely, with ranges of 38–87% reported for depressive and anxiety disorders relative to care-as-usual, waitlist, and pill placebo control groups (Cuijpers et al. 2021). Unfortunately, the public health impact of psychotherapy is limited by factors such as the costs, availability of trained therapists, and perceived inconvenience of traveling to weekly face-to-face sessions (Thase et al. 2020). Furthermore, the findings surrounding the benefit of psychotherapies reported in the research literature have been considerably overestimated, likely due to publication bias, sub-optimal quality of experimental trials, and the use of "nocebo" waiting list control groups, which may result in negative effects relative to the use of care-as-usual or pill placebo comparison groups (Cuijpers et al. 2016). The efficacy of psychological treatments may also depend on the quality of the practitioner (Stubbe 2018), the nature of the intervention (e.g., face-to-face, group-based, and Internet delivered) (Kambeitz-Ilankovic et al. 2022), and individual preferences (Steidtmann et al. 2012). Regardless, psychotherapy, like antidepressants, is effective for many people, helping them to manage their depressive symptoms (Cuijpers et al. 2013).

The availability of antidepressant medications and psychotherapy has increased markedly over the past three decades (Kovich et al. 2023). However, the general population prevalence of depression has not decreased over the same period of time (Ormel et al. 2022). This "treatment-prevalence paradox" has raised fundamental questions about the treatment of depression, including how to tailor the most effective treatment for each patient (i.e., personalized medicine) (Ormel et al. 2022). Despite an array of antidepressant and psychotherapy options, there is currently no empirically validated approach to match patients to optimal treatments, and many patients go on to develop treatment-resistant depression over time. As a result, clinical treatment for depression remains largely based on trial and error and the preferences of the primary care provider.

3 Exercise in the Management of Depression

Arguments for the use of alternative treatments for depression often include the unpleasant side effects of antidepressant medications (Cascade et al. 2009; Wang et al. 2018), costs of treatment (Arnaud et al. 2021), and difficulty accessing in-person treatment (Mohr et al. 2010) along with the limited efficacy of existing treatments. However, alternative treatment approaches for depression warrant the same level of evaluation as conventional treatments. Research conducted over the past 50 years supports a clinically meaningful impact of exercise and physical activity for the prevention and treatment of depression (Meyer and Schuch 2018; Blumenthal and Rozanski 2023; Cooney et al. 2013; Martinsen 1990). Although early cross-sectional and intervention studies focused on cardiorespiratory fitness and exercise per se (North et al. 1990; Greist et al. 1979), more recent studies have broadened the focus to include lifestyle physical activity, defined as any bodily movement produced by skeletal muscles that results in an increase in energy expenditure and includes recreational, occupational, or household-based physical activities that can be accumulated throughout the day (Caspersen et al. 1985; Farris and Abrantes 2020). Although exercise and physical activity have been recognized as important lifestyle behaviors for millenia (Tipton 2014), the seminal studies conducted by William P. Morgan in the late 1960s and early 1970s helped to bring contemporary scientific attention to the potential for exercise to prevent and treat depression. Morgan found that patients with depression tended to have lower cardiovascular fitness and physical capacity compared to their non-depressed counterparts (Morgan 1969, 1970). Since these early studies, there have been a large number of cross-sectional and epidemiological studies that support the utility of exercise as a stand-alone or adjunctive treatment for depression (Blumenthal and Rozanski 2023). There have also been an increasing number of RCTs conducted to compare exercise and physical activity to usual care, placebo controls, or established therapies in treating depression (Blumenthal and Rozanski 2023). Below, we present the evidence supporting the role of exercise and physical activity in the prevention and treatment of depression.

3.1 Exercise and Physical Activity for the Prevention of Depression

The potential for exercise and physical activity to prevent depression has been examined in numerous cross-sectional and prospective epidemiological studies. In one of the earliest cross-sectional studies, Farmer and colleagues (Farmer et al. 1988) examined the relationship between self-reported physical activity and depressive symptoms among 1,900 healthy adults between the ages of 25 and 77 years from the Epidemiologic Follow-up Study (1982–1984) to the first National Health and Nutrition Examination Survey (NHANES I). There was an almost twofold increase in the odds of experiencing increased depressive symptoms among individuals who reported little-to-no physical activity compared to those engaging in moderate or high activity. Goodwin (2003) similarly examined the association between self-reported physical activity and depression among 5,877 Americans aged between 15 and 54 years who participated in the National Comorbidity Survey (1990–1992), a nationally representative survey to better understand the prevalence, risk factors, and consequences of psychiatric morbidity and comorbidity. Physical activity was assessed using a single item that asked, "How often do you get physical exercise – either in your job or in a recreational activity?" with response choices of regularly, occasionally, rarely, and never. Depression status was confirmed using a structured interview (the World Health Organization Composite International Diagnostic Interview). Individuals who reported no regular physical activity were more likely to be depressed relative to those who reported regular physical activity engagement. The findings also suggested a potential dose-response relationship: depression was least prevalent among individuals who regularly engaged in physical activity (8.2%) compared to those who occasionally (11.6%), rarely (15.6%), or never (16.8%) engaged in physical activity. Interestingly, Goodwin noted that the observed association between physical activity and depression could be explained through reverse causation, with depression leading to less physical activity. The temporal nature of the relationship between physical activity and depression (i.e., does physical inactivity precede an increase in depressive symptoms, or vice versa) can be tested through prospective epidemiological studies.

In a prospective cohort study, Gudmundsson et al. (2015) examined the relationship between physical activity and depressive symptoms among 676 women followed over a period of 32 years (1974–2005, with follow-up assessments conducted in 1992, 2000, and 2005). The repeated assessments allowed the investigators to test the directionality of the relationship between physical activity and the symptom burden of depression. In 1974 at baseline, women who engaged in lower levels of physical activity also reported greater depressive symptoms. Across time, a decrease in physical activity was also associated with higher depression scores. In support of the reverse causality pathway mentioned by Goodwin (2003), higher depression scores at earlier assessments predicted lower levels of physical activity at subsequent examinations, but not vice versa. However, many additional prospective studies have been conducted to help answer the temporality question.

Mammen and Faulkner (2013) reviewed the available published prospective longitudinal studies to determine whether physical activity is protective against the onset of depression. A total of 25 of the 30 reviewed studies found a significant inverse relationship between baseline physical activity levels and subsequent depression, suggesting that physical activity can prevent the onset of depression. They also reported that engaging in as little as 10–29 min of daily physical activity could reduce the subsequent risk for depression by approximately 10%. Dishman et al. (2021) conducted a review and meta-regression analysis of 111 studies, including more than three million people, to determine whether the odds of depression vary according to the amount of physical activity or change in physical activity exposure across time. This evidence is important for clinical practice and for satisfying minimal criteria (e.g., strength of association, consistency, temporal sequence, and dose–response) in observational studies for a possible causal association between physical activity and depression. Lower odds for depression were associated with: (World Health Organization 2017) moderate-to-vigorous physical activity in line with public health guidelines; and (Brody et al. 2018) increases in physical activity across time.

Choi et al. (2019) examined physical activity and depression data from the UK Biobank Study to test the causal pathway between physical activity and depression. Using a Mendelian randomization approach, they reported that for each 1 SD unit increase in accelerometer-based physical activity there was a 26% reduction in risk for developing depression. This 1 SD increase in physical activity was equivalent to replacing sedentary time (e.g., sitting) with 15 min of vigorous physical activity, 60 min of moderate activity, or a combination of light and vigorous activities. No associations were observed between the pathway from depression to physical activity, providing support for the causal pathway of physical activity reducing the likelihood of depression. There is also evidence suggesting a potential dose–response relationship between physical activity and future depression. A review of 15 studies comprising 191,130 participants revealed an inverse dose–response association between physical activity and depression (Pearce et al. 2022). Relative to adults not reporting any activity, those who accumulated half of the recommended volume of physical activity had an 18% lower risk of depression while adults who accumulated the recommended volume of physical activity per week (equivalent to 2.5 h/week of brisk walking) had a 25% lower risk of depression. Based on an estimate, if less active adults achieved current physical activity recommendations, 11.5% of depression cases could be prevented. Although it is possible that the relationship between physical activity and depression is bidirectional in nature (Stavrakakis et al. 2012), the findings from these reviews support lifestyle physical activity for the prevention of depression and depressive symptoms.

3.2 Exercise and Physical Activity for the Treatment of Depression

The use of exercise as a treatment for depression has gained considerable traction, as evidenced by the inclusion of exercise in recent depression and mood disorder treatment guidelines published by the American Psychiatric Association (Rethorst and Trivedi 2013), the Canadian Network for Mood and Anxiety Treatments (Ravindran et al. 2016), and the Royal Australian and New Zealand College of

Psychiatrists (Malhi et al. 2015). However, despite this recognition, the American Psychiatric Association's clinical guidelines included the statement, "...if a patient with mild depression wishes to try exercise alone for several weeks as a first intervention, there is little to argue against it..." (Gelenberg 2010 (p. 30)). This statement, from one of the largest psychiatric organizations in the world, seems incompatible with the available research evidence on the use of exercise as a therapeutic lifestyle intervention for depression. Moreover, patients who may have the most to benefit from exercise (e.g., those who present with higher depressive symptoms, are inactive, or have comorbid health conditions) may be relatively uninformed about the benefits of exercise or less inclined to try exercise as a first-line therapy for depression. One way to provide convincing evidence for the therapeutic potential of exercise for depression is from gold standard RCTs.

In an early RCT, Blumenthal et al. (1999) randomized 156 adults with depression who were 50 years of age and older to a 4-month treatment of aerobic exercise, standard antidepressant treatment with sertraline, or a combination of aerobic exercise plus sertraline. For the exercise program, participants attended three supervised sessions per week in a group setting and were prescribed walking or jogging at 70%–85% of heart rate reserve (HRR), a range of vigorous intensity that accounts for the difference between maximal and resting heart rate. Although patients receiving sertraline monotherapy exhibited the fastest initial response relative to the exercise alone and combined treatment groups, all three groups showed comparable reduction in depressive symptoms by the end of the 16-week intervention. The rates of participants who no longer met criteria for depression (i.e., remission) were 60.4% in the exercise group, 65.5% in the medication group, and 68.8% in the combined treatment group. These findings suggest comparable improvements in depressive symptoms and remission from depression from exercise and antidepressant medication, even among patients with more severe depression.

Six months following the intervention, Babyak et al. (2000) found that individuals assigned to the exercise monotherapy group had significantly lower rates of depression (30%) compared to the sertraline (50%) or combined treatment (55%) groups, suggesting that exercise monotherapy may help to decrease depression relapse. During the follow-up period, physical activity levels were also associated with lower depressive symptoms, while the use of antidepressant medications was not associated with depression. Blumenthal and Rozanski (2023) noted that several questions remained following this initial RCT, including whether the improvements were the result of a placebo effect, spontaneous remission, or demand characteristics inherent to the study (e.g., expectation of benefit or the increased attention provided to participants by the study staff). Because participants exercised in a group setting, it is also possible that enhanced social support further enhanced the antidepressant benefits of exercise. To address these questions, Blumenthal et al. (2007) randomized 202 participants with depression to the following groups for 16 weeks: (a) supervised group exercise, (b) home-based aerobic exercise, (c) antidepressant medication (sertraline, 50–200 mg daily), or (d) a placebo pill. At end of treatment, 41% of the patients achieved remission, and all active treatment groups had greater remission rates compared to the placebo pill group (45% for supervised exercise, 40% for home-based exercise, and 47% for those receiving medication, compared with only 31% who received the pill placebo). Exercise had comparable efficacy to antidepressant medication, and both exercise and medication tended to be better than the placebo group. These two RCTs helped to establish the efficacy of exercise as a stand-alone and adjunctive treatment for depression, with comparable effects to antidepressant medications.

In addition to the question of overall efficacy, several issues remain to be studied within the context of treating depression with exercise and physical activity. For instance, prescribing exercise remains a challenging endeavor due in part to the number of program variables involved (e.g., frequency, intensity, and duration of exercise sessions as well as treatment length) and the type of exercise performed. For instance, although the early research evidence was primarily focused on aerobic forms of exercise, more recent studies—including RCTs—are beginning to focus on resistance exercise (Gordon et al. 2018; O'Sullivan et al. 2023). Gordon and colleagues conducted a meta-analysis of published RCTs and found that resistance exercise training programs were associated with a significant reduction in depressive symptoms, regardless of study participants' characteristics (age, sex, and health status) or features of the exercise stimulus (program duration, session duration, intensity, frequency, or total prescribed volume). Given the rising popularity of resistance exercise, particularly among younger adults who are at higher risk for depression (Mojtabai et al. 2016), these findings are promising for tailoring exercise prescriptions while considering individual preferences toward exercise.

Establishing the optimal or minimal dose of physical activity to produce health benefits has been a focus of scientific interest since the early 2000s (Kesaniemi et al. 2001). At the time, aerobic exercise training for 6–12 weeks was believed to be consistently associated with a reduction in depressive symptoms of a magnitude comparable to that observed following treatment with antidepressants; however, the evidence base at the time was insufficient to support a dose–response relationship between physical activity and depression. Indeed, an umbrella review of systematic reviews showed that exercise interventions that last ≤12 weeks have an effect size of 0.84, but this antidepressant response drops to 0.46 for exercise interventions lasting 12–23 weeks and further drops to 0.28 for exercise interventions lasting ≥24 weeks (Singh et al. 2023a). It is important to determine whether the initial antidepressant response from exercise can be sustained over time and whether exercise program variables (frequency, intensity, duration) moderate the long-term antidepressant response.

Dunn et al. (2005) randomized 80 adults aged 20–45 years diagnosed with mild-to-moderate depression to one of four aerobic exercise treatment groups that varied in total energy expenditure (7.0 kcal/kg/week or 17.5 kcal/kg/week) and frequency (3 days/week or 5 days/week) or to a control group that performed light stretching exercises 3 days/week for 15 to 20 min per session. The findings from this clinical trial supported a dose–response effect given that the public health dose of exercise (17.5 kcal/kg/week) led to significantly lower depression scores on the Hamilton Rating Scale for Depression (HRSD) compared to the low dose (7.0 kcal/kg/week) and stretching control groups. However, it is noteworthy that both the low-dose exercise and stretching exercise groups also led to significant reductions in depression scores. Further, the antidepressant response and remission rates following the public health dose are comparable to those observed following antidepressant medication and CBT. These findings suggest that the public health dose of exercise is likely to result in a significant and clinically meaningful reduction in depressive symptoms. Current recommendations suggest that adults should do at least 150–300 min per week of moderate intensity, or 75–150 min per week of vigorous-intensity aerobic physical activity, or an equivalent combination of moderate- and vigorous-intensity aerobic activity spread throughout the week. Adults should also do muscle-strengthening activities of at least moderate intensity that includes all major muscle groups on 2 or more days per week.

Pearce et al. (2022) reviewed the dose–response association between physical activity and incidence of depression among 15 studies, involving 191,130 participants. An inverse curvilinear association was observed, with steeper gradients at lower volumes of physical activity. Relative to those adults not reporting any activity, those accumulating half the recommended volume of physical activity (approximately 75 min of moderate-intensity activity per week) had an 18% lower risk of depression, while those who engaged in the recommended amount of physical activity (150 min of moderate-intensity activity per week) had 25% lower risk, with diminishing benefits beyond that exposure level. The greatest differences in risk for depression were observed between those who were totally sedentary versus those who were engaging in low doses of physical activity. At a population level, this suggests that moving people from being completely sedentary to engaging in low levels of physical activity may have a significant impact on the overall burden of depression. At present, the available evidence suggests that the USDHHS physical activity recommendations (Piercy et al. 2018) may be used as an initial prescription for depression and should be modified based on physical activity history and individual preferences.

One important consideration for treatment guidelines is that exercise may be an effective adjunct treatment to antidepressant medication and psychotherapies (Meyer et al. 2022). For instance, Legrand and Neff (2016) evaluated the effects of a short add-on aerobic exercise intervention in hospitalized patients with severe depression being treated with antidepressant medication (and no other form of therapy) for less than 2 weeks. Thirty-five participants were randomly assigned to aerobic exercise ($n = 14$), placebo exercise (stretching) ($n = 11$), or to a no-intervention control group ($n = 10$) for 10 days. The aerobic exercise group consisted of 30 min of daily brisk walking or jogging outdoors at 65–75% of age-predicted maximal heart rate and the stretching group consisted of 30 min of stretching per day for 10 consecutive days. Intent-to-treat analysis showed significant improvements in depression severity for both the aerobic exercise and stretching groups compared to the control group. However, aerobic exercise led to a significantly larger reduction in depressive symptoms from pre- to post-intervention compared to the control (Cohen's $d = -1.39$) group although the reductions relative to the stretching group (Cohen's $d = -1.06$) did not reach significance ($p = 0.082$). This study provides further support for the acceptability and safety of exercise, as well as its utility as an adjunctive treatment, even among patients with more severe forms of depression.

A Cochrane review reported a small effect favoring exercise compared to a no-treatment control in reducing depression, although no significant differences were reported between exercise and psychological or pharmacological interventions (Cooney et al. 2013). Although the clinical value of this systematic review and meta-analysis has been questioned (Honey 2015), exercise was found to improve numerous aspects of physical health, including cardiorespiratory fitness, body mass index, and heart rate variability. Further, there are far fewer side effects associated with exercise and patients with depression who have been randomized to exercise in RCTs are significantly less likely to drop out compared to participants who were allocated to the control arms in these trials (Stubbs et al. 2016). Exercise is also much more accessible and cost-effective in comparison to both antidepressant medications and psychotherapy (Singh et al. 2023b), making it a feasible option for patients with depression. While antidepressants and psychotherapy remain important treatments in the toolbox for depression, the available evidence supports the efficacy of exercise as a monotherapy or add on to traditional treatments. However, fewer adverse side effects and the accompanying physical health benefits may make exercise a preferable option.

4 Symptoms of Depression Undermine Exercise Behavior

Despite the numerous neurobiological benefits associated with exercise (Dishman et al. 2006), it is important to recognize that symptoms of depression may undermine exercise motivation. For instance, common affective symptoms of depression include depressed mood, fatigue, and a lack of motivation or initiative (apathy). Further, one of the core symptoms of depression (anhedonia) involves the loss of enjoyment or interest in hobbies and activities that were once viewed as pleasurable. These affective symptoms, combined with negative cognitions and reward system dysfunction in depression (Admon and Pizzagalli 2015), may contribute to exercise avoidance and intolerance (Glowacki et al. 2017). Firth et al. (Firth et al. 2016) examined motivating factors and barriers to exercise in 12 independent studies of 6,431 psychiatric patients and found that 91% of people with serious mental illness endorsed "improving health" as a reason to exercise, particularly for "losing weight" (83% of patients), "improving mood" (81%), and "reducing stress" (78%). However, low mood and high stress were identified as the most prominent barriers to exercise (61%), followed by "lack of support" (50%). In a scoping review (Glowacki et al. 2017), low mood, lack of energy, and being too tired were the most commonly reported barriers to physical activity among adults with depression, whereas being stressed, boredom, and lack of enjoyment were also frequently reported barriers. The authors highlighted that these emotional barriers are not typically addressed by traditional theories of behavior change (e.g., Theory of Planned Behavior, Self-Determination Theory, and Self-Efficacy Theory) and may be overlooked when trying to promote physical activity among adults with depression. Future behavior change research should specifically target these emotional barriers when attempting to help individuals with depression in adopting and maintaining physical activity and exercise.

Exercise has proven to be a challenging and difficult health behavior for most individuals to engage in, and this is particularly so for patients with depression. The approach to incorporating exercise into a depression treatment plan should prioritize ways to overcome common barriers to participation while considering motivating factors identified in the research literature (Busch et al. 2016). First, when it comes to developing an initial exercise prescription, the USDHHS guidelines of 150 min of moderate-intensity physical activity and 2 days per week of muscle-strengthening activity should be used as a starting point. This general prescription can be tailored based on the current activity level, physical capacity, and goals and exercise preferences of the individual. Patients should be counseled to begin slowly and choose an enjoyable and preferred form of physical activity. Individuals with depression have reported that walking is their most preferred form of exercise, followed by weightlifting, yoga, and dance (Busch et al. 2016). Men preferred weightlifting marginally more than women did, while women preferred yoga significantly more than men. It may be useful to initially focus on the frequency of physical activity, for instance, by setting a goal of three exercise sessions per week at a comfortable pace for a manageable length of time (e.g., 10 min). Once the patient becomes confident and comfortable with exercise frequency, then duration and intensity of exercise can be gradually increased (Craft and Perna 2004). The goal should be to establish a regular physical activity routine rather than being overly concerned about meeting an exercise dose (intensity and duration) for symptom reduction. Also, focusing on frequency may help to establish a behavioral habit and routine while addressing any symptoms that are serving as barriers to adherence. Next, increasing the duration of exercise may help to build endurance and exercise tolerance. Finally, increasing the intensity of the exercise can be effective for increasing cardiorespiratory fitness, reaching specific health goals, improving quality of life, and decreasing depression. Increasing exercise intensity too quickly, however, may lead to burnout, injury, reduced self-efficacy, or exercise nonadherence (Vetter and Symonds 2010). Therefore, it should not be the primary focus of exercise treatment until an individual has established the habit (i.e., frequency) and tolerance (i.e., duration) for regular physical activity. It is also important for clinical exercise specialists and health providers to provide basic education, give realistic expectations, and outline timelines for an anticipated antidepressant response (e.g., 4–6 weeks).

5 Tailored Exercise Prescriptions for Depression Management

Despite substantial evidence supporting the use of exercise in the treatment of depression, previous studies have not provided a clear indication of a standard or tailored dose of exercise for promoting antidepressant effects. Exercise benefits numerous aspects of physical and mental health, but the motivational effort required to adopt and maintain a regular exercise routine can quickly become overwhelming to depressed patients. It is therefore important to make exercise recommendations on

Table 1 Exercise prescription for depression

Program stage	Frequency (days/week)	Intensity	Duration (min)	Mode
Initial stage (weeks 1–4)	3–4	Low to moderate	20–30	Walking, light aerobic
Middle stage (weeks 5–8)	4–5	Moderate	30–45	Aerobic, light resistance
Later stage (weeks 9–12)	5	Moderate-to-vigorous	30–60	Aerobic, resistance

Table 2 An evidence-informed approach to increase exercise motivation and adherence

Before prescribing exercise
1. Provide brief psychoeducation about the recommended physical activity guidelines and benefits of exercise for mental health
2. Question patients about their current activity level and preferences
3. Perform motivational interviewing and question patients about their personal barriers and facilitators to exercise
4. Explain therapeutic expectations and realistic timelines for the exercise treatment plan
Prescribing exercise
1. Start low and go slow: Emphasize starting with lower intensity and duration, gradually increasing over time to prevent negative exercise experiences
2. Use an individualized approach tailored to the patient's abilities, preferences, and tolerance levels
3. Discuss strategies to increase adherence, including goal-setting, incorporating enjoyable activities, and social support
4. Acknowledge the potential impact of depressive symptoms on motivation and adherence. Encourage small, manageable steps and consider the use of incentives to aid in achieving goals
During the exercise program
1. Regularly monitor patient progress and mood responses to exercise
2. Adjust the prescription according to individual progress and preferences
3. Periodically assess depressive symptoms and provide feedback about progress

an individual level when possible, accounting for personal differences in exercise history, preferences, barriers, and facilitators. Acute bouts of exercise have been shown to be effective for depression management, and it is possible that the antidepressant response to chronic exercise is the result of daily symptom management through individual bouts of activity (Meyer et al. 2016). Like any clinical prescription, treatment with exercise should include the following key elements: (1) type or mode of exercise; (2) frequency and duration of exercise sessions; and (3) exercise intensity. Given the importance of behavioral adherence, additional key elements to include are: (4) duration of the intervention; and (5) psychoeducation about motivating factors, barriers, and monitoring of adherence (Rethorst and Trivedi 2013). These elements provide the core structure for an exercise prescription, while specific recommendations can then be individually tailored. An example exercise prescription plan for depression is summarized in Table 1. An interventional approach to addressing barriers to exercise and improving behavioral adherence to exercise is described in Table 2.

6 Conclusion

Depression is a prevalent and disabling disease and has become the leading cause of avoidable suffering worldwide. Clinically, depression is characterized by feelings of sadness and helplessness, loss of pleasure (anhedonia), and lack of motivation. Antidepressant medications and psychological therapies are effective treatments for some patients with depression; however, non-response, partial response, and small effects remain significant problems. Additionally, these treatments are associated with several barriers and many people with depression either never seek treatment or gravitate exclusively to medications (Beshai et al. 2019), which pose considerable side effects (Cascade et al. 2009; Wang et al. 2018). Exercise and physical activity are lifestyle behaviors that can be used to prevent and treat depression. Cross-sectional and prospective epidemiological studies indicate that engaging in regular physical activity is associated with a lowered risk of developing depression. Several large-scale RCTs have provided evidence for efficacy of exercise in treating depression, with effects that are comparable to those typically observed for antidepressant medications and psychotherapy. Systematic reviews and meta-analyses indicate that aerobic and resistance exercise training results in moderate-to-large reductions in depressive symptoms, and exercise can be used as an adjunct to antidepressants and psychotherapy. Although several organizations dedicated to psychiatric care, including the American Psychiatric Association and National Institute of Mental Health, have not specifically endorsed the use of exercise as a treatment for depression, statements from these organizations highlight the use of exercise as an "effective, cost-effective treatment for depression" and that "exercise helps keep your psyche fit." This is promising, considering the decades of research supporting the effects of exercise on depression. By incorporating some of the recommendations outlined in this chapter, we believe that exercise can be incorporated as an effective behavioral monotherapy or adjunctive treatment to advance treatment for depression.

References

Admon R, Pizzagalli DA (2015) Dysfunctional reward processing in depression. Curr Opin Psychol 4:114–118. https://doi.org/10.1016/j.copsyc.2014.12.011

American Psychiatric Association (2013) DSM-5 task force. Diagnostic and statistical manual of mental disorders: DSM-5™, 5th edn. American Psychiatric Publishing, Inc. https://doi.org/10.1176/appi.books.9780890425596

Arnaud A, Suthoff E, Tavares RM, Zhang X, Ravindranath AJ (2021) The increasing economic burden with additional steps of pharmacotherapy in major depressive disorder. Pharmacoeconomics 39(6):691–706. https://doi.org/10.1007/s40273-021-01021-w

Artigas F, Nutt DJ, Shelton R (2002) Mechanism of action of antidepressants. Psychopharmacol Bull 36(Suppl 2):123–132

Babyak M, Blumenthal JA, Herman S et al (2000) Exercise treatment for major depression: maintenance of therapeutic benefit at 10 months. Psychosom Med 62(5):633

Baune BT, Caniato RN, Arolt V, Berger K (2009) The effects of dysthymic disorder on health-related quality of life and disability days in persons with comorbid medical conditions in the general population. Psychother Psychosom 78(3):161–166. https://doi.org/10.1159/000206870

Beshai S, Watson LM, Meadows TJS, Soucy JN (2019) Perceptions of cognitive-behavioral therapy and antidepressant medication for depression after brief psychoeducation: examining shifts in attitudes. Behav Ther 50(5):851–863. https://doi.org/10.1016/j.beth.2019.01.001

Blumenthal JA, Rozanski A (2023) Exercise as a therapeutic modality for the prevention and treatment of depression. Prog Cardiovasc Dis 77:50–58. https://doi.org/10.1016/j.pcad.2023.02.008

Blumenthal JA, Babyak MA, Moore KA et al (1999) Effects of exercise training on older patients with major depression. Arch Intern Med 159(19):2349–2356. https://doi.org/10.1001/archinte.159.19.2349

Blumenthal JA, Babyak MA, Murali Doraiswamy P et al (2007) Exercise and pharmacotherapy in the treatment of major depressive disorder. Psychosom Med 69(7):587–596. https://doi.org/10.1097/PSY.0b013e318148c19a

Breedvelt JJF, Warren FC, Segal Z, Kuyken W, Bockting CL (2021) Continuation of antidepressants vs sequential psychological interventions to prevent relapse in depression: an individual participant data meta-analysis. JAMA Psychiatry 78(8):868–875. https://doi.org/10.1001/jamapsychiatry.2021.0823

Brody DJ, Pratt LA, Hughes JP (2018) Prevalence of depression among adults aged 20 and over: United States, 2013-2016. NCHS Data Brief 303:1–8

Busch AM, Ciccolo JT, Puspitasari AJ, Nosrat S, Whitworth JW, Stults-Kolehmainen MA (2016) Preferences for exercise as a treatment for depression. Ment Health Phys Act 10:68–72. https://doi.org/10.1016/j.mhpa.2015.12.004

Cartwright C, Gibson K, Read J, Cowan O, Dehar T (2016) Long-term antidepressant use: patient perspectives of benefits and adverse effects. Patient Prefer Adherence 10:1401–1407. https://doi.org/10.2147/PPA.S110632

Cascade E, Kalali AH, Kennedy SH (2009) Real-world data on SSRI antidepressant side effects. Psychiatry (Edgmont) 6(2):16–18

Caspersen CJ, Powell KE, Christenson GM (1985) Physical activity, exercise, and physical fitness: definitions and distinctions for health-related research. Public Health Rep 100(2):126–131

Castaldelli-Maia JM, Scomparini LB, Andrade AG, Bhugra D, de Toledo Ferraz Alves TC, D'Elia G (2011) Perceptions of and attitudes toward antidepressants. J Nerv Ment Dis 199(11): 866–871. https://doi.org/10.1097/NMD.0b013e3182388950

Choi KW, Chen CY, Stein MB et al (2019) Assessment of bidirectional relationships between physical activity and depression among adults: a 2-sample Mendelian randomization study. JAMA Psychiatry 76(4):399–408. https://doi.org/10.1001/jamapsychiatry.2018.4175

Cooney GM, Dwan K, Greig CA et al (2013) Exercise for depression. Cochrane Database Syst Rev:9. https://doi.org/10.1002/14651858.CD004366.pub6

Craft LL, Perna FM (2004) The benefits of exercise for the clinically depressed. Prim care companion. J Clin Psychiatry 6(3):104–111

Cuijpers P, Gentili C (2017) Psychological treatments are as effective as pharmacotherapies in the treatment of adult depression: a summary from randomized clinical trials and neuroscience evidence. Res Psychother 20(2):273. https://doi.org/10.4081/ripppo.2017.273

Cuijpers P, Sijbrandij M, Koole SL, Andersson G, Beekman AT, Reynolds CF III (2013) The efficacy of psychotherapy and pharmacotherapy in treating depressive and anxiety disorders: a meta-analysis of direct comparisons. World Psychiatry 12:137–148. https://doi.org/10.1002/wps.20038

Cuijpers P, Cristea IA, Karyotaki E, Reijnders M, Huibers MJH (2016) How effective are cognitive behavior therapies for major depression and anxiety disorders? A meta-analytic update of the evidence. World Psychiatry 15(3):245–258. https://doi.org/10.1002/wps.20346

Cuijpers P, Karyotaki E, Ciharova M, Miguel C, Noma H, Furukawa TA (2021) The effects of psychotherapies for depression on response, remission, reliable change, and deterioration: a meta-analysis. Acta Psychiatr Scand 144(3):288–299. https://doi.org/10.1111/acps.13335

DeRubeis RJ, Siegle GJ, Hollon SD (2008) Cognitive therapy versus medication for depression: treatment outcomes and neural mechanisms. Nat Rev Neurosci 9(10):788–796. https://doi.org/10.1038/nrn2345

Dishman RK, Berthoud H-R, Booth FW, Cotman CW, Edgerton VR, Fleshner MR, Gandevia SC, Gomez-Pinilla F, Greenwood BN, Hillman CH, Kramer AF, Levin BE, Moran TH, Russo-Neustadt AA, Salamone JD, van Hoomissen JD, Wade CE, York DA, Zigmond MJ (2006) Neurobiology of exercise. Obesity 2006(14):345–356. https://doi.org/10.1038/oby.2006.46

Dishman RK, McDowell CP, Herring MP (2021) Customary physical activity and odds of depression: a systematic review and meta-analysis of 111 prospective cohort studies. Br J Sports Med 55(16):926–934. https://doi.org/10.1136/bjsports-2020-103140

Dunn AL, Trivedi MH, Kampert JB, Clark CG, Chambliss HO (2005) Exercise treatment for depression: efficacy and dose response. Am J Prev Med 28(1):1. https://doi.org/10.1016/j.amepre.2004.09.003

Farmer ME, Locke BZ, Mościcki EK, Dannenberg AL, Larson DB, Radloff LS (1988) Physical activity and depressive symptoms: the NHANES I epidemiologic follow-up study. Am J Epidemiol 128(6):1340–1351. https://doi.org/10.1093/oxfordjournals.aje.a115087

Farris S, Abrantes A (2020) Mental health benefits from lifestyle physical activity interventions: a systematic review. Bull Menninger Clin 84:337–372. https://doi.org/10.1521/bumc.2020.84.4.337

Feighner JP (1999) Mechanism of action of antidepressant medications. J Clin Psychiatry 60(Suppl 4):4–11

Firth J, Rosenbaum S, Stubbs B, Gorczynski P, Yung AR, Vancampfort D (2016) Motivating factors and barriers towards exercise in severe mental illness: a systematic review and meta-analysis. Psychol Med 46(14):2869–2881. https://doi.org/10.1017/S0033291716001732

Fried EI, Nesse RM (2015) Depression is not a consistent syndrome: an investigation of unique symptom patterns in the STAR*D study. J Affect Disord 172:96–102. https://doi.org/10.1016/j.jad.2014.10.010

Gelenberg AJ (2010) A review of the current guidelines for depression treatment. J Clin Psychiatry 71(7):e15. https://doi.org/10.4088/jcp.9078tx1c

Glowacki K, Duncan MJ, Gainforth H, Faulkner G (2017) Barriers and facilitators to physical activity and exercise among adults with depression: a scoping review. Ment Health Phys Act 13: 108–119. https://doi.org/10.1016/j.mhpa.2017.10.001

Goodwin RD (2003) Association between physical activity and mental disorders among adults in the United States. Prev Med 36(6):698–703. https://doi.org/10.1016/S0091-7435(03)00042-2

Gordon BR, McDowell CP, Hallgren M, Meyer JD, Lyons M, Herring MP (2018) Association of efficacy of resistance exercise training with depressive symptoms: meta-analysis and meta-regression analysis of randomized clinical trials. JAMA Psychiatry 75(6):566–576. https://doi.org/10.1001/jamapsychiatry.2018.0572

Greenberg PE, Fournier AA, Sisitsky T, Pike CT, Kessler RC (2015) The economic burden of adults with major depressive disorder in the United States (2005 and 2010). J Clin Psychiatry 76(2):155–162. https://doi.org/10.4088/JCP.14m09298

Greist JH, Klein MH, Eischens RR, Faris J, Gurman AS, Morgan WP (1979) Running as treatment for depression. Compr Psychiatry 20:41–54

Gudmundsson P, Lindwall M, Gustafson DR et al (2015) Longitudinal associations between physical activity and depression scores in Swedish women followed 32 years. Acta Psychiatr Scand 132(6):451–458. https://doi.org/10.1111/acps.12419

Hirschfeld RM (2000) History and evolution of the monoamine hypothesis of depression. J Clin Psychiatry 61(Suppl 6):4–6

Honey EP (2015) I shrunk the pooled SMD! Guide to critical appraisal of systematic reviews and meta-analyses using the Cochrane review on exercise for depression as example. Ment Health Phys Act:8. https://doi.org/10.1016/j.mhpa.2014.12.001
Jesulola E, Micalos P, Baguley IJ (2018) Understanding the pathophysiology of depression: from monoamines to the neurogenesis hypothesis model – are we there yet? Behav Brain Res 341:79–90. https://doi.org/10.1016/j.bbr.2017.12.025
Judd LL, Rapaport MH, Paulus MP, Brown JL (1994) Subsyndromal symptomatic depression: a new mood disorder? J Clin Psychiatry 55(Suppl):18–28
Kambeitz-Ilankovic L, Rzayeva U, Völkel L, Wenzel J, Weiske J, Jessen F, Reininghaus U, Uhlhaas PJ, Alvarez-Jimenez M, Kambeitz J (2022) A systematic review of digital and face-to-face cognitive behavioral therapy for depression. NPJ Dig Med 5(1):144. https://doi.org/10.1038/s41746-022-00677-8
Kesaniemi YA, Danforth EJ, Jensen MD, Kopelman PG, Lefebvre P, Reeder BA (2001) Dose-response issues concerning physical activity and health: an evidence-based symposium. Med Sci Sports Exerc 33(6):S351–S358. https://doi.org/10.1097/00005768-200106001-00003
Kovich H, Kim W, Quaste AM (2023) Pharmacologic treatment of depression. Am Fam Physician 107(2):173–181
Legrand FD, Neff EM (2016) Efficacy of exercise as an adjunct treatment for clinically depressed inpatients during the initial stages of antidepressant pharmacotherapy: an open randomized controlled trial. J Affect Disord 191:139–144. https://doi.org/10.1016/j.jad.2015.11.047
Luo Y, Kataoka Y, Ostinelli EG, Cipriani A, Furukawa TA (2020) National Prescription Patterns of antidepressants in the treatment of adults with major depression in the US between 1996 and 2015: a population representative survey based analysis. Front Psych 11:35. https://doi.org/10.3389/fpsyt.2020.00035
Malhi GS, Bassett D, Boyce P et al (2015) Royal Australian and New Zealand College of Psychiatrists clinical practice guidelines for mood disorders. Aust N Z J Psychiatry. https://doi.org/10.1177/0004867415617657
Mammen G, Faulkner G (2013) Physical activity and the prevention of depression: a systematic review of prospective studies. Am J Prev Med 45(5):649–657. https://doi.org/10.1016/j.amepre.2013.08.001
Martinsen EW (1990) Benefits of exercise for the treatment of depression. Sports Med 9(6):380–389. https://doi.org/10.2165/00007256-199009060-00006
Mechanic D (2002) Removing barriers to care among persons with psychiatric symptoms. Health Aff 21(3):137–147. https://doi.org/10.1377/hlthaff.21.3.137
Meyer J, Schuch FB (2018) Chapter 1 – exercise for the prevention and treatment of depression. In: Stubbs B, Rosenbaum S (eds) Exercise-based interventions for mental illness. Academic Press, pp 1–18. https://doi.org/10.1016/B978-0-12-812605-9.00001-0
Meyer JD, Koltyn KF, Stegner AJ, Kim JS, Cook DB (2016) Influence of exercise intensity for improving depressed mood in depression: a dose-response study. Behav Ther 47(4):527–537. https://doi.org/10.1016/j.beth.2016.04.003
Meyer JD, Perkins SL, Brower CS et al (2022) Feasibility of an exercise and CBT intervention for treatment of depression: a pilot randomized controlled trial. Front Psych:13. Accessed 2 Dec 2023. https://www.frontiersin.org/articles/10.3389/fpsyt.2022.799600
Milea D, Guelfucci F, Bent-Ennakhil N, Toumi M, Auray JP (2010) Antidepressant monotherapy: a claims database analysis of treatment changes and treatment duration. Clin Ther 32(12):2057–2072. https://doi.org/10.1016/j.clinthera.2010.11.011
Mohr DC, Ho J, Duffecy J, Baron KG, Lehman KA, Jin L, Reifler D (2010) Perceived barriers to psychological treatments and their relationship to depression. J Clin Psychol 66(4):394–409. https://doi.org/10.1002/jclp.20659
Mojtabai R, Olfson M, Han B (2016) National trends in the prevalence and treatment of depression in adolescents and young adults. Pediatrics (Evanston) 138(6):1. https://doi.org/10.1542/peds.2016-1878

Morgan WP (1969) A pilot investigation of physical working capacity in depressed and nondepressed psychiatric males. Res Q Am Assoc Health Phys Educ Recreat 40(4):859–861. https://doi.org/10.1080/10671188.1969.10614932

Morgan WP (1970) Physical working capacity in depressed and non-depressed psychiatric females: a preliminary study. Am Correct Ther J 24:14–16

Moussavi S, Chatterji S, Verdes E et al (2007) Depression, chronic diseases, and decrements in health: results from the World Health Surveys. Lancet 370(9590):851–858. https://doi.org/10.1016/S0140-6736(07)61415-9

North TC, McCullagh P, Tran ZV (1990) Effect of exercise on depression. Exerc Sport Sci Rev 18(1):379–416

O'Sullivan D, Gordon BR, Lyons M, Meyer JD, Herring MP (2023) Effects of resistance exercise training on depressive symptoms among young adults: a randomized controlled trial. Psychiatry Res 326:115322–115322. https://doi.org/10.1016/j.psychres.2023.115322

Ormel J, Hollon SD, Kessler RC, Cuijpers P, Monroe SM (2022) More treatment but no less depression: the treatment-prevalence paradox. Clin Psychol Rev 91:102111. https://doi.org/10.1016/j.cpr.2021.102111

Orsolini L, Latini R, Pompili M et al (2020) Understanding the complex of suicide in depression: from research to clinics. Psychiatry Investig 17(3):207–221. https://doi.org/10.30773/pi.2019.0171

Pearce M, Garcia L, Abbas A et al (2022) Association between physical activity and risk of depression: a systematic review and meta-analysis. JAMA Psychiatry 79(6):550. https://doi.org/10.1001/jamapsychiatry.2022.0609

Piercy KL, Troiano RP, Ballard RM, Carlson SA, Fulton JE, Galuska DA et al (2018) The physical activity guidelines for Americans. JAMA 320(19):2020–2028

Pigott HE, Leventhal AM, Alter GS, Boren JJ (2010) Efficacy and effectiveness of antidepressants: current status of research. Psychother Psychosom 79(5):267–279. https://doi.org/10.1159/000318293

Ravindran AV, Balneaves LG, Faulkner G et al (2016) Canadian network for mood and anxiety treatments (CANMAT) 2016 clinical guidelines for the management of adults with major depressive disorder: section 5. Complementary and alternative medicine treatments. Can J Psychiatry. https://doi.org/10.1177/0706743716660290

Rethorst CD, Trivedi MH (2013) Evidence-based recommendations for the prescription of exercise for major depressive disorder. J Psychiatr Pract 19(3):204. https://doi.org/10.1097/01.pra.0000430504.16952.3e

Singh B, Olds T, Curtis R, Dumuid D, Virgara R, Watson A et al (2023a) Effectiveness of physical activity interventions for improving depression, anxiety and distress: an overview of systematic reviews. Br J Sports Med 57(18):1203–1209

Singh B, Olds T, Curtis R et al (2023b) Effectiveness of physical activity interventions for improving depression, anxiety and distress: an overview of systematic reviews. Br J Sports Med 57:1203–1209

Stavrakakis N, de Jonge P, Ormel J, Oldehinkel AJ (2012) Bidirectional prospective associations between physical activity and depressive symptoms. The TRAILS study. J Adolesc Health 50(5):503–508. https://doi.org/10.1016/j.jadohealth.2011.09.004

Steidtmann D, Manber R, Arnow BA, Klein DN, Markowitz JC, Rothbaum BO, Thase ME, Kocsis JH (2012) Patient treatment preference as a predictor of response and attrition in treatment for chronic depression. Depress Anxiety 29:896–905. https://doi.org/10.1002/da.21977

Stubbe DE (2018) The therapeutic alliance: the fundamental element of psychotherapy. Focus (American Psychiatric Publishing) 16(4):402–403. https://doi.org/10.1176/appi.focus.20180022

Stubbs B, Vancampfort D, Rosenbaum S et al (2016) Dropout from exercise randomized controlled trials among people with depression: a meta-analysis and meta regression. J Affect Disord 190: 457–466. https://doi.org/10.1016/j.jad.2015.10.019

Substance Abuse and Mental Health Services Administration (2021) Key Substance Use and Mental Health Indicators in the United States: results from the 2021 National Survey on Drug Use and Health

Thase ME, McCrone P, Barrett MS et al (2020) Improving cost-effectiveness and access to cognitive behavior therapy for depression: providing remote-ready, computer-assisted psychotherapy in times of crisis and beyond. Psychother Psychosom 89(5):307–313. https://doi.org/10.1159/000508143

Tipton CM (2014) The history of "Exercise Is Medicine" in ancient civilizations. Adv Physiol Educ 38(2):109–117. https://doi.org/10.1152/advan.00136.2013

Vetter RE, Symonds ML (2010) Correlations between injury, training intensity, and physical and mental exhaustion among college athletes. J Strength Cond Res 24(3):587–596. https://doi.org/10.1519/JSC.0b013e3181c7c2eb

Wang SM, Han C, Bahk WM, Lee SJ, Patkar AA, Masand PS, Pae CU (2018) Addressing the side effects of contemporary antidepressant drugs: a comprehensive review. Chonnam Med J 54(2):101–112. https://doi.org/10.4068/cmj.2018.54.2.101

World Health Organization (2017) Depression and other common mental disorders: Global Health estimates. World Health Organization

Exercise and Smoking Cessation

Lorra Garey, Jessica M. Thai, Michael J. Zvolensky, and Jasper A. J. Smits

Contents

Abstract Smoking is a public health crisis, leading to a multitude of health complications. Exercise is associated with numerous health benefits and is accepted by health professionals and smokers as a potentially effective smoking cessation aid. This chapter discusses the extant literature on the relation between exercise and smoking, including cross-sectional studies, experiments, and randomized clinical trials. There is robust evidence for exercise's efficacy in reducing cigarette craving, tobacco withdrawal symptoms, and negative affect. Further, exercise-based

L. Garey (✉)
Department of Psychology, University of Houston, Houston, TX, USA

HEALTH Institute, University of Houston, Houston, TX, USA
e-mail: llgarey@central.uh.edu

J. M. Thai
Department of Psychology, University of Houston, Houston, TX, USA

M. J. Zvolensky
Department of Psychology, University of Houston, Houston, TX, USA

HEALTH Institute, University of Houston, Houston, TX, USA

Department of Behavioral Science, The University of Texas MD Anderson Cancer Center, Houston, TX, USA

J. A. J. Smits
Department of Psychology, The University of Texas at Austin, Austin, TX, USA

Curr Topics Behav Neurosci (2024) 67: 177–198
https://doi.org/10.1007/7854_2024_497
Published Online: 2 August 2024

interventions appear to boost short-term abstinence yet may fall short of facilitating long-term abstinence. Methodological limitations of extant work are reviewed. We conclude with a discussion of the next steps in this line of work to fine-tune exercise interventions and their application for smoking cessation.

1 Introduction

According to the World Health Organization (WHO), tobacco use kills more than eight million people each year, with more than seven million of these deaths due to direct tobacco use, and about 1.2 million due to non-smokers being exposed to second-hand smoke (World Health Organization 2022). Additionally, smoking can lead to the development of chronic diseases, including cardiovascular disease, respiratory diseases, and cancer, as well as other health issues such as infertility, premature aging, and oral health problems (Bonnie et al. 2015).

Smoking cessation is the process of quitting smoking, and it is essential for improving health outcomes for smokers (Centers for Disease Control Prevention 2020). There are several efficacious approaches to smoking cessation, including pharmacological interventions (e.g., nicotine replacement therapy, bupropion, and varenicline) and behavioral interventions (e.g., counseling and support groups; García-Gómez et al. 2019). Although these interventions can be efficacious, many smokers struggle to quit smoking. Indeed, most established interventions are associated with relatively low short-term abstinence rates (15–35%), and long-term (≥6 months) abstinence rates rarely exceed 20–25%, even with the gold standard combination of psychotherapy and pharmacotherapy for smoking cessation (Simons et al. 2010). Accordingly, the most recent guidelines for clinical practice on treating tobacco dependence indicate a need for innovative, potent strategies for smoking cessation (Fiore et al. 2008).

Physical activity and exercise are known to have numerous health benefits, including reducing the risk of developing chronic diseases, improving mental health and well-being, and increasing overall longevity (Penedo and Dahn 2005). Therefore, it is unsurprising that exercise interventions have been considered as a possible aid in smoking cessation efforts. Indeed, exercise is routinely recommended as a cessation aid by smoking professionals (Everson et al. 2010) and is endorsed by the current Clinical Practice Guideline for Treating Tobacco Use and Dependence (Fiore et al. 2008). Many smokers themselves identify exercise as a potentially useful quitting strategy (Everson-Hock et al. 2010).

In this chapter, we will review the relation between smoking and exercise and exercise interventions for smoking cessation. We will also discuss putative mechanisms of exercise interventions for smoking cessation and special populations to consider regarding exercise interventions for smoking cessation. We will conclude the chapter by discussing future directions.

2 Smoking and Physical Activity

Extensive evidence has highlighted an inverse relation between smoking and physical activity (Boutelle et al. 2000; Picavet et al. 2011). Indeed, smoking is known to have several adverse effects on physical fitness and exercise capacity (Conway and Cronan 1992). Nicotine found in cigarettes acts as a vasoconstrictor, leading to reduced blood flow to muscles and other tissues (Benowitz and Burbank 2016) and reduced exercise capacity (McAllister et al. 1995). Smoking is also known to lower lung function, leading to a decreased ability to perform physical activities requiring exertion (Tantisuwat and Thaveeratitham 2014; Mohamed Abdelaal and Mohamed Mousa 2022). Smokers also have a higher risk of developing cardiovascular disease, which can further reduce exercise capacity (US Department of Health Human Services 2010). In part, these adverse health effects likely contribute to lower exercise rates among smokers relative to the general population (Furlanetto et al. 2014; Kaczynski et al. 2008).

Smoking can negatively impact one's ability and desire to exercise. Those persons who can overcome these challenges report better motivational and behavioral outcomes related to quitting smoking. Specifically, among smokers, being physically active has been positively associated with the intention to quit (Frith and Loprinzi 2018), initiating a quit attempt (DeRuiter et al. 2008; Gauthier et al. 2012; Haddock et al. 2000), confidence in maintaining smoking abstinence (King et al. 1996), smoking cessation (Abrantes et al. 2009; Derby et al. 1994; Loprinzi and Walker 2016; Paavola et al. 2001; Sedgwick et al. 1988), and reduced likelihood of smoking relapse (Emery et al. 2014). In part, the positive effect of exercise on smoking urges (Haasova et al. 2013; Abrantes et al. 2018; Kunicki et al. 2022; Roberts et al. 2012; Bock et al. 1999; Ussher et al. 2001; Daniel et al. 2004; Taylor et al. 2005) and withdrawal (Ruslan et al. 2023; Ussher et al. 2019) may contribute to these relations and help promote cessation. Indeed, recent work found that even an acute bout of exercise reduced smoking urges in both men and women smokers during an acute state of abstinence (Allen et al. 2018). Further, exercise engagement prior to quitting has been shown to relate to changes in smoking behavior and perception of use among women, including decreased puff duration, smoking satisfaction, psychological reward, enjoyment of respiratory tract sensations, craving, and increased puff flow (De Jesus and Prapavessis 2018). Taken together, there is evidence supporting the benefits of exercise as a component of tobacco dependence treatments. It should be highlighted that being physically active is associated with numerous health benefits on its own (Penedo and Dahn 2005) and can promote the health of the smoker even if cessation is not attempted or achieved (Ferrucci et al. 1999; Auer et al. 2014; Ussher et al. 2003; Bock et al. 2012). Nevertheless, the combination of smoking cessation and regular physical activity, relative to each behavior on its own, has the potential to lead to more significant health improvements among current smokers.

3 Exercise Interventions for Smoking Cessation

Exercise interventions for smoking cessation employ physical activity and exercise to promote smoking cessation. Several types of exercise approaches have been tested in the context of exercise interventions for smoking cessation. The most used approaches utilize aerobic exercise (i.e., cardiovascular based), resistance exercise (i.e., weight training), or a combination of the two (Whiteley et al. 2012). Exercise interventions for smoking cessation also have been tested in various formats, such as supervised exercise programs, home-based exercise programs, or a combination of both (Taylor et al. 1988). Additionally, exercise interventions have been studied as standalone smoking cessation interventions or as adjunct interventions that are provided in combination with traditional smoking cessation approaches (i.e., behavioral therapy and/or pharmacotherapy; (Chen et al. 2022; Prapavessis et al. 2007; Marcus et al. 2005).

The overall potential for exercise as a standalone treatment or adjunct treatment to standard smoking cessation has been summarized by a recent Cochrane review (see Ussher et al. 2019). This review focused on 21 (total number of participants = 6,607) studies that included an exercise intervention (with exercise alone or as an add-on) for smoking cessation. The outcome of interest was self-reported and biochemically verified abstinence at the final follow-up, ranging from 6 to 16 months post-quit. Based on the review, the authors concluded that there is limited supporting evidence that exercise in conjunction with smoking cessation support increased quit rates relative to smoking cessation support alone when measured at distal follow-ups. Indeed, only one study demonstrated a significant difference in smoking cessation at the 6-month follow-up among those who received an exercise intervention (but confounded by severity of sensitivity to somatic symptoms; Smits et al. 2016) and only one study supported a significant benefit of exercise at the 1-year follow-up (Marcus et al. 1999). Another recent meta-analysis found that yoga, in combination with standard smoking cessation treatment, was the only exercise type that increased the likelihood of smoking cessation (Klinsophon et al. 2017). This inference, however, was based on a single study.

Several notable explanations can be offered to explain the lack of findings among most studies (Ussher et al. 2019). First, several studies were underpowered to detect effects (Taylor et al. 1988; Abrantes et al. 2014; Hill 1985; Marcus et al. 1995; Patten et al. 2016; Russell et al. 1988; Vickers et al. 2009; Whiteley et al. 2007). Second, differences in methodology across studies, including timing and intensity of the smoking cessation and exercise programs offered, could impact the findings. Third, the review included studies that reported smoking cessation outcomes, but the operationalization of cessation differed across studies, which may have biased results. Fourth, the comparison group differed across studies, with some being more robust than others. Finally, adherence may have influenced findings. Indeed, some work indicated stronger adherence to the exercise condition, whereas other trials reported lower adherence or limited data on adherence. This is an important consideration, as several studies suggest that the degree of adherence to exercise programs is integral to the observed outcomes (Whiteley et al. 2012, 2007). Thus, although the review by Ussher et al. (2019) provides limited support for exercise interventions to assist with smoking cessation, these data must be considered in the context of the limitations and inconsistencies across the studies included.

It is possible that the efficacy of exercise interventions for smoking cessation may be specific to certain subgroups of smokers, exercise interventions, or their combinations. Indeed, recent studies have found promising support for exercise interventions to promote smoking cessation, particularly among specific populations. For example, Smits et al. (2021) conducted a randomized controlled trial in which they randomized 150 daily smokers with high anxiety sensitivity who were interested in quitting to (a) standard smoking cessation treatment through the Texas Tobacco Quitline, which consists of a 5-session counseling package, combined with nicotine replacement therapy and high-intensity aerobic exercise (60–85% of heart rate reserve) or (b) the same standard smoking cessation treatment through the Texas Tobacco Quitline with nicotine replacement therapy and low-intensity aerobic exercise (20–40% of heart rate reserve). The trial consisted of a 15-week intervention period and, importantly, was conducted in a community center (i.e., YMCA). Participants met with their personal trainers during the first week to complete three training sessions, establish an exercise training schedule, and set their target quit day for week 6. The next 14 weeks consisted of one supervised and two non-supervised exercise training sessions each week. Participants were connected to the Texas Tobacco Quitline in week 3 and were provided nicotine replacement therapy in weeks 6 through 14. The authors found that point prevalence abstinence was significantly higher at the 6-month follow-up in the high-intensity group relative to the low-intensity group (27.6% vs. 14.8%). This data highlights the potential of a supervised, high-intensity exercise program integrated into a community center to promote smoking cessation among smokers with elevated anxiety sensitivity.

Stockton et al. (2023) also conducted a study of an exercise-based intervention for smoking cessation protocol in the YCMA. Daily smokers ($N = 392$) interested in quitting were randomized into one of two study conditions: (1) a physical activity intervention consisting of an individualized exercise plan containing moderate and vigorous intensity exercise, with supervised activity sessions and behavioral counseling and supportive contacts during the post-cessation maintenance phase, or (2) a wellness intervention with curriculum surrounding multiple dimensions of wellness and health (e.g., healthy eating, injury reduction, and avoidance). The target quit date was set to week 3 after the intervention was initiated, and the treatment lasted for 7 weeks. Participants in the physical activity intervention participated in 16 face-to-face physical activity/cessation sessions, 11 telephone activity/cessation sessions, and received 11 supportive mailings (i.e., supportive contacts to boost adherence). The physical activity intervention consisted of supervised leisure exercises, "lifestyle activities" (i.e., manual labor that is performed in day-to-day life), and short bouts (>10 min) of exercise to manage urges and cravings. Participants started off engaging in physical activity three times per week. They gradually increased the frequency until they reached five to six times per week (i.e., 150–180 min of exercise per week). Participants achieved the weekly goal by engaging in any combination of the different exercises. Participants in the wellness intervention participated in 8 face-to-face wellness/cessation sessions, 12 telephone wellness/cessation sessions, and received 18 follow-up mailings. Participants in both treatments received nicotine replacement therapy in the form of transdermal nicotine patches and behavioral smoking cessation counseling. The Social Cognitive Theory informed all behavioral counseling to increase self-efficacy and address barriers to cessation. Biochemical verification of smoking was collected at 7 weeks, 6 months, and 1 year. The physical activity intervention did not result in a higher abstinence rate compared to the wellness intervention (19.6% and 25.4% at 7 weeks, 15.1% and 16.6% at 6 months, and 14.1% and 17.1% at 12 months, respectively). Despite the community-based physical activity intervention not improving long-term quit rates, the program was found to be acceptable among individuals interested in quitting smoking. Interestingly, unlike Smits et al. (2021), who enrolled smokers with high anxiety sensitivity, Stockton enrolled a general sample of smokers. Thus, considering the findings by Smits et al. (2021), community-based exercise interventions for smoking cessation may be most effective among emotionally vulnerable subgroups of the larger smoking population, such as those with elevated sensitivity to anxiety-related symptoms commonly experienced during exercise and withdrawal.

Another study (Abrantes et al. 2018) documented the effects of aerobic exercise on smoking abstinence through negative affect reduction. Female ($N = 57$) low-active daily smokers were randomized to either a group-based moderate-intensity aerobic exercise intervention or a health-education intervention. An exercise physiologist guided aerobic exercise to ensure moderate-intensity exercises (55–69% of age-predicted maximal heart rate). Based on fitness level and progress, weekly exercises were decided collaboratively between the exercise physiologist and the participant. Exercise sessions gradually increased in duration throughout the weeks. The aerobic exercises offered to participants included ellipticals, treadmills, and recumbent bicycles. Exercise sessions started as 15-min sessions each week and gradually increased to 40-min sessions by week 12 (51.7% of participants reached 40 min). Participants were also prescribed home-based exercises that would take place 2–4 additional times per week, with a goal of achieving 150 min of exercise per week. Home-based exercises were reported by the participant using self-monitored logs. The health-education intervention consisted of weekly 45–60-min health-education sessions that covered health topics (i.e., oral and sleep hygiene) but did not discuss physical activity. The information was conveyed through handouts, in-group exercises, lectures, and Internet resources. Both interventions were 12 weeks in length, and the target quit day was set to week 5. Standard cognitive-behavioral smoking cessation treatment (Abrams and Niaura 2003) was delivered to both groups via telephone starting at week 1 and continued throughout the intervention period. In addition to cognitive-behavioral smoking cessation treatment, both groups received nicotine replacement therapy in the form of transdermal patches beginning on the quit day. To avoid multiple onsets of behavioral changes that could affect outcomes (King et al. 1996), exercise was initiated 1 month prior to the quit day. The wellness group was not instructed to engage in any new behaviors prior to the quit day. Beginning at week 5, biochemical verification of smoking was collected at each weekly session until week 12. Among smokers, Abrantes et al. (2018) found a statistically significant ($p = 0.0019$) difference in pre- to post-cessation decreases in anxiety among those in the aerobic exercise intervention compared to those in the health-education intervention after an exercise session. Those in the aerobic exercise intervention also reported a statistically significant ($p = 0.025$) reduction in cravings at the 8-week follow-up period compared to those in the health-education intervention. Though this study did not find statistically significant differences in smoking cessation rates between groups, the findings suggest that aerobic exercise sessions prior to a quit date aid in alleviating anxiety and, in turn, could decrease fear and avoidance of arousal sensations (Smits et al. 2011), which are common and impactful barriers to cessation (Lasser et al. 2000; Piper et al. 2010).

Similarly, Bock et al. (2019) conducted a randomized clinical trial using yoga as a complementary therapy for adult smokers wanting to quit. Yoga was chosen over other types of exercise because it has been shown to improve mood and reduce stress through *asanas* (Yoga postures) and *pranayama* (breathing exercises; (Li and Goldsmith 2012; Shahab et al. 2013). Yoga has also been shown to enhance mindfulness, with increases in mindfulness associated with reductions in nicotine withdrawal, improvement in coping with cravings, and increases in cognitive deliberations required to avoid smoking (Gard et al. 2012; Ussher et al. 2006; McClernon et al. 2004). The study looked at 227 daily smokers and consisted of an 8-week yoga or general wellness program. Both groups had participants meeting for a 1-h weekly group-based smoking cessation counseling session using cognitive-behavior therapy, which was scheduled after their respective wellness or yoga class. The Social Cognitive Theory informed smoking cessation counseling. The quit date was scheduled for week 4. The type of Yoga practiced throughout the intervention was *Iyengar* Yoga (Iyengar 2007) due to its emphasis on posture (*asanas*) and the use of props to reduce injury and facilitate learning. Participants in the yoga intervention attended yoga classes twice weekly for 8 weeks. Certified *Iyengar* instructors would select a pattern of *asanas* appropriate for beginners and teach them during class. The classes consisted of 5 min of breathing exercises and seated medication, followed by 45 min of *asanas*, and concluded with 10 min of closing postures and a final set of seated meditation. To encourage continued practice, participants were given one free class voucher for weeks 9 through 20 after the 8-week intervention. The wellness group consisted of 1-h twice-weekly classes consisting of videos and lectures on health topics (i.e., oral and sleep hygiene). At the end of the 8 weeks, participants were sent health and wellness materials during weeks 9–20. In both interventions, participants completed and mailed in a monthly log to report either their yoga practice or health behaviors. Follow-ups were completed at 3- and 6 months. Results indicated that *Iyengar* Yoga was feasible and acceptable among smokers based on the strong attendance and retention rates at the 6-month follow-up (94.7% for both intervention groups). Participants in the Yoga group were also more likely to quit compared to those in the wellness group (46.9% vs. 17.4%, respectively), and lighter smokers were also over twice as likely to quit smoking if they were engaging in Yoga. The results highlight the potential of Yoga as a smoking cessation aid and may particularly be more beneficial for light smokers.

Onkcen et al. (2020) recently completed a trial among 301 postmenopausal women. Researchers were particularly interested in this group, given that postmenopausal women who smoke are at a much greater risk than their nonsmoking female counterparts to develop osteoporosis and fractures (Health UDo, Services H 2004; Alberg et al. 2014), cardiovascular disease, chronic obstructive lung disease, and cancer (Health UDo, Services H 2004). The study compared the effectiveness of a smoking cessation program with an aerobic and strength-based, moderate exercise (i.e., 50–69% of maximum heart rate) program to an identical smoking cessation program combined with a relaxation program as a comparison condition. The intervention lasted 6 months, and participants were followed up for a year post-intervention. Participants received varenicline, a smoking cessation pharmacological agent, during the first 12 weeks of the treatment. The study contained 3 phases: (1) Phase I included smoking cessation counseling with varenicline for 4 weeks, (2) Phase II included supervised exercise or relaxation two times per week with varenicline for 8 weeks, and (3) Phase III included supervised exercise or relaxation every week for 8 weeks and then every other week for 4 weeks. Quit day was set for the second week in phase I. In phase I, 1.5-h long supervised sessions were conducted weekly and introduced exercise and relaxation depending on the intervention group. Smoking cessation was emphasized to both groups during these sessions. The exercise condition was introduced to supervised isometric exercises, contract/relax exercises, breathing and stretching exercises, proper clothing to wear during exercise, hydration during exercise, and the Borg scale to determine the intensity of the exercises. The relaxation group was introduced to different relaxation techniques. Both groups were encouraged to practice what they had learned at home. In phase II, 1.5-h long sessions were conducted twice weekly and focused more on exercise and relaxation. Participants in the exercise intervention began a supervised 1-h twice weekly aerobic and strength training program, totaling up to 180 min of moderate exercise per week. The sessions for participants in the relaxation intervention followed a similar schedule and focused on breathing techniques and muscle relaxation. In phase III, 1-h sessions on exercise and relaxation were conducted every week for 8 weeks and then every other week for 4 weeks. Follow-ups occurred at 3-, 6-, 9-, and 12 months to assess smoking status. Exercise testing was conducted 3-, 6-, 9-, and 15 months after the quit date. Findings did not support the exercise intervention, compared to relaxation, to increase end-of-treatment prolonged abstinence (i.e., abstinence at the end of 60 weeks). There were no statistically significant effects observed in the exercise condition at 4 weeks (Wald χ 2 $=$ 1.85; $df = 292$; $p = 0.18$) or for prolonged abstinence (Wald χ 2 $= 0.08$, $df = 292$, $p > 0.78$). In part, the use of varenicline by participants in both conditions may have contributed to the non-significant effects. Further, as relaxation techniques exert a positive effect on negative affect and mood (Ng and Jeffery 2003), participants in the control group may have received an alternative yet equally effective treatment as those in the exercise-based treatment. Notably, the exercise intervention was associated with reduced cigarettes per day during the treatment period. Thus, this study suggests that continued exercise may relate to reduced smoking behavior but not result in cessation among postmenopausal women.

There is increased recognition that long-term (> 6 months) abstinence following a quit attempt is low (Stead et al. 2016). Focusing on such distal outcomes may mask the potential of exercise interventions to assist smoking cessation. Indeed, of the articles reviewed by Ussher et al. (2019), when considering abstinence at end-of-treatment or more proximal follow-up assessments, evidence was more favorable to exercise conditions, with several trials demonstrating significantly higher abstinence rates in an exercise condition relative to a control (Bock et al. 2012; Smits et al. 2016; Marcus et al. 1991, 1999; Patten et al. 2016; Thompson et al. 2015; Martin et al. 1997). Further, a recent meta-analytic review by Santos et al. (2021) found that aerobic exercise was better than usual care in promoting smoking cessation in the short-term (11 trials, risk ratio 0.79; 95% confidence interval, 0.66–0.94) but not in the medium- or long-term (i.e., after engaging in exercise, smoking cessation was achieved for a short period of time but did not last). Other reviews have found similar findings (Chen et al. 2022). Based on these collective data, exercise interventions for smoking cessation are efficacious at promoting cessation in the short-term. Including booster sessions to promote sustained effects of an exercise intervention on cessation following the end-of-treatment may be beneficial.

4 Mechanisms of Exercise Interventions for Smoking Cessation

Identifying treatment mechanisms may aid the efforts to optimize interventions (Kazdin 2007). At a biological level, aerobic exercise induces an increase in plasma β-endorphins (De Meirleir et al. 1986). This is important in the context of smoking cessation as higher levels of β-endorphin have been found to be associated with lesser smoking relapse after cessation (Shaw and al'Absi 2008). Yet, this may vary as a function of exercise regime, given that increases in β-endorphin depend on the intensity and duration of the exercise (Goldfarb et al. 1990). Other work has suggested that specific neurobiological mechanisms may be relevant to the effect of exercise at particular stages of addiction. For example, during drug use initiation and withdrawal, the efficacy of exercise may be related to its ability to facilitate dopaminergic transmission. Once addiction develops, its efficacy may be explained by its effects on glutamatergic and dopaminergic signaling and reverse drug-induced changes in chromatin via epigenetic interactions with brain-derived neurotrophic factor (BDNF) in the reward pathway (Lynch et al. 2013). A recent review of the neural mechanisms unpinning the effects of exercise on smoking-related processes and behavior also suggested that exercise leads to increased serum dopamine levels, which partially replaces the rewarding effect of smoking and reduces the desire to smoke in smokers (Ni et al. 2022). Although such work indicates a biological mechanism, additional research is needed to isolate the precise neurobiological factors and identify the conditions under which specific mechanisms may be engaged to their full potential.

Several studies also indicate that the effect of exercise on cognitive-affective processes may help explain how exercise affects smoking cessation. As noted above, exercise leads to reductions in smoking craving and nicotine withdrawal symptom severity and increases in positive affect and decreases in negative affect (Haasova et al. 2013, 2014; Roberts et al. 2012; Taylor and Katomeri 2007). These changes are essential, given increases in negative affect (Burgess et al. 2002; Piasecki et al. 2000; Kang and Lee 2010; Boden et al. 2010), decreases in positive affect (Cook et al. 2004; Lerman et al. 2002), and more severe craving and withdrawal symptoms are consistently associated with poorer cessation outcomes (Xian et al. 2003; Killen and Fortmann 1997; Allen et al. 2008; Kahler et al. 2002; Kenford et al. 1994; Brnstrm et al. 2010). In an effort to fully understand the potential for mood as a mechanism of exercise intervention efficacy on smoking cessation outcomes, Zvolensky et al. (2018) conducted a secondary analysis using data collected by Smits et al. (2016) to examine anxiety sensitivity and dysphoria (sadness; Covinsky et al. 2014) as unique explanatory mechanisms for the intervention effect on smoking cessation. Findings indicated that both anxiety sensitivity and dysphoria explained the effect of the exercise intervention on abstinence among smokers with high anxiety sensitivity. These two mechanisms may be particularly relevant to exercise treatment responsivity given that by exercising, participants may learn how to manage uncomfortable physical sensations better, thereby promoting reduced anxiety sensitivity and promoting cessation. Further, extensive evidence has highlighted the benefits of exercise on depression (Ross et al. 2023). As dysphoria is a cardinal feature of depression (Kovacs and Yaroslavsky 2014), these data indicate that exercise influences specific components of depression to promote cessation. There is a need for more research to understand how other aspects of depression, such as anhedonia (loss of interest in usually pleasurable activities; Covinsky et al. 2014), may be impacted by an exercise intervention to promote smoking cessation. Taken together, through its acute effect on craving and affect, including such affective vulnerability processes as anxiety sensitivity and dysphoria, aerobic exercise may play an important role in smokers' quit attempts.

5 Special Populations of Exercise Interventions for Smoking Cessation

Women also represent a unique group of smokers who may be apt for exercise-based smoking cessation trials. Although men are more likely to be smokers than women (20.6% vs. 12.8%; Creamer et al. 2019), the gender gap in smoking rates has closed considerably over the past 50 years, and some studies show that women have more difficulty quitting and remaining abstinent (for a review, see Smith et al. 2016). Weight control is a leading reason for smoking among women (Saarni et al. 2004) and may pose a barrier to cessation among women smokers (Berg et al. 2008; Levine et al. 2010). Weight gain has been observed in individuals who quit; however, women gain more weight than men (O'Hara et al. 1998; Williamson et al. 1991; Pisinger and Jorgensen 2007). Weight gain following cessation is related to relapse among women but not men, suggesting that a weight control component (e.g., physical activity) should be included in smoking cessation interventions for women (Kuo et al. 2022). Additionally, evidence has indicated that for postmenopausal women, in particular, smoking provides a weight control benefit. As a result, postmenopausal women are at a higher risk for relapse than premenopausal women (Caan et al. 1996; Copeland et al. 2006). Pregnancy is also an ideal time for cessation and is a preventable cause of poor health outcomes for both mother and child (Fingerhut et al. 1990; Linke et al. 2013). Moderate-intensity exercise has already been recommended to improve pregnant women's health (Obstetricians ACo, Gynecologists 2003; Bell and Dooley 2006). It presents another option for pregnant smokers who are reluctant to use nicotine replacement therapy (Ussher and West 2003) or fear post-cessation or postpartum weight gain (Pomerleau et al. 2000). A survey by Ussher et al. (2004) found that pregnant smokers are interested in personal exercise plans for smoking cessation, with 44% of participants expressing interest, and two studies have already demonstrated the feasibility of an exercise intervention for pregnant women (Ussher et al. 2008). Although exercise interventions for smoking cessation have enrolled both men and women, a high percentage of trials have enrolled only women (Bock et al. 2012; Whiteley et al. 2012; Prapavessis et al. 2007; Marcus et al. 1991, 1995, 1999, 2005; Russell et al. 1988; Vickers et al. 2009; Williams et al. 2010; Chaney and Sheriff 2008; Jung et al. 2010; Ussher et al. 2012; Kinnunen et al. 2008), as exercise theoretically targets weight-related barriers for women. Among women, acute bouts of exercise have been found to temporarily reduce cigarette cravings and withdrawal symptoms (Haasova et al. 2013; Taylor et al. 2007). Taken together, exercise interventions may alleviate weight and health concerns related to smoking cessation among women.

Individuals who experience elevated psychological distress or meet the criteria for a mental health disorder represent another group who may benefit from exercise interventions for smoking cessation. Extensive work has highlighted that regular physical activity is related to better mental health outcomes and that exercise interventions are effective at improving mental health conditions among those who struggle with such symptoms or syndromes (Smith and Merwin 2021). Smokers are more likely to experience elevated psychological distress (Jamal et al. 2018) and meet the criteria for a mental health disorder relative to the general population (Smith and Merwin 2021). Indeed, adults with anxiety and depressive symptoms also report a higher smoking prevalence (35.8%) compared with adults without such symptoms (14.7%; Jamal et al. 2018). Such data are concerning as smokers who experience mental health concerns have more difficulty quitting than those without mental health problems (Lê Cook et al. 2014). As such, an exercise intervention plus a smoking cessation program may benefit these individuals by (1) engaging neuromechanistic functions associated with improved mental health and (2) subsequently reducing mental health distress that promotes relapse. This has served as the theoretical foundation for several studies on exercise intervention for smoking cessation among those with mental health concerns. Specifically, based on a meta-analysis of 6 studies and 296 participants (see Schöttl et al. (2022) for review), evidence suggests that exercise interventions benefit smoking cessation in individuals with mental illness; however, results were not statistically significant at the 6-month follow-up, suggesting a short-term effect. Thus, there may be potential for physical activity interventions to support cessation among this group, but additional refinement is needed to improve efficacy and maintenance.

6 Future Directions

There is no consistency in "best methods" to prescribe an exercise intervention for smoking cessation. Across studies, there are considerable methodological differences regarding exercise modality, dose (i.e., duration and frequency), intensity (i.e., low, moderative and vigorous), and modality of implementation (i.e., supervised or unsupervised; Ussher et al. 2019). The design considerations within these areas have the potential to impact efficacy. Furthermore, there is considerable inconsistency for *when* to initiate the exercise intervention to support cessation. Specifically, some studies start the exercise intervention before the quit date (Whiteley et al. 2012; Prapavessis et al. 2007; Marcus et al. 1991, 1995, 1999, 2005; Smits et al. 2016; Abrantes et al. 2014; Patten et al. 2016; Kinnunen et al. 2008; Bernard et al. 2015; Bize et al. 2010; Prapavessis et al. 2016; Ussher et al. 2015), others begin on the quit date (Hill 1985; Martin et al. 1997; Ciccolo et al. 2011), and others initiate the exercise intervention after the quit date (Taylor et al. 1988; Russell et al. 1988; Hassandra et al. 2017; Maddison et al. 2014). There has been limited evidence suggesting that initiating exercise prior to a quit date is more efficacious; however, results have been mixed, and the effect of exercise on a self-initiated quit attempt has not been explored (Ussher et al. 2019). Thus, additional work is needed to parse the potential influence of specifics of an exercise program on cessation outcomes, particularly considering the potential mediational effects via changes in β-endorphin depending on the exercise regime. To achieve this goal, it may be fruitful for researchers to look at the study design of recently reported trials that found promising findings and conduct a dismantling study to isolate the precise aspects related to improved smoking cessation outcomes. Specifically, it may be prudent to look at the timing, i.e., whether the exercise is initiated pre-quit, at the same time as a quit attempt, or post-quit attempt. It may be interesting to explore whether the time between a quit attempt and initiation of exercise influences successful abstinence. Additionally, research could be usefully directed at different intensities or types of exercise, such as isometric versus aerobic exercises.

As this area of work continues to develop, researchers are urged to include smokers from racial or ethnic minority backgrounds. There is little to no understanding of how individuals from diverse racial or ethnic backgrounds may respond to exercise treatment for smoking cessation. The lack of diversity among participants enrolled in published trials on exercise for smoking cessation has contributed to this limitation in scientific knowledge. Indeed, nearly all published work on exercise as an aid to promote smoking cessation has been conducted with predominantly Non-Hispanic White samples (Ussher et al. 2019). Of note, one recently completed small-scale pilot trial for an exercise intervention to improve smoking cessation specifically targeted Black women (Williams et al. 2021). Although the data did not support differences in cessation between the exercise and a control group, the exercise intervention was deemed feasible and acceptable and resulted in high retention. Yet, a notable limitation was that exercise adherence was low, suggesting the need for procedural refinement to engage this group actively. Given the established, significant smoking-related health disparities that exist across racial and ethnic minority groups (Sakuma et al. 2021), targeted efforts are needed to examine the potential benefit, generalizability of findings, and protocol regimen for an exercise program to assist with smoking cessation among racial and ethnic minority groups.

Persons living with HIV (PLWH) are another unique group of smokers who may benefit from exercise-based smoking cessation treatment. Smoking rates are significantly higher among PLWH relative to the general population (33.6% vs. 12.5%; (Frazier et al. 2018; Cornelius et al. 2023). HIV infection also increases the susceptibility to the harmful effects of smoking (Calvo et al. 2015). For example, compared to uninfected smokers, PLWH who smoke are more likely to develop respiratory complications and disease than smokers without HIV (Reddy et al. 2017; Depp et al. 2016). Further, emerging evidence suggests that smoking contributes to more years of life lost than HIV-related illnesses among HIV-infected smokers when HIV is properly managed (Reddy et al. 2016). Notably, previous work has highlighted the benefits of exercise among PLWH (Jaggers and Hand 2016; Bopp et al. 2003; Heissel et al. 2019; O'Brien et al. 2010). Specifically, exercise helps with HIV and ART side effect symptom management, improves mental health, and leads to increased quality of life among PLWH. Yet, there has been limited work on exercise-based smoking cessation treatments among PLWH (Ledgerwood and Yskes 2016). Further work is encouraged to evaluate such smoking interventions among this group.

Electronic cigarette use (i.e., vaping) has quickly become a popular alternative to combustible cigarette smoking (Dutra et al. 2017). Electronic cigarette use has been associated with higher smoking cessation rates (Tackett et al. 2015; Rahman et al. 2015). However, recent studies have shown that vaping may cause similar detrimental effects to the lung and cardiovascular system as combustible cigarettes (Darabseh et al. 2020, 2021; Li et al. 2020). A recent meta-analysis by Darabseh et al. (2022) on the effects of aerobic exercise on vaping and smoking cessation found that there have been no clinical trials conducted on aerobic exercise and vaping cessation. Consequently, the effect of exercise on vaping cessation is relatively unknown. Vaping and smoking cessation have barrier overlaps, such as stress reduction as a reason for use (Sanchez et al. 2021). As such, exercise-based smoking cessation interventions targeted toward stress reduction may prove fruitful for vaping cessation (Abrantes et al. 2018; Smits et al. 2021). However, with vaping cessation, there are new barriers to consider, such as the enjoyment of flavors (i.e., e-liquids), convenience and discreetness, and lack of self-awareness regarding vaping behavior (Sanchez et al. 2021). Barriers such as e-liquid flavors may be targeted directly with exercise, as previous studies have shown that human taste preferences change after exercise (Horio and Kawamura 1998). Additionally, Helen et al. (2018) found that vaping behavior changes between different e-liquids, such as the amount of puffs and puff duration, can affect nicotine intake. Vaping and smoking consist of many overlapping components and barriers. Future studies should explore the implications of exercise-based treatments for vaping cessation.

Evidence is relatively robust for the end-of-treatment and short-term benefits of an exercise intervention on smoking cessation outcomes; however, the long-term effects of exercise on smoking cessation are limited. Indeed, data indicate that these benefits fade without continued intervention. Future research may consider the potential of 'booster sessions' related to the exercise intervention to promote sustained effects. Booster sessions have been shown to help maintain both smoking abstinence and exercise following treatment (Metz et al. 2007; Fleig et al. 2013). Yet, this work has not been applied to an exercise-based smoking cessation program. As continued efforts to develop a programmatic method for prescriptive exercise interventions for smoking cessation continue to be refined, research is needed to understand the clinical utility of booster sessions focused on continued exercise and smoking cessation to promote, maintain, or re-initiate cessation.

7 Summary

This chapter reviewed the extensive literature on smoking cessation and exercise, discussing different exercise interventions, mechanisms of exercise interventions, special populations of exercise interventions, and discusses future directions of exercise and smoking cessation. A vast literature showcases the inverse relationship between smoking and physical activity and that engagement in exercise can lead to long-lasting health improvements irrespective of successful cessation. Although smoking is a serious health concern, smoking cessation interventions have shown little success in prolonged abstinence rates, indicating a need for more potent and lasting strategies. Exercise smoking cessation interventions have been shown to be efficacious for short-term abstinence, indicating a need for more potent and lasting exercise strategies.

References

Abrams DB, Niaura R (2003) The tobacco dependence treatment handbook: a guide to best practices. Guilford Press

Abrantes AM, Strong DR, Lloyd-Richardson EE, Niaura R, Kahler CW, Brown RA (2009) Regular exercise as a protective factor in relapse following smoking cessation treatment. Am J Addict 18(1):100–101

Abrantes AM, Bloom EL, Strong DR, Riebe D, Marcus BH, Desaulniers J et al (2014) A preliminary randomized controlled trial of a behavioral exercise intervention for smoking cessation. Nicotine Tob Res 16(8):1094–1103
Abrantes AM, Farris SG, Minami H, Strong DR, Riebe D, Brown RA (2018) Acute effects of aerobic exercise on affect and smoking craving in the weeks before and after a cessation attempt. Nicotine Tob Res 20(5):575–582
Alberg AJ, Shopland DR, Cummings KM (2014) The 2014 surgeon general's report: commemorating the 50th anniversary of the 1964 Report of the Advisory Committee to the US Surgeon General and updating the evidence on the health consequences of cigarette smoking. Am J Epidemiol 179(4):403–412
Allen SS, Bade T, Hatsukami D, Center B (2008) Craving, withdrawal, and smoking urges on days immediately prior to smoking relapse. Nicotine Tob Res 10(1):35–45
Allen AM, Abdelwahab NM, Carlson S, Bosch TA, Eberly LE, Okuyemi K (2018) Effect of brief exercise on urges to smoke in men and women smokers. Addict Behav 77:34–37
Auer R, Vittinghoff E, Kiefe C, Reis JP, Rodondi N, Khodneva YA et al (2014) Change in physical activity after smoking cessation: the C oronary A rtery R isk D evelopment in Y oung A dults (CARDIA) study. Addiction 109(7):1172–1183
Bell B, Dooley M (2006) Exercise in pregnancy. Royal College of Obstetricians and Gynaecologists
Benowitz NL, Burbank AD (2016) Cardiovascular toxicity of nicotine: implications for electronic cigarette use. Trends Cardiovasc Med 26(6):515–523. Epub 2016/04/16.eng
Berg CJ, Park ER, Chang Y, Rigotti NA (2008) Is concern about post-cessation weight gain a barrier to smoking cessation among pregnant women? Nicotine Tob Res 10(7):1159–1163
Bernard P, Ninot G, Cyprien F, Courtet P, Guillaume S, Georgescu V et al (2015) Exercise and counseling for smoking cessation in smokers with depressive symptoms: a randomized controlled pilot trial. J Dual Diagn 11(3–4):205–216
Bize R, Willi C, Chiolero A, Stoianov R, Payot S, Locatelli I et al (2010) Participation in a population-based physical activity programme as an aid for smoking cessation: a randomised trial. Tob Control 19(6):488–494
Bock BC, Marcus BH, King TK, Borrelli B, Roberts MR (1999) Exercise effects on withdrawal and mood among women attempting smoking cessation. Addict Behav 24(3):399–410
Bock BC, Fava JL, Gaskins R, Morrow KM, Williams DM, Jennings E et al (2012) Yoga as a complementary treatment for smoking cessation in women. J Womens Health 21(2):240–248
Bock BC, Dunsiger SI, Rosen RK, Thind H, Jennings E, Fava JL et al (2019) Yoga as a complementary therapy for smoking cessation: results from BreathEasy, a randomized clinical trial. Nicotine Tob Res 21(11):1517–1523
Boden JM, Fergusson DM, Horwood LJ (2010) Cigarette smoking and depression: tests of causal linkages using a longitudinal birth cohort. Br J Psychiatry 196(6):440–446
Bonnie RJ, Stratton K, Kwan L (2015) Committee on the Public Health Implications of Raising the Minimum Age for Purchasing Tobacco Products. Public health implications of raising the minimum age of legal access to tobacco products Institute of Medicine (US). National Academies Press, Washington, pp 1–399
Bopp C, Phillips KD, Fulk LJ, Hand GA (2003) Clinical implications of therapeutic exercise in HIV/AIDS. J Assoc Nurses AIDS Care 73
Boutelle KN, Murray DM, Jeffery RW, Hennrikus DJ, Lando HA (2000) Associations between exercise and health behaviors in a community sample of working adults. Prev Med 30(3): 217–224
Brnstrm R, Penilla C, Prez-Stable EJ, Muoz RF (2010) Positive affect and mood management in successful smoking cessation. Am J Health Behav 34(5):553–562
Burgess ES, Brown RA, Kahler CW, Niaura R, Abrams DB, Goldstein MG et al (2002) Patterns of change in depressive symptoms during smoking cessation: who's at risk for relapse? J Consult Clin Psychol 70(2):356

Caan B, Coates A, Schaefer C, Finkler L, Sternfeld B, Corbett K (1996) Women gain weight 1 year after smoking cessation while dietary intake temporarily increases. J Am Diet Assoc 96(11): 1150–1155

Calvo M, Laguno M, Martínez M, Martínez E (2015) Effects of tobacco smoking on HIV-infected individuals. AIDS Rev 17(1)

Centers for Disease Control Prevention (2020) Smoking cessation-the role of healthcare professionals and health systems. Smoking and Tobacco use

Chaney SE, Sheriff S (2008) Weight gain among women during smoking cessation: testing the effects of a multifaceted program. AAOHN J 56(3):99–105

Chen H, Yang Y, Miyai H, Yi C, Oliver BG (2022) The effects of exercise with nicotine replacement therapy for smoking cessation in adults: a systematic review. Front Psych 13: 1053937

Ciccolo JT, Dunsiger SI, Williams DM, Bartholomew JB, Jennings EG, Ussher MH et al (2011) Resistance training as an aid to standard smoking cessation treatment: A pilot study. Nicotine Tob Res 13(8):756–760

Conway TL, Cronan TA (1992) Smoking, exercise, and physical fitness. Prev Med 21(6):723–734. Epub 1992/11/01.eng

Cook JW, Spring B, McChargue DE, Borrelli B, Hitsman B, Niaura R et al (2004) Influence of fluoxetine on positive and negative affect in a clinic-based smoking cessation trial. Psychopharmacology (Berl) 173:153–159

Copeland AL, Martin PD, Geiselman PJ, Rash CJ, Kendzor DE (2006) Smoking cessation for weight-concerned women: group vs. individually tailored, dietary, and weight-control follow-up sessions. Addict Behav 31(1):115–127

Cornelius ME, Loretan CG, Jamal A, Lynn BCD, Mayer M, Alcantara IC et al (2023) Tobacco product use among adults–United States, 2021. Morb Mortal Wkly Rep 72(18):475

Covinsky KE, Cenzer IS, Yaffe K, O'Brien S, Blazer DG (2014) Dysphoria and anhedonia as risk factors for disability or death in older persons: implications for the assessment of geriatric depression. Am J Geriatr Psychiatry 22(6):606–613

Creamer MR, Wang TW, Babb S, Cullen KA, Day H, Willis G et al (2019) Tobacco product use and cessation indicators among adults – United States, 2018. Morb Mortal Wkly Rep 68(45): 1013

Daniel J, Cropley M, Ussher M, West R (2004) Acute effects of a short bout of moderate versus light intensity exercise versus inactivity on tobacco withdrawal symptoms in sedentary smokers. Psychopharmacology (Berl) 174:320–326

Darabseh MZ, Selfe J, Morse CI, Degens H (2020) Is vaping better than smoking for cardiorespiratory and muscle function? Multidiscip Respir Med 15(1)

Darabseh MZ, Selfe J, Morse CI, Degens H (2021) Impact of vaping and smoking on maximum respiratory pressures and respiratory function. Int J Adolesc Youth 26(1):421–431

Darabseh MZ, Selfe J, Morse CI, Aburub A, Degens H (2022) Does aerobic exercise facilitate vaping and smoking cessation: a systematic review of randomized controlled trials with meta-analysis. Int J Environ Res Public Health 19(21):14034

De Jesus S, Prapavessis H (2018) Smoking behaviour and sensations during the pre-quit period of an exercise-aided smoking cessation intervention. Addict Behav 81:143–149

De Meirleir K, Naaktgeboren N, Van Steirteghem A, Gorus F, Olbrecht J, Block P (1986) Beta-endorphin and ACTH levels in peripheral blood during and after aerobic and anaerobic exercise. Eur J Appl Physiol Occup Physiol 55:5–8

Depp TB, McGinnis KA, Kraemer K, Akgün KM, Edelman EJ, Fiellin DA et al (2016) Risk factors associated with acute exacerbation of chronic obstructive pulmonary disease in HIV-infected and uninfected patients. AIDS (London, England) 30(3):455

Derby CA, Lasater TM, Vass K, Gonzalez S, Carleton RA (1994) Characteristics of smokers who attempt to quit and of those who recently succeeded. Am J Prev Med 10(6):327–334

DeRuiter WK, Faulkner G, Cairney J, Veldhuizen S (2008) Characteristics of physically active smokers and implications for harm reduction. Am J Public Health 98(5):925–931

Dutra LM, Grana R, Glantz SA (2017) Philip Morris research on precursors to the modern e-cigarette since 1990. Tob Control 26(e2):e97–e105
Emery RL, Levine MD, Cheng Y, Marcus MD (2014) Change in body weight does not mediate the relationship between exercise and smoking cessation among weight-concerned women smokers. Nicotine Tob Res 17(9):1142–1148
Everson ES, Taylor AH, Ussher M (2010) Determinants of physical activity promotion by smoking cessation advisors as an aid for quitting: support for the transtheoretical model. Patient Educ Couns 78(1):53–56
Everson-Hock ES, Taylor AH, Ussher M (2010) Readiness to use physical activity as a smoking cessation aid: a multiple behaviour change application of the transtheoretical model among quitters attending stop smoking clinics. Patient Educ Couns 79(2):156–159
Ferrucci L, Izmirlian G, Leveille S, Phillips CL, Corti M-C, Brock DB et al (1999) Smoking, physical activity, and active life expectancy. Am J Epidemiol 149(7):645–653
Fingerhut LA, Kleinman JC, Kendrick JS (1990) Smoking before, during, and after pregnancy. Am J Public Health 80(5):541–544
Fiore MC, Jaen CR, Baker TB, Bailey WC, Benowitz NL, Curry SJ et al (2008) Treating tobacco use and dependence: 2008 update US Public Health Service Clinical Practice Guideline executive summary. Respir Care 53(9):1217–1222
Fleig L, Pomp S, Schwarzer R, Lippke S (2013) Promoting exercise maintenance: how interventions with booster sessions improve long-term rehabilitation outcomes. Rehabil Psychol 58(4): 323
Frazier EL, Sutton MY, Brooks JT, Shouse RL, Weiser J (2018) Trends in cigarette smoking among adults with HIV compared with the general adult population, United States-2009–2014. Prev Med 111:231–234
Frith E, Loprinzi PD (2018) Exercise facilitates smoking cessation indirectly via intention to quit smoking: prospective cohort study among a national sample of young smokers. Am J Health Promot 32(5):1234–1238
Furlanetto KC, Mantoani LC, Bisca G, Morita AA, Zabatiero J, Proença M et al (2014) Reduction of physical activity in daily life and its determinants in smokers without airflow obstruction. Respirology 19(3):369–375
García-Gómez L, Hernández-Pérez A, Noé-Díaz V, Riesco-Miranda JA, Jiménez-Ruiz C (2019) Smoking cessation treatments: current psychological and pharmacological options. Rev Invest Clin 71(1):7–16
Gard T, Brach N, Hölzel BK, Noggle JJ, Conboy LA, Lazar SW (2012) Effects of a yoga-based intervention for young adults on quality of life and perceived stress: the potential mediating roles of mindfulness and self-compassion. J Posit Psychol 7(3):165–175
Gauthier AP, Snelling SJ, King M (2012) "Thinking Outside the Pack" examining physically active smokers and implications for practice among Ontario residents. Health Promot Pract 13(3): 395–403
Goldfarb AH, Hatfield B, Armstrong D, Potts J (1990) Plasma beta-endorphin concentration: response to intensity and duration of exercise. Med Sci Sports Exerc 22(2):241–244
Haasova M, Warren FC, Ussher M, Janse Van Rensburg K, Faulkner G, Cropley M et al (2013) The acute effects of physical activity on cigarette cravings: systematic review and meta-analysis with individual participant data. Addiction 108(1):26–37
Haasova M, Warren FC, Ussher M, Janse Van Rensburg K, Faulkner G, Cropley M et al (2014) The acute effects of physical activity on cigarette cravings: exploration of potential moderators, mediators and physical activity attributes using individual participant data (IPD) meta-analyses. Psychopharmacology (Berl) 231:1267–1275
Haddock CK, O'Byrne KK, Klesges RC, Talcott GW, Lando H, Peterson AL (2000) Relapse to smoking after basic military training in the US air force. Mil Med 165(11):884–888
Hassandra M, Lintunen T, Hagger MS, Heikkinen R, Vanhala M, Kettunen T (2017) An mHealth app for supporting quitters to manage cigarette cravings with short bouts of physical activity: a randomized pilot feasibility and acceptability study. JMIR Mhealth Uhealth 5(5):e6252

Health UDo, Services H (2004) Bone health and osteoporosis: a report of the surgeon general. US Department of Health and Human Services, Office of the Surgeon General, Rockville, p 87
Heissel A, Zech P, Rapp MA, Schuch FB, Lawrence JB, Kangas M et al (2019) Effects of exercise on depression and anxiety in persons living with HIV: A meta-analysis. J Psychosom Res 126: 109823
Helen GS, Shahid M, Chu S, Benowitz NL (2018) Impact of e-liquid flavors on e-cigarette vaping behavior. Drug Alcohol Depend 189:42–48
Hill JS (1985) Effect of a program of aerobic exercise on the smoking behaviour of a group of adult volunteers. Can J Public Health 76(3):183–186
Horio T, Kawamura Y (1998) Influence of physical exercise on human preferences for various taste solutions. Chem Senses 23(4):417–421
Iyengar BKS (2007) BKS Iyengar yoga: the path to holistic health: penguin
Jaggers JR, Hand GA (2016) Health benefits of exercise for people living with HIV: a review of the literature. Am J Lifestyle Med 10(3):184–192
Jamal A, Phillips E, Gentzke AS, Homa DM, Babb SD, King BA et al (2018) Current cigarette smoking among adults – United States, 2016. Morb Mortal Wkly Rep 67(2):53
Jung ME, Fitzgeorge L, Prapavessis H, Faulkner G, Maddison R (2010) The getting physical on cigarettes trial: rationale and methods. Ment Health Phys Act 3(1):35–44
Kaczynski AT, Manske SR, Mannell RC, Grewal K (2008) Smoking and physical activity: a systematic review. Am J Health Behav 32(1):93–110
Kahler CW, Brown RA, Ramsey SE, Niaura R, Abrams DB, Goldstein MG et al (2002) Negative mood, depressive symptoms, and major depression after smoking cessation treatment in smokers with a history of major depressive disorder. J Abnorm Psychol 111(4):670
Kang E-J, Lee J-H (2010) A longitudinal study on the causal association between smoking and depression. J Prev Med Public Health 43(3):193–204
Kazdin AE (2007) Mediators and mechanisms of change in psychotherapy research. Annu Rev Clin Psychol 3:1–27
Kenford SL, Fiore MC, Jorenby DE, Smith SS, Wetter D, Baker TB (1994) Predicting smoking cessation: who will quit with and without the nicotine patch. JAMA 271(8):589–594
Killen JD, Fortmann SP (1997) Craving is associated with smoking relapse: findings from three prospective studies. Exp Clin Psychopharmacol 5(2):137
King TK, Marcus BH, Pinto BM, Emmons KM, Abrams DB (1996) Cognitive–behavioral mediators of changing multiple behaviors: smoking and a sedentary lifestyle. Prev Med 25(6): 684–691
Kinnunen T, Leeman RF, Korhonen T, Quiles ZN, Terwal DM, Garvey AJ et al (2008) Exercise as an adjunct to nicotine gum in treating tobacco dependence among women. Nicotine Tob Res 10(4):689–703
Klinsophon T, Thaveeratitham P, Sitthipornvorakul E, Janwantanakul P (2017) Effect of exercise type on smoking cessation: a meta-analysis of randomized controlled trials. BMC Res Notes 10(1):1–21
Kovacs M, Yaroslavsky I (2014) Practitioner review: dysphoria and its regulation in child and adolescent depression. J Child Psychol Psychiatry 55(7):741–757
Kunicki ZJ, Hallgren M, Uebelacker LA, Brown RA, Price LH, Abrantes AM (2022) Examining the effect of exercise on the relationship between affect and cravings among smokers engaged in cessation treatment. Addict Behav 125:107156
Kuo C-W, Lin C-F, Chen C-Y, Wang R-H, Chou C-Y, Cheng H-J et al (2022) Body-weight gain in women during smoking cessation is a sex-specific predictor of 6-month abstinence: A retrospective cohort study. Front Public Health 10:872220
Lasser K, Boyd JW, Woolhandler S, Himmelstein DU, McCormick D, Bor DH (2000) Smoking and mental illness: a population-based prevalence study. JAMA 284(20):2606–2610
Lê Cook B, Wayne GF, Kafali EN, Liu Z, Shu C, Flores M (2014) Trends in smoking among adults with mental illness and association between mental health treatment and smoking cessation. JAMA 311(2):172–182

Ledgerwood DM, Yskes R (2016) Smoking cessation for people living with HIV/AIDS: a literature review and synthesis. Nicotine Tob Res 18(12):2177–2184

Lerman C, Roth D, Kaufmann V, Audrain J, Hawk L, Liu A et al (2002) Mediating mechanisms for the impact of bupropion in smoking cessation treatment. Drug Alcohol Depend 67(2):219–223

Levine MD, Marcus MD, Kalarchian MA, Houck PR, Cheng Y (2010) Weight concerns, mood, and postpartum smoking relapse. Am J Prev Med 39(4):345–351

Li AW, Goldsmith C-AW (2012) The effects of yoga on anxiety and stress. Altern Med Rev 17(1)

Li D, Sundar IK, McIntosh S, Ossip DJ, Goniewicz ML, O'Connor RJ et al (2020) Association of smoking and electronic cigarette use with wheezing and related respiratory symptoms in adults: cross-sectional results from the Population Assessment of Tobacco and Health (PATH) study, wave 2. Tob Control 29(2):140–147

Linke SE, Ciccolo JT, Ussher M, Marcus BH (2013) Exercise-based smoking cessation interventions among women. Womens Health 9(1):69–84

Loprinzi PD, Walker JF (2016) Association of longitudinal changes of physical activity on smoking cessation among young daily smokers. J Phys Act Health 13(1):1–5

Lynch WJ, Peterson AB, Sanchez V, Abel J, Smith MA (2013) Exercise as a novel treatment for drug addiction: a neurobiological and stage-dependent hypothesis. Neurosci Biobehav Rev 37(8):1622–1644

Maddison R, Roberts V, McRobbie H, Bullen C, Prapavessis H, Glover M et al (2014) Exercise counseling to enhance smoking cessation outcomes: the Fit2Quit randomized controlled trial. Ann Behav Med 48(2):194–204

Marcus BH, Albrecht AE, Niaura RS, Abrams DB, Thompson PD (1991) Usefulness of physical exercise for maintaining smoking cessation in women. Am J Cardiol 68(4):406–407

Marcus BH, Albrecht AE, Niaura RS, Taylor ER, Simkin LR, Feder SI et al (1995) Exercise enhances the maintenance of smoking cessation in women. Addict Behav 20(1):87–92

Marcus BH, Albrecht AE, King TK, Parisi AF, Pinto BM, Roberts M et al (1999) The efficacy of exercise as an aid for smoking cessation in women: a randomized controlled trial. Arch Intern Med 159(11):1229–1234

Marcus BH, Lewis BA, Hogan J, King TK, Albrecht AE, Bock B et al (2005) The efficacy of moderate-intensity exercise as an aid for smoking cessation in women: a randomized controlled trial. Nicotine Tob Res 7(6):871–880

Martin JE, Calfas KJ, Patten CA, Polarek M, Hofstetter CR, Noto J et al (1997) Prospective evaluation of three smoking interventions in 205 recovering alcoholics: one-year results of project SCRAP-tobacco. J Consult Clin Psychol 65(1):190

McAllister RM, Delp MD, Laughlin MH (1995) Thyroid status and exercise tolerance: cardiovascular and metabolic considerations. Sports Med 20:189–198

McClernon FJ, Westman EC, Rose JE (2004) The effects of controlled deep breathing on smoking withdrawal symptoms in dependent smokers. Addict Behav 29(4):765–772

Metz K, Flöter S, Kröger C, Donath C, Piontek D, Gradl S (2007) Telephone booster sessions for optimizing smoking cessation for patients in rehabilitation centers. Nicotine Tob Res 9(8): 853–863

Mohamed Abdelaal AA, Mohamed Mousa GS (2022) Impact of cigarettes smoking on undergraduates' lung health and functional performance: observational cross-sectional study. Int J Prev Med 13:59. Epub 2022/06/17.eng

Ng DM, Jeffery RW (2003) Relationships between perceived stress and health behaviors in a sample of working adults. Health Psychol 22(6):638

Ni Y, Li L, Qu Y, Chen S, Zhang H, Wang H (2022) Advances in the study of neural mechanisms associated with exercise to ameliorate nicotine addiction. Adv Physiol Educ 12(2):126–141

O'Brien K, Nixon S, Tynan AM, Glazier R (2010) Aerobic exercise interventions for adults living with HIV/AIDS. Cochrane Database Syst Rev (8)

Obstetricians ACo, Gynecologists (2003) Exercise during pregnancy and the postpartum period. Clin Obstet Gynecol 46(2):496–499

O'Hara P, Connett JE, Lee WW, Nides M, Murray R, Wise R (1998) Early and late weight gain following smoking cessation in the lung health study. Am J Epidemiol 148(9):821–830

Oncken C, Allen S, Litt M, Kenny A, Lando H, Allen A et al (2020) Exercise for smoking cessation in postmenopausal women: a randomized, controlled trial. Nicotine Tob Res 22(9):1587–1595

Paavola M, Vartiainen E, Puska P (2001) Smoking cessation between teenage years and adulthood. Health Educ Res 16(1):49–57

Patten CA, Bronars CA, Vickers Douglas KS, Ussher MH, Levine JA, Tye SJ et al (2016) Supervised, vigorous intensity exercise intervention for depressed female smokers: a pilot study. Nicotine Tob Res 19(1):77–86

Penedo FJ, Dahn JR (2005) Exercise and well-being: a review of mental and physical health benefits associated with physical activity. Curr Opin Psychiatry 18(2):189–193

Piasecki TM, Niaura R, Shadel WG, Abrams D, Goldstein M, Fiore MC et al (2000) Smoking withdrawal dynamics in unaided quitters. J Abnorm Psychol 109(1):74

Picavet HSJ, Wendel-Vos GW, Vreeken HL, Schuit AJ, Verschuren WMM (2011) How stable are physical activity habits among adults? The Doetinchem Cohort Study. Med Sci Sports Exerc 43(1):74–79

Piper ME, Smith SS, Schlam TR, Fleming MF, Bittrich AA, Brown JL et al (2010) Psychiatric disorders in smokers seeking treatment for tobacco dependence: relations with tobacco dependence and cessation. J Consult Clin Psychol 78(1):13

Pisinger C, Jorgensen T (2007) Waist circumference and weight following smoking cessation in a general population: The Inter99 study. Prev Med 44(4):290–295

Pomerleau CS, Brouwer RJN, Jones LT (2000) Weight concerns in women smokers during pregnancy and postpartum. Addict Behav 25(5):759–767

Prapavessis H, Cameron L, Baldi JC, Robinson S, Borrie K, Harper T et al (2007) The effects of exercise and nicotine replacement therapy on smoking rates in women. Addict Behav 32(7): 1416–1432

Prapavessis H, De Jesus S, Fitzgeorge L, Faulkner G, Maddison R, Batten S (2016) Exercise to enhance smoking cessation: the getting physical on cigarette randomized control trial. Ann Behav Med 50(3):358–369

Rahman MA, Hann N, Wilson A, Mnatzaganian G, Worrall-Carter L (2015) E-cigarettes and smoking cessation: evidence from a systematic review and meta-analysis. PLoS One 10(3): e0122544

Reddy KP, Parker RA, Losina E, Baggett TP, Paltiel AD, Rigotti NA et al (2016) Impact of cigarette smoking and smoking cessation on life expectancy among people with HIV: a US-based modeling study. J Infect Dis 214(11):1672–1681

Reddy KP, Kong CY, Hyle EP, Baggett TP, Huang M, Parker RA et al (2017) Lung cancer mortality associated with smoking and smoking cessation among people living with HIV in the United States. JAMA Intern Med 177(11):1613–1621

Roberts V, Maddison R, Simpson C, Bullen C, Prapavessis H (2012) The acute effects of exercise on cigarette cravings, withdrawal symptoms, affect, and smoking behaviour: systematic review update and meta-analysis. Psychopharmacology (Berl) 222:1–15

Ross RE, VanDerwerker CJ, Saladin ME, Gregory CM (2023) The role of exercise in the treatment of depression: biological underpinnings and clinical outcomes. Mol Psychiatry 28(1):298–328

Ruslan N-H, Yasin SM, Mohd Nasir N, Isa MR (2023) The impacts of supervised exercise intervention on tobacco withdrawal symptoms. Tobacco Use Insights 16:1179173X231179811

Russell PO, Epstein LH, Johnston JJ, Block DR, Blair E (1988) The effects of physical activity as maintenance for smoking cessation. Addict Behav 13(2):215–218

Saarni S, Silventoinen K, Rissanen A, Sarlio-Lähteenkorva S, Kaprio J (2004) Intentional weight loss and smoking in young adults. Int J Obes (Lond) 28(6):796–802

Sakuma K-LK, Pierce JP, Fagan P, Nguyen-Grozavu FT, Leas EC, Messer K et al (2021) Racial/ethnic disparities across indicators of cigarette smoking in the era of increased tobacco control, 1992–2019. Nicotine Tob Res 23(6):909–919

Sanchez S, Kaufman P, Pelletier H, Baskerville B, Feng P, O'Connor S et al (2021) Is vaping cessation like smoking cessation? A qualitative study exploring the responses of youth and young adults who vape e-cigarettes. Addict Behav 113:106687

Santos CP, Proença M, dos Santos GT, de Oliveira CBS, Tacao GY, Trevisan IB et al (2021) Effectiveness of aerobic exercise on smoking cessation in adults: a systematic review and meta-analysis. J Phys Act Health 18(2):230–242

Schöttl SE, Niedermeier M, Kopp-Wilfling P, Frühauf A, Bichler CS, Edlinger M et al (2022) Add-on exercise interventions for smoking cessation in people with mental illness: a systematic review and meta-analysis. BMC Sports Sci Med Rehabil 14(1):1–16

Sedgwick A, Davidsond A, Taplin R, Thomas D (1988) Effects of physical activity on risk factors for coronary heart disease in previously sedentary women: a five-year longitudinal study. Aust N Z J Med 18(4):600–605

Shahab L, Sarkar BK, West R (2013) The acute effects of yogic breathing exercises on craving and withdrawal symptoms in abstaining smokers. Psychopharmacology (Berl) 225:875–882

Shaw D, al'Absi M (2008) Attenuated beta endorphin response to acute stress is associated with smoking relapse. Pharmacol Biochem Behav 90(3):357–362

Simons JS, Dvorak RD, Batien BD, Wray TB (2010) Event-level associations between affect, alcohol intoxication, and acute dependence symptoms: effects of urgency, self-control, and drinking experience. Addict Behav 35(12):1045–1053

Smith PJ, Merwin RM (2021) The role of exercise in management of mental health disorders: an integrative review. Annu Rev Med 72:45–62

Smith PH, Bessette AJ, Weinberger AH, Sheffer CE, McKee SA (2016) Sex/gender differences in smoking cessation: a review. Prev Med 92:135–140

Smits JA, Tart CD, Rosenfield D, Zvolensky MJ (2011) The interplay between physical activity and anxiety sensitivity in fearful responding to CO2 challenge. Psychosom Med 73(6):498

Smits JA, Zvolensky MJ, Davis ML, Rosenfield D, Marcus BH, Church TS et al (2016) The efficacy of vigorous-intensity exercise as an aid to smoking cessation in adults with high anxiety sensitivity: a randomized controlled trial. Psychosom Med 78(3):354

Smits JA, Zvolensky MJ, Rosenfield D, Brown RA, Otto MW, Dutcher CD et al (2021) Community-based smoking cessation treatment for adults with high anxiety sensitivity: A randomized clinical trial. Addiction 116(11):3188–3197

Stead LF, Koilpillai P, Fanshawe TR, Lancaster T (2016) Combined pharmacotherapy and behavioural interventions for smoking cessation. Cochrane Database Syst Rev:3

Stockton MB, Ward KD, McClanahan BS, Vander Weg MW, Coday M, Wilson N et al (2023) The efficacy of individualized, community-based physical activity to aid smoking cessation: a randomized controlled trial. J Smok Cessat:2023

Tackett AP, Lechner WV, Meier E, Grant DM, Driskill LM, Tahirkheli NN et al (2015) Biochemically verified smoking cessation and vaping beliefs among vape store customers. Addiction 110(5):868–874

Tantisuwat A, Thaveeratitham P (2014) Effects of smoking on chest expansion, lung function, and respiratory muscle strength of youths. J Phys Ther Sci 26(2):167–170. Epub 2014/03/22.eng

Taylor A, Katomeri M (2007) Walking reduces cue-elicited cigarette cravings and withdrawal symptoms, and delays ad libitum smoking. Nicotine Tob Res 9(11):1183–1190

Taylor C, Houston-Miller N, Haskell WL, Debusk RF (1988) Smoking cessation after acute myocardial infarction: the effects of exercise training. Addict Behav 13(4):331–335

Taylor AH, Katomeri M, Ussher M (2005) Acute effects of self-paced walking on urges to smoke during temporary smoking abstinence. Psychopharmacology (Berl) 181:1–7

Taylor AH, Ussher MH, Faulkner G (2007) The acute effects of exercise on cigarette cravings, withdrawal symptoms, affect and smoking behaviour: a systematic review. Addiction 102(4): 534–543

Thompson TP, Greaves CJ, Ayres R, Aveyard P, Warren FC, Byng R et al (2015) An exploratory analysis of the smoking and physical activity outcomes from a pilot randomized controlled trial of an exercise assisted reduction to stop smoking intervention in disadvantaged groups. Nicotine Tob Res 18(3):289–297

US Department of Health Human Services (2010) How tobacco smoke causes disease: the biology and behavioral basis for smoking-attributable disease: a report of the surgeon general. US Department of Health and Human Services, Centers for Disease . . ., Atlanta

Ussher M, West R (2003) Interest in nicotine replacement therapy among pregnant smokers. Tob Control 12(1):108–109

Ussher M, Nunziata P, Cropley M, West R (2001) Effect of a short bout of exercise on tobacco withdrawal symptoms and desire to smoke. Psychopharmacology (Berl) 158:66–72

Ussher M, West R, McEwen A, Taylor A, Steptoe A (2003) Efficacy of exercise counselling as an aid for smoking cessation: a randomized controlled trial. Addiction 98(4):523–532

Ussher M, West R, Hibbs N (2004) A survey of pregnant smokers' interest in different types of smoking cessation support. Patient Educ Couns 54(1):67–72

Ussher M, West R, Doshi R, Sampuran AK (2006) Acute effect of isometric exercise on desire to smoke and tobacco withdrawal symptoms. Hum Psychopharmacol Clin Exp 21(1):39–46

Ussher M, Aveyard P, Coleman T, Straus L, West R, Marcus B et al (2008) Physical activity as an aid to smoking cessation during pregnancy: two feasibility studies. BMC Public Health 8(1):1–9

Ussher M, Aveyard P, Manyonda I, Lewis S, West R, Lewis B et al (2012) Physical activity as an aid to smoking cessation during pregnancy (LEAP) trial: study protocol for a randomized controlled trial. Trials 13(1):1–12

Ussher M, Lewis S, Aveyard P, Manyonda I, West R, Lewis B et al (2015) Physical activity for smoking cessation in pregnancy: randomised controlled trial. BMJ:350

Ussher MH, Faulkner GEJ, Angus K, Hartmann-Boyce J, Taylor AH (2019) Exercise interventions for smoking cessation. Cochrane Database Syst Rev 2019(10) Pubmed Central PMCID: PMC6819982 Patten 2017; Ussher 2003; Ussher 2015). AT: was involved with three of the included trials (Bernard 2015; Ussher 2003; Ussher 2015). GF: was involved with one of the included trials (Prapavessis 2016). JHB: none declared. KA: none declared. Epub 2019/11/07. eng

Vickers KS, Patten CA, Lewis BA, Clark MM, Ussher M, Ebbert JO et al (2009) Feasibility of an exercise counseling intervention for depressed women smokers. Nicotine Tob Res 11(8): 985–995

Whiteley JA, Napolitano MA, Lewis BA, Williams DM, Albrecht A, Neighbors CJ et al (2007) Commit to quit in the YMCAs: translating an evidence-based quit smoking program for women into a community setting. Nicotine Tob Res 9(11):1227–1235

Whiteley JA, Williams DM, Dunsiger S, Jennings EG, Ciccolo JT, Bock BC et al (2012) YMCA commit to quit: randomized trial outcomes. Am J Prev Med 43(3):256–262

Williams DM, Whiteley JA, Dunsiger S, Jennings EG, Albrecht AE, Ussher MH et al (2010) Moderate intensity exercise as an adjunct to standard smoking cessation treatment for women: a pilot study. Psychol Addict Behav 24(2):349

Williams CD, Taylor T, Stanton C, Makambi K, Hicks J, Adams-Campbell LL (2021) A feasibility study of smoking cessation utilizing an exercise intervention among black women: 'Quit and Fit'. J Natl Med Assoc 113(3):243–251

Williamson DF, Madans J, Anda RF, Kleinman JC, Giovino GA, Byers T (1991) Smoking cessation and severity of weight gain in a national cohort. N Engl J Med 324(11):739–745

World Health Organization (2022) World Health Organization, Tobacco

Xian H, Scherrer JF, Madden PA, Lyons MJ, Tsuang M, True WR et al (2003) The heritability of failed smoking cessation and nicotine withdrawal in twins who smoked and attempted to quit. Nicotine Tob Res 5(2):245–254

Zvolensky MJ, Rosenfield D, Garey L, Kauffman BY, Langdon KJ, Powers MB et al (2018) Does exercise aid smoking cessation through reductions in anxiety sensitivity and dysphoria? Health Psychol 37(7):647

Exercise and Anxiety

Kevin M. Crombie and Patrick J. O'Connor

Contents

Abstract Anxiety disorders are among the most commonly diagnosed mental health disorders in the United States – affecting over 40 million adults per year. Although anxiety disorders are commonly treated via psychotherapy and/or pharmacotherapy, there is also accumulating evidence to suggest that physical activity and exercise may play an important role in prevention and treatment. This chapter provides an extensive overview of literature examining the effects of physical activity, acute bouts of exercise, and chronic exercise training on several anxiety outcomes, primarily in adults. The collective evidence to date suggests that: (1) an acute bout of aerobic or resistance exercise generally results in reductions in state anxiety among adults with and without clinical anxiety disorders, (2) greater levels of physical activity are associated with fewer anxiety symptoms and a reduced

K. M. Crombie (✉)
Department of Kinesiology, The University of Alabama, Tuscaloosa, AL, USA

Department of Psychiatry and Behavioral Sciences, The University of Texas at Austin Dell Medical School, Austin, TX, USA
e-mail: kmcrombie@ua.edu

P. J. O'Connor
Department of Kinesiology, University of Georgia, Athens, GA, USA

Curr Topics Behav Neurosci (2024) 67: 199–222
https://doi.org/10.1007/7854_2024_498
Published Online: 31 July 2024

likelihood of developing an anxiety disorder, (3) exercise training (aerobic and resistance) appears to reduce general anxiety symptoms in adults without a clinical anxiety disorder and adults with a chronic illness (e.g., cardiovascular disease) and may reduce disorder-specific symptoms of anxiety in adults with a clinical anxiety disorder. Although the collective body of evidence is promising, there is a need for additional well-designed and adequately powered randomized controlled trials, especially among adults with clinical anxiety disorders.

Keywords State anxiety · Clinical anxiety disorders · Acute anxiolytic effects · Anxiety symptoms · Exercise training · Physical activity · Generalized anxiety disorder (GAD) · Panic disorder · Social anxiety disorder · Specific phobias · Obsessive compulsive disorder (OCD) · Posttraumatic stress disorder (PTSD)

1 Overview of Anxiety

Anxiety is a psychobiological state characterized by anticipatory cognitive, behavioral, and affective changes in response to uncertainty about a future event or potential threat (Grupe and Nitschke 2013). It is normal for individuals to experience occasional anxiety as they navigate life. This general transient fluctuation in anxiety is often referred to as state anxiety (i.e., how anxious one feels at a given moment). Some individuals are also more prone to appraise and respond to many situations with elevated levels of anxiety, which is often referred to as trait-anxiety (i.e., how anxious one generally feels) (Endler and Kocovski 2001; Spielberger 1966). In addition to state and trait anxiety, there are several types of anxiety disorders that individuals can develop when specific anxiety symptoms persist or worsen over a significant period of time and cause clinically significant impairment or distress (American Psychiatric Association 2013). Anxiety disorders all share some common affective and physiological features. However, each category of disordered anxiety differs based of the objects or situations that induce symptoms, and there is considerable variability in frequency and severity of symptoms between and within disorders (see Table 1).

Anxiety disorders are consistently among the most commonly diagnosed mental health disorders in the United States, affecting over 40 million adults per year (approximately 20% of US adult population). In fact, data from the National Comorbidity Survey Replication (NCS-R) study among American adults (N = 9,282) suggest that up to 30% of US adults meet criteria for one or more anxiety disorder(s) at some point during their lifetime (Blumenthal et al. 2021; Kessler et al. 2005; Stonerock et al. 2015). Prevalence rates for anxiety disorders are generally higher in women compared to men (McLean et al. 2011), with NCS-R data revealing past-year year prevalence rates of any anxiety disorder of 23.4% for women compared to 14.3% in men (Kessler et al. 2005). Anxiety disorders are most commonly treated with psychotherapy (e.g., cognitive behavioral therapy, CBT) or pharmacotherapies, including, but not limited to, selective serotonin reuptake

Table 1 General overview of prevalence, sex differences, primary characteristics, and basic diagnostic criteria for clinical anxiety disorders

Clinical anxiety disorders	Prevalence	Sex differences	Primary characteristics/basic diagnostic criteria
Generalized anxiety disorder	Affects 6.8 million adults (3.1% of US population)	Women are two times as likely to be affected as men	Excessive anxiety and worry that occurs more days than not for at least 6 months, about a number of events or activities (e.g., work or school performance) that is accompanied by at least three of the following symptoms: Restless or feeling keyed up or on edge, being easily fatigued, difficulty concentrating or mind going blank, irritability, muscle tension, and sleep disturbance (difficulty falling or staying asleep or restless unsatisfying sleep)
Panic disorder	Affects six million adults (2.7% of US population)	Women are two times as likely to be affected as men	Panic attacks involve abrupt surge of intense fear or intense discomfort that reaches a peak within minutes and during which time four or more of the following symptoms occur: accelerated heart rate, sweating, trembling or shaking, sensations of shortness of breath, feeling of choking, chest pain or discomfort, nausea or abdominal distress, feeling dizzy/lightheaded/faint, derealization or depersonalization, fear of losing control, fear of dying, paresthesia, or chills and hot flushes. Panic disorder involves recurrent and unexpected panic attacks, with at least one panic attack followed by 1 month or more of either persistent concern about additional attacks or their consequences and/or a significant maladaptive change in behavior related to the attacks
Social anxiety disorder	Affects 15 million adults (7.1% of US population)	No significant sex differences	Marked fear or anxiety about one or more social situations in which the individual is exposed to possible scrutiny by others (e.g., social interactions, being observed, and performing in front of others). The social situations almost always provoke fear or anxiety and are

(continued)

Table 1 (continued)

Clinical anxiety disorders	Prevalence	Sex differences	Primary characteristics/basic diagnostic criteria
			avoided or endured with intense fear or anxiety. The fear or anxiety is out of proportion to the actual threat posed by the social situation and to the sociocultural context. The fear, anxiety, or avoidance causes clinically significant distress or impairment in social, occupational, or other important areas of functioning for at least 6 months or more
Specific phobias	Affects 19.3 million adults (9.1% of US population)	Women are two times as likely to be affected as men	Marked fear or anxiety about a specific object or situation (e.g., flying, heights, animals, receiving an injection, seeing blood). The phobic object or situation almost always provokes immediate fear or anxiety and is actively avoided or endured with intense fear or anxiety. The fear or anxiety is out of proportion to the actual danger posed by the specific object or situation and to the sociocultural context. The fear, anxiety, or avoidance causes clinically significant distress or impairment in social, occupational, or other important areas of functioning and is persistent (usually lasting 6 months or more)
Obsessive compulsive disorder*	Affects 2.5 million adults (1.2% of the population)	Women are three times as likely to be affected as men	Presence of obsessions, compulsions, or both. The obsessions or compulsions are time consuming (e.g., take more than 1 h per day) or cause clinically significant distress or impairment in social, occupational, or other important areas of functioning. Obsessions refer to recurrent and persistent thoughts, urges, or images that are experienced, at some time during the disturbance, as intrusive and unwanted, and that in most individuals cause marked anxiety or distress. The individual attempts to ignore or suppress such thoughts, urges, or images or to neutralize them with some thought or action

(continued)

Table 1 (continued)

Clinical anxiety disorders	Prevalence	Sex differences	Primary characteristics/basic diagnostic criteria
			(i.e., by performing a compulsion). Compulsions refer to repetitive behaviors (e.g., hand washing, ordering checking) or mental acts (e.g., praying, counting, repeating words silently) that the person feels driven to perform in response to an obsession or according to the rules that must be applied rigidly. The behaviors or mental acts are aimed at preventing or reducing distress or preventing some dreaded event or situation. However, these behaviors or mental acts either are not connected in a realistic way with what they are designed to neutralize or prevent or are clearly excessive
Posttraumatic stress disorder*	Affects 7.7 million adults (3.6% of the population)	Women are five times more likely to be affected as men	The primary characteristics of PTSD include reexperiencing symptoms, avoidance of trauma-related stimuli, negative alterations in cognitions and mood, and alterations in arousal and reactivity – all of which must cause significant impairment for at least 1 month following exposure to a traumatic event or multiple traumatic events (actual or threatened death, serious injury, or sexual violence)

Note. Information reported in the table was obtained from the *National Institute of Mental Health* and the *Diagnostic and Statistical Manual of Mental Disorders, Fifth Edition, Text Revision (DSM-5-TR)*. Posttraumatic stress disorder (PTSD) and obsessive-compulsive disorder (OCD) are no longer classified as anxiety disorders (denoted *), but given that they were categorized as anxiety disorders up until the release of the DSM-V TR, they are still often discussed alongside anxiety disorders. As a result, studies examining individuals with OCD are discussed in this chapter, as well as some studies that examined individuals with PTSD (e.g., meta-analyses that grouped all anxiety disorders together and as a result included participants with PTSD). Please see Chap. 13 for a comprehensive overview of PTSD studies

inhibitors (SSRIs), serotonin-norepinephrine reuptake inhibitors (SNRIs), or benzodiazepines (Baldwin et al. 2014; Carpenter et al. 2018). Despite high efficacy rates for commonly administered treatments, less than 40% of adults with an anxiety disorder seek treatment (Herring 2018; Kroenke et al. 2007). This is unfortunate as anxiety is also a risk factor for several other adverse outcomes, such as lower health-related quality of life and increased risk of all-cause mortality and chronic health conditions, especially cardiovascular disease (Blumenthal et al. 2021; Stonerock et al. 2015; Allgulander 2016; Emdin et al. 2016; Roy-Byrne et al. 2008). In fact, individuals with an anxiety disorder are over 50% more likely to develop cardiovascular disease compared to those without an anxiety disorder (Batelaan et al. 2016; Kandola et al. 2018).

There has been a continued effort to incorporate physical activity and exercise with treatments for anxiety, in part as a result of the relationship between poorer physical health and anxiety. Given that none of the conventional treatments for anxiety (psychotherapy and/or pharmacotherapy) directly address physical health concerns, incorporating exercise into a comprehensive treatment plan makes intuitive sense. Moreover, there is also evidence to suggest that (1) acute bouts of exercise often reduce state anxiety in individuals with and without clinical anxiety disorders, (2) greater levels of regular physical activity are associated with fewer anxiety symptoms and lowered risk of developing a clinical anxiety disorder, and (3) exercise interventions may reduce general anxiety symptoms and anxiety disorder-specific symptoms among individuals with and without a clinical anxiety disorder. The following sections will present a brief overview of the evidence that forms the basis of these claims.

2 Acute Effects of Exercise on State Anxiety in Individuals With and Without Anxiety Disorders

The majority of early investigations into the anxiolytic effects of acute aerobic exercise examined individuals without a clinical diagnosis of an anxiety disorder. Small, but statistically significant, transient reductions in state anxiety (approximately 1/6–1/2 standard deviation) following acute bouts of aerobic exercise (compared to control conditions) have been consistently demonstrated in several quantitative reviews and meta-analyses (Ensari et al. 2015; McDonald and Hodgdon 1991; Petruzzello 2012; Petruzzello et al. 1991; Landers and Petruzzello 1994). One review quantified this change as approximately a 7.7–15.4% reduction in state anxiety scores (based on state anxiety norms from the commonly used State Trait Anxiety Inventory [STAI]) (Herring 2018). Results from some of the earliest investigations suggested that an acute bout of moderate-intensity aerobic exercise is as effective at reducing state anxiety as meditation (Bahrke and Morgan 1978) and quiet-rest (Bahrke and Morgan 1978; Breus and O'Connor 1998; Motl et al. 2004; Smith et al. 2002), with the duration of anxiolytic effects favoring exercise compared to the other aforementioned modalities. Reductions in state anxiety following aerobic exercise typically occur anywhere from 5- to 30-min post-exercise, with effects persisting up to 2 h in some instances (Raglin and Wilson 1996). There is evidence that a single session of yoga reduces feelings of anxiety, but the evidence that other mindful exercise modes, such as Tai Chi and Qigong, reduce anxiety is limited by the small number of studies conducted to date (Ray and O'Connor 2023; Yin et al. 2021). Additional research is needed to confirm if exercise training augments reductions in state anxiety among those with elevated anxiety or a diagnosed anxiety disorder (Lucibello et al. 2019).

It should be noted that a few experiments among adults without a clinical anxiety disorder have reported no change or an increase in state anxiety following an acute bout of aerobic exercise (Raglin and Wilson 1996; Broman-Fulks et al. 2015; Crombie et al. 2018; Koltyn et al. 1998; O'Connor et al. 1995). While this may seem contradictory to the aforementioned findings, methodological issues may explain these effects. For instance, the majority of trials reporting unchanged levels of state anxiety after exercise often tested individuals with baseline anxiety levels well-below normative values (Broman-Fulks et al. 2015; Crombie et al. 2018). Additionally, the trials that resulted in transient increases in state anxiety either administered a maximal exercise test (Koltyn et al. 1998; O'Connor et al. 1995) or more vigorous-intensity exercise (Raglin and Wilson 1996). It is important to note that increases in state anxiety following vigorous intensity aerobic exercise are typically only present when assessed shortly after the exercise bout, with reductions eventually occurring when assessed after a delay. For instance, a prior investigation revealed that state anxiety was increased 5 min following completion of a 20-min cycling bout at 70% VO_2 peak but decreased 60- and 120-min post-exercise (Raglin and Wilson 1996). This was in contrast to the less intense exercise conditions (40% and 60% VO_2 peak) for which reductions in anxiety were observed at all time points (Raglin and Wilson 1996).

Investigations examining the anxiolytic effect of aerobic exercise among clinical populations have lagged a bit. The delay is thought (in part) to be a result of misconceived interpretation of findings from a study published in the *New England Journal of Medicine* in 1967 (Pitts and McClure1967). In this study, 93% of participants with an anxiety disorder experienced a panic attack following infusion of sodium DL-lactate, which was not the case following glucose injection (Pitts and McClure 1967). Given that exercise increases lactate, this led many researchers and clinicians to speculate that individuals with anxiety disorders should avoid exercising. However, several rebuttals to this study (Grosz and Farmer 1972; Morgan 1979) suggested that the authors made an erroneous conclusion given that an acute bout of aerobic exercise is vastly different from acute lactate infusion. For example, infused lactate results in metabolic alkalosis and hyperventilation whereas exercise-induced elevations in lactate result in metabolic acidosis (Tappy et al. 1996). Furthermore, O'Connor et al. reviewed 15 exercise studies among patients with panic disorder ($n = 420$) and found that there were only 5 instances of panic attacks following 444 acute exercise bouts (O'Connor et al. 2000). O'Connor later demonstrated that not only was it safe for individuals with panic disorder to exercise, but these patients exhibit significant reductions in state anxiety following acute bouts of moderate- or high-intensity aerobic exercise compared to quiet-rest (O'Connor 2022). Additionally, Knapen et al. reported significant reductions in state anxiety following 20 min of prescribed (50% heart rate reserve) or preferred and preferred (self-selected intensity) aerobic exercise in a noncontrolled trial of men ($n = 19$) and women ($n = 29$) with one or more clinically diagnosed anxiety disorders (Knapen et al. 2009). More recently, investigations reported large reductions ($d = 0.98$) in state anxiety following an acute bout of moderate-intensity aerobic exercise (70–75% MHR) in adults with posttraumatic stress disorder (PTSD) (Crombie et al. 2018, 2019) and moderate reductions in state anxiety 10 and 30 min (hedges g = 0.59, g = 0.61, respectively) following an acute bout of moderate-intensity aerobic exercise (rating of 13 on Borg's Ratings of Perceived Exertion scale) compared to quiet-rest in adult women with major depressive disorder (Perkins et al. 2023). It should be noted that PTSD is no longer classified as an anxiety disorder in the DSM-V (Crombie et al. 2023a) but is discussed here given that it was previously categorized as an anxiety disorder and given that the dedicated chapter on PTSD focused on effects beyond state anxiety (see Chap. 13: Exercise and PTSD).

Overall, reductions in state anxiety following an acute bout of aerobic exercise are generally larger among individuals with a clinical anxiety disorder or higher levels of state anxiety compared to those without an anxiety disorder or those exhibiting lower levels of state anxiety. However, this is likely due (at least partially) to floor effects among individuals without a clinical anxiety disorder (Ensari et al. 2015). There is also evidence to suggest that an acute bout of aerobic exercise and greater habitual levels of physical activity contribute to reduced reactivity to stressors and anxiety- and fear-provoking stimuli (Broman-Fulks et al. 2015; Crombie et al. 2021; Esquivel et al. 2002, 2008; Rejeski et al. 1992; Smits et al. 2009, 2011; Ströhle et al. 2005, 2009). The reader is referred to Chap. 3: Multiple Sex- and Circuit-Specific Mechanisms Underlie Exercise-Induced Stress Resistance for a comprehensive overview of this literature.

An acute bout of resistance training (e.g., weightlifting) has also been shown to reduce state anxiety in nonclinical adults, although several investigations have revealed a delayed-onset effect (Focht and Koltyn 1999; O'Connor et al. 1993; Bartholomew and Linder 1998; Bibeau et al. 2010). Similar to aerobic exercise, the anxiolytic effect of resistance exercise is typically greater for individuals with higher baseline levels of state and trait-anxiety (Hale and Raglin 2002). One of the earliest investigations found no significant effect of weight training on state anxiety when assessed shortly after training (Raglin et al. 1993). Broman-Fulks et al. reported no significant change in state anxiety scores when assessed 5 min following an acute resistance exercise session (Broman-Fulks et al. 2015). However, as noted above, mean baseline state anxiety values for this study were well below normative values (Broman-Fulks et al. 2015). A recent investigation also reported nonsignificant reductions in state anxiety in untrained young adults (n = 62) when assessed 10 min after acute resistance exercise sessions administered as part of a resistance exercise training program (Gordon et al. 2022). However, several investigations found that reductions in state anxiety occur after weight training when assessed 15–30 min after exercise (Bartholomew and Linder 1998; Bibeau et al. 2010) and in some cases up to 90–180 min post-exercise (Focht and Koltyn 1999; O'Connor et al. 1993). It is worth nothing that a prior investigation reported reductions in state anxiety after only a 10-min delay, although participants in this study were collegiate athletes administered a combination of aerobic exercise (30 min of cycling at 70% heart rate reserve) and resistance training (6–8 sets of 8–10 repetitions targeting upper and lower body muscles in a 30-min period) (Hale et al. 2002). Relatedly, a recent narrative review concluded that vigorous (e.g., high-intensity) resistance exercise worsens acute anxiety symptoms (Cavarretta et al. 2019). However, this conclusion warrants further examination. Although state anxiety has been reported to increase shortly after (5 and 15 min after) completing 20 min of high-intensity resistance exercise training (75–85% 1RM) (Bartholomew and Linder 1998), other investigations have reported significant reductions in state anxiety 20 and 40 min following high-intensity (80–85% 1RM) and moderate-intensity (50–55% 1RM) resistance exercise sessions with varying (short and long) rest periods in between sets (Bibeau et al. 2010).

3 Physical Activity and Exercise and the Risk for Developing Anxiety Symptoms or Anxiety Disorders

Relationships between physical activity/exercise and anxiety symptoms have been thoroughly investigated. Several large cross-sectional and population-based studies have demonstrated that higher rates of physical activity are associated with less severe anxiety symptoms (De Moor et al. 2006; Goodwin 2003; Stephens 1988; Strine et al. 2008; Ströhle et al. 2007; Thorsen et al. 2005). One of the earliest cross-sectional survey studies found that more active individuals (i.e., those who self-reported they expended five or more kilocalories/kilogram of body weight per day during leisure time physical activities) reported fewer symptoms of anxiety compared to less active individuals (i.e., those who self-reported they expended fewer than 2 kcal/kg of body weight per day during leisure time physical activities) (Stephens 1988). Additionally, Renee Goodwin examined self-report data from a nationally representative sample of Americans ($n = 5{,}877$) between the ages of 18 and 54 and found that individuals who regularly exercised had lower rates (OR = 0.64–0.78) of a past year diagnosis of agoraphobia, social anxiety, specific phobias, panic disorder, and generalized disorder. All of these effects, other than the generalized anxiety disorder effect, held after controlling for age, sex, race, marital status, education, income, and comorbid physical conditions and mental health disorders (Goodwin 2003). More broadly, World Health Survey data among individuals from 46 low- and middle-income countries ($n = 237{,}023$) revealed that low levels of physical activity are significantly associated with greater anxiety symptoms alone (OR = 1.37, 95% CI = 1.23–1.53) and comorbid anxiety and depression symptoms (OR = 1.75; 95% CI–1.52 to 2.01). In this study, self-reported low physical activity (obtained via short form of the International Physical Activity Questionnaire (Craig et al. 2003)) was defined as meeting one or more of the following criteria: (1) three days of vigorous activity of at least 20 min per day, (2) five days of moderate-intensity activity or walking at least 30 min per day (minimum of 10-min bouts), or (3) five days of any combination of walking, moderate-intensity, or vigorous-intensity activities achieving at least 600 MET-minutes per week (Ma et al. 2023). Additionally, the researchers of this study created a four-category variable for anxiety and depression categories (no anxiety, no depression, depression without anxiety, anxiety without depression, anxiety and depression) (Ma et al. 2023).

A significant limitation with the aforementioned cross-sectional studies is that they do not address temporal relationships and therefore cannot establish causality. However, evidence from prospective cohort studies can improve causal inferences by examining physical activity/exercise levels prior to the onset of an anxiety disorder. As a result, prospective cohort studies provide some insight into whether exercise may offer protection against the development of anxiety disorders. In one such investigation, Ströhle et al. examined a cohort ($N = 2{,}548$) of adolescent and young adults (aged 14–24) and found that individuals that self-reported that they engaged in regular physical exercise were 48% less likely (OR = 0.52, 95% CI = 0.37–0.74) to be diagnosed with any anxiety disorder over a four-year follow-up period, compared to those who self-reported that they were inactive (Ströhle et al. 2007). Several other prospective cohort studies conducted outside the United States have also demonstrated that regular participation in physical activity is associated with a 25–57% reduction in the risk of developing an anxiety disorder (Beard et al. 2007; Jonsdottir et al. 2010; McDowell et al. 2018; Sanchez-Villegas et al. 2008). For instance, meeting physical activity guidelines (based on self-report) compared to not meeting physical activity guidelines was associated with 57% reduced odds (OR = 0.43; 95% CI = 0.19–0.99) of developing generalized anxiety disorder over a two-year period in adult men and women ($n = 3{,}950$) that participated in *The Irish Longitudinal Study on Ageing* (McDowell et al. 2018). Lastly, participants in an endurance snow ski race in Sweden ($n = 197{,}685$, median age of 36 years, 38% women) were 60% less likely to develop an anxiety disorder during the >20-year follow-up when compared to nonskiers (Svensson et al. 2021).

McDowell et al. conducted a systematic review and meta-analysis of prospective cohort studies that examined relationships between physical activity/exercise and self-reported anxiety symptoms, a diagnosis of any anxiety disorder, and a diagnosis of generalized anxiety disorder (McDowell et al. 2019). Twenty-four studies were reviewed for the systematic review (median follow-up of 4.75 years), and 13 studies were included in the meta-analysis. The researchers found that the odds of elevated anxiety symptoms (OR = 0.87, 95% CI = 0.77–0.99), any anxiety disorder diagnosis (OR = 0.66, 95% CI = 0.54–0.82), and generalized anxiety disorder (OR = 0.54, 95% CI = 0.32–0.92) were significantly lower for those that were physically active compared to those who were not (McDowell et al. 2019). Another meta-analysis published in that same year ($n = 69{,}037$ participants from 11 cohorts) similarly reported that greater levels of physical activity significantly reduced the incidence of anxiety symptoms (OR = 0.74; 95% CI = 0.63–0.89) or anxiety disorders (OR = 0.75; 95% CI = 0.61–0.92) over a 3.5-year follow-up period compared to those with lower levels of physical activity (Schuch et al. 2019). The researchers also reported that those who reported accumulating at least 150 min of moderate- to vigorous-intensity physical activity had reduced odds of developing anxiety symptoms (OR = 0.71, 95% CI = 0.54, 0.94) over time compared to those who did not meet the aforementioned activity threshold (Schuch et al. 2019).

Although the collective body of evidence suggests that engaging in physical activity offers protection against increased anxiety symptoms and the development of anxiety disorder(s), our understanding of this relationship could be strengthened by using device-based measures of physical activity (e.g., accelerometers) to corroborate the commonly administered self-report measures. Future research should also consider assessing cardiorespiratory fitness, as some investigations have reported that greater cardiorespiratory fitness is associated with a lower risk of developing a clinical anxiety disorder (Baumeister et al. 2017; Kandola et al. 2019; Nyberg et al. 2018), while others have not (Kandola et al. 2019; Shigdel et al. 2019).

4 Physical Activity and Exercise Interventions in Individuals Without an Anxiety Disorder

Evidence has demonstrated that exercise training (aerobic and resistance) significantly reduces (small to moderate effect size reduction) anxiety symptoms among adults without a clinical anxiety disorder (Petruzzello et al. 1991; Conn 2010; Gordon et al. 2018; Long and van Stavel 1995; Rebar et al. 2015; Wipfli et al. 2008). Small to large effect size reductions in anxiety have also been reported among children and adolescents (Petruzzello et al. 1991; Wipfli et al. 2008; Ahn and Fedewa 2011; Calfas and Taylor 1994; Larun et al. 2006). Several recent meta-analyses have also demonstrated that exercise training contributes to reductions in anxiety symptoms (small to moderate effects) among adults with comorbid chronic disease and illnesses, particularly among individuals with cardiovascular disease and fibromyalgia (Herring et al. 2010; Kugler et al. 1994; McDowell et al. 2017; Rossy et al. 1999). For instance, Matthew Herring, Pat O'Connor, and Rod Dishman conducted an extensive review of over 40 exercise training randomized controlled trials (with 75 effects) and found a small but significant reduction in anxiety symptoms (mean effect size $\Delta = 0.29$) across all chronically-ill patient groups, with small to moderate reductions across each of the individual groups, including multiple sclerosis (mean effect size $\Delta = 0.19$), cancer (mean effect size $\Delta = 0.19$), fibromyalgia (mean effect size $\Delta = 0.29$), chronic pain (mean effect size $\Delta = 0.30$), cardiovascular disease (mean effect size $\Delta = 0.32$), chronic obstructive pulmonary disease (mean effect size $\Delta = 0.40$), and others (mean effect size $\Delta = 0.47$) (Herring et al. 2010). Importantly, meta-regression analyses revealed that the largest effects were found for exercise training programs that involved at least 30-min exercise bouts and lasted at least 12 weeks long, with a time frame of anxiety assessment of more than 1 week in the past (Herring 2018; Herring et al. 2010).

In addition to aerobic exercise, a recent meta-analysis reported a small but significant reduction (mean effect size $\Delta = 0.31$) in anxiety symptoms following participation in resistance exercise training interventions (Gordon et al. 2017). This effect is comparable to several previously published meta-analyses that reported reductions in anxiety symptoms among adults without a clinical anxiety disorder following aerobic exercise interventions (Wipfli et al. 2008; Herring et al. 2010; Herring et al. 2014a). In contrast to meta-analytic evidence in chronically ill patients (Herring et al. 2010), the effects were not moderated based on intervention length, time frame of anxiety assessment, and exercise bout duration (in addition to age, sex, and control condition). Although the aforementioned meta-analysis (Gordon et al. 2018) analyzed 31 effects from 16 different randomized controlled trials, only one of the trials examined individuals with a clinical anxiety disorder diagnosis (generalized anxiety disorder), highlighting the need for further investigations of the anxiolytic effect of resistance training in clinical populations. At least three randomized controlled trials tested the short-term influence of yoga compared to relaxation and found large mean benefits favoring the yoga interventions (Cramer et al. 2018).

Several quantitative reviews have also concluded that exercise training reduces anxiety symptoms in individuals with elevated anxiety symptoms (compared to normative data) but not necessarily with a clinical anxiety disorder diagnosis (Petruzzello et al. 1991; Wipfli et al. 2008; Aylett et al. 2018). Several of these investigations examined adults with high levels of anxiety sensitivity (i.e., fear of anxiety-related sensations), which plays a role in the etiology and maintenance of panic attacks and panic disorder (Broman-Fulks and Storey 2008; Ehlers 1995; Schmidt et al. 2006). For instance, Smits et al. randomly assigned adult men and women with elevated levels of anxiety sensitivity to either a two-week aerobic exercise intervention (six, 20-min treadmill sessions at 70% MHR), a two-week exercise plus cognitive restructuring intervention, or a waitlist control condition. Results revealed that both exercise conditions resulted in significant reductions in anxiety sensitivity and general anxiety symptoms (measured via Beck Anxiety Inventory) at 1 and 3 weeks after the intervention compared to the waitlist control condition (Smits et al. 2008). Another study published in that same year among men and women undergraduates ($n = 24$) with elevated levels of anxiety sensitivity reported a significant reduction in self-reported fears of anxiety sensations (i.e., anxiety sensitivity) immediately after and one-week following six, 20-min treadmill sessions at 60–90% MHR over a two-week period, compared to a waitlist control condition (Broman-Fulks and Storey 2008).

5 Physical Activity and Exercise in the Treatment of Anxiety Disorders

Several recent reviews and meta-analyses concluded that exercise training appears to be an effective stand-alone and adjunct treatment for adults with clinical anxiety disorders (Wipfli et al. 2008; Herring et al. 2014a; Aylett et al. 2018; Asmundson et al. 2013; Jayakody et al. 2014; Ramos-Sanchez et al. 2021; Stubbs et al. 2017). In fact, one of the most recent meta-analyses examined the influence of exercise training in adults ($n = 262$) with an anxiety disorder (and PTSD) from six randomized controlled trials (Broocks et al. 1998; Gaudlitz et al. 2015; Herring et al. 2012; Merom et al. 2008; Powers et al. 2015; Rosenbaum et al. 2015) and found a significant, moderate effect size improvement in anxiety symptoms (Stubbs et al. 2017). It should be noted that there was considerable variability in the magnitude of the effect among studies included in this meta-analysis (Stubbs et al. 2017), with standardized mean differences ranging from 0.012 (Gaudlitz et al. 2015) to 1.418 (Broocks et al. 1998) for anxiety disorders and up to 2.648 for a study among adults with PTSD (Powers et al. 2015). Similarly, a recent random-effect meta-analysis of 13 randomized controlled trials (consisting of 376 adults with an anxiety disorder or PTSD randomly assigned to an exercise condition and 355 adults with an anxiety disorder or PTSD randomly assigned to a control condition) found that exercise interventions resulted in a small, but statistically significant effect (standardized mean difference $= -0.425$) reduction in anxiety symptoms compared to a control condition (Ramos-Sanchez et al. 2021).

5.1 Effectiveness of Exercise Training Only (or Waitlist Control)

Egil Martinsen and colleagues conducted some of the earliest investigations examining the effect of exercise training on anxiety symptoms in individuals with an anxiety disorder. One of their first investigations was a noncontrolled trial that found significant reductions in anxiety symptoms for adults ($n = 44$) with a clinical anxiety disorder (except for social phobia) following an exercise intervention consisting of walking or jogging five times a week (60 min per session) for a duration of 8 weeks (Martinsen et al. 1989b). A second noncontrolled investigation revealed significant (and comparable) reductions in anxiety following random assignment to either an aerobic or nonaerobic exercise intervention, both of which involved three sessions a week (60 min per session) for 8 weeks. Importantly, aerobic capacity was assessed as part of this intervention, which revealed that reductions in anxiety occurred independently of changes in aerobic capacity (Martinsen et al. 1989a). Similarly, another noncontrolled early study among inpatients ($n = 52$) reported significant (and comparable) reductions in anxiety for up to 6 months following 8 weeks of either walking or jogging (Sexton et al. 1989).

There have been a few recent studies that examined the effect of exercise training on anxiety symptoms among individuals with a specific anxiety disorder. For instance, Herring et al. compared the effect of 6 weeks of resistance exercise training (two sessions per week of lower body weightlifting at 50% of 1RM, which progressed by 5% weekly) to 6 weeks of aerobic exercise training (two weekly sessions per week) and a waitlist control condition on generalized anxiety disorder symptoms and remission in women ($n = 30$) with a clinically confirmed diagnosis of generalized anxiety disorder (Herring et al. 2012). A strength of this study was that researchers selected cycling to match the primary muscle groups engaged during the resistance exercise intervention, along with matching for the work completed during the sessions and the actual time engaged in exercise (as opposed to simply matching for the duration of the exercise session [e.g., including rest periods]). Results revealed that remission rates were higher following resistance exercise training (60% remission) and aerobic exercise training (40% remission) compared to the weight-list control (30% remission). Additionally, in comparison to the waitlist control condition, the exercise training groups exhibited significant reductions in worry symptoms (hallmark of generalized anxiety disorder) and moderate reductions in trait-anxiety (among other secondary outcomes including depression, feeling of energy and fatigue, and pain) (Herring et al. 2012). Although not directly compared as part of this study, the reported remission rates for the exercise groups were comparable to other commonly administered treatments (e.g., pharmacotherapies, psychotherapies, relaxation therapy) for generalized anxiety disorder (Chessick et al. 2006; Hackett et al. 2003; Hofmann et al. 2010, 2012; Springer et al. 2018). In addition to generalized anxiety disorder, a noncontrolled aerobic exercise trial among adults with obsessive compulsive disorder (OCD) resulted in a mean reduction in OCD symptoms (obtained via Yale-Brown Obsessive Compulsive Scale) from the severe to moderate range for the 11 out of 16 individuals that completed the intervention (Lancer et al. 2007). It should be noted that OCD (along with PTSD) is no longer classified as an anxiety disorder in the DSM-V but is discussed here given that it was previously categorized as an anxiety disorder and there is not a dedicated chapter on OCD (which is the case for PTSD; see Chap. 13: Exercise and PTSD).

5.2 *Effectiveness of Exercise Training in Comparison to Other Treatments*

Several investigations among adults with varying anxiety disorders have also examined the effect of exercise training in conjunction with or in comparison to other treatment modalities. The strongest research designs have been implemented in investigations examining adults with panic disorder. Broocks et al. examined anxiety symptoms among adults with a clinical diagnosis of moderate-to-severe panic disorder (with or without agoraphobia) following random assignment to either an aerobic exercise intervention (3–4 weekly sessions of walking or running a four-mile route for 10 weeks), antidepressant medication (112.5 mg daily dose of a tricyclic antidepressant [clomipramine]), or placebo (Broocks et al. 1998). Although the antidepressant treatment contributed to quicker reductions in anxiety symptoms, both the exercise and medications groups exhibited a significant reduction (large effect size reductions) in anxiety symptoms compared to placebo (Broocks et al. 1998). A separate randomized controlled trial implemented a slightly different design in which adults with panic disorder ($n = 75$) were randomized to either 10 weeks of (1) aerobic exercise and antidepressant medication (40 mg daily dose of a selective serotonin reuptake inhibitor [paroxetine]), (2) aerobic exercise and placebo, (3) relaxation and antidepressant medication (as described above), or (4) relaxation and placebo (Wedekind et al. 2010). Results revealed that all four groups exhibited significant reductions (large effect size improvements) in self-report and clinician-administered assessments, which in part suggests that antidepressant medication did not enhance the efficacy of exercise in reducing anxiety symptoms (Wedekind et al. 2010).

In addition to medication, a few trials among adults with panic disorder have also compared the effects of aerobic exercise to cognitive behavioral therapy. Hovland et al. reported that a 12-week group-based multicomponent (cardiorespiratory fitness and muscular strength) exercise intervention (3 sessions per week) resulted in significant and comparable reductions in anxiety compared to 12 weeks of CBT (one session per week) in a sample of adult men and women ($n = 36$; 81% women) with panic disorder (Hovland et al. 2013). In a separate trial (Gaudlitz et al. 2015), researchers found that an aerobic exercise intervention (three weekly sessions of treadmill running at 70% VO_2 max) in conjunction with CBT resulted in significant and similar reductions in anxiety symptoms compared to a control (very-low-intensity flexibility training) and CBT condition in adults with panic disorder ($n = 47$). However, there was a moderate effect size difference favoring the exercise group at a seven-month follow-up. Unfortunately, it is difficult to draw strong conclusions from this study as there was no control group in this trial and due to the fact that physical activity and/or exercise levels were not assessed during the follow-up period (Gaudlitz et al. 2015).

In addition to panic disorder, there have been a few exercise trials conducted in adults with OCD. The first trial conducted was a quasi-experimental pilot study that reported large reductions ($d = 1.69$) in OCD symptoms after the intervention and 6-months later for participants ($n = 15$) after they completed a 12-week aerobic exercise intervention (3–4 sessions per week of 20–40 min of moderate-intensity aerobic exercise). Importantly, all of the participants in this study previously received at least 12 weeks of either pharmacotherapy or psychotherapy and had not experienced a significant remission of symptoms (Brown et al. 2007). Relatedly, another uncontrolled small pilot study ($n = 11$) reported large improvements ($d = 2.55$) in OCD symptoms following a 12-week aerobic exercise intervention administered in conjunction with CBT (Rector et al. 2015). The exercise intervention for this pilot study (Rector et al. 2015) consisted of three weekly sessions at 40–60% MHR for 15–30 min for 4 weeks and progressed to 60–80% MHR for 30–45 min for an additional 8 weeks. More recently, Ana Abrantes and colleagues conducted a randomized controlled trial that examined the efficacy of a 12-week aerobic exercise intervention in comparison to 12 weeks of health education sessions among physically inactive adults ($n = 56$) currently receiving treatment for OCD with either pharmacotherapy or CBT (Abrantes et al. 2017). The aerobic exercise intervention consisted of a weekly supervised moderate-intensity session on either a treadmill, elliptical machine, or recumbent cycle (participant's choice). Participants were also encouraged to exercise on their own in between weekly sessions, building up to a goal of obtaining 150 min of moderate-intensity aerobic exercise per week. Results revealed that both groups exhibited significant reductions in anxiety and

OCD-specific symptom severity at posttreatment, with no significant differences between groups (exercise group $d = 1.16$; health education $d = 0.63$). However, there was a trend toward greater treatment response (i.e., at least a 35% reduction in symptoms from baseline) among participants in the aerobic exercise group. Specifically, 30.4% of participants from the aerobic exercise group exhibited a 35% reduction in symptom severity from baseline (i.e., treatment responders), which was in contrast to the health education condition, in which only two individuals (7.7%) exhibited such a response (Abrantes et al. 2017).

There have been even fewer investigations among individuals with a primary diagnosis of social anxiety disorder. One such trial found moderate reductions in anxiety symptoms among unmedicated patients ($n = 56$) following random assignment to either 8 weeks of an aerobic exercise intervention (two individual and one group-based exercise session per week) or a mindfulness-based stress reduction condition (a weekly 150-min group class, a 1-day retreat, and daily practice) (Jazaieri et al. 2012). These reported anxiety reductions were comparable between groups as evident via nonsignificant group differences (Jazaieri et al. 2012). Lastly, prescription of a home-based aerobic exercise program that included 30-min bouts of moderate-intensity walking with a goal of 150 min per week, in addition to 8–10 weeks of group-based CBT among adults with a diagnosis of social anxiety disorder, generalized anxiety disorder, or panic disorder, resulted in significant, large reductions in anxiety symptoms compared to a CBT-only condition (Merom et al. 2008).

More recently, researchers at the Duke University Medical Center conducted a randomized controlled trial in which adult men and women ($n = 128$) with coronary heart disease and a diagnosed anxiety disorder (any disorder, 73% of sample) or high levels of anxiety (Hospital Anxiety and Depression-Anxiety Subscale [HADS-A] score of 8 or higher) were randomly assigned to receive an aerobic exercise intervention, escitalopram (up to 20 mg per day), or a placebo (Blumenthal et al. 2021). The 12-week aerobic exercise intervention consisted of three weekly supervised moderate-intensity aerobic exercise sessions (35 min at 70–85% heart rate reserve) of either walking, cycling, or jogging. Researchers found that the exercise intervention and escitalopram significantly reduced anxiety symptoms (HADS-A scores) to a greater degree than the placebo group, with greater reductions reported for the escitalopram group (Blumenthal et al. 2021). Additionally, participants from the exercise and medication groups exhibited significantly lower state anxiety scores compared to the placebo group at posttreatment. Interestingly, weekly reductions in state anxiety occurred most rapidly in the escitalopram group; reductions primarily occurred in the last 6 weeks for the exercise group (Blumenthal et al. 2021). However, a follow-up study with 94% of the original sample revealed that participants from the escitalopram group exhibited greater reductions in anxiety scores (HADS-A) at a six-month follow-up compared to both the exercise and placebo conditions (Blumenthal et al. 2022).

6 Conclusions and Future Directions

To date, the collective evidence suggests that (1) acute bouts of aerobic or resistance exercise generally results in small reductions in state anxiety among adults with and without clinical anxiety disorders, (2) greater levels of physical activity are associated with fewer anxiety symptoms and reduced likelihood of developing an anxiety disorder, (3) exercise training (aerobic and resistance) appears to reduce general anxiety symptoms in adults without a clinical anxiety disorder, including adults with a chronic illness (e.g., cardiovascular disease), and often reduces anxiety disorder-specific symptoms and general anxiety in adults with a clinical anxiety disorder.

Continued research efforts are needed to provide further insight into the potential role for exercise as an affordable and easily accessible treatment modality for individuals with clinical anxiety disorders. Our current understanding of the relationship between physical activity, exercise, and anxiety is restricted by several common methodological limitations often found in the literature (e.g., cross-sectional designs, small sample sizes, lack of long-term follow-up and/or monitoring of activity levels during follow-up periods). There is also an urgent need for additional well-designed and adequately powered randomized controlled trials, especially among individuals (including children and adolescents) with clinical anxiety disorders. Such trials will help to improve our understanding of the efficacy of exercise of varying modalities, intensities, and doses as a potential treatment for anxiety. Additionally, it is important for future research to continue to examine potential and plausible mechanisms that may be at least partially responsible for the anxiolytic effects of acute and chronic exercise. Several mechanisms have been proposed in the literature (and discussed in several other chapters in this series), including but not limited to improvements in self-efficacy (Katula et al. 1999), self-esteem (Herring 2018; Herring et al. 2014b), adaptations to relevant neurotransmitter/neuromodulatory systems and signaling (e.g., GABA, norepinephrine, serotonin, dopamine, endocannabinoid) (Dishman 1997; Dishman et al. 2006; Dishman and O'Connor 2009; Greenwood et al. 2011; Greenwood and Fleshner 2011; Crombie et al. 2023b), and expectancy and/or placebo effects (see Chap. 19: The Placebo Effect in Exercise and Mental Health Research) (Herring et al. 2023; Lindheimer et al. 2015, 2020a, b).

References

Abrantes AM, Brown RA, Strong DR, McLaughlin N, Garnaat SL, Mancebo M et al (2017) A pilot randomized controlled trial of aerobic exercise as an adjunct to OCD treatment. Gen Hosp Psychiatry 49:51–55

Ahn S, Fedewa AL (2011) A meta-analysis of the relationship between children's physical activity and mental health. J Pediatr Psychol 36(4):385–397

Allgulander C (2016) Anxiety as a risk factor in cardiovascular disease. Curr Opin Psychiatry 29(1): 13–17

American Psychiatric Association (2013) Diagnostic and statistical manual of mental disorders, fifth edition (DSM-5). American Psychiatric Publishing, Washington

Asmundson GJG, Fetzner MG, Deboer LB, Powers MB, Otto MW, Smits JAJ (2013) Let's get physical: a contemporary review of the anxiolytic effects of exercise for anxiety and its disorders. Depress Anxiety 30(4):362–373

Aylett E, Small N, Bower P (2018) Exercise in the treatment of clinical anxiety in general practice – a systematic review and meta-analysis. BMC Health Serv Res 18(1):559

Bahrke MS, Morgan WP (1978) Anxiety reduction following exercise and meditation. Cogn Ther Res 2:323–333

Baldwin DS, Anderson IM, Nutt DJ, Allgulander C, Bandelow B, den Boer JA et al (2014) Evidence-based pharmacological treatment of anxiety disorders, post-traumatic stress disorder and obsessive-compulsive disorder: a revision of the 2005 guidelines from the British Association for Psychopharmacology. J Psychopharmacol 28(5):403–439

Bartholomew JB, Linder DE (1998) State anxiety following resistance exercise: the role of gender and exercise intensity. J Behav Med 21(2):205–219

Batelaan NM, Seldenrijk A, Bot M, van Balkom AJLM, Penninx BWJH (2016) Anxiety and new onset of cardiovascular disease: critical review and meta-analysis. Br J Psychiatry 208(3): 223–231

Baumeister SE, Leitzmann MF, Bahls M, Dörr M, Schmid D, Schomerus G et al (2017) Associations of leisure-time and occupational physical activity and cardiorespiratory fitness with incident and recurrent major depressive disorder, depressive symptoms, and incident anxiety in a general population. J Clin Psychiatry 78(1):e41–e47

Beard JR, Heathcote K, Brooks R, Earnest A, Kelly B (2007) Predictors of mental disorders and their outcome in a community based cohort. Soc Psychiatry Psychiatr Epidemiol 42(8):623–630

Bibeau WS, Moore JB, Mitchell NG, Vargas-Tonsing T, Bartholomew JB (2010) Effects of acute resistance training of different intensities and rest periods on anxiety and affect. J Strength Cond Res 24(8):2184–2191

Blumenthal JA, Smith PJ, Jiang W, Hinderliter A, Watkins LL, Hoffman BM et al (2021) Effect of exercise, escitalopram, or placebo on anxiety in patients with coronary heart disease: the understanding the benefits of exercise and escitalopram in anxious patients with coronary heart disease (UNWIND) randomized clinical trial. JAMA Psychiatry 78(11):1270–1278

Blumenthal JA, Smith PJ, Jiang W, Hinderliter A, Watkins LL, Hoffman BM et al (2022) Longer term benefits of exercise and escitalopram in the treatment of anxiety in patients with coronary heart disease: six month follow-up of the UNWIND randomized clinical trial. Am Heart J 251: 91–100

Breus MJ, O'Connor PJ (1998) Exercise-induced anxiolysis: a test of the "time out" hypothesis in high anxious females. Med Sci Sports Exerc 30(7):1107–1112

Broman-Fulks JJ, Storey KM (2008) Evaluation of a brief aerobic exercise intervention for high anxiety sensitivity. Anxiety Stress Coping 21(2):117–128

Broman-Fulks JJ, Kelso K, Zawilinski L (2015) Effects of a single bout of aerobic exercise versus resistance training on cognitive vulnerabilities for anxiety disorders. Cogn Behav Ther 44(4): 240–251

Broocks A, Bandelow B, Pekrun G, George A, Meyer T, Bartmann U et al (1998) Comparison of aerobic exercise, clomipramine, and placebo in the treatment of panic disorder. Am J Psychiatry 155(5):603–609

Brown RA, Abrantes AM, Strong DR, Mancebo MC, Menard J, Rasmussen SA et al (2007) A pilot study of moderate-intensity aerobic exercise for obsessive compulsive disorder. J Nerv Ment Dis 195(6):514–520

Calfas KJ, Taylor WC (1994) Effects of physical activity on psychological variables in adolescents. Pediatr Exerc Sci 6(4):406–423

Carpenter JK, Andrews LA, Witcraft SM, Powers MB, Smits JAJ, Hofmann SG (2018) Cognitive behavioral therapy for anxiety and related disorders: a meta-analysis of randomized placebo-controlled trials. Depress Anxiety 35(6):502–514

Cavarretta DJ, Hall EE, Bixby WR (2019) The acute effects of resistance exercise on affect, anxiety, and mood – practical implications for designing resistance training programs. Int Rev Sport Exerc Psychol 12:295–324
Chessick CA, Allen MH, Thase M, Miralha B, da Cunha ABC, Kapczinski FFK, de Lima MSML et al (2006) Azapirones for generalized anxiety disorder. Cochrane Database Syst Rev 3: CD006115
Conn VS (2010) Anxiety outcomes after physical activity interventions: meta-analysis findings. Nurs Res 59(3):224–231
Craig CL, Marshall AL, Sjöström M, Bauman AE, Booth ML, Ainsworth BE et al (2003) International physical activity questionnaire: 12-country reliability and validity. Med Sci Sports Exerc 35(8):1381–1395
Cramer H, Lauche R, Anheyer D, Pilkington K, de Manincor M, Dobos G et al (2018) Yoga for anxiety: a systematic review and meta-analysis of randomized controlled trials. Depress Anxiety 35(9):830–843
Crombie KM, Brellenthin AG, Hillard CJ, Koltyn KF (2018) Psychobiological responses to aerobic exercise in individuals with posttraumatic stress disorder. J Trauma Stress 31(1):134–145
Crombie KM, Leitzelar BN, Brellenthin AG, Hillard CJ, Koltyn KF (2019) Loss of exercise- and stress-induced increases in circulating 2-arachidonoylglycerol concentrations in adults with chronic PTSD. Biol Psychol 145:1–7
Crombie KM, Cisler JM, Hillard CJ, Koltyn KF (2021) Aerobic exercise reduces anxiety and fear ratings to threat and increases circulating endocannabinoids in women with and without PTSD. Ment Health Phys Act 20:100366
Crombie KM, Cisler JM, Nemeroff CB (2023a) Posttraumatic stress disorder (PTSD) and trauma. In: Reference module in neuroscience and biobehavioral psychology. Elsevier
Crombie KM, Adams TG, Dunsmoor JE, Greenwood BN, Smits JA, Nemeroff CB et al (2023b) Aerobic exercise in the treatment of PTSD: an examination of preclinical and clinical laboratory findings, potential mechanisms, clinical implications, and future directions. J Anxiety Disord 94:102680
De Moor MHM, Beem AL, Stubbe JH, Boomsma DI, De Geus EJC (2006) Regular exercise, anxiety, depression and personality: a population-based study. Prev Med 42(4):273–279
Dishman RK (1997) Brain monoamines, exercise, and behavioral stress: animal models. Med Sci Sports Exerc 29(1):63–74
Dishman RK, O'Connor PJ (2009) Lessons in exercise neurobiology: the case of endorphins. Ment Health Phys Act 2:4–9
Dishman RK, Berthoud HR, Booth FW, Cotman CW, Edgerton VR, Fleshner MR et al (2006) Neurobiology of exercise. Obesity (Silver Spring) 14(3):345–356
Ehlers A (1995) A 1-year prospective study of panic attacks: clinical course and factors associated with maintenance. J Abnorm Psychol 104(1):164–172
Emdin CA, Odutayo A, Wong CX, Tran J, Hsiao AJ, Hunn BHM (2016) Meta-analysis of anxiety as a risk factor for cardiovascular disease. Am J Cardiol 118(4):511–519
Endler NS, Kocovski NL (2001) State and trait anxiety revisited. J Anxiety Disord 15(3):231–245
Ensari I, Greenlee TA, Motl RW, Petruzzello SJ (2015) Meta-analysis of acute exercise effects on state anxiety: an update of randomized controlled trials over the past 25 years. Depress Anxiety 32(8):624–634
Esquivel G, Schruers K, Kuipers H, Griez E (2002) The effects of acute exercise and high lactate levels on 35% CO2 challenge in healthy volunteers. Acta Psychiatr Scand 106(5):394–397
Esquivel G, Díaz-Galvis J, Schruers K, Berlanga C, Lara-Muñoz C, Griez E (2008) Acute exercise reduces the effects of a 35% CO2 challenge in patients with panic disorder. J Affect Disord 107(1–3):217–220
Focht BC, Koltyn KF (1999) Influence of resistance exercise of different intensities on state anxiety and blood pressure. Med Sci Sports Exerc 31(3):456–463
Gaudlitz K, Plag J, Dimeo F, Ströhle A (2015) Aerobic exercise training facilitates the effectiveness of cognitive behavioral therapy in panic disorder. Depress Anxiety 32(3):221–228

Goodwin RD (2003) Association between physical activity and mental disorders among adults in the United States. Prev Med 36(6):698–703

Gordon BR, McDowell CP, Lyons M, Herring MP (2017) The effects of resistance exercise training on anxiety: a meta-analysis and meta-regression analysis of randomized controlled trials. Sports Med 47(12):2521–2532

Gordon BR, McDowell CP, Hallgren M, Meyer JD, Lyons M, Herring MP (2018) Association of efficacy of resistance exercise training with depressive symptoms: meta-analysis and meta-regression analysis of randomized clinical trials. JAMA Psychiatry 75(6):566–576

Gordon BR, McDowell CP, Lyons M, Herring MP (2022) The effects of acute resistance exercise among young adults: a randomized controlled trial. J Affect Disord 299:102–107

Greenwood BN, Fleshner M (2011) Exercise, stress resistance, and central serotonergic systems. Exerc Sport Sci Rev 39(3):140–149

Greenwood BN, Foley TE, Le TV, Strong PV, Loughridge AB, Day HEW et al (2011) Long-term voluntary wheel running is rewarding and produces plasticity in the mesolimbic reward pathway. Behav Brain Res 217(2):354–362

Grosz HJ, Farmer BB (1972) Pitts' and McClure's lactate-anxiety study revisited. Br J Psychiatry 120(557):415–418

Grupe DW, Nitschke JB (2013) Uncertainty and anticipation in anxiety: an integrated neurobiological and psychological perspective. Nat Rev Neurosci 14(7):488–501

Hackett D, Haudiquet V, Salinas E (2003) A method for controlling for a high placebo response rate in a comparison of venlafaxine XR and diazepam in the short-term treatment of patients with generalised anxiety disorder. Eur Psychiatry 18(4):182–187

Hale BS, Raglin JS (2002) State anxiety responses to acute resistance training and step aerobic exercise across eight weeks of training. J Sports Med Phys Fitness 42(1):108–112

Hale BS, Koch KR, Raglin JS (2002) State anxiety responses to 60 minutes of cross training. Br J Sports Med 36(2):105–107

Herring MP (2018) Chapter 2 – Exercise for the Management of Anxiety and Stress-Related Disorders. In: Stubbs B, Rosenbaum S (eds) Exercise-based interventions for mental illness: physical activity as part of clinical treatment, 1st edn. Academic Press, pp 19–52

Herring MP, O'Connor PJ, Dishman RK (2010) The effect of exercise training on anxiety symptoms among patients: a systematic review. Arch Intern Med 170(4):321–331

Herring MP, Jacob ML, Suveg C, Dishman RK, O'Connor PJ (2012) Feasibility of exercise training for the short-term treatment of generalized anxiety disorder: a randomized controlled trial. Psychother Psychosom 81(1):21–28

Herring MP, Lindheimer JB, O'Connor PJ (2014a) The effects of exercise training on anxiety. Am J Lifestyle Med 8(6):388–403

Herring MP, O'Connor PJ, Dishman RK (2014b) Self-esteem mediates associations of physical activity with anxiety in college women. Med Sci Sports Exerc 46(10):1990–1998

Herring MP, Gordon BR, Murphy J, Lyons M, Lindheimer JB (2023) The interplay between expected psychological responses to exercise and physical activity in analogue generalized anxiety disorder: a cross-sectional study. Int J Behav Med 30(2):221–233

Hofmann SG, Sawyer AT, Witt AA, Oh D (2010) The effect of mindfulness-based therapy on anxiety and depression: a meta-analytic review. J Consult Clin Psychol 78(2):169–183

Hofmann SG, Asnaani A, Vonk IJJ, Sawyer AT, Fang A (2012) The efficacy of cognitive behavioral therapy: a review of meta-analyses. Cognit Ther Res 36(5):427–440

Hovland A, Nordhus IH, Sjøbø T, Gjestad BA, Birknes B, Martinsen EW et al (2013) Comparing physical exercise in groups to group cognitive behaviour therapy for the treatment of panic disorder in a randomized controlled trial. Behav Cogn Psychother 41(4):408–432

Jayakody K, Gunadasa S, Hosker C (2014) Exercise for anxiety disorders: systematic review. Br J Sports Med 48(3):187–196

Jazaieri H, Goldin PR, Werner K, Ziv M, Gross JJ (2012) A randomized trial of MBSR versus aerobic exercise for social anxiety disorder. J Clin Psychol 68(7):715–731

Jonsdottir IH, Rödjer L, Hadzibajramovic E, Börjesson M, Ahlborg G (2010) A prospective study of leisure-time physical activity and mental health in Swedish health care workers and social insurance officers. Prev Med 51(5):373–377

Kandola A, Vancampfort D, Herring M, Rebar A, Hallgren M, Firth J et al (2018) Moving to beat anxiety: epidemiology and therapeutic issues with physical activity for anxiety. Curr Psychiatry Rep 20(8):63

Kandola A, Ashdown-Franks G, Stubbs B, Osborn DPJ, Hayes JF (2019) The association between cardiorespiratory fitness and the incidence of common mental health disorders: a systematic review and meta-analysis. J Affect Disord 257:748–757

Katula JA, Blissmer BJ, McAuley E (1999) Exercise intensity and self-efficacy effects on anxiety reduction in healthy, older adults. J Behav Med 22(3):233–247

Kessler RC, Chiu WT, Demler O, Merikangas KR, Walters EE (2005) Prevalence, severity, and comorbidity of 12-month DSM-IV disorders in the National Comorbidity Survey Replication. Arch Gen Psychiatry 62(6):617–627

Knapen J, Sommerijns E, Vancampfort D, Sienaert P, Pieters G, Haake P et al (2009) State anxiety and subjective well-being responses to acute bouts of aerobic exercise in patients with depressive and anxiety disorders. Br J Sports Med 43(10):756–759

Koltyn KF, Lynch NA, Hill DW (1998) Psychological responses to brief exhaustive cycling exercise in the morning and the evening. Int J Sport Psychol 29(2):145–156

Kroenke K, Spitzer RL, Williams JBW, Monahan PO, Löwe B (2007) Anxiety disorders in primary care: prevalence, impairment, comorbidity, and detection. Ann Intern Med 146(5): 317–325

Kugler J, Seelbach H, Krüskemper GM (1994) Effects of rehabilitation exercise programmes on anxiety and depression in coronary patients: a meta-analysis. Br J Clin Psychol 33(3):401–410

Lancer R, Motta R, Lancer D (2007) The effect of aerobic exercise on obsessive-compulsive disorder, anxiety, and depression: a preliminary investigation. Behav Ther 30(3):57–62

Landers DM, Petruzzello SJ (1994) Physical activity, fitness, and anxiety. In: Physical activity, fitness, and health: international proceedings and consensus statement. Human Kinetics Publishers, Champaign, pp 868–882

Larun L, Nordheim LV, Ekeland E, Hagen KB, Heian F (2006) Exercise in prevention and treatment of anxiety and depression among children and young people. Cochrane Database Syst Rev 3:CD004691

Lindheimer JB, O'Connor PJ, Dishman RK (2015) Quantifying the placebo effect in psychological outcomes of exercise training: a meta-analysis of randomized trials. Sports Med 45(5):693–711

Lindheimer JB, Szabo A, Raglin JS, Beedie C (2020a) Advancing the understanding of placebo effects in psychological outcomes of exercise: lessons learned and future directions. Eur J Sport Sci 20(3):326–337

Lindheimer JB, Szabo A, Raglin JS, Beedie C, Carmichael KE, O'Connor PJ (2020b) Reconceptualizing the measurement of expectations to better understand placebo and nocebo effects in psychological responses to exercise. Eur J Sport Sci 20(3):338–346

Long BC, van Stavel R (1995) Effects of exercise training on anxiety: a meta-analysis: journal of applied sport psychology. J Appl Sport Psychol 7(2):167–189

Lucibello KM, Parker J, Heisz JJ (2019) Examining a training effect on the state anxiety response to an acute bout of exercise in low and high anxious individuals. J Affect Disord 247:29–35

Ma R, Romano E, Vancampfort D, Firth J, Stubbs B, Koyanagi A (2023) Association between physical activity and comorbid anxiety/depression in 46 low- and middle-income countries. J Affect Disord 320:544–551

Martinsen EW, Hoffart A, Solberg ØY (1989a) Aerobic and non-aerobic forms of exercise in the treatment of anxiety disorders. Stress Med 5(2):115–120

Martinsen EW, Sandvik L, Kolbjørnsrud OB (1989b) Aerobic exercise in the treatment of nonpsychotic mental disorders: an exploratory study. Nord Psykiatr Tidsskr 43(6):521–529

McDonald DG, Hodgdon JA (1991) The psychological effects of aerobic fitness training: research and theory. Springer, New York

McDowell CP, Cook DB, Herring MP (2017) The effects of exercise training on anxiety in fibromyalgia patients: a meta-analysis. Med Sci Sports Exerc 49(9):1868–1876

McDowell CP, Dishman RK, Vancampfort D, Hallgren M, Stubbs B, MacDonncha C et al (2018) Physical activity and generalized anxiety disorder: results from the Irish longitudinal study on ageing (TILDA). Int J Epidemiol 47(5):1443–1453

McDowell CP, Dishman RK, Gordon BR, Herring MP (2019) Physical activity and anxiety: a systematic review and meta-analysis of prospective cohort studies. Am J Prev Med 57(4): 545–556

McLean CP, Asnaani A, Litz BT, Hofmann SG (2011) Gender differences in anxiety disorders: prevalence, course of illness, comorbidity and burden of illness. J Psychiatr Res 45(8): 1027–1035

Merom D, Phongsavan P, Wagner R, Chey T, Marnane C, Steel Z et al (2008) Promoting walking as an adjunct intervention to group cognitive behavioral therapy for anxiety disorders – a pilot group randomized trial. J Anxiety Disord 22(6):959–968

Morgan WP (1979) Anxiety reduction following acute physical activity. Psychiatr Ann 9(3):36–45

Motl RW, O'connor PJ, Dishman RK (2004) Effects of cycling exercise on the soleus H-reflex and state anxiety among men with low or high trait anxiety. Psychophysiology 41(1):96–105

Nyberg J, Henriksson M, Åberg MAI, Rosengren A, Söderberg M, Åberg ND et al (2018) Cardiovascular fitness in late adolescent males and later risk of serious non-affective mental disorders: a prospective, population-based study. Psychol Med 48(3):416–425

O'Connor PJ (2022) State anxiety is reduced after maximal and submaximal exercise among people with panic disorder. Int J Sport Exerc Psychol 3(4):501–508

O'Connor PJ, Bryant CX, Veltri JP, Gebhardt SM (1993) State anxiety and ambulatory blood pressure following resistance exercise in females. Med Sci Sports Exerc 25(4):516–521

O'Connor PJ, Petruzzello SJ, Kubitz KA, Robinson TL (1995) Anxiety responses to maximal exercise testing. Br J Sports Med 29(2):97–102

O'Connor PJ, Smith JC, Morgan WP (2000) Physical activity does not provoke panic attacks in patients with panic disorder: a review of the evidence. Anxiety Stress Coping 13(4):333–353

Perkins SL, Cook DB, Herring MP, Meyer JD (2023) Dose-response effects of acute exercise intensity on state anxiety among women with depression. Front Psych 14:1090077

Petruzzello SJ (2012) The ultimate tranquilizer? Exercise and its influence on anxiety. In: Acevedo EO (ed) The Oxford handbook of exercise psychology. Oxford University Press, New York, pp 37–54

Petruzzello SJ, Landers DM, Hatfield BD, Kubitz KA, Salazar W (1991) A meta-analysis on the anxiety-reducing effects of acute and chronic exercise. Outcomes and mechanisms. Sports Med 11(3):143–182

Pitts FN, McClure JN (1967) Lactate metabolism in anxiety neurosis. N Engl J Med 277(25): 1329–1336

Powers MB, Medina JL, Burns S, Kauffman BY, Monfils M, Asmundson GJG et al (2015) Exercise augmentation of exposure therapy for PTSD: rationale and pilot efficacy data. Cogn Behav Ther 44(4):314–327

Raglin JS, Wilson M (1996) State anxiety following 20 minutes of bicycle ergometer exercise at selected intensities. Int J Sports Med 17(6):467–471

Raglin JS, Turner PE, Eksten F (1993) State anxiety and blood pressure following 30 min of leg ergometry or weight training. Med Sci Sports Exerc 25(9):1044–1048

Ramos-Sanchez CP, Schuch FB, Seedat S, Louw QA, Stubbs B, Rosenbaum S et al (2021) The anxiolytic effects of exercise for people with anxiety and related disorders: an update of the available meta-analytic evidence. Psychiatry Res 302:114046

Ray LN, O'Connor PJ (2023) Session of yoga, with and without slow (Ujjayi) breathing, reduces anxiety; no change on acute pain sensitivity and endogenous pain modulation. Explore (NY) 19 (3):362–370

Rebar AL, Stanton R, Geard D, Short C, Duncan MJ, Vandelanotte C (2015) A meta-meta-analysis of the effect of physical activity on depression and anxiety in non-clinical adult populations. Health Psychol Rev 9(3):366–378
Rector NA, Richter MA, Lerman B, Regev R (2015) A pilot test of the additive benefits of physical exercise to CBT for OCD. Cogn Behav Ther 44(4):328–340
Rejeski WJ, Thompson A, Brubaker PH, Miller HS (1992) Acute exercise: buffering psychosocial stress responses in women. Health Psychol 11(6):355–362
Rosenbaum S, Sherrington C, Tiedemann A (2015) Exercise augmentation compared with usual care for post-traumatic stress disorder: a randomized controlled trial. Acta Psychiatr Scand 131(5):350–359
Rossy LA, Buckelew SP, Dorr N, Hagglund KJ, Thayer JF, McIntosh MJ et al (1999) A meta-analysis of fibromyalgia treatment interventions. Ann Behav Med 21(2):180–191
Roy-Byrne PP, Davidson KW, Kessler RC, Asmundson GJG, Goodwin RD, Kubzansky L et al (2008) Anxiety disorders and comorbid medical illness. Gen Hosp Psychiatry 30(3):208–225
Sanchez-Villegas A, Ara I, Guillén-Grima F, Bes-Rastrollo M, Varo-Cenarruzabeitia JJ, Martínez-González MA (2008) Physical activity, sedentary index, and mental disorders in the SUN cohort study. Med Sci Sports Exerc 40(5):827–834
Schmidt NB, Zvolensky MJ, Maner JK (2006) Anxiety sensitivity: prospective prediction of panic attacks and Axis I pathology. J Psychiatr Res 40(8):691–699
Schuch FB, Stubbs B, Meyer J, Heissel A, Zech P, Vancampfort D et al (2019) Physical activity protects from incident anxiety: a meta-analysis of prospective cohort studies. Depress Anxiety 36(9):846–858
Sexton H, Maere A, Dahl NH (1989) Exercise intensity and reduction in neurotic symptoms. A controlled follow-up study. Acta Psychiatr Scand 80(3):231–235
Shigdel R, Stubbs B, Sui X, Ernstsen L (2019) Cross-sectional and longitudinal association of non-exercise estimated cardiorespiratory fitness with depression and anxiety in the general population: the HUNT study. J Affect Disord 252:122–129
Smith JC, O'Connor PJ, Crabbe JB, Dishman RK (2002) Emotional responsiveness after low- and moderate-intensity exercise and seated rest. Med Sci Sports Exerc 34(7):1158–1167
Smits JAJ, Berry AC, Rosenfield D, Powers MB, Behar E, Otto MW (2008) Reducing anxiety sensitivity with exercise. Depress Anxiety 25(8):689–699
Smits JAJ, Meuret AE, Zvolensky MJ, Rosenfield D, Seidel A (2009) The effects of acute exercise on CO(2) challenge reactivity. J Psychiatr Res 43(4):446–454
Smits JAJ, Tart CD, Rosenfield D, Zvolensky MJ (2011) The interplay between physical activity and anxiety sensitivity in fearful responding to carbon dioxide challenge. Psychosom Med 73(6):498–503
Spielberger CD (1966) Chapter 14 – the effects of anxiety on complex learning and academic achievement. In: Spielberger CD (ed) Anxiety and behavior. Academic Press, pp 361–398
Springer KS, Levy HC, Tolin DF (2018) Remission in CBT for adult anxiety disorders: a meta-analysis. Clin Psychol Rev 61:1–8
Stephens T (1988) Physical activity and mental health in the United States and Canada: evidence from four population surveys. Prev Med 17(1):35–47
Stonerock GL, Hoffman BM, Smith PJ, Blumenthal JA (2015) Exercise as treatment for anxiety: systematic review and analysis. Ann Behav Med 49(4):542–556
Strine TW, Mokdad AH, Balluz LS, Gonzalez O, Crider R, Berry JT et al (2008) Depression and anxiety in the United States: findings from the 2006 behavioral risk factor surveillance system. Psychiatr Serv 59(12):1383–1390
Ströhle A, Feller C, Onken M, Godemann F, Heinz A, Dimeo F (2005) The acute antipanic activity of aerobic exercise. Am J Psychiatry 162(12):2376–2378
Ströhle A, Höfler M, Pfister H, Müller AG, Hoyer J, Wittchen HU et al (2007) Physical activity and prevalence and incidence of mental disorders in adolescents and young adults. Psychol Med 37(11):1657–1666

Ströhle A, Graetz B, Scheel M, Wittmann A, Feller C, Heinz A et al (2009) The acute antipanic and anxiolytic activity of aerobic exercise in patients with panic disorder and healthy control subjects. J Psychiatr Res 43(12):1013–1017
Stubbs B, Vancampfort D, Rosenbaum S, Firth J, Cosco T, Veronese N et al (2017) An examination of the anxiolytic effects of exercise for people with anxiety and stress-related disorders: a meta-analysis. Psychiatry Res 249:102–108
Svensson M, Brundin L, Erhardt S, Hållmarker U, James S, Deierborg T (2021) Physical activity is associated with lower long-term incidence of anxiety in a population-based, large-scale study. Front Psych 12:714014
Tappy L, Cayeux MC, Chioléro R (1996) Effects of sodium lactate on ventilation and acid-base balance in healthy humans. Clin Physiol 16(4):393–401
Thorsen L, Nystad W, Stigum H, Dahl O, Klepp O, Bremnes RM et al (2005) The association between self-reported physical activity and prevalence of depression and anxiety disorder in long-term survivors of testicular cancer and men in a general population sample. Support Care Cancer 13(8):637–646
Wedekind D, Broocks A, Weiss N, Engel K, Neubert K, Bandelow B (2010) A randomized, controlled trial of aerobic exercise in combination with paroxetine in the treatment of panic disorder. World J Biol Psychiatry 11(7):904–913
Wipfli BM, Rethorst CD, Landers DM (2008) The anxiolytic effects of exercise: a meta-analysis of randomized trials and dose-response analysis. J Sport Exerc Psychol 30(4):392–410
Yin J, Tang L, Dishman RK (2021) The effects of a single session of mindful exercise on anxiety: a systematic review and meta-analysis. Ment Health Phys Act 21:100403

Adaptive and Maladaptive Exercise in Eating Disorders

Katherine Schaumberg, Lauren Pictor, and Max Frank

Contents

Abstract While exercise is generally associated with positive health outcomes, in the context of eating disorders, exercise has high potential to become maladaptive. Maladaptive exercise is compelled or compulsive in nature for the purposes of weight and shape control or to obtain/avoid other eating disorder-relevant consequences. A transdiagnostic eating disorder feature with moderate-to-high prevalence across restrictive- and bulimic-spectrum eating disorders, maladaptive exercise is often associated with negative mental and physical health sequalae. Several proposed threat- and reward-related biobehavioral mechanisms may initiate or perpetuate maladaptive exercise. While exercise is generally contraindicated during periods of acute medical concern, adaptive forms of exercise are also present among those with eating disorders, and facilitation of adaptive exercise has potential to promote physical and mental health benefits during eating disorder

K. Schaumberg (✉), L. Pictor, and M. Frank
Department of Psychiatry, University of Wisconsin, Madison, WI, USA
e-mail: kschaumberg@wisc.edu; pictor@wisc.edu; mbfrank@wisc.edu

Curr Topics Behav Neurosci (2024) 67: 223–240
https://doi.org/10.1007/7854_2024_499
Published Online: 24 July 2024

recovery. Detailed assessment and targeted interventions are needed to address the clinical conundrum of how and when to integrate exercise into eating disorder treatment.

Keywords Anorexia Nervosa · Bulimia Nervosa · Eating Disorders · Maladaptive Exercise

1 Introduction

Exercise (i.e., planned, structured physical activity carried out to improve one or more fitness domains) is commonly associated with positive mental health outcomes (Deslandes et al. 2009; Vankim and Nelson 2013; Chekroud et al. 2018) and is increasingly prescribed as a component of mental health treatment (Herring et al. 2010; Carek et al. 2011). However, within the context of an eating disorder, exercise presents a mental health paradox, as it also has high potential to become a maladaptive behavior with compulsive qualities, which can interfere with both quality of life and eating disorder recovery (Davis and Kaptein 2006; Shroff et al. 2006; Cook et al. 2014; Davis et al. 1997). Priorities for exercise-related research in the context of eating disorders include clarifying when and for whom exercise is adaptive vs. maladaptive along with designing interventions that promote adaptive exercise in eating disorder recovery while mitigating risk for maladaptive use.

1.1 Defining and Understanding Maladaptive Exercise in Eating Disorders

Exercise-related research in eating disorders has historically focused on characterizing and understanding maladaptive forms of exercise that can arise during an active eating disorder. While maladaptive qualities of exercise may arise with any form or type of exercise, and a growing literature notes the presentation of maladaptive forms of resistance training that are common among boys and men with eating disorders (Griffiths et al. 2015; Martenstyn et al. 2021; Murray et al. 2013), the most commonly reported type of maladaptive exercise is aerobic (Nagata et al. 2017). Defining maladaptive exercise has led to substantial debate and myriad terminology, including compulsive exercise, obligatory exercise, compensatory exercise, exercise addiction, excessive exercise, and driven exercise. We use the term "maladaptive exercise" to refer to exercise which, broadly defined, fits into any of these categories and note that definitions of these terms and their measurement components overlap substantially (see Table 1).

Studies that have examined multiple definitions simultaneously find substantial overlap in endorsement of compulsive and addictive exercise, with compensatory exercise, which refers to exercise that occurs specifically to counteract the effects of

Table 1 Maladaptive exercise constructs

Construct	Cognitive features	Affective features	Behavioral features	Example assessment items
Compensatory exercise	NA	Reduces feelings of guilt about overeating	Occurs after overeating or binge eating	"Have you exercised as a means of controlling your weight, altering your shape or amount of fat, or burning off calories?" (Fairburn and Bèglin 1994) "Have you ever used [exercise] to compensate for episodes of binge eating or over eating?" (Thornton et al. 2018)
Compulsive exercise	Weight/shape and/or affect regulation motivation	Mood improvement Guilt if missed Lack of enjoyment Avoidance of negative affect	Exercise rigidity Rule-driven behavior (habit)	"I follow a set routine for my exercise sessions" (Meyer et al. 2016)
Excessive exercise	Weight/shape motivation	NA	Daily or almost daily, 1 h or more	"I exercise more than 3 days per week" (Steffen and Brehm 1999) "I spend most of my free time exercising" (Hausenblas and Downs 2002)
Driven exercise	Weight/shape and/or affect regulation motivation	Feeling "compelled" or "driven" to complete exercise	Interferes with life activities	How many times have you exercised in a "driven" or "compulsive" way as a means of controlling your weight, shape, or amount of fat or to burn off calories? (Fairburn and Bèglin 1994)
Exercise addiction	Weight/shape and/or affect regulation motivation	Needing to exercise more to produce a similar effect over time	Exercising more than intended Inability to reduce Interferes with life activities	"I continually increase my exercise intensity to achieve the desired effects/benefits" (Hausenblas and Downs 2002) "I exercise despite persistent physical problems" (Hausenblas and Downs 2002)

(continued)

Table 1 (continued)

Construct	Cognitive features	Affective features	Behavioral features	Example assessment items
Obligatory exercise	Preoccupation with exercise	Intense negative affect when unable to exercise	Exercise rigidity Rule-driven behavior (habit) Continued exercise despite negative consequences	"If I miss an exercise session, I will try and make up for it when I next exercise (Meyer et al. 2016)" "I will engage in other forms of exercise if I am unable to engage in my usual form of exercise" (Steffen and Brehm 1999) "Sometimes, I find that my mind wanders to thoughts about exercising" (Pasman and Thompson 1988)

overeating or binge eating, showing some distinction from other maladaptive exercise definitions (Holland et al. 2014; Scharmer et al. 2020). Though some debate persists regarding an optimal definition of maladaptive exercise, consistent evidence indicates that some individuals, can *adaptively* exercise at very high levels, without negative consequence (Juwono et al. 2022). Among athlete samples (e.g., ultramarathon runners and other competitive runners, elite dancers, competitive body builders, triathletes, and other mixed groups of athletes), exercise addiction or dependence was estimated between 3% and 42% (Juwono et al. 2022). Further, individuals may experience negative sequalae of maladaptive exercise even when exercise is present at moderate levels (Adkins and Keel 2005; Young et al. 2017). For example, average metabolic equivalents (METS) per week was weakly associated with compulsive exercise in a sample of adolescents ($r = 0.15$) (Goodwin et al. 2011), and time spent in driven exercise on the eating disorder examination questionnaire (EDEQ) was moderately associated ($r = 0.42$) with compulsive exercise total score in a clinical sample of individuals with eating disorders (Young et al. 2017). Further, a study of university students found no difference in self-reported exercise mins/week among those who reported high levels of appearance motivation (M = 241.44, SD = 217.75 min), vs. those who reported no appearance motivations for exercise (M = 299.29, SD = 277.42 min); however, those who reported high appearance motivations for exercise reported significantly higher levels of eating pathology (Adkins and Keel 2005). Defining features that are common across maladaptive exercise constructs include a problematic *psychological relationship* with exercise (e.g., exercise feels driven or compulsive, feelings of guilt when missing exercise, overreliance on exercise to manage affect and/or concerns about shape and weight) along with *life interfering symptoms* (e.g., inability to reduce exercise, exercising despite illness or injury, exercise routines interfering with work, school, or social functioning).

Maladaptive exercise occurs transdiagnostically across eating disorders but at somewhat varying rates based on eating disorder presentation (Davis and Kaptein 2006; Shroff et al. 2006; Cook et al. 2014; Davis et al. 1997). Definitional components of specific eating disorders and estimations of maladaptive exercise prevalence within these subpopulations is presented in Table 2.

Overall rates of maladaptive exercise are highest among individuals with anorexia nervosa (AN) followed by bulimia nervosa (BN) and other specified feeding or eating disorder (OSFED); however, maladaptive exercise is also present among individuals with BED. Little research has investigated whether maladaptive exercise presents in those with avoidant/restrictive food intake disorder. Other disorder-relevant characteristics and demographic factors associated with increased risk for maladaptive exercise include lower body mass index reached during illness, lower novelty, higher levels of persistence (Shroff et al. 2006), suicidal behavior (Smith et al. 2013), obsessive-compulsive traits and symptoms (Davis and Kaptein 2006; Shroff et al. 2006), higher levels of anxiety (Shroff et al. 2006; Holtkamp et al. 2004), perfectionism (Shroff et al. 2006), body dissatisfaction (Brewerton et al. 1995), reward dependence (Dalle Grave et al. 2008), dietary restraint (Dalle Grave et al. 2008; Holtkamp et al. 2004), and anhedonia (Davis and Woodside 2002). When present in an eating disorder, maladaptive exercise is associated with poor treatment outcome, prognosis, and quality of life (Dalle Grave et al. 2008), along with hypoleptinemia (Holtkamp et al. 2003, 2006). Further, among epidemiological samples, maladaptive exercise is *prospectively* associated with eating disorder symptom severity, and eating disorder cognitions in middle adolescence are associated with greater odds of engaging in maladaptive exercise across adolescence and young adulthood (Schaumberg et al. 2022, 2023). Additionally, while the current chapter focuses on exercise in the context of eating disorders, it is also relevant to note that exercise can develop maladaptive qualities outside of eating disorders, which may present similar treatment challenges. Some work on the construct of exercise addiction, for instance, distinguishes between *primary* exercise addiction, which includes addictive qualities but occurs outside of a co-occurring mental health condition, and *secondary* exercise addiction, which occurs secondary to a mental health condition, most often an eating disorder (Cook et al. 2014). The odds ratio for exercise addiction in populations with vs. without indicated eating disorders has been estimated at 3.71, indicating substantial increased risk for maladaptive exercise among those with an eating disorder (Trott et al. 2021).

1.2 Defining and Understanding Adaptive Exercise in Eating Disorders

While most research on exercise in eating disorders has focused on maladaptive forms of exercise, it is well established that exercise can promote positive mental health outcomes, including among those with eating disorders. In the context of

Table 2 Maladaptive exercise prevalence estimates across eating disorders

Eating disorder	Basic definitional components	Maladaptive exercise prevalence estimate during active eating disorders
Anorexia nervosa (AN)	Significant restriction of energy intake, resulting in low body weight, intense fear of gaining weight, and a distorted body image. AN primarily affects adolescents and young adults and is most prevalent in females	Adult: 38–80% (Shroff et al. 2006; Brewerton et al. 1995); AN with purging subtype has demonstrated a higher prevalence of maladaptive exercise than other AN subtypes (Shroff et al. 2006) Adolescent: 23–80% (Davis et al. 1997; Stiles-Shields et al. 2011)
Bulimia nervosa (BN)	Recurrent episodes of binging followed by compensatory behaviors (e.g., self-induced vomiting and laxative misuse). Individuals with BN have a preoccupation with body shape and weight and may experience feelings of guilt and loss of control	Adult: 23–51% (Shroff et al. 2006; Brewerton et al. 1995; Monell et al. 2018) Adolescent: ~66% (Stiles-Shields et al. 2015)
Binge eating disorder (BED)	Recurrent episodes of consuming large amounts of food in a short period accompanied by a sense of loss of control and marked distress. Unlike BN, individuals with binge eating disorder do not engage in compensatory behaviors	Adult: 13–43% (Monell et al. 2018; Lampe et al. 2021) Adolescent: No current estimates
Atypical AN	Similar to AN, with significant weight loss and restrictive eating behaviors, but individuals may maintain a weight within or above the normal range. Despite not meeting the low weight criteria of typical anorexia nervosa, individuals with atypical anorexia nervosa experience similar psychological and physical consequences to those with AN	Adult: No current estimates Adolescent: 73% (Sawyer et al. 2016)
Other specified feeding and eating disorders (OSFED)	A diagnosis used when an individual presents with significant eating disorder symptoms that do not meet the full criteria. OSFED encompasses a range of eating disorder presentations, such as subthreshold BN, BED, and purging disorder	Adult: 21–55% (Shroff et al. 2006; Monell et al. 2018; Dalle Grave et al. 2008) Adolescent: ~17% (Machado et al. 2007)
Avoidant/restrictive food intake disorder (ARFID)	Persistent avoidance or restriction of food intake, often driven by sensory sensitivity, lack of interest in eating, or fear of negative consequences. Unlike other eating disorders, individuals with ARFID do not necessarily have body image concerns or a desire for weight loss	Current research does not support the presentation of maladaptive exercise with ARFID (Nakai et al. 2017)

eating disorder treatment, structured, adaptive exercise (i.e., exercise that is noncompensatory and not motivated by intention to control shape or weight) has the potential to promote muscle health, prevent loss of bone density, enhance cognitive functioning and self-regulation skills, and facilitate social engagement and mood regulation. Further, as mood and anxiety disorders are highly comorbid with eating disorders (Bühren et al. 2014), exercise may improve outcomes for these comorbid conditions (Carek et al. 2011; Asmundson et al. 2013; Heyman et al. 2012). Some research has successfully distinguished adaptive from maladaptive exercise engagement in individuals with eating disorders. For example, among adults with binge-spectrum eating disorders, ecological momentary assessment protocols have identified that these individuals report a mix of maladaptive and nonmaladaptive exercise bouts, with some individuals reporting exclusively maladaptive exercise, others reporting exclusively nonmaladaptive exercise, and others reporting both types of exercise (Lampe et al. 2021). Further, among this sample, nonmaladaptive exercise promoted positive affect after an exercise bout, while maladaptive exercise did not (Lampe et al. 2023a), suggesting potential specificity of adaptive exercise in promoting positive affect among those with eating disorders. Outside of the exercise behavior itself, participation in sports has the potential to exacerbate eating disorder risk (Atkinson 2011; Martinsen and Sundgot-Borgen 2013) but may also facilitate social skills and connection, motivation for recovery, and support identity development outside of an eating disorder. Overall, research supports that (1) not all exercise behavior among individuals with eating disorders is maladaptive and (2) inclusion of intervention components that promote adaptive exercise, while largely untested, holds promise for facilitating eating disorder recovery.

2 The Biobehavioral Underpinnings of Maladaptive Exercise in Eating Disorders

2.1 Biological Reinforcement of Maladaptive Exercise

Models of maladaptive exercise propose that, for individuals with eating disorders, overengagement of biological reward and threat, which are mediated by cognitive processes, may lead to exercise that is habitual and compulsive (Davis and Kaptein 2006; Shroff et al. 2006; Cook et al. 2014; Adams 2009; Adams and Kirkby 2002; Bamber et al. 2003; Bratland Sanda et al. 2011; Davis and Claridge 1998; Guarda et al. 2015; Herring et al. 2014; Lichtenstein et al. 2017). Rodent research has specifically identified a model, the "activity-based anorexia" (ABA) model, in which free access to voluntary exercise during a period of low food availability can trigger an anorexia-like syndrome. Changes in leptin, a hormone that modulates reward circuitry, is specifically noted to predict hyperactivity in the ABA model (Holtkamp et al. 2003; Holtkamp et al. 2006). In addition to research supporting leptin as a mechanism of action in the ABA model (Hillebrand et al. 2008), exercise has demonstrated an association with reduced plasma leptin post-exercise in a small sample with AN but not in healthy control women (Dostalova et al. 2007). Further, plasma leptin levels have been associated with maladaptive exercise (Ehrlich et al. 2009) and serum leptin with motor restlessness (Holtkamp et al. 2006) in AN samples. Further investigations of the ABA model suggest mesolimbic activity, and related dopamine signaling regulates exercise behavior (Foldi et al. 2017a, b) and that disrupting these processes can impact wheel running activity (Klenotich et al. 2015; Verhagen et al. 2009). As with the preponderance of female (vs. male) humans who present with AN, female adolescent mice are particularly susceptible to experiencing ABA (Chowdhury et al. 2015). From an evolutionary perspective, ABA mechanisms may include that exercise reward is heightened during periods of starvation to motivate foraging (Adan et al. 2010) or that exercise may increase body temperature when metabolism drops due to starvation (Hillebrand et al. 2005). Altogether, the ABA model supports that food restriction and hyperactivity may drive one another and that rewarding aspects of exercise may entrench anorexia-like behaviors in rodents, with translational implications.

2.2 *Behavioral Reinforcement of Maladaptive Exercise*

Behavioral reinforcement and observations of exercise effects among individuals with EDs further support both reward- and threat-related mechanisms as relevant to the onset and maintenance of maladaptive exercise. Exercise may be a particularly salient reward for individuals with eating disorders due to the positive affect experienced in response to exercise coupled with a Western cultural contextualization which supports high valuation of exercise behavior (Coniglio et al. 2022). For example, a study using ecological momentary assessment in a population of individuals with AN found that, over the course of a week, physical activity level predicted positive affect and positive affect also predicted physical activity within the same hour and within the next hour (Karr et al. 2017). Behavioral research that has examined willingness to work for exercise among individuals with AN additionally found that individuals with greater commitment to exercise worked harder on a progressive-ratio computer task for the opportunity to exercise as compared to those with less commitment to exercise, a finding that persisted after patients were weight restored. Further, a sample of individuals recovered from AN demonstrated more motivation to work for exercise in a behavioral task compared to healthy controls, even when dopamine was depleted (O'Hara et al. 2016). Together, findings highlight (1) a high reward value of exercise reflective of willingness to work for exercise among those with AN and (2) variability among the reinforcement value of exercise which is potentially reflective of risk for maladaptive exercise (Klein et al. 2010).

In addition to its rewarding effects, exercise also demonstrates substantial potential impact related to reduction of negative affect in the context of eating disorders. Initial experience sampling studies support the notion that a primary function of maladaptive exercise includes reducing momentary negative affect and weight-related fears – for example, individuals with EDs report reductions in negative affect post-exercise (Engel et al. 2013; Kolar et al. 2020; Reichert et al. 2020). Among those with binge-spectrum eating disorders, negative affect is likely to occur immediately prior to exercise episodes characterized by feeling driven and intended to influence shape/weight (Lampe et al. 2023b). Further, the most central ED fears in a network analysis study predicted excessive exercise (but not other ED behaviors) at a two-month follow-up (Levinson and Williams 2020), suggesting potential importance of negative reinforcement patterns to maladaptive exercise maintenance.

2.3 *Potential Genetic Underpinnings of Maladaptive Exercise*

Both twin and molecular genetic studies indicate that exercise propensity is both variable and heritable (Herring et al. 2014; de Geus et al. 2014; Klimentidis et al. 2018; Flack et al. 2019), and accelerometer-assessed physical activity in a sample of adults in midlife demonstrates a positive genetic correlation with AN ($r_g = 0.17$) (Anorexia Nervosa Genetics Initiative, Eating Disorders Working Group of the Psychiatric Genomics Consortium et al. 2019). Thus, individuals who are genetically at risk for eating disorders, and more specifically AN, may also be prone to experiencing high levels of reinforcement from exercise behavior. As previously mentioned, not all individuals with eating disorders engage in maladaptive exercise; thus, a relevant question in identifying those at highest risk for maladaptive exercise includes distinguishing what factors, beyond eating disorder risk itself, coincide with engagement in maladaptive exercise. At a phenotypic level, both compulsive and addictive traits and symptoms associate with maladaptive exercise among individuals with EDs (Davis and Kaptein 2006; Shroff et al. 2006). Genetic findings further the narrative that risk for obsessive-compulsive disorder may be specifically related to maladaptive exercise. In a large, epidemiological sample of youth, for instance, maladaptive exercise was the only eating disorder behavior predicted by both an obsessive-compulsive and an anorexia nervosa polygenic risk score (Yilmaz et al. 2022). Overall, genetic links between obsessive compulsive disorder, anorexia nervosa, and physical activity suggest that some shared genetic liability among these traits may influence the degree to which exercise behavior is experienced as reinforcing. At extreme levels or in the context of an active eating disorder, a tendency to experience exercise as a potent modulator of affective experience may lead to exercise patterns which are rigid, compulsive, and maladaptive.

3 Advancing an Understanding of Maladaptive Exercise

Figure 1 presents a model of maladaptive exercise, based on the literature reviewed above. Overall, biobehavioral theories contend that maladaptive exercise may serve varying primary functions, to increase positive affect (reward) and decrease negative

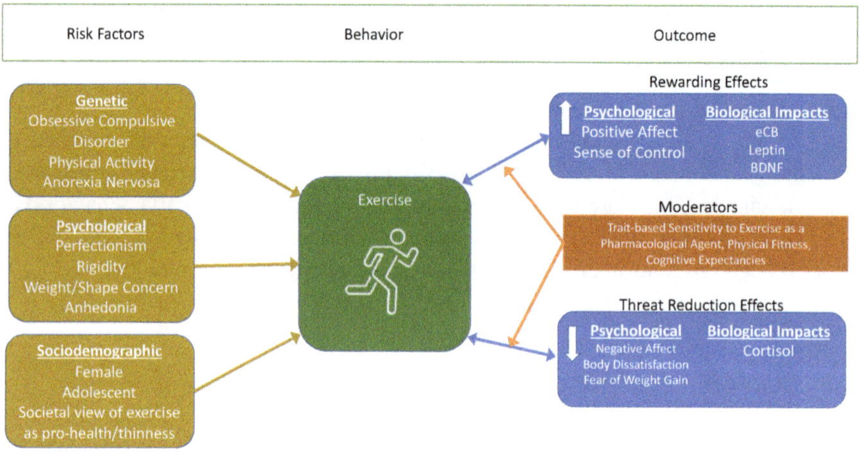

Fig. 1 A model of maladaptive exercise in the context of eating disorders. *Note*: This model of maladaptive exercise is based on the literature reviewed in Sect. 2. Biobehavioral theories contend that maladaptive exercise may serve varying primary functions, to increase positive affect (reward) and decrease negative affect (negative reinforcement), or as a rigid and compulsive (habitual) behavior. These potential functions of exercise implicate divergent neurobiological mechanisms in the development and maintenance of maladaptive exercise, which may vary across individuals, vary within individuals across time, and/or co-occur in a single behavioral instance. eCB, endocannabinoids; BDNF, brain-derived neurotrophic factor. Evidence for impacts of exercise on biological variables are reviewed in-depth throughout several chapters of this volume. Briefly, and with a focus on eating disorder-relevant effects: BDNF, expression in the mesolimbic reward system is altered in the rodent model of AN after wheel running (Ho et al. 2016). In humans, BNDF impacts hippocampal functioning: relating to reward sensations, emotions, and mood (Kumar et al. 2019; Severinsen and Pedersen 2020). Cortisol: elevated cortisol is consistently associated with threat processing, including in samples of individuals with anorexia nervosa (Guarda et al. 2015). In patients with AN show blunted reactivity to social stressors (Schmalbach et al. 2020). eCB: Exercise-induced increases in eCBs are associated with central mood state improvements in human clinical and nonclinical populations (Brellenthin et al. 2017; Crombie et al. 2018, 2021; Meyer et al. 2019) and reductions in anxiety and fear ratings during a threat task (Crombie et al. 2021). Leptin: Acute exercise associated with reduced plasma leptin postexercise in a small sample with AN but not in healthy control women. Plasma leptin levels show association with driven exercise (Ehrlich et al. 2009) and serum leptin with motor restlessness (Holtkamp et al. 2006) in AN

affect (negative reinforcement), or as a rigid and compulsive (habitual) behavior. These potential functions of exercise implicate divergent neurobiological mechanisms in the development and maintenance of maladaptive exercise, which may vary across individuals, vary within individuals across time, or co-occur in a single behavioral instance. At present, we do not fully understand mechanistic factors that contribute to the variability in presence vs. absence of maladaptive exercise among those with eating disorders—why some individuals present with maladaptive exercise and others do not or how and why exercise serves various functions among those with eating disorders. Greater attunement to moderators across levels of analysis (genetics, preexercise biomarker availability, exercise training status, exercise expectancies, neural circuitry differences, eating disorder illness factors) will clarify accounts of exercise-related risk across individuals. Further, differential functions of exercise impact treatment outcomes. Ultimately, improved understanding of varying maladaptive exercise maintenance factors – and how to reduce them – may better guide treatment programs.

4 Exercise in Eating Disorder Treatment

4.1 Reducing Maladaptive Exercise

There are growing calls for action within the eating disorder field (Cook et al. 2016; Cook and Leininger 2017; Hay et al. 2018; Hausenblas et al. 2008) to address exercise in eating disorder treatment. Maladaptive exercise patterns often present early in the course of an ED (Davis et al. 1997; Achamrah et al. 2016) and can mark poor treatment prognosis, suggesting that intervention strategies that directly address this symptom are essential. The most dominant treatments for eating disorders do not have specific guidance related to targeting maladaptive exercise: enhanced cognitive behavioral therapy (CBT-E), for instance, notes that exercise should be monitored and the patient should be educated about its potential effects, stating that "Unless the exercising has other functions than weight control, it generally decreases during the course of CBT-E as patients become less concerned about their eating, shape and weight," and encouragement to use more general affect management strategies to address exercise which primarily functions to reduce negative affect (Fairburn 2008). In more controlled settings (e.g., family-based treatment, residential and inpatient treatment facilities), exercise is often minimized during weight restoration (Lock and Le Grange 2015). Ultimately, a failure to adequately assess and address exercise behavior in treatment can pose risk to patients by perpetuating suboptimal treatment. Even if access to exercise is substantially reduced in acute treatment settings, these settings represent only a first phase of recovery for the subset of patients who access this care. Patients in free-living environments are navigating ongoing choices about when, whether, and how much to exercise at different points during eating disorder recovery, often with little empirical guidance. Intervention development which is targeted toward maladaptive exercise is underway, specifically a cognitive-behavioral based intervention for those with restrictive eating disorders (compulsive exercise activity therapy, LEAP) (Hay et al. 2018). As models of maladaptive exercise are further tested and refined, treatment development can include investigation of the ways that psychological agents (e.g., exposure and response prevention, distress tolerance) and exercise dose (e.g., frequency, type, intensity, and duration) might be optimally incorporated into eating disorder treatment and how prescriptions may be effectively tailored based on the function of exercise for individual patients along with their exercise-related history and risk.

4.2 Promoting Adaptive Exercise

Ultimately, a goal of eating disorder recovery includes a return to adaptive exercise. In addition to myriad benefits of adaptive exercise engagement for more general physical and mental health, adaptive exercise also holds potential to be used as a tool for improving eating disorder-specific symptoms, including reducing driven exercise behaviors, drive for thinness, binge eating frequency, compensatory behaviors, and body dissatisfaction (Lampe et al. 2022; Mathisen et al. 2020; Vancampfort et al. 2017). Nascent research investigating potential exercise-targeted interventions in eating disorders include an intervention for binge eating and bulimia which incorporates psychoeducation, supervised exercise, and recommendations for healthy exercise engagement (Mathisen et al. 2018, 2020), which could support increases in adaptive exercise. Research on standard cognitive behavioral therapy for those with binge-spectrum disorders indicates that there is no effect of this therapy on adaptive exercise (Wons et al. 2023), highlighting the potential for novel interventions to specifically target adaptive exercise with greater potency. Further, at present, exercise professionals are rarely integrated into eating disorder treatment teams; however, there may be great benefit for incorporating such professionals in eating disorder treatment, particularly for patients who have significant history of maladaptive exercise and those who aim to engage in high levels of adaptive exercise as they progress in ED recovery (e.g., athletes). Ultimately, a recovery goal could include return to adaptive activity that includes a psychologically flexible relationship with exercise and may even be at high levels for some individuals.

5 Conclusion

Exercise poses a complex challenge in the realm of mental health and the treatment of eating disorders, presenting a paradox and clinical dilemma (Moola et al. 2013; Mathisen et al. 2023). During the course of an eating disorder, maladaptive exercise patterns may develop and be reinforced when intertwined with obsessive-compulsive traits and preoccupation with weight and shape (Coniglio et al. 2022). Maladaptive exercise is a prevalent symptom of eating disorders, and effective intervention and treatment approaches that address exercise-related issues are still under development. While the presentation of maladaptive exercise in eating disorders is common, exercise also has high potential to contribute to positive mental well-being and may even aid recovery (Vancampfort et al. 2013, 2014). Current evidence indicates that exercise can have beneficial effects when incorporated thoughtfully into eating disorder treatment (Vancampfort et al. 2013, 2014; Rosa-Caldwell et al. 2023). Suggested guidelines include employing a team of experts and monitoring medical status; screening for exercise-related psychopathology; creating a written contract, including psychoeducation and focusing on positive reinforcement in exercise programming with debriefs after exercise sessions; and developing a graduated set of recommendations that starts slow and is tailored to the individual (Cook et al. 2016). In addition, careful assessment and differentiation between adaptive and maladaptive exercise may be relevant to supporting more adaptive exercise over time (Wons et al. 2023). Moving forward, close attention to the functions of maladaptive exercise, such as mood regulation and compensatory functions, may improve precision of interventions which aim to reduce maladaptive and increase adaptive exercise engagement.

References

Achamrah N, Coëffier M, Déchelotte P (2016) Physical activity in patients with anorexia nervosa. Nutr Rev 74(5):301–311

Adams J (2009) Understanding exercise dependence. J Contemp Psychother 39:231–240

Adams J, Kirkby RJ (2002) Excessive exercise as an addiction: a review. Addict Res Theory

Adan RAH, Hillebrand JJG, Danner. . . UN. (2010) Neurobiology driving hyperactivity in activity-based anorexia. Neurobiol Eat

Adkins EC, Keel PK (2005) Does "excessive" or "compulsive" best describe exercise as a symptom of bulimia nervosa. Int J Eat Disord 38:24–29

Anorexia Nervosa Genetics Initiative, Eating Disorders Working Group of the Psychiatric Genomics Consortium, Watson HJ, Yilmaz Z, Thornton LM, Hübel C, et al. Genome-wide association study identifies eight risk loci and implicates metabo-psychiatric origins for anorexia nervosa. Nat Genet [Internet]. 2019 Jul 15 [cited 2019 Jul 24]; Available from: http://www.nature.com/articles/s41588-019-0439-2

Asmundson GJG, Fetzner MG, DeBoer LB, Otto MW, Smits JAJ (2013) Let's get physical: a contemporary review of the anxiolytic effects of exercise for anxiety and its disorders. Depress Anxiety 30(4):362–373

Atkinson M (2011) Male athletes and the cult (ure) of thinness in sport. Deviant Behav 32(3): 224–256

Bamber DJ, Cockerill IM, Rodgers S, Carroll D (2003) Diagnostic criteria for exercise dependence in women. Br J Sports Med 37:393–400

Bratland Sanda S, Martinsen EW, Rosenvinge JH, Rø Ø, Hoffart A, Sundgot BJ (2011) Exercise dependence score in patients with longstanding eating disorders and controls: the importance of affect regulation and physical activity intensity. Eur Eat Disord Rev 19:249–255

Brellenthin AG, Crombie KM, Hillard CJ, Koltyn KF (2017) Endocannabinoid and mood responses to exercise in adults with varying activity levels. Med Sci Sports Exerc 49(8):1688–1696

Brewerton TD, Stellefson EJ, Hibbs N, Hodges EL, Cochrane CE (1995) Comparison of eating disorder patients with and without compulsive exercising. Int J Eat Disord 17(4):413–416

Bühren K, Schwarte R, Fluck F, Timmesfeld N, Krei M, Egberts K et al (2014) Comorbid psychiatric disorders in female adolescents with first-onset anorexia nervosa: comorbidity in adolescent anorexia nervosa. Eur Eat Disord Rev 22(1):39–44

Carek PJ, Laibstain SE, Carek SM (2011) Exercise for the treatment of depression and anxiety. Int J Psychiatry Med 41(1):15–28

Chekroud SR, Gueorguieva R, Zheutlin AB, Paulus M, Krumholz HM, Krystal JH et al (2018) Association between physical exercise and mental health in 1.2 million individuals in the USA between 2011 and 2015: a cross-sectional study. Lancet. Psychiatry 5(9):739–746

Chowdhury TG, Chen YW, Aoki C (2015) Using the activity-based anorexia rodent model to study the neurobiological basis of anorexia nervosa. J Vis Exp 22(105):e52927

Coniglio KA, Cooper M, Selby EA (2022) Behavioral reinforcement of pathological exercise in anorexia nervosa. Int J Eat Disord [Internet]. [cited 2021 Dec 6]; Available from: https://onlinelibrary.wiley.com. https://doi.org/10.1002/eat.23626
Cook B, Leininger L (2017) The ethics of exercise in eating disorders: can an ethical principles approach guide the next generation of research and clinical practice? J Sport Health Sci 6(3): 295–298
Cook B, Hausenblas H, Freimuth M (2014) Exercise addiction and compulsive exercising: Relationship to eating disorders, substance use disorders, and addictive disorders. In: Eating disorders, addictions, and substance use disorders. pp 127–144. (Eating Disorders)
Cook B, Wonderlich SA, Mitchell JE, Thompson R, Sherman R, McCallum K (2016) Exercise in eating disorders treatment: systematic review and proposal of guidelines. Med Sci Sports Exerc 48(7):1408–1414
Crombie KM, Brellenthin AG, Hillard CJ, Koltyn KF (2018) Psychobiological responses to aerobic exercise in individuals with posttraumatic stress disorder: psychobiological responses to exercise in PTSD. J Trauma Stress 31(1):134–145
Crombie KM, Sartin-Tarm A, Sellnow K, Ahrenholtz R, Lee S, Matalamaki M et al (2021) Exercise-induced increases in anandamide and BDNF during extinction consolidation contribute to reduced threat following reinstatement: preliminary evidence from a randomized controlled trial. Psychoneuroendocrinology 132:105355
Dalle Grave R, Calugi S, Marchesini G (2008) Compulsive exercise to control shape or weight in eating disorders: prevalence, associated features, and treatment outcome. Compr Psychiatry 49(4):346–352
Davis C, Claridge G (1998) The eating disorders as addiction: a psychobiological perspective. Addict Behav 23(4):463–475
Davis C, Kaptein S (2006) Anorexia nervosa with excessive exercise: a phenotype with close links to obsessive-compulsive disorder. Psychiatry Res 142(2–3):209–217
Davis C, Woodside DB (2002) Sensitivity to the rewarding effects of food and exercise in the eating disorders. Compr Psychiatry 43(3):189–194
Davis C, Katzman DK, Kaptein S, Kirsh C, Brewer H, Kalmbach K et al (1997) The prevalence of high-level exercise in the eating disorders: etiological implications. Compr Psychiatry 38(6): 321–326
de Geus E, Bartels M, Kaprio J, Lightfoot J, Thomis M (2014) Genetics of regular exercise and sedentary behaviors. Twin Res Hum Genet 17(4):262–271
Deslandes A, Moraes H, Ferreira C, Veiga H, Silveira H, Mouta R et al (2009) Exercise and mental health: many reasons to move. Neuropsychobiology 59(4):191–198
Dostalova I, Bartak V, Papezova H, Nedvidkova J (2007) The effect of short-term exercise on plasma leptin levels in patients with anorexia nervosa. Metabolism 56(4):497–503
Ehrlich S, Burghardt R, Schneider N, Broecker-Preuss M, Weiss D, Merle JV et al (2009) The role of leptin and cortisol in hyperactivity in patients with acute and weight-recovered anorexia nervosa. Prog Neuropsychopharmacol Biol Psychiatry 33(4):658–662
Engel SG, Wonderlich SA, Crosby RD, Mitchell JE, Crow S, Peterson CB et al (2013) The role of affect in the maintenance of anorexia nervosa: Evidence from a naturalistic assessment of momentary behaviors and emotion. J Abnorm Psychol 122(3):709
Fairburn C (2008) Eating disorders: the transdiagnostic view and the cognitive behavioral theory. Guilford Press, New York
Fairburn C, Bèglin SJ (1994) Assessment of eating disorders: interview or self-report questionnaire? Int J Eat Disord
Flack KD, Pankey C, Ufholz K, Johnson L, Roemmich JN (2019) Genetic variations in the dopamine reward system influence exercise reinforcement and tolerance for exercise intensity. Behav Brain Res 375:112148
Foldi CJ, Milton LK, Oldfield BJ (2017a) A focus on reward in anorexia nervosa through the lens of the activity-based anorexia rodent model. J Neuroendocrinol 29(10):e12479

Foldi CJ, Milton LK, Oldfield BJ (2017b) The role of mesolimbic reward neurocircuitry in prevention and rescue of the activity-based anorexia (ABA) phenotype in rats. Neuropsychopharmacology 42(12):2292–2300

Goodwin H, Haycraft E, Taranis L, Meyer C (2011) Psychometric evaluation of the compulsive exercise test (CET) in an adolescent population: links with eating psychopathology. Eur Eat Disord Rev 19(3):269–279

Griffiths S, Mond JM, Murray SB, Touyz S (2015) Positive beliefs about anorexia nervosa and muscle dysmorphia are associated with eating disorder symptomatology. Aust N Z J Psychiatry 49(9):812–820

Guarda AS, Schreyer CC, Boersma GJ, Tamashiro KL, Moran TH (2015) Anorexia nervosa as a motivated behavior: relevance of anxiety, stress, fear and learning. Physiol Behav 152 (B):466–472

Hausenblas HA, Downs DS (2002) How much is too much? The development and validation of the exercise dependence scale. Psychol Health 17(4):387–404

Hausenblas HA, Cook BJ, Chittester NI (2008) Can exercise treat eating disorders. Exerc Sport Sci Rev 36:43–47

Hay P, Touyz S, Arcelus J, Pike K, Attia E, Crosby RD et al (2018) A randomized controlled trial of the compuLsive exercise activity TheraPy (LEAP): a new approach to compulsive exercise in anorexia nervosa. Int J Eat Disord 51(8):999–1004

Herring M, O'Connor PJ, Dishman RK (2010) The effect of exercise training on anxiety symptoms among patients: a systematic review. Arch Intern Med 170(4):321–331

Herring M, Sailors M, Bray M (2014) Genetic factors in exercise adoption, adherence and obesity. Obes Rev 15(1):29–39

Heyman E, Gamelin FX, Goekint M, Piscitelli F, Roelands B, Leclair E et al (2012) Intense exercise increases circulating endocannabinoid and BDNF levels in humans – possible implications for reward and depression. Psychoneuroendicrinology 37(6):844–851

Hillebrand JJG, de Rijke CE, Brakkee JH, Kas MJH, Adan RAH (2005) Voluntary access to a warm plate reduces hyperactivity in activity-based anorexia. Physiol Behav 85(2):151–157

Hillebrand JJG, Kas MJH, van Elburg AA, Hoek HW, Adan R, a. H. (2008) Leptin's effect on hyperactivity: potential downstream effector mechanisms. Physiol Behav 94(5):689–695

Ho EV, Klenotich SJ, McMurray MS, Dulawa SC (2016) Activity-based anorexia alters the expression of BDNF transcripts in the mesocorticolimbic reward circuit. PLoS One 11(11): e0166756

Holland LA, Brown TA, Keel PK (2014) Defining features of unhealthy exercise associated with disordered eating and eating disorder diagnoses. Psychol Sport Exerc 15(1):116–123

Holtkamp K, Herpertz-Dahlmann B, Mika C, Heer M, Heussen N, Fichter M et al (2003) Elevated physical activity and low leptin levels co-occur in patients with anorexia nervosa. J Clin Endocrinol Metab 88(11):5169–5174

Holtkamp K, Hebebrand J, Herpertz-Dahlmann B (2004) The contribution of anxiety and food restriction on physical activity levels in acute anorexia nervosa. Int J Eat Disord 36(2):163–171

Holtkamp K, Herpertz-Dahlmann B, Hebebrand K, Mika C, Kratzsch J, Hebebrand J (2006) Physical activity and restlessness correlate with leptin levels in patients with adolescent anorexia nervosa. Biol Psychiatry 60(3):311–313

Juwono ID, Tolnai N, Szabo A (2022) Exercise addiction in athletes: a systematic review of the literature. Int J Ment Health Addict 20(5):3113–3127

Karr TM, Cook B, Zunker C, Cao L, Crosby RD, Wonderlich SA et al (2017) Examining physical activity and affect using objective measures: a pilot study of anorexia nervosa. Sport J:20

Klein DA, Schebendach JE, Gershkovich M, Bodell LP, Foltin RW, Walsh BT (2010) Behavioral assessment of the reinforcing effect of exercise in women with anorexia nervosa: further paradigm development and data. Int J Eat Disord 43(7):611–618

Klenotich SJ, Ho EV, McMurray MS, Server CH, Dulawa SC (2015) Dopamine D2/3 receptor antagonism reduces activity-based anorexia. Transl Psychiatry 4(5):e613

Klimentidis YC, Raichlen DA, Bea J, Garcia DO, Wineinger NE, Mandarino LJ et al (2018) Genome-wide association study of habitual physical activity in over 377,000 UK Biobank participants identifies multiple variants including CADM2 and APOE. Int J Obes (Lond) 42(6): 1161–1176

Kolar DR, Neumayr C, Roth M, Voderholzer U, Perthes K, Schlegl S (2020) Testing an emotion regulation model of physical activity in adolescents with anorexia nervosa: a pilot ecological momentary assessment. Eur Eat Disord Rev 28(2):170–183

Kumar A, Pareek V, Faiq MA, Ghosh SK, Kumari C (2019) Adult neurogenesis in humans: a review of basic concepts, history, current research, and clinical implications. Innov Clin Neurosci 16(5–6):30–37

Lampe EW, Trainor C, Presseller EK, Michael ML, Payne-Reichert A, Juarascio AS et al (2021) Characterizing reasons for exercise in binge-spectrum eating disorders. Eat Behav 1(43):101558

Lampe EW, Forman EM, Juarascio AS, Manasse SM (2022) Feasibility, acceptability, and preliminary target engagement of a healthy physical activity promotion intervention for bulimia nervosa: development and evaluation via case series design. Cogn Behav Pract 29(3):598–613

Lampe EW, Gorrell S, Smith K, Payne-Reichert AM, Juarascio AS, Manasse SM (2023a) Divergent trajectories of positive affect following maladaptive and non-maladaptive exercise among individuals with binge-spectrum eating disorders. Int J Eat Disord 56(5):1001–1010

Lampe EW, Presseller EK, Abber SR, Sonnenblick RM, Juarascio AS, Manasse SM (2023b) Affective profiles of exercise episodes are associated with maladaptive and adaptive motivations for exercise. Eur Eat Disord Rev:erv.3012

Levinson CA, Williams BM (2020) Eating disorder fear networks: identification of central eating disorder fears. Int J Eat Disord:eat.23382

Lichtenstein MB, Hinze CJ, Emborg B, Thomsen F, Hemmingsen SD (2017) Compulsive exercise: links, risks and challenges facedent. Psychol Res Behav Manag 10:85–95

Lock J, Le Grange D (2015) Treatment manual for anorexia nervosa: a family-based approach, 2nd edn. Guildford Press, New York

Machado PPP, Machado BC, Gonçalves S, Hoek HW (2007) The prevalence of eating disorders not otherwise specified. Int J Eat Disord 40(3):212–217

Martenstyn JA, Touyz S, Maguire S (2021) Treatment of compulsive exercise in eating disorders and muscle dysmorphia: protocol for a systematic review. J Eat Disord 9(1):19

Martinsen M, Sundgot-Borgen J (2013) Higher prevalence of eating disorders among adolescent elite athletes than controls. Med Sci Sports Exerc 45(6):1188–1197

Mathisen TF, Bratland-Sanda S, Rosenvinge JH, Friborg O, Pettersen G, Vrabel KA et al (2018) Treatment effects on compulsive exercise and physical activity in eating disorders. J Eat Disord 6(1):43

Mathisen TF, Rosenvinge JH, Friborg O, Vrabel K, Bratland-Sanda S, Pettersen G et al (2020) Is physical exercise and dietary therapy a feasible alternative to cognitive behavior therapy in treatment of eating disorders? A randomized controlled trial of two group therapies. Int J Eat Disord 53(4):574–585

Mathisen TF, Hay P, Bratland-Sanda S (2023) How to address physical activity and exercise during treatment from eating disorders: a scoping review. Curr Opin Psychiatry

Meyer C, Plateau CR, Taranis L, Brewin N, Wales J, Arcelus J (2016) The compulsive exercise test: confirmatory factor analysis and links with eating psychopathology among women with clinical eating disorders. J Eat Disord 4:22

Meyer JD, Crombie KM, Cook DB, Hillard CJ, Koltyn KF (2019) Serum endocannabinoid and mood changes after exercise in major depressive disorder. Med Sci Sports Exerc 51(9): 1909–1917

Monell E, Levallius J, Forsén Mantilla E, Birgegård A (2018) Running on empty – a nationwide large-scale examination of compulsive exercise in eating disorders. J Eat Disord 6(1):11

Moola FJ, Gairdner SE, Amara CE (2013) Exercise in the care of patients with anorexia nervosa: a systematic review of the literature. Ment Health Phys Act 6(2):59–68

Murray SB, Rieger E, Karlov L, Touyz SW (2013) An investigation of the transdiagnostic model of eating disorders in the context of muscle dysmorphia. Eur Eat Disord Rev 21(2):160–164
Nagata JM, Carlson JL, Kao JM, Golden NH, Murray SB, Peebles R (2017) Characterization and correlates of exercise among adolescents with anorexia nervosa and bulimia nervosa. Int J Eat Disord 50(12):1394–1403
Nakai Y, Nin K, Noma S, Hamagaki S, Takagi R, Teramukai S et al (2017) Clinical presentation and outcome of avoidant/restrictive food intake disorder in a Japanese sample. Eat Behav 1(24): 49–53
O'Hara C, Keyes A, Renwick B, Leyton M, Campbell I, Schmidt U (2016) The effects of acute dopamine precursor depletion on the reinforcing value of exercise in anorexia nervosa. PLoS One 11(1)
Pasman L, Thompson JK (1988) Body image and eating disturbance in obligatory runners, obligatory weightlifters, and sedentary individuals. Int J Eat Disord 7(6):759–769
Reichert M, Schlegel S, Jagau F, Timm I, Wieland L, Ebner-Priemer UW et al (2020) Mood and dysfunctional cognitions constitute within-subject antecedents and consequences of exercise in eating disorders. Psychother Psychosom 89(2):119–121
Rosa-Caldwell ME, Eddy KT, Rutkove SB, Breithaupt L (2023) Anorexia nervosa and muscle health: a systematic review of our current understanding and future recommendations for study. Int J Eat Disord 56(3):483–500
Sawyer SM, Whitelaw M, Le Grange D, Yeo M, Hughes EK (2016) Physical and psychological morbidity in adolescents with atypical anorexia nervosa. Pediatrics 137(4):e20154080
Scharmer C, Gorrell S, Schaumberg K, Anderson D (2020) Compulsive exercise or exercise dependence? Clarifying conceptualizations of exercise in the context of eating disorder pathology. J Clin Sport Psychol 46:101586
Schaumberg K, Bulik C, Micali N (2022) Patterns of maladaptive exercise behavior from ages 14-24 in a longitudinal cohort [Internet]. PsyArXiv; [cited 2023 Mar 31]. Available from: https://psyarxiv.com/htx5z/
Schaumberg K, Bulik CM, Micali N (2023) Patterns of maladaptive exercise behavior from ages 14-24 in a longitudinal cohort. J Child Psychol Psychiatry
Schmalbach I, Herhaus B, Pässler S, Runst S, Berth H, Wolff-Stephan S et al (2020) Cortisol reactivity in patients with anorexia nervosa after stress induction. Transl Psychiatry 10(1):275
Severinsen MCK, Pedersen BK (2020) Muscle-organ crosstalk: the emerging roles of myokines. Endocr Rev 41(4):bnaa016
Shroff H, Reba L, Thornton LM, Tozzi F, Klump KL, Berrettini WH et al (2006) Features associated with excessive exercise in women with eating disorders. Int J Eat Disord 39(6): 454–461
Smith AR, Fink EL, Anestis MD, Ribeiro JD, Gordon KH, Davis H et al (2013) Exercise caution: over-exercise is associated with suicidality among individuals with disordered eating. Psychiatry Res 206(2–3):246–255
Steffen JJ, Brehm BJ (1999) The dimensions of obligatory exercise. Eat Disord 7:219–226
Stiles-Shields EC, Goldschmidt AB, Boepple L, Glunz C, Le Grange D (2011) Driven exercise among treatment-seeking youth with eating disorders. Eat Behav 12(4):328–331
Stiles-Shields C, DclinPsy BB, Lock J, Le Grange D (2015) The effect of driven exercise on treatment outcomes for adolescents with anorexia and bulimia nervosa. Int J Eat Disord 48(4): 392–396
Thornton LM, Munn-Chernoff MA, Baker JH, Juréus A, Parker R, Henders AK et al (2018) The anorexia nervosa genetics initiative (ANGI): overview and methods. Contemp Clin Trials 74: 61–69
Trott M, Jackson SE, Firth J, Jacob L, Grabovac I, Mistry A et al (2021) A comparative meta-analysis of the prevalence of exercise addiction in adults with and without indicated eating disorders. Eat Weight Disord Stud Anorex Bulim Obes 26(1):37–46

Vancampfort D, Vanderlinden J, Hert M, Adámkova Segard M, Skjaerven L, Catalan D et al (2013) A systematic review on physical therapy interventions for patients with binge eating disorder. Disabil Rehabil:35

Vancampfort D, Vanderlinden J, De Hert M, Soundy A, Adámkova M, Skjaerven LH et al (2014) A systematic review of physical therapy interventions for patients with anorexia and bulemia nervosa. Disabil Rehabil 36(8):628–634

Vancampfort D, Firth J, Schuch FB, Rosenbaum S, Mugisha J, Hallgren M et al (2017) Sedentary behavior and physical activity levels in people with schizophrenia, bipolar disorder and major depressive disorder: a global systematic review and meta-analysis. World Psychiatry 16(3): 308–315

Vankim NA, Nelson TF (2013) Vigorous physical activity, mental health, perceived stress, and socializing among college students. Am J Health Promot 28:7

Verhagen LAW, Luijendijk MCM, Hillebrand JJG, Adan RAH (2009) Dopamine antagonism inhibits anorectic behavior in an animal model for anorexia nervosa. Eur Neuropsychopharmacol J Eur Coll Neuropsychopharmacol 19(3):153–160

Wons O, Lampe E, Patarinski AG, Schaumberg K, Juarascio A (2023) Change in adaptive and maladaptive exercise and objective physical activity throughout CBT for individuals with eating disorders. Eat Weight Disord Stud Anorex Bulim Obes 28(1):40

Yilmaz Z, Schaumberg K, Halvorsen M, Goodman EL, Brosof LC, Crowley JJ et al (2022) Predicting eating disorder and anxiety symptoms using disorder-specific and transdiagnostic polygenic scores for anorexia nervosa and obsessive-compulsive disorder. Psychol Med 4:1–15

Young S, Touyz S, Meyer C, Arcelus J, Rhodes P, Madden S et al (2017) Validity of exercise measures in adults with anorexia nervosa: the EDE, compulsive exercise test and other self-report scales. Int J Eat Disord 50(5):533–541

Exercise and PTSD

Laura M. Kinsman, Holden J. Norrie, Geoffrey S. Rachor, and Gordon J. G. Asmundson

Contents

Abstract Evidence indicating that exercise benefits mental health symptoms across a range of mental health diagnoses spans decades of scientific literature; however, fewer studies have examined the impact of exercise on posttraumatic stress disorder (PTSD). Exercise is an accessible, cost-effective, and scalable treatment option that has the potential to improve both physiological and psychological symptoms among individuals with PTSD. The purpose of this chapter is to review empirical literature on the role of exercise in the treatment of PTSD. Researchers have demonstrated that exercise improves PTSD symptoms as both a stand-alone treatment and as an adjunct to cognitive behavioral and trauma-focused therapies. Additional research is needed to clarify mechanisms that account for the impacts of exercise on PTSD and to identify which components of exercise (e.g., type of exercise, dose, intensity, frequency) are the most beneficial.

Keywords posttraumatic stress disorder · exercise · physical activity

L. M. Kinsman, H. J. Norrie, G. S. Rachor, and G. J. G. Asmundson (✉)
Department of Psychology, University of Regina, Regina, SK, Canada
e-mail: gordon.asmundson@uregina.ca

Curr Topics Behav Neurosci (2024) 67: 241–262
https://doi.org/10.1007/7854_2024_500
Published Online: 8 August 2024

1 Introduction

The physiological and psychological benefits of exercise in treating various mood- and anxiety-related disorders are well documented across the scientific literature. There is a growing interest in documenting the benefits of exercise among individuals with posttraumatic stress disorder (PTSD). The purpose of the current chapter is to review the empirical literature on the role of exercise in the treatment of PTSD. An overview of the research evidence for treating PTSD with exercise is provided, including recommended types of exercise, dose, intensity, and frequency. The chapter concludes with recommendations for further research efforts in elucidating the role of exercise in the treatment of PTSD.

2 PTSD

2.1 *What Is PTSD?*

PTSD is characterized by the sudden or delayed onset of psychological distress following exposure to one or more traumatic events (American Psychiatric Association 2022). Approximately 70.4% of individuals worldwide have experienced at least one traumatic event in their lifetime (Benjet et al. 2016; Kessler et al. 2017a). Interpersonal violence (e.g., sexual violence, physical violence) is among the most common potentially traumatic experiences, while the unexpected death of a loved one is among the greatest risk factors for developing PTSD (Kessler et al. 2017b). Approximately 6.8% of the adult population in the United States experience PTSD in their lifetime (American Psychiatric Association 2022). The risk of developing PTSD is elevated among females, veterans, active military members, public safety personnel (e.g., correctional officers, firefighters, paramedics, police), and survivors of sexual violence or traumatic injury (American Psychiatric Association 2022). PTSD often cooccurs with other disorders, including depressive, bipolar, anxiety, substance use disorders, and chronic pain conditions (e.g., migraine, back pain, arthritis, and fibromyalgia). For example, 25% ($N = 1761$) of Canadian Forces personnel who met diagnostic criteria for PTSD also reported chronic pain conditions (Vun et al. 2018). Moreover, individuals who are diagnosed with PTSD and a co-occurring disorder are more likely than those with PTSD alone to develop more severe comorbid symptoms, impairments, and suicide attempts and completions (Walter et al. 2018).

2.2 *Established PTSD Treatments*

Several evidence-based psychological and pharmacological interventions are available for treating PTSD, of which cognitive behavioral therapies (CBT) are regarded as the gold standard (Clinical practice guideline for the treatment of posttraumatic stress disorder (PTSD) in adults: (501872017-001) 2017; Schnurr 2017). Trauma-focused CBT, including cognitive therapy, cognitive processing therapy, and prolonged exposure (PE), exhibits short- and long-term efficacy in reducing symptoms of PTSD and is strongly recommended by the American Psychological Association (APA) (Clinical practice guideline for the treatment of posttraumatic stress disorder (PTSD) in adults: (501872017-001) 2017; Schnurr 2017; Asmundson et al. 2019; Bradley et al. 2005; Cusack et al. 2016; Kline et al. 2018; Olthuis et al. 2016; Van Etten and Taylor 1998; Watkins et al. 2018). Researchers have identified – using meta-analyses – that treatment gains may be sustained up to 20 months posttreatment (Kline et al. 2018). However, challenges such as treatment nonresponse, residual symptoms, and attrition may reduce positive treatment outcomes (Fonzo et al. 2020; Larsen et al. 2019; Najavits 2015; Koek et al. 2016; Schottenbauer et al. 2008). The APA also conditionally recommends psychological therapies such as brief eclectic therapy, eye movement and desensitization and reprocessing therapy, narrative exposure therapy, and pharmacological treatments such as selective serotonin reuptake inhibitors (SSRIs) and selective serotonin norepinephrine reuptake inhibitors (Clinical practice guideline for the treatment of posttraumatic stress disorder (PTSD) in adults: (501872017-001) 2017; Brady et al. 2000; Lindauer et al. 2005; Marshall et al. 2001; Schauer et al. 2011; Shapiro 2007). Additional research is necessary to determine the efficacy of conditionally recommended treatments among different populations.

Treatment planning and delivery may be influenced by factors such as access to services, stigma regarding mental health services utilization, and client preferences (Chen et al. 2013; Markowitz et al. 2016). For example, stigma regarding accessing mental health treatment may prevent individuals from seeking treatment for PTSD (Wade et al. 2015), in which case alternative treatment options such as pharmacological interventions may be considered or used in conjunction with psychotherapy. However, pharmacological treatments may confer adverse outcomes that are not associated with psychotherapeutic interventions (e.g., sexual dysfunction, insomnia) (Edinoff et al. 2021). The positive effects of pharmacological treatments may also be diminished following discontinuation of treatment, whereas treatment gains following trauma-focused psychotherapies generally remain stable following treatment completion (Kline et al. 2018).

2.3 *The Role of Exercise in Treatment*

As noted above, limited access to first-line treatments, stigmatization associated with utilizing mental health services, and side effects associated with pharmacotherapy may pose significant barriers to effective PTSD treatment. Conversely, while not entirely free from barriers, exercise is generally not associated with stigma or adverse side effects and tends to be lower in cost (Hegberg et al. 2019); however, limited financial resources may still pose a barrier to engaging in exercise (e.g., purchasing a gym membership or exercise equipment). Other barriers to exercise may include time constraints, low motivation, and reduced self-efficacy (Hegberg et al. 2019; Zen et al. 2012).

Exercise interventions have the potential to improve both psychological and physical sequelae associated with PTSD, which are explored further in subsequent sections of the chapter. Researchers have identified an association between PTSD and physiological changes that adversely impact multiple organ systems (e.g., central nervous system, autoimmune and inflammatory, neuroendocrine, metabolic, and cardiovascular disorders) (Koenen et al. 2017; McFarlane et al. 2017; Pacella et al. 2013; Qureshi et al. 2009; Ryder et al. 2018). Indeed, individuals with PTSD are at risk for poor health outcomes such as cardiovascular disease, diabetes, hypertension, metabolic syndrome, pain, and stroke (Asmundson et al. 2002; Boscarino 2004; Burg and Soufer 2016; Goodwin and Davidson 2005; Rosman et al. 2019; Rosenbaum et al. 2015a; Sareen et al. 2005). Researchers have suggested the potential for exercise to improve hypothalamic-pituitary axis (HPA) functioning and reduce inflammatory markers as a rationale for implementing exercise interventions for PTSD, as improvements in HPA functioning and inflammation may lead to improved stress response (e.g., reduced cortisol) and sleep regulation (Hegberg et al. 2019; de Kloet et al. 2006; Santos et al. 2007). Given the potential for exercise to improve physiological and psychological functioning among individuals with PTSD, exercise is a promising treatment option for PTSD, if not at least an adjunct to psychotherapy. Physical health conditions may also increase risk for the onset of PTSD symptoms (Carmassi et al. 2020), suggesting that exercise may also prevent PTSD.

3 Exercise and Mental Health

3.1 Psychological Impacts of Exercise

A well-documented short-term psychological benefit of exercise is improved over time following engaging in exercise (Guérin et al. 2013; Reed and Buck 2009). A meta-analysis of 158 lab-based studies identified an increase in positive affect (e.g., energy, joy, pleasant mood, vigor) following acute aerobic exercise with a medium effect size ($d = 0.47$) (Reed and Ones 2006). The benefits of exercise on positive affect also extend to naturalistic field-based studies, as evidenced by a systematic review that found enhanced positive affect and feelings of vitality following physical exercise (Liao et al. 2015). Factors associated with greater increases in positive affect following exercise interventions include more negative baseline preexercise affect, lower intensity exercise, durations of exercise spanning 7–35 min, and low to moderately intense doses of exercise (Reed and Ones 2006). These findings were corroborated by a study of individuals in a weight training class who were randomly assigned to one of four conditions of resistance exercise with varying intensity and rest time (i.e., low-intensity and long rest time, low-intensity and short rest time, high-intensity and long rest time, high-intensity and short rest-time) or a no-exercise control (Bibeau et al. 2010). Positive affect improved significantly among all exercise conditions, with the largest increase in positive affect observed in the group who exercised at a low intensity for a long period of time (Bibeau et al. 2010). Exercise enjoyment also impacts affective states, with one study finding that greater enhancement in affect was reporting following individuals engaging in their most preferred compared to their least preferred form of exercise (Lane et al. 2005). Studies have also shown that individuals with PTSD reported increased positive affect after engaging in moderate-intensity aerobic exercise (Crombie et al. 2018, 2019, 2021a, b).

Researchers have also demonstrated that negative affect (e.g., anxiety, stress) decreases following exercise (Basso and Suzuki 2017; Petruzzello et al. 1991). A meta-analysis of 16 randomized controlled trials determined that acute bouts of exercise can produce reductions in self-reported state anxiety with a small but statistically significant effect size (Ensari et al. 2015). Researchers have identified that reductions in negative affect persist for up to 24 h postexercise (Reed and Ones 2006; Basso and Suzuki 2017; Petruzzello et al. 1991; Ensari et al. 2015; Maroulakis and Zervas 1993; Raglin and Morgan 1987). One form of exercise that is associated with worsened negative affect is high-intensity interval training (HIIT), a type of exercise consisting of brief, intermittent bouts of exercise followed by periods of rest or reduced activity (Karlsen et al. 2017). HIIT may have a negative impact on affect by enhancing the experience of negative emotions and perceptions of physical exertion, as well as decreasing feelings of pleasure (Hall et al. 2020; Jung et al. 2014; Kilpatrick et al. 2007; Parfitt et al. 2006; Saanijoki et al. 2015). Researchers speculate that negative feelings following HIIT may be attributed to the strong link between exercise intensity and adverse emotions (Saanijoki et al. 2015; Ekkekakis 2003; Oliveira et al. 2013). However, in a study where eight males were randomly assigned to complete either HIIT running or moderate-intensity continuous running, researchers demonstrated that HIIT running was rated as more enjoyable than moderate intensity running (Bartlett et al. 2011).

Accumulating evidence demonstrates that exercise improves mental health through reduced symptoms of anxiety among individuals with and without an anxiety disorder diagnosis (Asmundson et al. 2013; Bartley et al. 2013; Herring 2018; Herring et al. 2014; Jayakody et al. 2014; Ramos-Sanchez et al. 2021; Stonerock et al. 2015; Stubbs et al. 2017). In a recent meta-analysis of 13 randomized controlled trials, researchers confirmed that exercise is efficacious in reducing anxiety symptoms (e.g., worry, nervousness) with a small-medium effect size compared to control groups (SMD = −0.43 among individuals with anxiety and related disorders) (Ramos-Sanchez et al. 2021). Both aerobic exercise and resistance training have received the most empirical support for reducing anxiety symptoms; however, other forms of low-intensity exercise such as tai chi, walking, and yoga have also been associated with lowered anxiety (Asmundson et al. 2013; Ramos-Sanchez et al. 2021).

3.2 Physiological Impacts of Exercise

Past research indicates that changes in psychological states are associated with changes in measures of immunity and inflammation (Weisse 1992). More specifically, psychological states like depression and stress are correlated to decreases in immune system response (Herbert and Cohen 1993a, b), and meta-analytic findings indicate that depression may be associated with inflammation (Zorrilla et al. 2001). Studies also show that anxiety is associated with increased inflammation, leading to a higher risk of developing inflammatory disease (O'Donovan et al. 2010). The increase in inflammation due to anxiety is theorized to be related to HPA activation, as the states of anxiety are significantly associated with the inflammatory markers interleukin-6 (IL-6) and high sensitivity C-reactive protein (hs-CRP) and decreased cortisol (O'Donovan et al. 2010). Individuals with increased levels of inflammatory cytokines are also more likely to develop PTSD, and the increased inflammatory response is theorized to influence health declines indirectly related to PTSD (Gill et al. 2009).

Researchers have suggested that the enhancing effects of exercise on cognition and mood are mediated by brain-derived neurotrophic factor (BDNF) (Szuhany et al. 2015). Past studies have shown that BDNF is correlated with increases in neurogenesis, potentiation of neurons, and dendrite maintenance and development (Szuhany et al. 2015). In studies examining the impact of exercise and depression in rats, exercise was shown to lessen the decrease of BDNF mRNA expression in comparison to the nonexercise group, 24 h after the chronic unpredictable stress was discontinued (Zheng et al. 2006). Levels of cortisol in the nonexercise group were also observed to be higher in comparison to the exercise group. These results suggest that exercise may lead to more flexible HPA responses to stress and a faster return to baseline levels of cortisol following chronic unpredictable stress (Zheng et al. 2006). Additionally, the group of rats that exercised exhibited less depressive behaviors.

Exercise has also been noted as an effective method for increasing endocannabinoids, dopamine, and levels of serotonin, similar to SSRIs (Desai et al. 2022; Wipfli et al. 2011; Crombie et al. 2021c; Heyman et al. 2012). The endocannabinoid system exists within the central nervous system and is involved in synaptic plasticity and several biological processes (e.g., cognition, eating, inflammation, metabolism, pain, reproduction) (Desai et al. 2022; Zou and Kumar 2018). Recent studies have shown that aerobic exercise maintains and increases circulating concentrations of endocannabinoids (Desai et al. 2022). The increase in endocannabinoids has also been associated with increases in BDNF, which, in turn, further supports cognitive enhancement and reductions in depression (Crombie et al. 2021c; Heyman et al. 2012). In a study comparing the effects of exercise and static stretching in a seven-week intervention, participants who were assigned to the exercise group were found to have lower levels of depression and decreased levels of serum serotonin in comparison to the stretching group (Wipfli et al. 2011). These results mimic similar results to using pharmaceutical interventions like SSRIs (Hughes et al. 1996). Researchers have hypothesized that the effectiveness of antidepressants is partly based on increases in dopaminergic synapses (Muscat et al. 1992). In animal studies examining the influence of exercise on dopaminergic circuits, researchers found that regular exercise increased endurance and plasticity within dopaminergic neurocircuits (Wipfli et al. 2011). In turn, results showed that the aerobic exercise group was found to synthesize monoamine dopamine faster than the sedentary group.

The HPA is responsible for regulating physiological reactions to stress through the secretion of corticotrophin-releasing hormone (CRH) (Hegberg et al. 2019). Secretion of CRH stimulates the pituitary gland to release adrenocorticotropic hormone (ACTH) in response to stress. Previous research has shown that individuals diagnosed with PTSD have increases in CRH and excessive HPA axis activation (Daskalakis et al. 2016). Frequent aerobic exercise has been shown to be an effective method for normalizing the HPA (Childs and de Wit 2014). More specifically, moderate to vigorous physical activity is a more efficient intensity for normalizing the HPA in comparison to low-intensity aerobic exercise (Hill et al. 2008). Exercise has also been noted as an effective method for decreasing pro-inflammatory responses (i.e., interleukin-1, IL-6) that are associated with PTSD (Archer et al. 2011; Passos et al. 2015). Specifically, studies utilizing moderate to vigorous aerobic exercise interventions have shown that regular aerobic exercise decreases pro-inflammatory responses, leading to an increased quality of sleep and decreased symptoms of PTSD (Santos et al. 2007; Archer et al. 2011).

4 Exercise and PTSD

4.1 Overview of General Research Findings

Researchers have identified associations between exercise behaviors and reduced PTSD symptoms via observational, cross-sectional research (Hegberg et al. 2019; Vancampfort et al. 2017; Whitworth et al. 2019a; b). One study of American Veterans ($N = 38{,}883$, 77% of male), 96% of whom did not have PTSD at baseline, found that participants who engaged in vigorous exercise (≥20 min, two or more times per week), but not light- or moderate-intensity exercise, had significantly reduced odds of developing new PTSD symptoms (OR 50.58, [95% CI: 0.49, 0.70]) or reporting persistent PTSD symptoms at 3–5-year follow-ups (OR 50.59, [95% CI: 0.42, 0.83]) (LeardMann et al. 2011). An additional cross-sectional study of American Veterans identified a U-shaped association between self-reported exercise frequency and probable PTSD ($N = 2{,}832$, 90% male) (Adams et al. 2020). Veterans with probable PTSD were more likely to report no exercise or exercising daily and were less likely to report exercising 1–6 days a week compared to those without probable PTSD. Indeed, exercising three or more days a week was associated with approximately 49% lower odds of probable PTSD compared to exercising less than 3 days per week [95% CI: 0.21, 0.70). No exercise was associated with more severe emotional numbing symptoms and less severe anxious arousal symptoms than those who exercised 1–6 days a week, while daily exercise was associated with greater severity of reexperiencing symptoms compared with participants who exercised 1–6 days per week. A recent follow-up study of American Veterans replicated these findings ($N = 3,875$, 9.4% women), showing that self-reported moderate levels of physical activity were associated with the lowest likelihood of probable PTSD (3.8% adjusted probability), whereas greater levels of physical activity were associated with the highest likelihood of screening positive for PTSD (7.8% adjusted probability) (Adams et al. 2022). Intrusion symptoms (e.g., nightmares) and anhedonia symptoms were associated with significantly lower probability of reporting moderate physical activity. Moreover, researchers using a national sample of American Veterans ($N = 501,161$, 95.9% male) found that the odds of reporting no regular exercise were the highest among Veterans with PTSD (OR = 1.26, [95% CI: 1.20, 1.32), as well as schizophrenia and bipolar disorder compared to Veterans without these diagnoses (Chwastiak et al. 2011). Another study of male Croatian Veterans ($N = 478$, 269 with PTSD as determined using the diagnostic criteria of the *International Classification of Diseases, 10th revision*, 209 without PTSD) found that 27.9% of Veterans with PTSD reported engaging in any exercise compared to 48.8% of Veterans without PTSD (Kozaric-Kovacic et al. 2009).

Researchers have also examined associations between exercise and PTSD symptoms among community samples. A longitudinal investigation of community-recruited adults who screened positive for self-reported PTSD symptoms ($N = 182$, 72% female) found that strenuous-intensity exercise (e.g., vigorous running or cycling) was associated with reduced avoidance, numbing, and hyperarousal symptoms at a three-month follow-up (Whitworth et al. 2017). However, reporting both strenuous and moderate (e.g., nonexhaustive sports, jogging, or weight training) intensity exercise was only associated with reduced avoidance/numbing symptoms. A significant association between exercise and reduced PTSD symptoms at three-month follow-up did not emerge among participants who only reported engaging in moderate-intensity exercise (Whitworth et al. 2017). A cross-sectional study investigated the association between exercise intensity (i.e., light, moderate, vigorous) and posttraumatic stress symptom cluster severity (i.e., reexperiencing, avoidance/numbing, and hyperarousal) in a nonclinical community-based sample exposed to a DSM-IV PTSD Criterion A traumatic lifetime event ($N = 108$, 54.6% female) (Harte et al. 2015). Results showed a significant, inverse association between self-reported vigorous-intensity exercise, but not light- or moderate-intensity exercise, and PTSD hyperarousal symptom severity after adjusting for gender and lifetime number of trauma exposures (Harte et al. 2015). However, vigorous-intensity exercise uniquely accounted for a small portion of the variance in hyperarousal symptoms (4%). No significant associations emerged between exercise of any intensity and reexperiencing or avoidance/numbing symptom clusters. As this was a nonclinical sample that did not endorse comorbid mental health disorders, these findings may not generalize to those with more severe PTSD symptoms and comorbid psychopathology (Harte et al. 2015).

Other researchers have failed to find an association between exercise and PTSD symptoms. One study of American Veterans completing a 90-day rehabilitation program for PTSD ($N = 346$, 81% male) found that PTSD symptoms were not associated with exercise (i.e., participants rated how frequently they engaged in exercise on a scale ranging from not at all to every day) but that higher levels of exercise were associated with lower levels of depressive symptoms and higher sleep quality. Additionally, decreased depressive symptoms and improved sleep quality were each associated with reduced suicide risk (Davidson et al. 2013). Another study of males with combat-related PTSD ($N = 55$) did not find a significant difference in the severity of PTSD symptoms between participants who reported regular physical activity versus those who were not active (Arnson et al. 2007). A study of Veterans ($N = 76$, 82% male) found a significant association between engagement in exercise at baseline (i.e., light, moderate, vigorous) and improved sleep quality at one-year follow-up but failed to find a significant relationship between exercise engagement at baseline and PTSD symptoms at one-year follow-up (Bosch et al. 2017). Similarly, a study of Veterans ($N = 239$, 91.1% male) did not observe a significant main effect of physical activity on PTSD symptom severity but found that exercise moderated the relationship between PTSD and pain, with participants who reported being active and endorsing high levels of pain severity or interference reporting less PTSD symptoms (Bourn et al. 2016).

Due to the cross-sectional nature of these studies, causal inferences cannot be made (e.g., those with lower PTSD symptoms may exercise more frequently or vigorously) (Hegberg et al. 2019). The use of self-report questionnaires may also influence results due to the high potential for inaccurate reporting or misunderstanding questions (Hegberg et al. 2019). As previously described, other observational studies have failed to identify a significant relationship between exercise and PTSD symptomatology (Davidson et al. 2013; Arnson et al. 2007; Bosch et al. 2017; Bourn et al. 2016), but this may be attributed to methodological limitations (Hegberg et al. 2019; Harte et al. 2015). For example, some studies measured exercise with a single-item questionnaire or variable, which does not allow for assessment of frequency, intensity, duration, and type of exercise and may not accurately reflect how exercise impacts PTSD symptoms (Hegberg et al. 2019).

4.2 Randomized Controlled Trials

Uncontrolled studies offered initial support for exercise as an effective intervention for PTSD. For example, one study with a small sample ($N = 9$) of adults with PTSD found that PTSD symptoms, anxiety, and depression reduced following a 12-week aerobic exercise program (Manger and Motta 2005). In another study, adolescent females ($N = 15$) completed an eight-week aerobic exercise program (three 20-min weekly exercise sessions) (Newman and Motta 2007). Results indicated that PTSD, depression, and anxiety were significantly reduced from pre- to postintervention and treatment gains remained at one-month follow-up. More recently, researchers evaluated a 12-week moderate-intensity exercise program (i.e., brisk walking) in a sample of Veteran women ($N = 16$) and found significant improvements in PTSD and depression symptoms by the end of the intervention (Shivakumar et al. 2017). However, the aforementioned findings may only be considered preliminary due to the lack of control groups, unblinded assessors, and small sample sizes.

A small but growing number of randomized controlled trials have investigated the impact of exercise interventions on PTSD symptoms across a variety of settings, populations, and exercise modalities (Björkman and Ekblom 2022). For example, a sample of American Veterans ($N = 47$, 81% male) were randomized to receive an integrative exercise program (3-h group sessions of aerobic and resistance training combined with mindfulness-based practices for 12 weeks) or a waitlist control condition (Goldstein et al. 2018). Participants in the integrative exercise group displayed greater reductions in PTSD symptom severity ($d = 0.90$) and greater improvements in psychological quality of life ($d = 0.53$). Analyses of PTSD symptom clusters also indicated that symptoms of hyperarousal were significantly more reduced among the integrative exercise conditions compared to the control group ($d = 0.80$). While there was a moderate effect of improvement in symptoms of reexperiencing and avoidance/numbing that favored the integrative exercise group, these effects did not reach statistical significance. Another study examined the impact of cycling on PTSD symptoms (Fetzner and Asmundson 2015). Adults with PTSD or subsyndromal PTSD ($N = 33$, 76% female) were assigned into three groups with varying attentional focus: (1) attention to somatic arousal (i.e., participants were promoted to direct their attention to the interoceptive effects of exercise), (2) distraction from somatic arousal (i.e., participants watched a nature documentary), and (3) exercise with no distractions or interceptive prompts. Reductions in PTSD symptoms and anxiety were observed among all groups, with large effect sizes, indicating that aerobic exercise has a therapeutic effect on PTSD and PTSD-related symptoms regardless of attentional focus. However, it is necessary to note that this study did not employ a nonexercise control group. A study using the same data found that greater reductions in PTSD symptoms were observed among participants with lower levels of baseline fitness, suggesting that individuals with lower fitness may experience greater benefits from aerobic exercise (LeBouthillier et al. 2016).

Studies have also examined the impact of resistance training on PTSD symptoms (Whitworth et al. 2019a, b). Researchers randomly assigned 30 (73.3% female) nontreatment-seeking, urban-dwelling adults who screened positive for PTSD to a three-week resistance exercise intervention or a time-matched control condition (Whitworth et al. 2019a). Total self-reported PTSD scores at follow-up did not significantly differ between groups; however, large effect sizes for avoidance ($d = 1.26$) and hyperarousal symptoms ($d = 0.90$), sleep quality ($d = 1.31$), and hazardous alcohol use ($d = 0.99$) were observed at follow-up between the intervention group compared to the control group. Significant differences did not emerge between groups for posttraumatic intrusions or mood and cognitive symptoms. In a separate study, researchers randomized community-dwelling adults with probable PTSD ($N = 22$, 82% women) to complete a three-week, nine-session resistance exercise training intervention or a time-matched attention control (Whitworth et al. 2019b). Results indicated that the intervention group reported significantly improved sleep quality ($d = -0.93$), sleep latency (i.e., the length of time it takes an individual to fall asleep; $d = -1.61$), and reduced anxiety ($d = -0.68$) compared to the control group; however, both groups reported similar beneficial effects on PTSD and depression symptoms. Nonsignificant effects of resistance training on overall PTSD severity may be explained by low statistical power, or it may also be that resistance training only has beneficial effects on certain PTSD symptom clusters (e.g., hyperarousal).

Randomized controlled trials designed to determine whether yoga improves PTSD symptoms have produced mixed findings (Cramer et al. 2018; Kysar-Moon et al. 2021; van der Kolk et al. 2014; Davis et al. 2020). One study randomized women with treatment-resistant PTSD ($N = 64$) to participate in weekly trauma-informed hatha yoga or supportive women's health education classes for 10 weeks (van der Kolk et al. 2014). Both groups improved in terms of reduction in PTSD diagnostic status; however, the improvement of the yoga group ($d = 1.07$) was significantly greater than the control group ($d = 0.66$). Moreover, both groups improved from pretreatment to the mid-treatment point, but only the yoga group maintained gains at post-treatment, suggesting yoga may offer more durable benefits than a women's health education class. Another study ($N = 209$, 66% male) found that participants who completed a 16-week yoga intervention reported significantly greater reductions in PTSD symptom severity compared with participants in a wellness program (i.e., walking, health education) at posttreatment ($d = 0.43$), but that these improvements were not sustained at seven-month follow-up (van der Kolk et al. 2014). Contrarily, other research has found marginally significant to no effects of yoga on PTSD and depression symptoms in a sample of women (Kysar-Moon et al. 2021).

In addition to examining the impact of exercise as a stand-alone treatment for PTSD, researchers have assessed how adding exercise to trauma-focused therapy impacts treatment outcomes and have identified mixed findings (see (Crombie et al. 2023) for a full review). One pilot study with a small sample ($N = 9$) randomized participants with PTSD to receive either 12 weeks of PE alone or PE plus a 30-min session of moderate-intensity treadmill exercise prior to each PE session (Powers et al. 2015). Individuals in the PE plus exercise group exhibited improvements in PTSD symptoms ($d = 2.65$); however, significance tests were not used to directly compare changes in PTSD symptoms between groups due to the small sample size. A recent study of adults with PTSD ($N = 130$, 61% women) randomly assigned participants to receive either nine weekly sessions of PE with aerobic exercise or PE with passive stretching (Bryant et al. 2023). Participants in the PE with aerobic exercise group exhibited greater reductions in PTSD scores compared to those in the PE with stretching group at the six-month follow-up ($d = 0.6$). An additional randomized controlled trial compared the effectiveness of physical activity added to an intensive trauma-focused treatment (i.e., daily PE, EMDR, and psychoeducation) for PTSD with nonphysical control activities (e.g., crafting activities with paint, clay, and crochet) (Voorendonk et al. 2023). Significant decreases in both clinician and self-reported PTSD symptoms were observed among both conditions with large effect sizes (pre- to posttreatment: $d = 2.28$, pretreatment to -month follow-up: $d = 2.20$), but significant differences between conditions were not found. Further, 80.0% of participants in the physical activity condition and 82.7% of participants in the control condition no longer met diagnostic criteria for PTSD at posttreatment (Voorendonk et al. 2023).

Another randomized controlled trial examined the impact of exercise added to treatment as usual for PTSD (Rosenbaum et al. 2015b). Participants ($N = 81$) were randomly assigned to receive either usual care (i.e., psychotherapy, pharmaceutical interventions, group therapy) or exercise (i.e., resistance training and walking) in addition to usual care. Participants in the exercise group showed significant reductions in PTSD symptoms compared with the usual care group, with a medium effect size (MD = −5.4, partial eta squared = 0.07, [95% CI: −13.7 to −0.60]). Participants in the intervention group also reported greater improvements in depressive symptoms, sleep quality, and sedentary time than those in the usual care group.

Other studies have failed to find evidence that treatment outcomes are improved by the addition of exercise protocols (Nordbrandt et al. 2020; Young-McCaughan et al. 2022). A recent study of active duty service members ($N = 72$) with clinically significant PTSD symptoms investigated the impact of adding aerobic exercise to imaginal exposure for PTSD (Young-McCaughan et al. 2022). Participants were randomly assigned to one of four conditions: exercise only, imaginal exposure only, imaginal exposure plus exercise, and a self-care control condition. Participants exercised five times a week for 8 weeks and chose which type of exercise they would do to keep their heart rate above 60% of their individual heart rate reserve for 20–25 min, with the majority of participants electing to run outside or on a treadmill. While PTSD symptom severity decreased from pre- to postintervention with large effect sizes among all three experimental conditions, there were no significant differences between the experimental conditions, suggesting that adding aerobic exercise before completing an imaginal exposure protocol may not confer additional benefits than imaginal exposure alone. However, this study is limited by the lack of a true control condition, as participants assigned to the self-care control group may have participated in exercise, online self-help, or other interventions that may have targeted PTSD symptoms. Participants were also instructed to engage in their preferred form of aerobic exercise at >60% of their individually determined heart rate reserve for 20–25 min but did not provide data on exercise intensity (e.g., heart rate), which may have had an impact on participants' physiological response to exercise and treatment outcomes. This finding also stands in contrast to previous research findings that found exercise augmented PTSD treatment outcomes (Bryant et al. 2023). Another study randomized refugees with PTSD ($N = 318$, 53% women) to one of three groups: (1) basic body awareness therapy, (2) individual mixed physical activity for 1 h per week for 20 weeks plus treatment as usual, or (3) treatment as usual (Nordbrandt et al. 2020). Reductions in PTSD symptoms were observed among all groups with no significant group differences, indicating that adding basic body awareness therapy or mixed physical activity did not augment treatment as usual.

Researchers have also examined how exercise may impact PTSD symptoms in older adults. One pilot trial randomly assigned older adult Veterans ($N = 54$, 90.7% male) to a supervised exercise condition (i.e., aerobic, strength training, balance, and flexibility exercises) or a waitlist condition (Hall et al. 2020). The authors reported what appears to be clinically meaningful reductions in PTSD symptoms (i.e., negative cognitions and mood, reexperiencing) among the exercise condition; however, the effects were not statistically significant. Worsened PTSD symptoms were also observed among a significant minority of participants in both groups, suggesting that additional research is needed to investigate individual variability in response to exercise protocols among older adults (Hall et al. 2020).

Results from meta-analyses have found additional support for exercise interventions reducing PTSD symptoms (Stubbs et al. 2017; Björkman and Ekblom 2022; Rosenbaum et al. 2015c). A recent meta-analysis of 11 randomized controlled trials ($N = 605$) examining the impact of exercise on PTSD symptoms found a small to medium effect size of exercise on PTSD symptom severity relative to control conditions (ES 0.29 [95% CI: 0.10, 0.49]) (Björkman and Ekblom 2022). A small but significant effect of exercise on depression symptoms (ES 0.20, [95% CI: 0.01, 0.38]) and a moderate effect of exercise on sleep symptoms were also identified (ES 0.51, [95% CI: 0.29 to 0.73]). Additionally, moderate effects were found for the effect of exercise on substance use (alcohol and drugs combined; ES 0.52, [95% CI: 0.06, 0.98]) and quality of life (ES 0.51, [95% CI: 0.34, 0.69]). Contrary to previous meta-analytic findings (Stubbs et al. 2017), exercise was not found to have a significant effect on anxiety symptoms (ES = 0.18, [95% CI: −0.15 to 0.51]). Evidence for publication bias was not detected. Treatment moderators such as volume of exercise (i.e., >20 h in total vs. ≤20 h in total), dose of training (i.e., high or low), type of exercise (e.g., aerobic exercise, resistance training), and type of trauma (e.g., military trauma) were also examined. The effect of exercise on PTSD symptoms was more robust when exercise interventions consisted of more than 20 h of exercise, indicating that greater amounts of exercise may be more beneficial. Exercise type, training dose, and type of trauma did not influence treatment effects.

In sum, these results suggest that exercise can be a beneficial addition to PTSD treatment, and larger amounts of exercise may be more beneficial. Adverse effects or negative experiences associated with engaging in exercise were not reported, confirming the feasibility of exercise interventions for PTSD. Additional research is needed to elucidate which type of exercise is most effective at reducing PTSD symptoms. Studies investigating the impact of exercise on PTSD are limited by factors such as small sample sizes and lack of follow-up data (Björkman and Ekblom 2022; Rosenbaum et al. 2015c). Many studies have also been conducted using homogenous samples (e.g., Veterans) or convenience samples, which limits generalization to the general population. Additionally, some studies included restrictions of age, ethnicity, and gender, which may also reduce generalizability to other demographics with PTSD. Several studies did not include a nonexercise control group, which may influence outcomes and limit the conclusions that can be drawn. Researchers have suggested that other aspects of participating in exercise, such as social support or receiving increased attention, may contribute to decreases in PTSD symptoms (Voorendonk et al. 2023).

5 Summary and Conclusions

Exercise is recognized to exert a beneficial effect on physiological and psychological symptoms across several mood- and anxiety-related disorders. Research to date indicates that various forms of exercise (i.e., aerobic, resistance training) effectively reduce symptoms of PTSD when applied as a stand-alone intervention or adjunct to cognitive behavioral and trauma-focused therapies. Evidence further indicates that greater amounts of exercise may produce greater benefits. Cross-sectional, observational studies suggest that moderate levels of exercise are associated with a lower likelihood of probable PTSD compared to low and high levels of exercise. Longitudinal findings indicated that strenuous/vigorous-intensity exercise is associated with greater reductions in PTSD symptoms than moderate- and low-intensity exercise. Associations between exercise and PTSD symptom clusters have been mixed in cross-sectional and longitudinal studies.

More research is needed to clarify which dose, intensity, and type of exercise are most effective at improving PTSD symptoms and to elucidate which prescriptions of exercise interventions are most effective for different types of trauma. Studies using larger, more diverse sample sizes are also needed, as is additional research investigating mediators and mechanisms that are responsible for the therapeutic effects of exercise on PTSD symptoms. Notwithstanding, the available evidence indicates that exercise is an accessible, cost-effective, and scalable treatment option that has the potential to improve both physiological and psychological symptoms among individuals with PTSD.

References

Adams TG, Forte J, Fogle BM, Southwick SM, Pietrzak RH (2020) Self-reported exercise frequency and PTSD: results from the National Health and Resilience in Veterans Study. Acta Psychiatr Scand 142(6):486–495. https://doi.org/10.1111/acps.13234

Adams TG, Fenlon E, Penn C, Hubert T, Pietrzak RH (2022) A significant U-shaped association between physical activity level and posttraumatic stress disorder in US military veterans: results from the 2019–2020 National Health and Resilience in Veterans Study. Mental Health Physical Activity 23:100460

American Psychiatric Association (2022) Diagnostic and statistical manual of mental disorders, 5th edn, TR. American Psychiatric Association, Washington

Archer T, Fredriksson A, Schütz E, Kostrzewa RM (2011) Influence of physical exercise on neuroimmunological functioning and health: aging and stress. Neurotox Res 20(1):69–83. https://doi.org/10.1007/s12640-010-9224-9

Arnson Y, Amital D, Fostick L, Silberman A, Polliack ML, Zohar J, Rubinow A, Amital H (2007) Physical activity protects male patients with post-traumatic stress disorder from developing severe fibromyalgia. Clin Exp Rheumatol 25(4):529–533

Asmundson GJG, Coons MJ, Taylor S, Katz J (2002) PTSD and the experience of pain: research and clinical implications of shared vulnerability and mutual maintenance models. Can J Psychiatr 47(10):930–937. https://doi.org/10.1177/070674370204701004

Asmundson GJG, Fetzner MG, Deboer LB, Powers MB, Otto MW, Smits JA (2013) Let's get physical: a contemporary review of the anxiolytic effects of exercise for anxiety and its disorders. Depress Anxiety 30(4):362–373. https://doi.org/10.1002/da.22043

Asmundson GJG, Thorisdottir AS, Roden-Foreman JW, Baird SO, Witcraft SM, Stein AT, Smits JAJ, Powers MB (2019) A meta-analytic review of cognitive processing therapy for adults with posttraumatic stress disorder. Cogn Behav Ther 48(1):1–14. https://doi.org/10.1080/16506073.2018.1522371

Bartlett JD, Close GL, MacLaren DP, Gregson W, Drust B, Morton JP (2011) High-intensity interval running is perceived to be more enjoyable than moderate-intensity continuous exercise: implications for exercise adherence. J Sports Sci 29(6):547–553. https://doi.org/10.1080/02640414.2010.545427

Bartley CA, Hay M, Bloch MH (2013) Meta-analysis: aerobic exercise for the treatment of anxiety disorders. Prog Neuro-Psychopharmacol Biol Psychiatry 45:34–39. https://doi.org/10.1016/j.pnpbp.2013.04.016

Basso JC, Suzuki WA (2017) The effects of acute exercise on mood, cognition, neurophysiology, and neurochemical pathways: a review. Brain Plast 2(2):127–152. https://doi.org/10.3233/BPL-160040

Benjet C, Bromet E, Karam EG, Kessler RC, McLaughlin KA, Ruscio AM, Shahly V, Stein DJ, Petukhova M, Hill E, Alonso J, Atwoli L, Bunting B, Bruffaerts R, Caldas-de-Almeida JM, de Girolamo G, Florescu S, Gureje O, Huang Y, Lepine JP, Kawakami N, Kovess-Masfety V, Medina-Mora ME, Navarro-Mateu F, Piazza M, Posada-Villa J, Scott KM, Shalev A, Slade T, ten Have M, Torres Y, Viana MC, Zarkov Z, Koenen KC (2016) The epidemiology of traumatic event exposure worldwide: results from the World Mental Health Survey Consortium. Psychol Med 46(2):327–343. https://doi.org/10.1017/S0033291715001981

Bibeau WS, Moore JB, Mitchell NG, Vargas-Tonsing T, Bartholomew JB (2010) Effects of acute resistance training of different intensities and rest periods on anxiety and affect. J Strength Cond Res 24(8):2184–2191. https://doi.org/10.1519/JSC.0b013e3181ae794b

Björkman F, Ekblom Ö (2022) Physical exercise as treatment for PTSD: a systematic review and meta-analysis. Mil Med 187(9–10):e1103–e1113. https://doi.org/10.1093/milmed/usab497

Boscarino JA (2004) Posttraumatic stress disorder and physical illness: results from clinical and epidemiologic studies. Ann N Y Acad Sci 1032:141–153. https://doi.org/10.1196/annals.1314.011

Bosch J, Weaver TL, Neylan TC, Herbst E, McCaslin SE (2017) Impact of engagement in exercise on sleep quality among veterans with posttraumatic stress disorder symptoms. Mil Med 182(9):e1745–e1750. https://doi.org/10.7205/MILMED-D-16-00385

Bourn LE, Sexton MB, Porter KE, Rauch SA (2016) Physical activity moderates the association between pain and PTSD in treatment-seeking veterans. Pain Med 17(11):2134–2141. https://doi.org/10.1093/pm/pnw089

Bradley R, Greene J, Russ E, Dutra L, Westen D (2005) A multidimensional meta-analysis of psychotherapy for PTSD. Am J Psychiatry 162(2):214–227. https://doi.org/10.1176/appi.ajp.162.2.214

Brady K, Pearlstein T, Asnis GM, Baker D, Rothbaum B, Sikes CR, Farfel GM (2000) Efficacy and safety of sertraline treatment of posttraumatic stress disorder: a randomized controlled trial. JAMA 283(14):1837–1844. https://doi.org/10.1001/jama.283.14.1837

Bryant RA, Dawson KS, Azevedo S, Yadav S, Cahill C, Kenny L, Maccallum F, Tran J, Rawson N, Tockar J, Garber B, Keyan D (2023) Augmenting trauma-focused psychotherapy for post-

traumatic stress disorder with brief aerobic exercise in Australia: a randomised clinical trial. Lancet Psychiatry 10(1):21–29. https://doi.org/10.1016/S2215-0366(22)00368-6
Burg MM, Soufer R (2016) Post-traumatic stress disorder and cardiovascular disease. Curr Cardiol Rep 18(10):94. https://doi.org/10.1007/s11886-016-0770-5
Carmassi C, Cordone A, Pedrinelli V, Dell'Osso L (2020) PTSD and cardiovascular disease: a bidirectional relationship. Brain Heart Dynam 26(2):355–376
Chen JA, Keller SM, Zoellner LA, Feeny NC (2013) "How will it help me?" Reasons underlying treatment preferences between sertraline and prolonged exposure in posttraumatic stress disorder. J Nerv Ment Dis 201(8):691–697. https://doi.org/10.1097/NMD.0b013e31829c50a9
Childs E, de Wit H (2014) Regular exercise is associated with emotional resilience to acute stress in healthy adults. Front Physiol 5:161. https://doi.org/10.3389/fphys.2014.00161
Chwastiak LA, Rosenheck RA, Kazis LE (2011) Association of psychiatric illness and obesity, physical inactivity, and smoking among a national sample of veterans. Psychosomatics 52(3): 230–236. https://doi.org/10.1016/j.psym.2010.12.009
Clinical practice guideline for the treatment of posttraumatic stress disorder (PTSD) in adults: (501872017-001). 2017. Available from: http://doi.apa.org https://doi.org/10.1037/e501872017-001
Cramer H, Anheyer D, Saha FJ, Dobos G (2018) Yoga for posttraumatic stress disorder – a systematic review and meta-analysis. BMC Psychiatry 18(1):72. https://doi.org/10.1186/s12888-018-1650-x
Crombie KM, Brellenthin AG, Hillard CJ, Koltyn KF (2018) Psychobiological responses to aerobic exercise in individuals with posttraumatic stress disorder. J Trauma Stress 31(1):134–145. https://doi.org/10.1002/jts.22253
Crombie KM, Leitzelar BN, Brellenthin AG, Hillard CJ, Koltyn KF (2019) Loss of exercise- and stress-induced increases in circulating 2-arachidonoylglycerol concentrations in adults with chronic PTSD. Biol Psychol 145:1–7. https://doi.org/10.1016/j.biopsycho.2019.04.002
Crombie KM, Cisler JM, Hillard CJ, Koltyn KF (2021a) Aerobic exercise reduces anxiety and fear ratings to threat and increases circulating endocannabinoids in women with and without PTSD. Ment Health Phys Act 20:100366. https://doi.org/10.1016/j.mhpa.2020.100366
Crombie KM, Sartin-Tarm A, Sellnow K, Ahrenholtz R, Lee S, Matalamaki M, Adams TG, Cisler JM (2021b) Aerobic exercise and consolidation of fear extinction learning among women with posttraumatic stress disorder. Behav Res Ther 142:103867. https://doi.org/10.1016/j.brat.2021.103867
Crombie KM, Sartin-Tarm A, Sellnow K, Ahrenholtz R, Lee S, Matalamaki M et al (2021c) Exercise-induced increases in anandamide and BDNF during extinction consolidation contribute to reduced threat following reinstatement: preliminary evidence from a randomized controlled trial. Psychoneuroendocrinology 132:105355–105355. https://doi.org/10.1016/j.psyneuen.2021.105355
Crombie KM, Adams TG, Dunsmoor JE, Greenwood BN, Smits JA, Nemeroff CB, Cisler JM (2023) Aerobic exercise in the treatment of PTSD: an examination of preclinical and clinical laboratory findings, potential mechanisms, clinical implications, and future directions. J Anxiety Disord 94:102680. https://doi.org/10.1016/j.janxdis.2023.102680
Cusack K, Jonas DE, Forneris CA, Wines C, Sonis J, Middleton JC, Feltner C, Brownley KA, Olmsted KR, Greenblatt A, Weil A, Gaynes BN (2016) Psychological treatments for adults with posttraumatic stress disorder: a systematic review and meta-analysis. Clin Psychol Rev 43:128–141. https://doi.org/10.1016/j.cpr.2015.10.003
Daskalakis NP, McGill MA, Lehrner A, Yehuda R (2016) Endocrine aspects of PTSD: hypothalamic-pituitary-adrenal (HPA) axis and beyond. In: Martin CR, Preedy VR, Patel VB (eds) Comprehensive guide to post-traumatic stress disorders. Springer, Cham, pp 245–260. Available from: https://link.springer.com. https://doi.org/10.1007/978-3-319-08359-9_130
Davidson CL, Babson KA, Bonn-Miller MO, Souter T, Vannoy S (2013) The impact of exercise on suicide risk: examining pathways through depression, PTSD, and sleep in an inpatient sample of veterans. Suicide Life Threat Behav 43(3):279–289. https://doi.org/10.1111/sltb.12014

Davis LW, Schmid AA, Daggy JK, Yang Z, O'Connor CE, Schalk N, Do AL, Maric D, Lazarick D, Knock H (2020) Symptoms improve after a yoga program designed for PTSD in a randomized controlled trial with veterans and civilians. Psychol Trauma 12(8):904–912. https://doi.org/10.1037/tra0000564

de Kloet CS, Vermetten E, Geuze E, Kavelaars A, Heijnen CJ, Westenberg HG (2006) Assessment of HPA-axis function in posttraumatic stress disorder: pharmacological and non-pharmacological challenge tests, a review. J Psychiatr Res 40(6):550–567. https://doi.org/10.1016/j.jpsychires.2005.08.002

Desai S, Borg B, Cuttler C, Crombie KM, Rabinak CA, Marusak HA (2022) A systematic review and meta-analysis on the effects of exercise on the endocannabinoid system. Cannabis Cannabinoid Res 7(4):388–408. https://doi.org/10.1089/can.2021.0113

Edinoff AN, Akuly HA, Hanna TA, Ochoa CO, Patti SJ, Ghaffar YA, Kaye AD, Viswanath O, Urits I, Boyer AG, Cornett EM, Kaye AM (2021) Selective serotonin reuptake inhibitors and adverse effects: a narrative review. Neurol Int 13(3):387–401. https://doi.org/10.3390/neurolint13030038

Ekkekakis P (2003) Pleasure and displeasure from the body: perspectives from exercise. Cogn Emot 17(2):213–239. https://doi.org/10.1080/02699930302292

Ensari I, Greenlee TA, Motl RW, Petruzzello SJ (2015 Aug) Meta-analysis of acute exercise effects on state anxiety: an update of randomized controlled trials over the past 25 years. Depress Anxiety 32(8):624–634. https://doi.org/10.1002/da.22370

Fetzner MG, Asmundson GJ (2015) Aerobic exercise reduces symptoms of posttraumatic stress disorder: a randomized controlled trial. Cogn Behav Ther 44(4):301–313. https://doi.org/10.1080/16506073.2014.916745

Fonzo GA, Federchenco V, Lara A (2020) Predicting and managing treatment non-response in posttraumatic stress disorder. Curr Treat Options Psychiatry 7(2):70–87. https://doi.org/10.1007/s40501-020-00203-1

Gill JM, Saligan L, Woods S, Page G (2009) PTSD is associated with an excess of inflammatory immune activities. Perspect Psychiatr Care 45(4):262–277. https://doi.org/10.1111/j.1744-6163.2009.00229.x

Goldstein LA, Mehling WE, Metzler TJ, Cohen BE, Barnes DE, Choucroun GJ, Silver A, Talbot LS, Maguen S, Hlavin JA, Chesney MA, Neylan TC (2018) Veterans group exercise: a randomized pilot trial of an integrative exercise program for veterans with posttraumatic stress. J Affect Disord 227:345–352. https://doi.org/10.1016/j.jad.2017.11.002

Goodwin RD, Davidson JR (2005) Self-reported diabetes and posttraumatic stress disorder among adults in the community. Prev Med 40(5):570–574. https://doi.org/10.1016/j.ypmed.2004.07.013

Guérin E, Fortier MS, Sweet SN (2013) An experience sampling study of physical activity and positive affect: investigating the role of situational motivation and perceived intensity across time. Health Psychol Res 1(2):e21. https://doi.org/10.4081/hpr.2013.e21

Hall KS, Morey MC, Bosworth HB, Beckham JC, Pebole MM, Sloane R, Pieper CF (2020) Pilot randomized controlled trial of exercise training for older veterans with PTSD. J Behav Med 43(4):648–659. https://doi.org/10.1007/s10865-019-00073-w

Harte CB, Vujanovic AA, Potter CM (2015) Association between exercise and posttraumatic stress symptoms among trauma-exposed adults. Eval Health Prof 38(1):42–52. https://doi.org/10.1177/0163278713494774

Hegberg NJ, Hayes JP, Hayes SM (2019) Exercise intervention in PTSD: a narrative review and rationale for implementation. Front Psych 10:133. https://doi.org/10.3389/fpsyt.2019.00133

Herbert TB, Cohen S (1993a) Depression and immunity: a meta-analytic review. Psychol Bull 113(3):472–486. https://doi.org/10.1037/0033-2909.113.3.472

Herbert TB, Cohen S (1993b) Stress and immunity in humans: a meta-analytic review. Psychosom Med 55(4):364–379. https://doi.org/10.1097/00006842-199307000-00004

Herring MP (2018) Exercise for the management of anxiety and stress-related disorders. In: Exercise-based interventions for mental illness. Elsevier, pp 19–52. Available from: https://linkinghub.elsevier.com/retrieve/pii/B9780128126059000022

Herring MP, Lindheimer JB, O'Connor PJ (2014) The effects of exercise training on anxiety. Am J Lifestyle Med 8(6):388–403. https://doi.org/10.1177/1559827613508542

Heyman E, Gamelin FX, Goekint M, Piscitelli F, Roelands B, Leclair E et al (2012) Intense exercise increases circulating endocannabinoid and BDNF levels in humans – possible implications for reward and depression. Psychoneuroendocrinology 37(6):844–851. https://doi.org/10.1016/j.psyneuen.2011.09.017

Hill EE, Zack E, Battaglini C, Viru M, Viru A, Hackney AC (2008) Exercise and circulating cortisol levels: the intensity threshold effect. J Endocrinol Investig 31(7):587–591. https://doi.org/10.1007/BF03345606

Hughes CW, Petty F, Sheikha S, Kramer GL (1996) Whole-blood serotonin in children and adolescents with mood and behavior disorders. Psychiatry Res 65(2):79–95. https://doi.org/10.1016/s0165-1781(96)02932-0

Jayakody K, Gunadasa S, Hosker C (2014) Exercise for anxiety disorders: systematic review. Br J Sports Med 48(3):187–196. https://doi.org/10.1136/bjsports-2012-091287

Jung ME, Bourne JE, Little JP (2014) Where does HIT fit? An examination of the affective response to high-intensity intervals in comparison to continuous moderate- and continuous vigorous-intensity exercise in the exercise intensity-affect continuum. PLoS One 9(12):e114541. https://doi.org/10.1371/journal.pone.0114541

Karlsen T, Aamot IL, Haykowsky M, Rognmo Ø (2017) High intensity interval training for maximizing health outcomes. Prog Cardiovasc Dis 60(1):67–77. https://doi.org/10.1016/j.pcad.2017.03.006

Kessler RC, Aguilar-Gaxiola S, Alonso J, Benjet C, Bromet EJ, Cardoso G, Degenhardt L, de Girolamo G, Dinolova RV, Ferry F, Florescu S, Gureje O, Haro JM, Huang Y, Karam EG, Kawakami N, Lee S, Lepine JP, Levinson D, Navarro-Mateu F, Pennell BE, Piazza M, Posada-Villa J, Scott KM, Stein DJ, Ten Have M, Torres Y, Viana MC, Petukhova MV, Sampson NA, Zaslavsky AM, Koenen KC (2017a) Trauma and PTSD in the WHO World Mental Health Surveys. Eur J Psychotraumatol 8:1353383. https://doi.org/10.1080/20008198.2017.1353383

Kessler RC, Aguilar-Gaxiola S, Alonso J, Benjet C, Bromet EJ, Cardoso G et al (2017b) Trauma and PTSD in the WHO World Mental Health Surveys. Eur J Psychotraumatol 8:1353383

Kilpatrick M, Kraemer R, Bartholomew J, Acevedo E, Jarreau D (2007) Affective responses to exercise are dependent on intensity rather than total work. Med Sci Sports Exerc 39(8):1417–1422. https://doi.org/10.1249/mss.0b013e31806ad73c

Kline AC, Cooper AA, Rytwinksi NK, Feeny NC (2018) Long-term efficacy of psychotherapy for posttraumatic stress disorder: a meta-analysis of randomized controlled trials. Clin Psychol Rev 59:30–40. https://doi.org/10.1016/j.cpr.2017.10.009

Koek RJ, Schwartz HN, Scully S, Langevin JP, Spangler S, Korotinsky A, Jou K, Leuchter A (2016) Treatment-refractory posttraumatic stress disorder (TRPTSD): a review and framework for the future. Prog Neuro-Psychopharmacol Biol Psychiatry 70:170–218. https://doi.org/10.1016/j.pnpbp.2016.01.015

Koenen KC, Ratanatharathorn A, Ng L, McLaughlin KA, Bromet EJ, Stein DJ, Karam EG, Meron Ruscio A, Benjet C, Scott K, Atwoli L, Petukhova M, Lim CCW, Aguilar-Gaxiola S, Al-Hamzawi A, Alonso J, Bunting B, Ciutan M, de Girolamo G, Degenhardt L, Gureje O, Haro JM, Huang Y, Kawakami N, Lee S, Navarro-Mateu F, Pennell BE, Piazza M, Sampson N, Ten Have M, Torres Y, Viana MC, Williams D, Xavier M, Kessler RC (2017) Posttraumatic stress disorder in the World Mental Health Surveys. Psychol Med 47(13):2260–2274. https://doi.org/10.1017/S0033291717000708

Kozaric-Kovacic D, Ilic MG, Romic Z, Vidovic A, Jendricko T, Pivac N (2009) Body mass index in male Caucasian veterans with or without posttraumatic stress disorder. Prog Neuro-Psychopharmacol Biol Psychiatry 33(8):1447–1450. https://doi.org/10.1016/j.pnpbp.2009.07.026

Kysar-Moon A, Vasquez M, Luppen T (2021) Trauma-sensitive yoga interventions and posttraumatic stress and depression outcomes among women: a systematic review and analysis of randomized control trials. Int J Yoga Therap 31(1) Article_23. https://doi.org/10.17761/2021-D-20-00005

Lane AM, Jackson A, Terry PC (2005) Preferred modality influences on exercise-induced mood changes. J Sports Sci Med 4(2):195–200

Larsen SE, Fleming CJE, Resick PA (2019) Residual symptoms following empirically supported treatment for PTSD. Psychol Trauma 11(2):207–215. https://doi.org/10.1037/tra0000384

LeardMann CA, Kelton ML, Smith B, Littman AJ, Boyko EJ, Wells TS, Smith TC, Millennium Cohort Study Team (2011) Prospectively assessed posttraumatic stress disorder and associated physical activity. Public Health Rep 126(3):371–383. https://doi.org/10.1177/003335491112600311

LeBouthillier DM, Fetzner MG, Asmundson GJG (2016) Lower cardiorespiratory fitness is associated with greater reduction in PTSD symptoms and anxiety sensitivity following aerobic exercise. Ment Health Phys Act 10:33–39

Liao Y, Shonkoff ET, Dunton GF (2015) The acute relationships between affect, physical feeling states, and physical activity in daily life: a review of current evidence. Front Psychol 6:1975. https://doi.org/10.3389/fpsyg.2015.01975

Lindauer RJ, Gersons BP, van Meijel EP, Blom K, Carlier IV, Vrijlandt I, Olff M (2005) Effects of brief eclectic psychotherapy in patients with posttraumatic stress disorder: randomized clinical trial. J Trauma Stress 18(3):205–212. https://doi.org/10.1002/jts.20029

Manger TA, Motta RW (2005) The impact of an exercise program on posttraumatic stress disorder, anxiety, and depression. Int J Emerg Ment Health 7(1):49–57

Markowitz JC, Meehan KB, Petkova E, Zhao Y, Van Meter PE, Neria Y, Pessin H, Nazia Y (2016) Treatment preferences of psychotherapy patients with chronic PTSD. J Clin Psychiatry 77(3):363–370. https://doi.org/10.4088/JCP.14m09640

Maroulakis E, Zervas Y (1993) Effects of aerobic exercise on mood of adult women. Percept Mot Skills 76(3 Pt 1):795–801. https://doi.org/10.2466/pms.1993.76.3.795

Marshall RD, Beebe KL, Oldham M, Zaninelli R (2001) Efficacy and safety of paroxetine treatment for chronic PTSD: a fixed-dose, placebo-controlled study. Am J Psychiatry 158(12):1982–1988. https://doi.org/10.1176/appi.ajp.158.12.1982

McFarlane AC, Lawrence-Wood E, Van Hooff M, Malhi GS, Yehuda R (2017) The need to take a staging approach to the biological mechanisms of PTSD and its treatment. Curr Psychiatry Rep 19(2):10. https://doi.org/10.1007/s11920-017-0761-2

Muscat R, Papp M, Willner P (1992) Antidepressant-like effects of dopamine agonists in an animal model of depression. Biol Psychiatry 31(9):937–946. https://doi.org/10.1016/0006-3223(92)90119-k

Najavits LM (2015) The problem of dropout from "gold standard" PTSD therapies. F1000Prime Rep 7:43. https://doi.org/10.12703/P7-43

Newman CL, Motta RW (2007) The effects of aerobic exercise on childhood PTSD, anxiety, and depression. Int J Emerg Ment Health 9(2):133–158

Nordbrandt MS, Sonne C, Mortensen EL, Carlsson J (2020) Trauma-affected refugees treated with basic body awareness therapy or mixed physical activity as augmentation to treatment as usual - a pragmatic randomised controlled trial. PLoS One 15(3):e0230300. https://doi.org/10.1371/journal.pone.0230300

O'Donovan A, Hughes BM, Slavich GM, Lynch L, Cronin MT, O'Farrelly C, Malone KM (2010) Clinical anxiety, cortisol and interleukin-6: evidence for specificity in emotion-biology relationships. Brain Behav Immun 24(7):1074–1077. https://doi.org/10.1016/j.bbi.2010.03.003

Oliveira BR, Slama FA, Deslandes AC, Furtado ES, Santos TM (2013) Continuous and high-intensity interval training: which promotes higher pleasure? PLoS One 8(11):e79965. https://doi.org/10.1371/journal.pone.0079965

Olthuis JV, Wozney L, Asmundson GJ, Cramm H, Lingley-Pottie P, McGrath PJ (2016) Distance-delivered interventions for PTSD: a systematic review and meta-analysis. J Anxiety Disord 44: 9–26. https://doi.org/10.1016/j.janxdis.2016.09.010

Pacella ML, Hruska B, Delahanty DL (2013) The physical health consequences of PTSD and PTSD symptoms: a meta-analytic review. J Anxiety Disord 27(1):33–46. https://doi.org/10.1016/j.janxdis.2012.08.004

Parfitt G, Rose EA, Burgess WM (2006) The psychological and physiological responses of sedentary individuals to prescribed and preferred intensity exercise. Br J Health Psychol 11: 39–53. https://doi.org/10.1348/135910705X43606

Passos IC, Vasconcelos-Moreno MP, Costa LG, Kunz M, Brietzke E, Quevedo J, Salum G, Magalhães PV, Kapczinski F, Kauer-Sant'Anna M (2015) Inflammatory markers in post-traumatic stress disorder: a systematic review, meta-analysis, and meta-regression. Lancet Psychiatry 2(11):1002–1012. https://doi.org/10.1016/S2215-0366(15)00309-0

Petruzzello SJ, Landers DM, Hatfield BD, Kubitz KA, Salazar W (1991) A meta-analysis on the anxiety-reducing effects of acute and chronic exercise. Outcomes and mechanisms. Sports Med 11(3):143–182. https://doi.org/10.2165/00007256-199111030-00002

Powers MB, Medina JL, Burns S, Kauffman BY, Monfils M, Asmundson GJ, Diamond A, McIntyre C, Smits JA (2015) Exercise augmentation of exposure therapy for PTSD: rationale and pilot efficacy data. Cogn Behav Ther 44(4):314–327. https://doi.org/10.1080/16506073.2015.1012740

Qureshi SU, Pyne JM, Magruder KM, Schulz PE, Kunik ME (2009) The link between post-traumatic stress disorder and physical comorbidities: a systematic review. Psychiatry Q 80(2): 87–97. https://doi.org/10.1007/s11126-009-9096-4

Raglin JS, Morgan WP (1987) Influence of exercise and quiet rest on state anxiety and blood pressure. Med Sci Sports Exerc 19(5):456–463

Ramos-Sanchez CP, Schuch FB, Seedat S, Louw QA, Stubbs B, Rosenbaum S, Firth J, van Winkel R, Vancampfort D (2021) The anxiolytic effects of exercise for people with anxiety and related disorders: an update of the available meta-analytic evidence. Psychiatry Res 302: 114046. https://doi.org/10.1016/j.psychres.2021.114046

Reed J, Buck S (2009) The effect of regular aerobic exercise on positive-activated affect: a meta-analysis. Psychol Sport Exerc 10(6):581–594

Reed J, Ones DS (2006) The effect of acute aerobic exercise on positive activated affect: a meta-analysis. Psychol Sport Exerc 7:477–514

Rosenbaum S, Stubbs B, Ward PB, Steel Z, Lederman O, Vancampfort D (2015a) The prevalence and risk of metabolic syndrome and its components among people with posttraumatic stress disorder: a systematic review and meta-analysis. Metabolism 64(8):926–933. https://doi.org/10.1016/j.metabol.2015.04.009

Rosenbaum S, Sherrington C, Tiedemann A (2015b) Exercise augmentation compared with usual care for post-traumatic stress disorder: a randomized controlled trial. Acta Psychiatr Scand 131(5):350–359. https://doi.org/10.1111/acps.12371

Rosenbaum S, Vancampfort D, Steel Z, Newby J, Ward PB, Stubbs B (2015c) Physical activity in the treatment of post-traumatic stress disorder: a systematic review and meta-analysis. Psychiatry Res 230(2):130–136. https://doi.org/10.1016/j.psychres.2015.10.017

Rosman L, Sico JJ, Lampert R, Gaffey AE, Ramsey CM, Dziura J, Chui PW, Cavanagh CE, Brandt C, Haskell S, Burg MM (2019) Posttraumatic stress disorder and risk for stroke in young and middle-aged adults: a 13-year cohort study. Stroke 50(11):2996–3003. https://doi.org/10.1161/STROKEAHA.119.026854

Ryder AL, Azcarate PM, Cohen BE (2018) PTSD and physical health. Curr Psychiatry Rep 20(12): 116. https://doi.org/10.1007/s11920-018-0977-9

Saanijoki T, Nummenmaa L, Eskelinen JJ, Savolainen AM, Vahlberg T, Kalliokoski KK, Hannukainen JC (2015) Affective responses to repeated sessions of high-intensity interval training. Med Sci Sports Exerc 47(12):2604–2611. https://doi.org/10.1249/MSS.0000000000000721

Santos RV, Tufik S, De Mello MT (2007) Exercise, sleep and cytokines: is there a relation? Sleep Med Rev 11(3):231–239. https://doi.org/10.1016/j.smrv.2007.03.003

Sareen J, Cox BJ, Clara I, Asmundson GJ (2005) The relationship between anxiety disorders and physical disorders in the U.S. National Comorbidity Survey. Depress Anxiety 21(4):193–202. https://doi.org/10.1002/da.20072

Schauer M, Neuner F, Elbert T (2011) Narrative exposure therapy: a short-term treatment for traumatic stress disorders. Hogrefe Publishing, Newburyport

Schnurr PP (2017) Focusing on trauma-focused psychotherapy for posttraumatic stress disorder. Curr Opin Psychol 14:56–60. https://doi.org/10.1016/j.copsyc.2016.11.005

Schottenbauer MA, Glass CR, Arnkoff DB, Tendick V, Gray SH (2008) Nonresponse and dropout rates in outcome studies on PTSD: review and methodological considerations. Psychiatry 71(2): 134–168. https://doi.org/10.1521/psyc.2008.71.2.134

Shapiro FEMDR (2007) Adaptive information processing, and case conceptualization. J EMDR Pract Res 1(2):68–87

Shivakumar G, Anderson EH, Surís AM, North CS (2017) Exercise for PTSD in women veterans: a proof-of-concept study. Mil Med 182(11):e1809–e1814. https://doi.org/10.7205/MILMED-D-16-00440

Stonerock GL, Hoffman BM, Smith PJ, Blumenthal JA (2015) Exercise as treatment for anxiety: systematic review and analysis. Ann Behav Med 49(4):542–556. https://doi.org/10.1007/s12160-014-9685-9

Stubbs B, Vancampfort D, Rosenbaum S, Firth J, Cosco T, Veronese N, Salum GA, Schuch FB (2017) An examination of the anxiolytic effects of exercise for people with anxiety and stress-related disorders: a meta-analysis. Psychiatry Res 249:102–108. https://doi.org/10.1016/j.psychres.2016.12.020

Szuhany KL, Bugatti M, Otto MW (2015) A meta-analytic review of the effects of exercise on brain-derived neurotrophic factor. J Psychiatr Res 60:56–64. https://doi.org/10.1016/j.jpsychires.2014.10.003

van der Kolk BA, Stone L, West J, Rhodes A, Emerson D, Suvak M, Spinazzola J (2014) Yoga as an adjunctive treatment for posttraumatic stress disorder: a randomized controlled trial. J Clin Psychiatry 75(6):e559–e565. https://doi.org/10.4088/JCP.13m08561

Van Etten ML, Taylor S (1998) Comparative efficacy of treatments for post-traumatic stress disorder: a meta-analysis. In: Database of abstracts of reviews of effects (DARE): quality-assessed reviews [Internet}. Available from: https://www.ncbi.nlm.nih.gov/books/NBK67406/

Vancampfort D, Stubbs B, Richards J, Ward PB, Firth J, Schuch FB, Rosenbaum S (2017) Physical fitness in people with posttraumatic stress disorder: a systematic review. Disabil Rehabil 39(24): 2461–2467. https://doi.org/10.1080/09638288.2016.1226412

Voorendonk EM, Sanches SA, Tollenaar MS, Hoogendoorn EA, de Jongh A, van Minnen A (2023) Adding physical activity to intensive trauma-focused treatment for post-traumatic stress disorder: results of a randomized controlled trial. Front Psychol 14:1215250. https://doi.org/10.3389/fpsyg.2023.1215250

Vun E, Turner S, Sareen J, Mota N, Afifi TO, El-Gabalawy R (2018) Prevalence of comorbid chronic pain and mental health conditions in Canadian Armed Forces active personnel: analysis of a cross-sectional survey. CMAJ Open 6(4):E528–E536. https://doi.org/10.9778/cmajo.20180093

Wade NG, Vogel DL, Armistead-Jehle P, Meit SS, Heath PJ, Strass HA (2015) Modeling stigma, help-seeking attitudes, and intentions to seek behavioral healthcare in a clinical military sample. Psychiatr Rehabil J 38(2):135–141. https://doi.org/10.1037/prj0000131

Walter KH, Levine JA, Highfill-McRoy RM, Navarro M, Thomsen CJ (2018) Prevalence of posttraumatic stress disorder and psychological comorbidities among U.S. active duty service members, 2006-2013. J Trauma Stress 31(6):837–844. https://doi.org/10.1002/jts.22337

Watkins LE, Sprang KR, Rothbaum BO (2018) Treating PTSD: a review of evidence-based psychotherapy interventions. Front Behav Neurosci 12:258. https://doi.org/10.3389/fnbeh.2018.00258

Weisse CS (1992) Depression and immunocompetence: a review of the literature. Psychol Bull 111(3):475–489. https://doi.org/10.1037/0033-2909.111.3.475

Whitworth JW, Craft LL, Dunsiger SI, Ciccolo JT (2017) Direct and indirect effects of exercise on posttraumatic stress disorder symptoms: a longitudinal study. Gen Hosp Psychiatry 49:56–62. https://doi.org/10.1016/j.genhosppsych.2017.06.012

Whitworth JW, Nosrat S, SantaBarbara NJ, Ciccolo JT (2019a) Feasibility of resistance exercise for posttraumatic stress and anxiety symptoms: a randomized controlled pilot study. J Trauma Stress 32(6):977–984. https://doi.org/10.1002/jts.22464

Whitworth JW, Nosrat S, SantaBarbara NJ, Ciccolo JT (2019b) High intensity resistance training improves sleep quality and anxiety in individuals who screen positive for posttraumatic stress disorder: a randomized controlled feasibility trial. Mental Health Physical Activity 16:43–49

Wipfli B, Landers D, Nagoshi C, Ringenbach S (2011) An examination of serotonin and psychological variables in the relationship between exercise and mental health. Scand J Med Sci Sports 21(3):474–481. https://doi.org/10.1111/j.1600-0838.2009.01049.x

Young-McCaughan S, Peterson AL, Mintz J, Hale WJ, Dondanville KA, Borah EV, Blount TH, Blankenship AE, Fina BA, Hall-Clark BN, Hernandez AM, Jacoby VM, Malach SL, Williams JM, Compton KE, Bingham MO, Vriend CA, Inman AW, Brundige A, Arzola SM, Lapiz-Bluhm MD, Williamson DE, Litz BT, Hembree EA, Roache JD, Taylor DJ, Pruiksma KE, Borah AM, Yarvis JS (2022) Testing the role of aerobic exercise in the treatment of posttraumatic stress disorder (PTSD) symptoms in U.S. active duty military personnel: a pilot study. Cogn. Behav Ther 51(4):309–325. https://doi.org/10.1080/16506073.2021.2001689

Zen AL, Whooley MA, Zhao S, Cohen BE (2012) Post-traumatic stress disorder is associated with poor health behaviors: findings from the heart and soul study. Health Psychol 31(2):194–201. https://doi.org/10.1037/a0025989

Zheng H, Liu Y, Li W, Yang B, Chen D, Wang X, Jiang Z, Wang H, Wang Z, Cornelisson G, Halberg F (2006) Beneficial effects of exercise and its molecular mechanisms on depression in rats. Behav Brain Res 168(1):47–55. https://doi.org/10.1016/j.bbr.2005.10.007

Zorrilla EP, Luborsky L, McKay JR, Rosenthal R, Houldin A, Tax A, McCorkle R, Seligman DA, Schmidt K (2001) The relationship of depression and stressors to immunological assays: a meta-analytic review. Brain Behav Immun 15(3):199–226. https://doi.org/10.1006/brbi.2000.0597

Zou S, Kumar U (2018) Cannabinoid receptors and the endocannabinoid system: signaling and function in the central nervous system. Int J Mol Sci 19(3):833. https://doi.org/10.3390/ijms19030833

Exercise and Protection from Age-Related Cognitive Decline

J. Carson Smith, Daniel D. Callow, Gabriel S. Pena, Yash Kommula, Naomi Arnold-Nedimala, Junyeon Won, and Kristy A. Nielson

Contents

Abstract In this chapter, we review the cross-sectional evidence in healthy human subjects for physical activity and cardiorespiratory fitness to offer neuroprotection and moderate cognitive decline in older age. The role of exercise training on cognition in healthy older adults and those diagnosed with mild cognitive impairment (MCI) is also discussed, including the evidence from neuroimaging studies that document changes to brain structure and function after a period of exercise training and improved fitness. Finally, in reference to animal models, the potential neurophysiological mechanisms for physical activity and exercise to impact human brain health are highlighted.

J. Carson Smith (✉), D. D. Callow, G. S. Pena, Y. Kommula, and N. Arnold-Nedimala
University of Maryland, College Park, MD, USA
e-mail: carson@umd.edu

J. Won
Institute for Exercise and Environmental Medicine, Texas Health Presbyterian Hospital, Dallas, TX, USA

K. A. Nielson
Marquette University, Milwaukee, WI, USA

Curr Topics Behav Neurosci (2024) 67: 263–280
https://doi.org/10.1007/7854_2024_501
Published Online: 31 July 2024

Keywords Aging · Brain Health · Cardiorespiratory Fitness · Cognition · Executive Function · Memory · Neuroimaging · Physical Activity

1 Significance and Impact of Age-Related Cognitive Decline

The prevalence of age-related cognitive impairment among older adults in the USA is estimated to be over 20%, and approximately 6.7 million people over the age of 65 have been diagnosed with Alzheimer's disease (AD) in the USA. Women have nearly twice the incidence of AD compared to men, and African Americans are at greater risk than European Americans. With an aging US population, cognitive impairment and AD constitute alarming public health problems. For example, the number of older adults with AD is expected to increase to over 13 million by 2060. The brunt of the burden and cost for cognitive impairment and AD overwhelmingly falls on caregivers, who experience increased stress and mental health concerns, and expend untold financial and physical resources, which have been estimated at over $330 billion in annual unpaid wage (Alzheimer's Association 2023).

Modifiable lifestyle behaviors such as exercise and leisure-time physical activity (PA), which even modestly delay the onset of cognitive impairment or help maintain cognitive function in older adults at greater risk, could slow the development of AD and have a major public health impact (Woodard et al. 2012). For example, from data collected in the ABC Health Study from 2,509 well-functioning older adults, those who maintained weekly moderate-to-vigorous PA had a 31% lower odds of cognitive decline over a period of 8 years (Yaffe et al. 2009). Moreover, it has been further estimated that physical *inactivity* is attributable to 13% of AD cases worldwide and nearly 21% of AD cases in the USA. If physical inactivity rates were reduced by 25%, this could result in a prevention of nearly one million cases of AD worldwide and over 230,000 cases in the USA (Barnes and Yaffe 2011). Indeed, physical exercise training (exercise training) has received considerable attention as an intervention (Geda et al. 2010; Kramer et al. 2006; Lautenschlager et al. 2008; Smith et al. 2013a), primarily because it has relatively few side effects, a key factor in implementing a treatment in a population with comorbid conditions (ACSM 2006; IOM 2015). A physically active lifestyle and greater cardiorespiratory fitness has long been associated with reduced all-cause mortality (Samitz et al. 2011) and also with reduced mortality attributable to dementia (Liu et al. 2012).

1.1 Definitions: PA, Cardiorespiratory Fitness, and Exercise

Although PA, cardiorespiratory fitness (CRF), and exercise represent distinct constructs, they are often used interchangeably in the literature. First, PA refers to any kind of skeletal muscle-related bodily movements that increase energy expenditure above resting energy expenditure (Caspersen et al. 1985). PA includes leisure time activities as well as structured health-related activities. Self-report questionnaire or actigraphy are widely used to measure amount of PA (see Chapter "Introduction").

CRF is cardiovascular ability to supply and utilize oxygen to large muscle groups during dynamic PA of moderate-to-vigorous intensity for prolonged periods of time (Caspersen et al. 1985). The integrated functions of musculoskeletal, respiratory, and cardiovascular systems constitute the level of CRF. It has been well documented that higher CRF is associated with lower risk of cardiovascular disease, all-cause mortality, and improvement in other health outcomes (Samitz et al. 2011; Blair et al. 1989; Paffenbarger Jr. et al. 1986). The gold standard method to assess CRF is to measure the maximal rate of oxygen consumption ($\dot{V}O_{2max}$) using open circuit spirometry. $\dot{V}O_{2max}$ reflects the highest rate of oxygen uptake that an individual can attain during sustained exercise and is typically defined as a plateau of oxygen consumption despite increased work rate that is combined with attainment of other criteria for maximal effort (such as a near maximal subjective rating of perceived exertion, heart rate near age-predicted maximum, respiratory exchange ratio greater than 1.0) and represents maximal aerobic capacity. For physically inactive older participants or those with known cardiovascular disease, a modified submaximal test is often used due to safety concerns. Although regular participation in moderate-to-vigorous aerobic exercise largely influences $\dot{V}O_{2max}$, there are other factors (e.g., training status, age, and sex) that may moderate $\dot{V}O_{2max,}$ which also are strongly influenced by genetic inheritance (Bouchard et al. 2011). Thus, it is imperative to factor in these potential moderators into the analyses to investigate the associations with CRF.

Exercise is a specific subtype of PA comprising planned and structured PA to enhance physical fitness or health. Aerobic exercise (also referred to as cardiovascular exercise) refers to planned and structured large-muscle dynamic activity that engages the cardiovascular system, such as walking, running, swimming, or cycling (ACSM 2006; Caspersen et al. 1985). While resistance exercise (also referred to as strength training or weight lifting) also involves the cardiovascular system, it is distinguished by providing resistance to concentric or eccentric contractions of specific muscle groups, usually repeated for a limited number of repetitions and sets of repetitions in order to increase strength or ability to produce force (ACSM 2006; Caspersen et al. 1985). In clinical research, acute exercise is a single bout of exercise typically administered using cycle ergometer or treadmill or resistance exercise (e.g., a single session of cardiovascular exercise for 10–60 min) and then undergo experimental conditions at various timepoints (immediately to minutes to several hours) afterward. Exercise training refers to repeated participation in exercise for a certain period. Most exercise training studies are conducted for ≥3 months, and participants typically undergo experimental conditions at baseline and postintervention. Randomized controlled trials are commonly used in exercise training investigations. It is common for randomized controlled trials to randomize participants into either aerobic exercise or nonaerobic exercise control (e.g., balance, stretching, and toning) groups. See Chapter "Introduction" for a comprehensive overview of key physical activity and exercise-related definitions and concepts.

1.2 *Definitions: Cognitive Domains, Age-Related Cognitive Decline, and Mild Cognitive Impairment (Leading to Alzheimer's Disease)*

Cognitive functioning includes an array of mental faculties that encompass experience of the world around and within and the acquisition of knowledge, skills, and understanding, with maximum speed and accuracy. The major domains of cognition include sensory-perceptual processing, mental and motor speed, attention, learning, memory, language, reasoning, problem-solving, and executive functioning. Memory generally encompasses a wide array of encoding and retrieval functions including knowledge and facts (i.e., semantic memory), temporally specific information and experiences (i.e., episodic memory), skills (i.e., procedural memory), and unconscious/automated learning (i.e., implicit memory). Executive functioning is an umbrella term encompassing high-level (i.e., "fluid") cognitive skills including attentional and inhibitory control, cognitive flexibility, and working memory (i.e., holding, updating, and manipulating information in current consciousness).

Cognitive functions develop capacity and efficiency throughout development into the second decade, followed by incremental but continuous decline in most cognitive domains thereafter (McArdle et al. 2002). Although the degree and trajectory of change vary by domain, both cross-sectional and longitudinal data show this general, modest pattern of decline (Salthouse 2019; Singh-Manoux et al. 2012). Importantly, there are also spared cognitive domains in aging, such as in low-load and nondeclarative (e.g., implicit, procedural) memory tasks, naming, and the acquisition and retrieval of vocabulary and general knowledge (Salthouse 2019; Cohen et al. 2019; Zec et al. 2005). Distinctions between sparing and decline often go unnoticed until much later in life and decline may be offset by growth in expertise and skills into later life (Haaland et al. 2003). However, there is clear evidence that cognitive functions that require effortful and/or time-sensitive processing peak in younger adulthood and decline quite consistently throughout the rest of life (Salthouse 2019). Additionally, older adults have more difficulty ignoring irrelevant information, solving problems, learning new things, and most especially retrieving previously learned information (Craik and Salthouse 2007). There are multiple theories that attempt to explain cognitive aging processes, ranging from single-process, all-cause theories that rely on behavioral findings (e.g., processing speed, inhibitory deficit, binding in memory) to the most contemporary models that explain cognitive decline and resilience in terms of neural changes and individual adaptation to those changes (e.g., compensation, scaffolding) (Dennis et al. 2020). Adaptation to change and areas of cognitive sparing are given less attention in the literature, but they are of particular importance as they demonstrate that functional cognitive decline is not an inevitable consequence of aging and that cognitive plasticity can be leveraged for targeted interventions that can protect and enhance functioning into old age.

Cognitive impairment, as distinguished from age-related cognitive decline, is typically considered performance that is more than 1.5 standard deviations below the normative mean for the domain assessed (Petersen and Morris 2005). Dementia is a heterogenous major neurocognitive disorder that is characterized by a significant impairment and progressive decline of function in multiple cognitive domains, including attention, memory, processing speed, and executive function, and the disruption of normal activities of daily living (American Psychiatric Association 2022). AD, characterized by widespread, progressive brain atrophy and beta-amyloid and tau neuropathology, is by far the most prevalent form of dementia (Alzheimer's Association 2023). Juxtaposed between typical age-related cognitive decline and AD-related cognitive impairment is a "preclinical," transitional phase termed "mild cognitive impairment" (MCI). While 1–2% of healthy elders progress to dementia annually, 10–15% of those diagnosed with MCI will do so (Peltz et al. 2011). As such, this transitional MCI stage is an ideal, perhaps crucial time to initiate interventions to slow or reverse the dementia process.

MCI is currently characterized by cognitive complaints (by the individual or informant) accompanied by verified cognitive impairment in one or more cognitive domains that is not severe enough to meet the criteria for dementia and intact independence of daily activities (Albert et al. 2011). Various subtypes exist, most simply those that categorize MCI by amnestic or nonamnestic and by single or multiple domains of impairment. Episodic memory is commonly impaired, particularly in those who later convert to AD (Albert et al. 2011), as is executive functioning (Corbo and Casagrande 2022). The threshold for impairment in MCI, by definition less precise than in dementia because deficits are mild, is typically 1–1.5 standard deviations below the age- and education-corrected norms (Albert et al. 2011). In addition, MCI does not necessarily progress as consistently as it does in dementia. Instead, there is sometimes "reversion" over time, where impairment resolves or improves for a time before declining again (Malek-Ahmadi 2016).

2 Leisure-Time PA and Protection from Age-Related Cognitive Decline

Including regular PA as part of a healthy lifestyle offers numerous health-related benefits to include disease prevention, maintenance of independence, and improved quality of life particularly in older adults (Lautenschlager et al. 2008; Sun et al. 2013). In the USA, approximately 10% of premature mortality and an annual healthcare cost of an estimated $117 billion are associated with inadequate PA (Piercy et al. 2018). A large-scale, eight-year longitudinal study found that a moderate amount of additional daily PA (as little as 15 min) decreased mortality rates by 4% (Sun et al. 2013). The PA Guidelines for Americans recommends adults to engage in 150–300 min a week of moderate-intensity or 75–150 min a week of vigorous-intensity aerobic PA or an equivalent combination of moderate- and vigorous-intensity aerobic activity, including muscle-strengthening activities on two or more days a week (Piercy et al. 2018). Additionally, older adults should do multicomponent PA that includes balance training as well as aerobic and muscle-strengthening activities.

In recent years, focus on the impact of leisure-time PA in reducing risk for age-related cognitive decline in older adults has become an emerging area of research (Smith et al. 2013b). For example, a population-based case-control longitudinal study from the Mayo Clinic Study of Aging investigated the association between PA and risk of the development of MCI in older adults. They found that engagement in moderate intensity PA at any frequency greater than zero days per week in midlife (at ages 50–65 years) and late-life (70–89 years, less than 1 year prior to the cognitive assessment) was associated with an approximately 40% and 30% reduction, respectively, in risk for a diagnosis of MCI (Geda et al. 2010). The largest protective effect at midlife was found for moderate PA at a frequency of 5–6 days per week, which provided a 60% reduction in risk for MCI after age 70. The findings from this study suggest that at midlife, and even later in life, frequent engagement in PA may help to preserve cognitive abilities and may provide protection from the onset of MCI. In another longitudinal study, 3,903 dementia-free older adults (age 55+) were followed in Ebersberg, Bavaria, for 2 years. Those who engaged in moderate (<3 times per week) or high levels (≥ 3 times per week) of PA had significantly reduced odds for developing incident cognitive impairment at follow-up compared to those who did not engage in PA (Etgen et al. 2010). In a 1,740-person study of healthy, cognitively intact older adults (age 65+), with a mean follow-up of 6.2 years, the incidence rate of dementia was 13.0 per 1,000 person-years for those who were physically active greater than or equal to 3 times per week compared with a rate of 19.7 per 1,000-person years for those who exercised less than 3 times per week (Larson et al. 2006). Moreover, the risk reduction associated with PA was greatest in those who were less physically active at baseline.

In summary, leisure time PA is associated with lower incidence of cognitive decline. These findings support the adoption of PA for older adults as a method to attenuate the rate of cognitive decline with age, even when the adoption of an active lifestyle begins mid- to late-life or with a history of very little leisure-time PA. These hypothesized associations are depicted in Fig. 1. Note, however, that it is not possible to determine if these associations are linear with age or have a greater or lesser slope at younger or older ages. Moreover, very little is known regarding the comparative effectiveness of different modes and intensities of PA/exercise in relation to cognitive outcomes or symptoms of AD.

3 Cardiorespiratory Fitness and Protection from Age-Related Cognitive Decline

CRF, a facet of cardiovascular health, has been positively correlated with better overall and domain-specific cognitive function in healthy older adults (Pentikainen et al. 2019). A meta-analysis of 18 aerobic fitness training studies reveals that CRF in

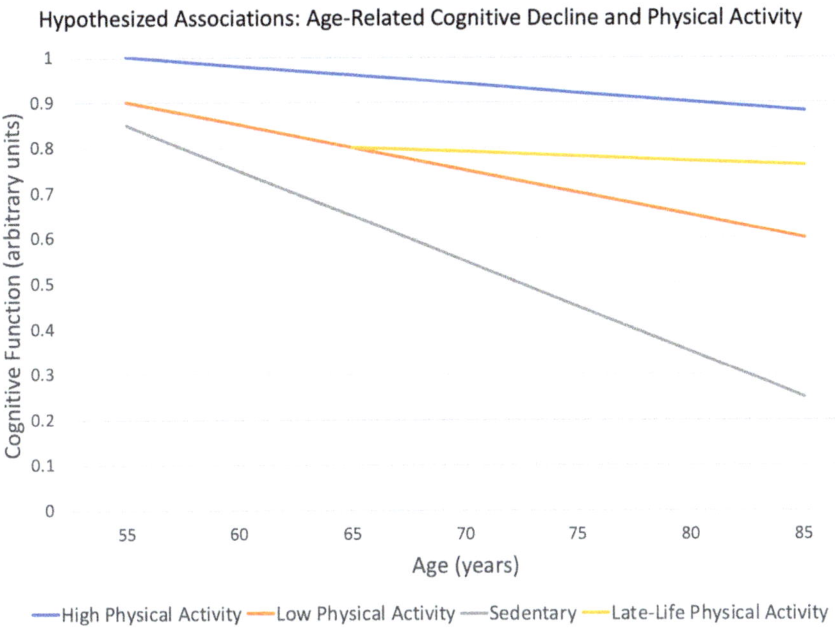

Fig. 1 Hypothesized associations between PA and the trajectory of cognitive decline, with protective effects even when started in late life

healthy older adults may bolster executive function most strongly, when compared to other key cognitive functions such as processing speed, visuospatial processing, and effortful-controlled processes that, unlike executive functions, can become automatic without need of constant mediation by a central executor over time (Colcombe and Kramer 2003). Cross-sectional studies consistently support that higher $\dot{V}O_{2max}$ is related to better executive functioning, as assessed by tasks such as the Wisconsin Card Sorting Test, the Trail-Making Tests (part B), and the Stroop task (Gauthier et al. 2015). Meanwhile, longitudinal research examining cognitive function has identified that higher baseline $\dot{V}O_{2max}$ in older adults is positively associated with better executive function 6 years later (Barnes et al. 2003). Although the mechanism of fitness benefits on executive function in healthy aging is unconfirmed, recent research suggests that higher global neural efficiency and lower local neural efficiency might play a role (Kawagoe et al. 2017; Callow and Smith 2023). Higher global efficiency reflects that the whole brain is more well integrated, whereas lower local efficiency implies that the brain is less well functionally segregated.

CRF has also frequently been associated with memory in healthy aging. For instance, one study has supported that high-fit older adults outperform low-fit older adults on episodic memory tasks such as the face-name memory task (Hayes et al. 2016). A study considering the Rey Auditory Verbal Learning Test as the episodic memory outcome measure, however, showed that CRF was associated with Rey

Auditory Verbal Learning Test performance in older men but not older women (Dougherty et al. 2017). Still, this study revealed a significant correlation between CRF in older women and hippocampal volume, which is encouraging given the classically noted relationship between the hippocampus and episodic memory (Squire and Wixted 2011). The degree of sex specificity in the relationship between fitness and episodic memory should be clarified in future research. Other forms of memory such as working memory have also correlated with CRF in healthy older adults, as research supports that CRF increase across a five-year moderate aerobic exercise training is positively associated with better working memory performance as assessed by the Digit Span Backwards test (Sokolowski et al. 2021). Working memory, such as visual or verbal working memory, is vital for processing and using limited information to support everyday reasoning, comprehension, and problem-solving, among other processes.

Like executive function and memory, language function is considered integral for older adults' functional independence. Challenges with communication can burden older adults during routine life activities, such as interaction with community, caregivers, and healthcare professionals. One study showed that after a 12-week spin cycling intervention in healthy, previously sedentary older adults, improvement in estimated VO_{2max} was correlated positively with increase in semantic fluency (Nocera et al. 2015). Similarly, higher aerobic fitness levels have been found to associate with decreased likelihood of tip-of-the-tongue states in healthy older adults, even when the effect of age is controlled (Segaert et al. 2018). Tip-of-the-tongue states represent a common cognitive difficulty in healthy older adults and occur when a speaker knows the desired word but is unable to produce it. One study assessing language function with the Boston Naming Test also supported that greater estimated cardiorespiratory fitness in healthy older adults is correlated with better language function, as well as functions within other cognitive domains (Velazquez-Diaz et al. 2023).

CRF in healthy older adults has also been associated with other vital cognitive processes, including processing speed, pattern separation, and visuospatial reasoning (Sokolowski et al. 2021; Espana-Irla et al. 2021). Apart from increased efficiency of neural networks, other candidate mechanisms may entail preservation of brain regions key for cognitive function. For instance, one study found that the relationship between CRF and executive function task performance was mediated by volume of regions within the prefrontal cortex (Weinstein et al. 2012). It may also be that better CRF improves sleep or attenuates the effects of poor sleep (Kuhn and Heisz 2022). Indeed, sleep quality declines with aging (Mander et al. 2017; Moraes et al. 2014), and worsened sleep quality is associated with poor cognition (Lim et al. 2012) and accelerated cognitive decline (Altena et al. 2010).

3.1 Exercise Training in Healthy Older Adults

In humans, aerobic exercise training interventions have been associated with better cognitive function in healthy older adults (Kramer et al. 2006). In a meta-analysis from 18 randomized-controlled trials, exercise interventions (cardiovascular exercise, resistance exercise, or combined cardiovascular and resistance exercise) had a significant effect on multiple domains of cognitive function compared to a nonexercise control condition, including executive function, controlled processing, visuospatial performance, processing speed, and episodic and short-term memory (Colcombe and Kramer 2003). Moreover, in their meta-analysis, these effects were shown to be larger in studies with a greater proportion of women, in participants greater than 65 years old (compared to ages 55–65), when the intervention lasts at least 6 months and when it consists of a combination of cardiovascular and resistance exercise training (Colcombe and Kramer 2003). Notably, the effects did not depend on the duration of each individual exercise session (up to 60 min long). Subsequent research has also indicated that one of the most consistent effects of exercise training compared with other types of interventions is improved processing speed (e.g., simple reaction time or other timed tasks, such as Trail-Making part A) (Callow and Smith 2023). Improved processing speed also positively impacts performance in other cognitive domains, particularly in executive function. The other finding among healthy older adults after exercise training compared to any other type of active intervention is improved immediate recall verbal memory performance (Angevaren et al. 2008). However, research has been less consistent regarding effects in other domains, including visual and auditory attention and visual memory (Young et al. 2015).

3.2 *Exercise Training in MCI*

Despite the efforts of the medical and research communities, there is insufficient evidence to support the use of pharmaceutical agents or dietary supplements to prevent cognitive decline or AD (Daviglus et al. 2010). Even the most recently FDA-approved medication, lecanemab, has been shown to only modestly attenuate (but not prevent) the rate of cognitive decline in patients diagnosed with early Alzheimer's disease (van Dyck et al. 2023), an effect that may or may not be clinically meaningful (Valenzuela and Pascual-Leone 2023; Zeng et al. 2023; Qiao et al. 2023), Antiamyloid infusions, including lecanemab, also have potential adverse and life-threatening side effects, particularly among those who are at increased genetic risk for AD (Qiao et al. 2023; Cummings et al. 2023). This has stimulated efforts to identify preclinical biomarkers to predict future cognitive decline and AD more effectively and to facilitate targeted intervention and prevention trials. One approach is to initiate trials at a much earlier stage of disease progression to slow or delay disease progression once a diagnosis of MCI due to AD has been determined.

Results from the Fitness for the Ageing Brain Study, a randomized trial of older adults with subjective memory complaints showed improved scores on the Alzheimer Disease Assessment Scale-Cognitive subscale in those assigned to a home-based PA program compared to the usual care condition (Lautenschlager et al. 2008). Another six-month randomized clinical trial in 33 adults diagnosed with MCI compared aerobic exercise training to a stretching control group (Baker et al. 2010). They found that exercise training resulted in significant improvement on verbal fluency, with the strongest effects for category fluency, an index of semantic memory retrieval. In the Mental Activity and Exercise trial (Barnes et al. 2013), older adults with cognitive complaints did 12 weeks of exercise plus mental activity or control conditions, and all groups improved on a composite index of cognition, which suggests that the effects of exercise may not be independent from other cognitively and/or socially stimulating interventions. Another study reported increased efficiency of semantic memory activation and improved verbal learning after 12 weeks of walking exercise (Smith et al. 2013a). Relatedly, another study showed that 6 months of aerobic exercise in women 70–80 years old with subjective memory complaints resulted in stable hippocampal volume compared to hippocampal atrophy measured in a control group that performed balance and toning exercise (Ten Brinke et al. 2015). While there is evidence for the efficacy of exercise to modestly improve cognition, it is unknown if exercise training among individuals diagnosed with MCI leads to a reduced incidence of, or progression to, AD. Accordingly, there is a need to conduct an exercise clinical trial earlier in the disease process among those who are at greatest risk, including those with increased genetic risk.

4 Neuroimaging Evidence for Cognitive Aging

Magnetic resonance imaging (MRI) has been crucial in understanding the effect of exercise on the brain and cognition in older adults. The importance of brain imaging has grown with technological advancements in the development and implementation of various MRI approaches, which researchers now use to investigate numerous structural and functional changes in the brain that may result from exercise interventions and facilitate healthy cognitive aging. For example, structural MRI approaches suggest that exercise may reduce age-related atrophy of brain structures essential for higher-order cognitive processes, such as memory and executive function. Additionally, perfusion MRI suggests that exercise can influence cognitive aging by altering levels of brain tissue perfusion with blood, oxygen, and essential nutrients. Thus, neuroimaging provides a noninvasive means of investigating neuroanatomical and neurophysiological changes in the human brain that may account for improvements in cognitive function previously documented by PA and exercise interventions.

Structural MRI provided some of the first evidence for the effects of exercise on cognitive aging in humans by observing how exercise might alter the relative size and trajectories of different brain regions with age, particularly gray matter tissue. Changes in brain volume are well characterized throughout healthy and pathological aging and suggest that the majority of cortical and subcortical gray matter tissue begins to shrink in mid-adulthood, with atrophy hastening in older age. Notably, the prefrontal cortex and the hippocampus are specific gray matter brain regions that consistently shrink approximately 1–2% a year in older age, with more rapid declines experienced after the onset of cognitive impairment. Given that these two structures are critical for memory and executive function, their rapid deterioration in older age may be related to the often pronounced declines in fluid cognitive abilities associated with aging (O'Shea et al. 2016).

Two foundational exercise intervention studies (Colcombe et al. 2006; Erickson et al. 2011) found cardiorespiratory fitness improvements from a six-month and one-year exercise training program in healthy older adults led to increases in frontal lobe and anterior hippocampal gray matter volume. Consistent with these early findings, recent observational studies suggested that PA levels and CRF are related to preserved gray matter volume in older adults, particularly in the frontal lobes and hippocampus (Koblinsky et al. 2021). Notably the benefits of PA and fitness for gray matter preservation may be most pronounced for those genetically predisposed to cognitive impairment and hippocampal deterioration. Specifically, higher PA levels can ameliorate the negative effect of the Apolipoprotein-E ε4 allele – a genetic polymorphism associated with AD risk – on hippocampal atrophy in old age (Fraser et al. 2022; Smith et al. 2014). However, many randomized controlled trials for exercise interventions conducted in the last decade fail to establish that exercise training can increase gray matter or hippocampal volume. For example, a meta-analysis of exercise intervention randomized controlled trials in adults only noted a retention of left hippocampal volume but not bilateral volume (Firth et al. 2018). Meanwhile, a recent meta-analysis of 14 exercise randomized controlled trials focusing exclusively on older adults found that exercise training was not associated with differences in gray matter volume compared to control conditions (Gogniat et al. 2021). Several potential reasons exist for inconsistencies in the reported relationships between exercise and gray matter volume, providing a framework and direction for future studies. Thus, future studies should differentiate between exercise responders and nonresponders by reporting changes in fitness and consider potential moderating factors. Additional MRI-based measures of brain tissue integrity and microstructure can proceed volumetric decline and thus should be explored within the context of exercise and cognitive aging (Callow et al. 2021). High-resolution structural imaging approaches focusing on the hippocampus are needed, as animal and recent human studies indicate the benefits of exercise training and fitness for cognitive aging may be specific to the preservation and integration of new neurons in particular hippocampal subfields (Duzel et al. 2016; Frodl et al. 2020). Finally, future studies must determine what cognitive domains and processes exercise-related preservation of gray matter tissue specifically mediates.

Researchers have also used perfusion MRI approaches to discern the cerebrovascular effects of exercise on cognitive aging. Notably, aging is often associated with cardiovascular dysfunction, which may contribute to consistent age-related declines in gray matter cerebral blood flow (CBF) (Leidhin et al. 2021; Tarumi and Zhang 2018). Additionally, longitudinal studies have linked age-related CBF declines to poorer fluid cognitive abilities such as executive function and processing speed (Staffaroni et al. 2019). Given the strong ties between exercise and cardiovascular health, understanding exercise and PA's effects on CBF in older adults may provide additional insight into how exercise may promote healthy cognitive aging.

Cross-sectional studies suggest that a sedentary lifestyle is associated with reduced resting cerebral perfusion in the frontal lobe. In contrast, greater light and moderate intensity PA levels have a positive, dose-dependent association with resting frontal lobe perfusion in older adults (Zlatar et al. 2019). Meanwhile, a small study of ten masters athletes found that 10 days of exercise cessation was associated with a decline in resting hippocampal CBF (Alfini et al. 2016). Furthermore, a within-subject 8-week fully supervised personalized and progressive aerobic exercise training program on a cycle ergometer in 17 older sedentary individuals found that in addition to improved CRF and executive functioning, exercise training was associated with higher resting CBF in the frontal lobe but lower resting CBF in the medial temporal lobe (Kleinloog et al. 2019). Finally, a 12-week progressive walking aerobic exercise intervention in healthy older adults and those with MCI found that exercise ameliorated elevated insular CBF at baseline in those with MCI and increased CBF in the cingulate cortex of the healthy older adults (Alfini et al. 2019). Both groups of older adults also saw improvements in executive function and memory following the intervention (Alfini et al. 2019). Notably, a lack of exercise intervention randomized controlled trials with adequate sample sizes, proper control groups, and that specifically test for relationships between exercise-induced CBF changes and cognition in older adults makes it challenging to interpret how exercise may affect CBF and cognitive aging. However, limited evidence indicates that these relationships may depend on individual factors, the specific brain regions evaluated, and the cognitive tasks analyzed.

In summary, greater PA, CRF, and exercise training programs appear to influence the trajectory of age-related changes across various structural and physiological neuroimaging domains and mechanisms. However, there is a notable lack of well-designed exercise intervention randomized controlled trials in older adults that test for exercise-induced MRI-based changes and subsequently link these changes to alterations in cognitive performance. Future studies must target this gap to improve our understanding of how these various neurophysiological mechanisms mediate the positive relationship between exercise and healthy cognitive aging.

5 Potential Mechanisms from Animal Models

The molecular mechanisms behind exercise-related neuroprotection are challenging to address in humans given that circulating peripheral biomarkers do not always reflect the neural milieu (Calverley et al. 2020). Nevertheless, animal models of biological and pathological aging suggest that exercise is neuroprotective through the upregulation of neurotrophins and growth factors that, over time, promote the proliferation and differentiation of quiescent adult neural stem cells in a process termed adult neurogenesis (Cotman et al. 2007). In the context of human health, animal studies have shown that adult neurogenesis takes place in the subgranular zone of the dentate gyrus in the hippocampus, a subcortical structure that is particularly susceptible to neurodegeneration and pathologies such as AD (Henneman et al. 2009). As such, understanding how exercise may lead to adult neurogenesis is a promising way to delay cognitive decline and neurodegeneration.

Neurotrophins are growth factors widely expressed throughout the central nervous system that play fundamental roles in neural plasticity and the development and maintenance of neurons (Noble et al. 2011). Among these, brain-derived neurotrophic factor (BDNF) has been shown to regulate adult neurogenesis and be upregulated in response to aerobic exercise (Cotman et al. 2007). In this context, aerobic exercise acutely upregulates mRNA levels of BDNF's high-affinity receptor TrkB as well as mRNA levels for proteins that facilitate neurotransmitter release and synaptic plasticity such as synapsin I (Gomez-Pinilla et al. 2002). Moreover, aerobic exercise increases phosphorylation of BDNF-related signal transduction pathways such as the MAPK ERK 1/MAPK ERK 2 pathway as well as the upregulation of transcription factors such as cyclic AMP response element that aid neuroplasticity (Gomez-Pinilla et al. 2002). Insulin growth factor 1 (IGF-1) and vascular endothelial growth factor (VEGF) are additional growth factors that play key roles in the neuroprotective benefits of aerobic exercise (Cotman et al. 2007). IGF-1 is a growth factor that is essential for cellular growth and metabolism, and it is intracellular signaling cascades that have been implicated in neurogenesis and cognitive health (Frater et al. 2018; Pardo et al. 2018). For example, in aging rats, IGF-1 upregulation is accompanied by higher expression of genes involved in synaptic plasticity such as synaptophysin-like protein 2 concomitant to improved performance in spatial memory tasks (Pardo et al. 2018). Importantly, exercise has been noted to upregulate the secretion of IGF-1 by neurons and peripheral tissues such as the skeletal muscle and liver (Cotman et al. 2007). This upregulation in central and circulating IGF-1 is associated with higher IGF-1 uptake by brain hippocampal neurons and synergistic interactions with BDNF that facilitate long-term potentiation and neuroplasticity via the activation of cellular signaling pathways such as MAPK (Cotman et al. 2007). Lastly, VEGF is an angiogenic growth factor that is upregulated by aerobic exercise, and VEGF in the brain may serve as an important regulator of adult neurogenesis via VEGF2/Filk 1 receptor signaling and subsequent phosphorylation of cell survival mediators such as PI3K/AKT1 (Fabel et al. 2003; Cao et al. 2004). Thus, although not exclusive, neurotrophins and growth factors are among the most well-documented molecular mechanisms by which PA and exercise can benefit long-term brain and cognitive health. It is important to note here, however, that this is a growing field of research, and recent evidence suggest that the molecular mediators of exercise-related neuroprotection may also involve proteins secreted from peripheral tissues such as skeletal muscle. Thus, there is continued need to employ animal models that can clarify the complex molecular neuroprotective responses upregulated during and following exercise.

6 Summary and Conclusions

Leisure-time PA, greater cardiorespiratory fitness, and exercise training interventions all have been associated with attenuation of age-related cognitive decline. These associations have been manifest in neurocognitive test performance outcomes, particularly executive function and cognitive processing speed, as well as indices of neural function and structure derived from MRI scans. There is some evidence that the protective effects of exercise and PA on cognition are greater among females, but additional research is needed due to the tendency for these studies to include more females than males. While there currently is no evidence to suggest that the cognitive benefits of PA do not extend to people from all racial and ethnic backgrounds, additional research with more diverse samples is needed. The complex mechanistic causes for these neuroprotective effects have not been conclusively demonstrated but likely involve neurotrophic pathways that stimulate growth and/or protection for neural atrophy within and between neural networks. Additional research in humans using cutting-edge neuroimaging methods will also help to demonstrate whether or not acute and chronic exercise modifies the neural networks that govern multiple cognitive domains and provide the type of mechanistic findings that may help to establish and promote evidence-based prescription of exercise and PA for protection against age-related cognitive decline.

References

ACSM (2006) American College of Sports Medicine Guidelines for Exercise Testing and Prescription, 7th edn. Lippincott, Williams and Wilkins

Albert MS, DeKosky ST, Dickson D, Dubois B, Feldman HH, Fox NC et al (2011) The diagnosis of mild cognitive impairment due to Alzheimer's disease: recommendations from the National Institute on Aging-Alzheimer's Association workgroups on diagnostic guidelines for Alzheimer's disease. Alzheimers Dement 7(3):270–279

Alfini AJ, Weiss LR, Leitner BP, Smith TJ, Hagberg JM, Smith JC (2016) Hippocampal and cerebral blood flow after exercise cessation in master athletes. Front Aging Neurosci 8:184

Alfini AJ, Weiss LR, Nielson KA, Verber MD, Smith JC (2019) Resting cerebral blood flow after exercise training in mild cognitive impairment. J Alzheimers Dis 67(2):671–684

Altena E, Ramautar JR, Van Der Werf YD, Van Someren EJ (2010) Do sleep complaints contribute to age-related cognitive decline? Prog Brain Res 185:181–205

Alzheimer's Association (2023) Alzheimer's disease facts and figures. Alzheimers Dement 2023

American Psychiatric Association (2022) Diagnostic and statistical manual of mental disorders. American Psychiatric Association Publishing, Washington

Angevaren M, Aufdemkampe G, Verhaar HJ, Aleman A, Vanhees L (2008) Physical activity and enhanced fitness to improve cognitive function in older people without known cognitive impairment. Cochrane Database Syst Rev (3):CD005381

Baker LD, Frank LL, Foster-Schubert K, Green PS, Wilkinson CW, McTiernan A et al (2010) Effects of aerobic exercise on mild cognitive impairment: a controlled trial. Arch Neurol 67(1): 71–79

Barnes DE, Yaffe K (2011) The projected effect of risk factor reduction on Alzheimer's disease prevalence. Lancet Neurol 10(9):819–828

Barnes DE, Yaffe K, Satariano WA, Tager IB (2003) A longitudinal study of cardiorespiratory fitness and cognitive function in healthy older adults. J Am Geriatr Soc 51(4):459–465
Barnes DE, Santos-Modesitt W, Poelke G, Kramer AF, Castro C, Middleton LE et al (2013) The mental activity and eXercise (MAX) trial: a randomized controlled trial to enhance cognitive function in older adults. JAMA Intern Med 173(9):797–804
Blair SN, Kohl HW 3rd, Paffenbarger RS Jr, Clark DG, Cooper KH, Gibbons LW (1989) Physical fitness and all-cause mortality. A prospective study of healthy men and women. JAMA 262(17): 2395–2401
Bouchard C, Sarzynski MA, Rice TK, Kraus WE, Church TS, Sung YJ et al (2011) Genomic predictors of the maximal O(2) uptake response to standardized exercise training programs. J Appl Physiol 110(5):1160–1170
Callow DD, Smith JC (2023) Physical fitness, cognition, and structural network efficiency of brain connections across the lifespan. Neuropsychologia 182:108527
Callow DD, Won J, Pena GS, Jordan LS, Arnold-Nedimala NA, Kommula Y et al (2021) Exercise training-related changes in cortical gray matter diffusivity and cognitive function in mild cognitive impairment and healthy older adults. Front Aging Neurosci 13:645258
Calverley TA, Ogoh S, Marley CJ, Steggall M, Marchi N, Brassard P et al (2020) HIITing the brain with exercise: mechanisms, consequences and practical recommendations. J Physiol 598(13): 2513–2530
Cao L, Jiao X, Zuzga DS, Liu Y, Fong DM, Young D et al (2004) VEGF links hippocampal activity with neurogenesis, learning and memory. Nat Genet 36(8):827–835
Caspersen CJ, Powell KE, Christenson GM (1985) Physical activity, exercise, and physical fitness: definitions and distinctions for health-related research. Public Health Rep 100(2):126–131
Cohen RA, Marsiske MM, Smith GE (2019) Neuropsychology of aging. Handb Clin Neurol 167: 149–180
Colcombe S, Kramer AF (2003) Fitness effects on the cognitive function of older adults: a meta-analytic study. Psychol Sci 14(2):125–130
Colcombe SJ, Erickson KI, Scalf PE, Kim JS, Prakash R, McAuley E et al (2006) Aerobic exercise training increases brain volume in aging humans. J Gerontol 61(11):1166–1170
Corbo I, Casagrande M (2022) Higher-level executive functions in healthy elderly and mild cognitive impairment: a systematic review. J Clin Med 11(5)
Cotman CW, Berchtold NC, Christie LA (2007) Exercise builds brain health: key roles of growth factor cascades and inflammation. Trends Neurosci 30(9):464–472
Craik FI, Salthouse TA (2007) The handbook of aging and cognition, 3rd edn. Psychology Press, New York
Cummings J, Apostolova L, Rabinovici GD, Atri A, Aisen P, Greenberg S et al (2023) Lecanemab: Appropriate Use Recommendations. J Prev Alzheimers Dis 10(3):362–377
Daviglus ML, Bell CC, Berrettini W, Bowen PE, Connolly ES, Cox NJ et al (2010) NIH state-of-the-science conference statement: preventing Alzheimer's disease and cognitive decline. NIH Consens State Sci Statements 27(4):1–30
Dennis N, Gutchess A, Thomas A (eds) (2020) Overview of models of cognitive aging. Cambridge University Press, Cambridge
Dougherty RJ, Schultz SA, Boots EA, Ellingson LD, Meyer JD, Van Riper S et al (2017) Relationships between cardiorespiratory fitness, hippocampal volume, and episodic memory in a population at risk for Alzheimer's disease. Brain Behav 7(3):e00625
Duzel E, van Praag H, Sendtner M (2016) Can physical exercise in old age improve memory and hippocampal function? Brain 139(Pt 3):662–673
Erickson KI, Voss MW, Prakash RS, Basak C, Szabo A, Chaddock L et al (2011) Exercise training increases size of hippocampus and improves memory. Proc Natl Acad Sci U S A 108(7): 3017–3022
Espana-Irla G, Gomes-Osman J, Cattaneo G, Albu S, Cabello-Toscano M, Solana-Sanchez J et al (2021) Associations between cardiorespiratory fitness, cardiovascular risk, and cognition are mediated by structural brain health in midlife. J Am Heart Assoc 10(18):e020688

Etgen T, Sander D, Huntgeburth U, Poppert H, Forstl H, Bickel H (2010) Physical activity and incident cognitive impairment in elderly persons: the INVADE study. Arch Intern Med 170(2): 186–193

Fabel K, Fabel K, Tam B, Kaufer D, Baiker A, Simmons N et al (2003) VEGF is necessary for exercise-induced adult hippocampal neurogenesis. Eur J Neurosci 18(10):2803–2812

Firth J, Stubbs B, Vancampfort D, Schuch F, Lagopoulos J, Rosenbaum S et al (2018) Effect of aerobic exercise on hippocampal volume in humans: a systematic review and meta-analysis. Neuroimage 166:230–238

Fraser MA, Walsh EI, Shaw ME, Anstey KJ, Cherbuin N (2022) Longitudinal effects of physical activity change on hippocampal volumes over up to 12 years in middle and older age community-dwelling individuals. Cereb Cortex 32(13):2705–2716

Frater J, Lie D, Bartlett P, McGrath JJ (2018) Insulin-like growth factor 1 (IGF-1) as a marker of cognitive decline in normal ageing: a review. Ageing Res Rev 42:14–27

Frodl T, Strehl K, Carballedo A, Tozzi L, Doyle M, Amico F et al (2020) Aerobic exercise increases hippocampal subfield volumes in younger adults and prevents volume decline in the elderly. Brain Imaging Behav 14(5):1577–1587

Gauthier CJ, Lefort M, Mekary S, Desjardins-Crepeau L, Skimminge A, Iversen P et al (2015) Hearts and minds: linking vascular rigidity and aerobic fitness with cognitive aging. Neurobiol Aging 36(1):304–314

Geda YE, Roberts RO, Knopman DS, Christianson TJ, Pankratz VS, Ivnik RJ et al (2010) Physical exercise, aging, and mild cognitive impairment: a population-based study. Arch Neurol 67(1): 80–86

Gogniat MA, Robinson TL, Miller LS (2021) Exercise interventions do not impact brain volume change in older adults: a systematic review and meta-analysis. Neurobiol Aging 101:230–246

Gomez-Pinilla F, Ying Z, Roy RR, Molteni R, Edgerton VR (2002) Voluntary exercise induces a BDNF-mediated mechanism that promotes neuroplasticity. J Neurophysiol 88(5):2187–2195

Haaland KY, Price L, Larue A (2003) What does the WMS-III tell us about memory changes with normal aging? J Int Neuropsychol Soc 9(1):89–96

Hayes SM, Forman DE, Verfaellie M (2016) Cardiorespiratory fitness is associated with cognitive performance in older but not younger adults. J Gerontol B Psychol Sci Soc Sci 71(3):474–482

Henneman WJ, Sluimer JD, Barnes J, van der Flier WM, Sluimer IC, Fox NC et al (2009) Hippocampal atrophy rates in Alzheimer disease: added value over whole brain volume measures. Neurology 72(11):999–1007

IOM (2015) Cognitive aging: Progress in understanding and opportunities for action, Washington

Kawagoe T, Onoda K, Yamaguchi S (2017) Associations among executive function, cardiorespiratory fitness, and brain network properties in older adults. Sci Rep 7:40107

Kleinloog JPD, Mensink RP, Ivanov D, Adam JJ, Uludag K, Joris PJ (2019) Aerobic exercise training improves cerebral blood flow and executive function: a randomized, controlled cross-over trial in sedentary older men. Front Aging Neurosci 11:333

Koblinsky ND, Meusel LC, Greenwood CE, Anderson ND (2021) Household physical activity is positively associated with gray matter volume in older adults. BMC Geriatr 21(1):104

Kramer AF, Erickson KI, Colcombe SJ (2006) Exercise, cognition, and the aging brain. J Appl Physiol 101(4):1237–1242

Kuhn T, Heisz J (2022) Cardiorespiratory fitness may protect memory for poorer sleepers. Front Psychol 13:793875

Larson EB, Wang L, Bowen JD, McCormick WC, Teri L, Crane P et al (2006) Exercise is associated with reduced risk for incident dementia among persons 65 years of age and older. Ann Intern Med 144(2):73–81

Lautenschlager NT, Cox KL, Flicker L, Foster JK, van Bockxmeer FM, Xiao J et al (2008) Effect of physical activity on cognitive function in older adults at risk for Alzheimer disease: a randomized trial. JAMA 300(9):1027–1037

Leidhin CN, McMorrow J, Carey D, Newman L, Williamson W, Fagan AJ et al (2021) Age-related normative changes in cerebral perfusion: data from the Irish longitudinal study on ageing (TILDA). Neuroimage 229:117741

Lim AS, Yu L, Costa MD, Leurgans SE, Buchman AS, Bennett DA et al (2012) Increased fragmentation of rest-activity patterns is associated with a characteristic pattern of cognitive impairment in older individuals. Sleep 35(5):633–40B

Liu R, Sui X, Laditka JN, Church TS, Colabianchi N, Hussey J et al (2012) Cardiorespiratory fitness as a predictor of dementia mortality in men and women. Med Sci Sports Exerc 44(2):253–259

Malek-Ahmadi M (2016) Reversion from mild cognitive impairment to normal cognition: a meta-analysis. Alzheimer Dis Assoc Disord 30(4):324–330

Mander BA, Winer JR, Walker MP (2017) Sleep and Human Aging. Neuron 94(1):19–36

McArdle JJ, Ferrer-Caja E, Hamagami F, Woodcock RW (2002) Comparative longitudinal structural analyses of the growth and decline of multiple intellectual abilities over the life span. Dev Psychol 38(1):115–142

Moraes W, Piovezan R, Poyares D, Bittencourt LR, Santos-Silva R, Tufik S (2014) Effects of aging on sleep structure throughout adulthood: a population-based study. Sleep Med 15(4):401–409

Noble EE, Billington CJ, Kotz CM, Wang C (2011) The lighter side of BDNF. Am J Physiol Regul Integr Comp Physiol 300(5):R1053–R1069

Nocera JR, McGregor KM, Hass CJ, Crosson B (2015) Spin exercise improves semantic fluency in previously sedentary older adults. J Aging Phys Act 23(1):90–94

O'Shea A, Cohen RA, Porges EC, Nissim NR, Woods AJ (2016) Cognitive aging and the hippocampus in older adults. Front Aging Neurosci 8:298

Paffenbarger RS Jr, Hyde RT, Wing AL, Hsieh CC (1986) Physical activity, all-cause mortality, and longevity of college alumni. N Engl J Med 314(10):605–613

Pardo J, Abba MC, Lacunza E, Ogundele OM, Paiva I, Morel GR et al (2018) IGF-I gene therapy in aging rats modulates hippocampal genes relevant to memory function. J Gerontol 73(4): 459–467

Peltz CB, Corrada MM, Berlau DJ, Kawas CH (2011) Incidence of dementia in oldest-old with amnestic MCI and other cognitive impairments. Neurology 77(21):1906–1912

Pentikainen H, Savonen K, Ngandu T, Solomon A, Komulainen P, Paajanen T et al (2019) Cardiorespiratory fitness and cognition: longitudinal associations in the FINGER study. J Alzheimers Dis 68(3):961–968

Petersen RC, Morris JC (2005) Mild cognitive impairment as a clinical entity and treatment target. Arch Neurol 62(7):1160–1163. discussion 7

Piercy KL, Troiano RP, Ballard RM, Carlson SA, Fulton JE, Galuska DA et al (2018) The physical activity guidelines for Americans. JAMA 320(19):2020–2028

Qiao Y, Chi Y, Zhang Q, Ma Y (2023) Safety and efficacy of lecanemab for Alzheimer's disease: a systematic review and meta-analysis of randomized clinical trials. Front Aging Neurosci 15: 1169499

Salthouse TA (2019) Trajectories of normal cognitive aging. Psychol Aging 34(1):17–24

Samitz G, Egger M, Zwahlen M (2011) Domains of physical activity and all-cause mortality: systematic review and dose-response meta-analysis of cohort studies. Int J Epidemiol 40(5): 1382–1400

Segaert K, Lucas SJE, Burley CV, Segaert P, Milner AE, Ryan M et al (2018) Higher physical fitness levels are associated with less language decline in healthy ageing. Sci Rep 8(1):6715

Singh-Manoux A, Kivimaki M, Glymour MM, Elbaz A, Berr C, Ebmeier KP et al (2012) Timing of onset of cognitive decline: results from Whitehall II prospective cohort study. BMJ 344:d7622

Smith JC, Nielson KA, Antuono P, Lyons J-A, Hanson RJ, Butts AM et al (2013a) Semantic memory functional MRI and cognitive function after exercise intervention in mild cognitive impairment. J Alzheimers Dis 37(1):197–215

Smith JC, Nielson KA, Woodard JL, Seidenberg M, Rao SM (2013b) Physical activity and brain function in older adults at increased risk for Alzheimer's disease. Brain Sci 3:54–83

Smith JC, Nielson KA, Woodard JL, Seidenberg M, Durgerian S, Hazlett KE et al (2014) Physical activity reduces hippocampal atrophy in elders at genetic risk for Alzheimer's disease. Front Aging Neurosci 6(61):1–7

Sokolowski DR, Hansen TI, Rise HH, Reitlo LS, Wisloff U, Stensvold D et al (2021) 5 years of exercise intervention did not benefit cognition compared to the physical activity guidelines in older adults, but higher cardiorespiratory fitness did. A generation 100 substudy. Front Aging Neurosci 13:742587

Squire LR, Wixted JT (2011) The cognitive neuroscience of human memory since H.M. Annu Rev Neurosci 34:259–288

Staffaroni AM, Cobigo Y, Elahi FM, Casaletto KB, Walters SM, Wolf A et al (2019) A longitudinal characterization of perfusion in the aging brain and associations with cognition and neural structure. Hum Brain Mapp 40(12):3522–3533

Sun F, Norman IJ, While AE (2013) Physical activity in older people: a systematic review. BMC Public Health 13:449

Tarumi T, Zhang R (2018) Cerebral blood flow in normal aging adults: cardiovascular determinants, clinical implications, and aerobic fitness. J Neurochem 144(5):595–608

Ten Brinke LF, Bolandzadeh N, Nagamatsu LS, Hsu CL, Davis JC, Miran-Khan K et al (2015) Aerobic exercise increases hippocampal volume in older women with probable mild cognitive impairment: a 6-month randomised controlled trial. Br J Sports Med 49(4):248–254

Valenzuela MJ, Pascual-Leone A (2023) Lecanemab in early Alzheimer's disease. N Engl J Med 388(17):1630

van Dyck CH, Swanson CJ, Aisen P, Bateman RJ, Chen C, Gee M et al (2023) Lecanemab in early Alzheimer's disease. N Engl J Med 388(1):9–21

Velazquez-Diaz D, Cadenas-Sanchez C, Molina-Guzman FA, Saenz-Carrasco JA, Gonzalez-Rosa JJ, Erickson KI et al (2023) A new set of estimated cardiorespiratory fitness equations are associated with cognitive performance in older adults. Geroscience

Weinstein AM, Voss MW, Prakash RS, Chaddock L, Szabo A, White SM et al (2012) The association between aerobic fitness and executive function is mediated by prefrontal cortex volume. Brain Behav Immun 26(5):811–819

Woodard JL, Sugarman MA, Nielson KA, Smith JC, Seidenberg M, Durgerian S et al (2012) Lifestyle and genetic contributions to cognitive decline and hippocampal structure and function in healthy aging. Curr Alzheimer Res 9(4):436–446

Yaffe K, Fiocco AJ, Lindquist K, Vittinghoff E, Simonsick EM, Newman AB et al (2009) Predictors of maintaining cognitive function in older adults: the health ABC study. Neurology 72(23):2029–2035

Young J, Angevaren M, Rusted J, Tabet N (2015) Aerobic exercise to improve cognitive function in older people without known cognitive impairment. Cochrane Database Syst Rev (4):CD005381

Zec RF, Markwell SJ, Burkett NR, Larsen DL (2005) A longitudinal study of confrontation naming in the "normal" elderly. J Int Neuropsychol Soc 11(6):716–726

Zeng BS, Tseng PT, Liang CS (2023) Lecanemab in early Alzheimer's disease. N Engl J Med 388(17):1630

Zlatar ZZ, Hays CC, Mestre Z, Campbell LM, Meloy MJ, Bangen KJ et al (2019) Dose-dependent association of accelerometer-measured physical activity and sedentary time with brain perfusion in aging. Exp Gerontol 125:110679

Physical Activity and Fatigue Symptoms: Neurotypical Adults and People with Chronic Multisymptom Illnesses

Alex Boruch, Grace Branchaw, Patrick J. O'Connor, and Dane B. Cook

Contents

Abstract For neurotypical adults, a single bout of low-to-moderate intensity physical activity usually transiently improves feelings of energy. Similar bouts of exercise have the opposite effect of increased feelings of fatigue when performed by samples with chronic multisymptom illnesses (CMIs) such as Long-COVID, Gulf War Illness (GWI), or Myalgic Encephalomyelitis/Chronic Fatigue Syndrome (ME/CFS). The short-term adoption of regular moderate intensity physical activity (typical experiments are 1 to 6 months) among neurotypical adults results in small-to-moderate improvements in self-reported feelings of fatigue, energy, and vitality.

A. Boruch, G. Branchaw, and D. B. Cook (✉)
Research Service, William S. Middleton Memorial Veterans Hospital, Madison, WI, USA

Department of Kinesiology, University of Wisconsin–Madison, Madison, WI, USA
e-mail: dane.cook@wisc.edu

P. J. O'Connor
Department of Kinesiology, University of Georgia, Athens, GA, USA

Curr Topics Behav Neurosci (2024) 67: 281–308
https://doi.org/10.1007/7854_2024_502
Published Online: 23 July 2024

Small improvements in these feelings, or no change at all, occur for CMIs, but limited data precludes strong conclusions. The mechanisms of exercise effects on fatigue, whether acute or chronic, are poorly understood but likely involve multiple neural circuits and associated transmitters. For CMIs, the mechanisms of acute worsening of fatigue with exercise may be driven by the yet unknown pathophysiological mechanisms of the disease (perhaps involving brain, immune and autonomic system dysfunction, and their interactions). Likewise, fatigue improvements may depend on whether chronic physical activity is a disease-modifying treatment.

Keywords Energy · Exercise · Post-exertional malaise

1 Introduction

The primary purpose of this chapter is to overview research that has examined relationships between physical activity and symptoms of fatigue. The information in this chapter is preceded by background about fatigue, information about its measurement, and comments about the scope and impact of the problem of fatigue. Subsequent sub-sections focus on both acute and chronic physical activity in both neurotypical adults and atypical adults such as those with diseases that are defined by chronic fatigue and associated symptoms.

The Oxford English Dictionary provides a definition of the noun fatigue, dating back to the seventeenth century, as "lassitude or weariness resulting from either bodily or mental exertion" (Proffitt 2023). Much has been learned about fatigue since that time, especially about moderators of fatigue. For instance, factors such as one's genetics (Wang et al. 2017), acute drug use, including caffeine, modafinil, and methylphenidate (Qu et al. 2016; Ward-Ritacco et al. 2021), and environmental exposures such as bright light (Xiao et al. 2022) or viral pathogens (Islam et al. 2020) can moderate fatigue even though these variables are independent of physical or mental exertion.

Fatigue has multiple complex meanings. This chapter will focus on feelings of fatigue, which is synonymous with symptoms of fatigue. There is evidence that symptoms of fatigue have small-to-moderate negative relationships with feelings of energy yet are independent from feelings of energy (Boolani et al. 2019; Loy et al. 2018). Feelings of fatigue can covary with fatigue induced by skeletal muscle actions (Monroe et al. 2016). This is because high intensity or long duration exercise produces both feelings of fatigue and a transient reduction in the ability to produce power output. The criterion measure of symptoms of fatigue is self-reported feelings, often assessed with reference to the duration, frequency, or intensity of such feelings. There are dozens of measures of fatigue symptoms that have been used by researchers and clinicians. Most of these measures have been supported by at least one study providing at least one type of evidence to support the validity of the scores as a measure of fatigue (O'Connor 2004). The most widely used measures of fatigue symptoms, however, have thousands of references providing comprehensive supportive evidence for the validity of the measure (e.g., fatigue and vigor scales of the Profile of Mood States questionnaire; vitality scale of the SF-36 health survey) (Curran et al. 1995; Ware 2000). The National Institutes of Health supported the development of the Patient Reported Outcomes Measurement Information System (PROMIS®) which includes a measure of fatigue symptoms (Cella et al. 2016). PROMIS, based on Item Response Theory, uses a computer adaptive approach to determine fatigue levels. One key under-appreciated assessment issue is that some measures focus solely on the concept of fatigue (e.g., State Trait Energy and Fatigue Scales) (Boolani and Manierre 2019) while other measurement tools aim to simultaneously assess both fatigue and aspects of the affective, behavioral and/or cognitive consequences of fatigue (e.g., European Organization for Research and Treatment of Cancer – Quality of Life Questionnaire) (Aaronson et al. 1993). While this latter approach has the advantage of more comprehensively surveying fatigue-related experiences, it also has the potential disadvantage of introducing substantial construct-irrelevant variance into scores, which adds error to the measure of fatigue (Eignor 2013).

The prevalence of fatigue symptoms is not adequately represented by a single number but depends on several factors including the group studied, such as a community sample compared to samples with a chronic illness. The timeframe (e.g., point prevalence versus lifetime prevalence) and how fatigue is defined (e.g., elevated symptoms above a cut point versus meeting criteria for a diagnosis of a fatigue disorder) are also critical factors in determining prevalence of fatigue symptoms. For example, in 2017, the prevalence of prior month persistent fatigue was reported as 13.6% for the Japanese general population (Aritake et al. 2015). Eight years later a comprehensive worldwide review of studies from population-based samples identified 76 data points from investigations that measured general fatigue of less than six months duration or of an unspecified duration (i.e., not specified as chronic or lifetime) primarily using the Chalder Fatigue Scale. The prevalence of fatigue in adults was 20.4% (with a 95% Confidence Interval (CI) of 16.7–25.0). In the same review, the prevalence of chronic fatigue (fatigue lasting more than 6 months) in adults was 10.1% (95% CI, 8.2–12.5) (Yoon et al. 2023).

The prevalence of fatigue is typically higher in those characterized by a medical illness compared to community samples. A meta-analysis of 129 studies of over 71,000 individuals, primarily residing in Western countries, found that the prevalence of fatigue was 49% among cancer patients based on symptom scores elevated above cut-offs on multiple measures of fatigue and primarily 3 months after treatment (Al Maqbali et al. 2021). The prevalence of usual or prior month fatigue was 46% in a sample of 458 non-hospitalized adults with a positive polymerase chain reaction test for COVID-19 in the prior 1.5 to 6 months (Stavem et al. 2021). The prevalence of fatigue, using various time frames, has been reported in meta-analyses to be elevated on average for multiple types of chronic illnesses, including post-stroke (48%), multiple sclerosis (36.5% to 78%), irritable bowel syndrome (47%), and many others (D'Silva et al. 2022; Oliva Ramirez et al. 2021; Alghamdi et al. 2021). Also, the odds of severe fatigue increases as the number of chronic conditions increases; in one study, the odds of severe fatigue was 1.6 (95% CI,1.5–1.7) in those with one chronic condition and the odds increased to 5.5 (95% CI, 4.5–6.7) for those with four chronic conditions (Goërtz et al. 2021). When strict diagnostic criteria are used, such as with the Diagnostic and Statistical Manual of Mental Disorders, Fifth Edition (DSM-5) (Regier et al. 2013), the prevalence of fatigue is typically lower. For example, a systematic review of epidemiological studies of Myalgic Encephalomyelitis/Chronic Fatigue Syndrome (ME/CFS) in Europe identified only a few studies, two of adults and one of adolescents (Estévez-López et al. 2020). This research reported that the prevalence of ME/CFS ranged from 0.1% to 2.2% (Estévez-López et al. 2020). A separate review of ME/CFS studies with broader inclusion criteria reported a mean prevalence of 1.4 (± 1.6%) based on data from 46 studies that involved over one million participants (Lim et al. 2020a). Gulf War illness (GWI), a disease involving multiple and diverse symptoms similar to ME/CFS, has been estimated to affect ~30% of the nearly 700,000 Veterans deployed to the Persian Gulf War (United States Department of Veterans Affairs 2008). Based on the make-up of active-duty U.S. military personnel and those deployed to the Persian Gulf (~7% women), GWI affects mainly men who were deployed during Operations Dessert Shield and Storm. The global prevalence of post-acute sequelae of SARS CoV-2 infection (i.e., Long-COVID) has been estimated at 43% (95% CI: 39% to 46%) based on data from 41 studies that involved over one million COVID-19 positive patients. Among the symptoms assessed, fatigue was the most prevalent with an estimate of 23% (95% CI: 17% to 30%) (Chen et al. 2022).

Regardless of the timeframe, the prevalence of fatigue is usually higher in samples of women compared to men (Stavem et al. 2021; Reeves et al. 2007; Friedberg et al. 2015). A recent study found that the odds of suffering from fatigue was 1.4 (95% CI, 1.3–1.6) times higher for women than men and this magnitude of group difference is generally consistent with other studies. Elevated symptoms of fatigue are associated with an increased risk of mortality; for example, a 20-year prospective population study of more than 18,000 European men and women from 40 to 79 years found a 1.4 (95% CI, 1.25–1.56) times greater probability of death during the follow-up period compared to those reporting low fatigue symptoms (Basu et al. 2016).

The healthcare, societal, personal, and financial costs of fatigue are substantial. The economic impact of ME/CFS in the United States alone has been estimated to be 36 to 51 billion dollars annually (Jason and Mirin 2021). The economic impact of Long-COVID has yet to be fully determined, although a recent update estimated that the total costs (i.e., healthcare costs, lost wages, reduced quality of life) were upwards of $3.7 trillion, an increase from the previous estimate of $2.6 trillion (Cutler 2022). Affordable and accessible techniques with few adverse side effects to improve fatigue symptoms, such as engagement in physical activity, could have a considerable, positive public health impact.

2 Acute Physical Activity and Fatigue Symptoms in Neurotypical Adults

2.1 Summary of Acute Studies with Neurotypical Participants

Most of the investigations into the effect of a single bout of exercise on feelings of energy and fatigue have tested college students before and after performing a single cycle ergometry session for about 20 to 40 min. These exercise bouts are usually in a laboratory setting with comparisons made to a seated control condition involving either little cognitive activity (quiet rest) or a range of cognitive engagements such as coloring, playing videos, or listening to a lecture (Cooper et al. 2022; Arent et al. 2007). A meta-analysis of studies of this type found that low-to-moderate intensity aerobic type exercise typically results in transient increases in feelings of energy with a standardized mean effect of 0.47 (95% CI, 0.39–0.56) (Loy et al. 2013). Other experiments have examined the influence of weightlifting on feelings of energy and fatigue and found similar results (Herring and O'Connor 2009; Bartholomew et al. 2005). When a sample of pregnant women reported symptoms before and after 24 separate low-to-moderate intensity weight lifting sessions, perceptions of physical and mental energy were consistently increased while perceptions of physical and mental fatigue were consistently decreased (Ward-Ritacco et al. 2016). Overall, the evidence suggests that low-to-moderate intensity acute exercise improves feelings of energy and fatigue transiently for about 85% of individuals tested and across about 85% of workouts. The size of the effect is typically larger for measures of feelings of energy compared to measures of feelings of fatigue.

There appears to be no single experiment in which intensity and duration have been systematically manipulated for the purpose of examine interactions between intensity and duration. Acute exercise intensity and duration both affect feelings of energy and fatigue. A single bout of high intensity exercise can produce increased feelings of fatigue and reduced feelings of energy. These changes appear to depend on the relative exercise intensity such that greater fatigue is produced when the intensity exceeds anaerobic threshold and approaches an individual's peak capacity (Monroe et al. 2016). Intensities above anaerobic threshold cannot be maintained, consequently it is not possible to test long duration high intensity efforts through the same range of durations that are possible with low intensity exercise which can be endured longer (Yoshida 1984). Fatigue does not appear to accrue if the exercise intensity is high but the duration is quite short, such as with a single bout of high intensity cycling lasting 30 s or less (Monroe et al. 2016). However, competitive and recreational exercisers who engage in single workouts that include multiple, repeated near-maximal sprints, often from 30 s to several minutes duration, with short rest periods do show large increases in feelings of fatigue (Costello et al. 2022). In general, as the intensity is maintained at a high level and the total duration is lengthened or the rest interval is reduced, there is an increased likelihood of elevated feelings of fatigue. For example, studies comparing a bout of high intensity interval exercise to a bout of moderate intensity continuous exercise find that there is lower fatigue during the moderate intensity exercise (Greene et al. 2023). In popular recreational and competitive events, such as a 5-K, 10-K, and 42-K runs, anecdotal evidence supports that feelings of fatigue are elevated and gradually increase as the duration of the activity lengthens if a high effort is maintained. Also, we hypothesize the feelings of fatigue can take longer to return to the pre-exercise baseline level with higher intensities and longer durations when a high effort is maintained as has been shown with feelings of effort and pain during recovery (Cook et al. 1997).

Individual characteristics may moderate the influence of acute exercise on feelings of energy and fatigue. Post-exercise improvements in feelings of energy and fatigue were larger for females compared to males after a 30-min bout of vigorous treadmill running compared to a quiet rest control condition (McDowell et al. 2016). In a sample of patients with a depressive disorder, treadmill running for 25 min at a moderate-to-high intensity increased feelings of fatigue 30-min post-exercise compared to controls. The authors contended, if the findings are replicable and generalizable, this mood shift could contribute to the known reduced physical activity levels among depressed patients compared to controls (Weinstein et al. 2010). In contrast, 20 min of both moderate and preferred intensity cycling exercise reduced feelings of fatigue at both 10- and 30-min post-exercise in women with major depressive disorder (Meyer et al. 2019). Vigorous cycling for 20 min induced large increases in feelings of energy compared to a sitting control condition in a sample of young men characterized by elevated ADHD symptoms. The differences between exercise and control were large when assessed 3-min post-exercise, moderate when assessed 40-min post-exercise, and small when assessed 75-min after exercise (Fritz and O'Connor 2016). The findings above from single experiments above require replication before strong conclusions about individual characteristic moderators can be drawn.

The potential moderating role of the environment on the extent to which acute exercise influences feelings of energy and fatigue has been inadequately explored. A recent review stated that definitive conclusions could not be drawn regarding the effect of most countermeasures against mental fatigue, including exposures to elements that might co-occur during exercise such as music or odors (Proost et al. 2022). Another example is whether exercising outdoors enhances the positive mood-altering effects compared to indoor exercise. These comparisons are challenging to interpret because of the multitude of outdoor elements that could influence feelings of energy and fatigue (e.g., observers, traffic, safety concerns, pollution, noise, variations in sunlight). To date, results from a limited number of studies have been mixed with some showing approximately the same change in feelings of energy or fatigue occurring after outdoor exercise compared to indoor exercise (Legrand et al. 2018; Focht 2009), while other studies finding small-to-moderate augmentation in improvements in feelings of energy or fatigue following exercise performed outdoors or indoors during bright light exposure (LaCaille et al. 2004; O'Brien and O'Connor 2000).

2.2 Evidence Regarding Plausible Mechanisms

Relatively few studies have focused on the potential mechanisms by which a single bout of exercise could influence feelings of fatigue; however, there is strong evidence that acute exercise has widespread effects on the brain, which could support exercise-induced changes in symptom states. Evidence from microdialysis studies have revealed that acute exercise in rodents results in increased brain levels of monoaminergic transmitters such as norepinephrine and dopamine (Meeusen et al. 2001), and some of these observations have been confirmed in humans using radiolabeled non-invasive brain imaging techniques (Boecker and Drzezga 2016). These findings are relevant, given that animal and human psychopharmacology investigations suggest that feelings of energy and fatigue can change in response to alterations in neural circuits involving monoamine, acetylcholine, and histamine neurotransmission (Meeusen et al. 2001; Loy and O'Connor 2016; Stahl 2002). Anatomically, there are diffuse norepinephrine, dopamine, acetylcholine, serotonin, and histamine projections to the cortex that have been hypothesized to regulate feelings of fatigue. Many other variables could be involved, including inhibitory neurotransmitters such as gamma-aminobutyric acid (GABA), neurotrophic factors such as brain-derived neurotrophic factor (BDNF), neuropeptides (endorphins, galanin), gases (nitric oxide), and other signaling molecules (Dishman and O'Connor 2009; Basso and Suzuki 2017).

Several investigations have explored relationships between exercise-induced changes in brain electrocortical signals and changes in feelings of fatigue (Petruzzello et al. 2001; Woo et al. 2009). One of these experiments examined acute exercise across weeks 1, 3, and 6 of a 6-week training program in a sample of young adults with elevated symptoms of fatigue. Fatigue symptoms were improved after low intensity exercise on weeks 3 and 6 while feelings of energy were improved at all three measurement periods. About 50% of the improvement in feelings of energy with acute exercise was accounted for by changes in EEG in the theta frequency range measured in the posterior part of the brain (Dishman et al. 2010).

One approach to understanding the potential role of specific neurotransmitters in causing exercise-induced changes in fatigue symptoms is to administer drugs that specifically target neurotransmitters of interest. One experiment had female participants consume doxepin, a drug that blocks histamine H1 receptors, prior to cycling at a low intensity for 30 min. On a separate occasion, the participants consumed a placebo prior to an identical exercise bout. Feelings of energy changed similarly in both conditions while mental fatigue increased only for the doxepin condition. It was concluded that exercise-induced changes in brain histamine binding play a role in post-exercise changes in feelings of fatigue but not energy (Loy and O'Connor 2016).

Neuromodulators also could play a role in causing exercise-induced changes in feelings of fatigue. The endocannabinoid system is a large network of chemical signals and receptors that is involved in the regulation of many bodily functions, including sleep, pain, immune responses, and mood (Zou and Kumar 2018; Raichlen et al. 2012; Brellenthin et al. 2017; Crombie et al. 2018). At least one investigation has implicated endocannabinoids in positive improvements in feelings of fatigue after exercise. This study reported improvements in feelings of fatigue 10 to 30 min after a single bout of cycling exercise at a moderate and self-selected preferred intensity were significantly associated with changes in serum endocannabinoid levels in a sample of depressed women (Meyer et al. 2019). A separate study among people with post-traumatic stress disorder and elevated depression symptoms failed to find a significant relationship between endocannabinoids and fatigue (Botsford et al. 2023).

Improvements in self-reported feelings of fatigue after a single session of exercise might be partly a placebo response (Szabo 2013) (see Chapter 19: "*The Placebo Effect in Exercise and Mental Health Research*" for comprehensive overview). This idea has been tested experimentally at least once (Lindheimer et al. 2017a). Participants completed very low intensity exercise for 25 min on a semi-recumbent cycle ergometer. Active cycling was compared to passive cycling (the "exercise placebo") in which equipment moved participants legs for them. Some participants were told by a research confederate prior to exercise about the expected psychological benefits of acute exercise while others were not. Active cyclists told about the benefits of exercise showed larger improvements in feelings of energy after exercise compared to the active cyclists not given this information. Also, the improvements in feelings of energy after exercise were smaller after the placebo condition despite the conditions being matched on energy expenditure. While these effects were not statistically significant, arguably because of the low exercise intensity (35%), the findings support pursuing further research aimed at understanding fatigue symptom improvements after acute exercise as partially placebo responses.

3 Acute Physical Activity and Fatigue Symptoms: ME/CFS, GWI, and Long-COVID

3.1 Summary of Acute Studies: E.g., Extent of Post-Exertional Malaise

When fatigue is the defining feature of a disease, the relationships between physical activity and fatigue feelings change. Unlike the fatigue reducing and energy enhancing potential of acute exercise in neurotypical adults, fatigue is more often worsened along with other disease defining symptoms including pain, concentration/memory (i.e., brain fog), and unrefreshing sleep. This is the case for diseases such as ME/CFS, GWI, and Long-COVID (Chu et al. 2018; Khan et al. 2004; Nehme et al. 2023; Thaweethai et al. 2023). Collectively, ME/CFS, GWI, and Long-COVID can be broadly considered as chronic multisymptom illnesses (CMIs) (Fukuda et al. 1998), although the heterogeneity within and between these distinct diagnostic entities is high and each disease has unique case definitions (Thaweethai et al. 2023; Carruthers et al. 2003; Fukuda et al. 1994; Clayton 2015; Shine et al. 2014; Steele 2000).

The phenomenon of symptom worsening (e.g., increased fatigue) consequent to physical and/or mental challenge has been termed post-exertional malaise (PEM) (Clayton 2015) and is considered a cardinal feature of CMIs (Clayton 2015; Steele 2000; Brown and Jason 2020; Davis et al. 2021; Fappiano and Baraniuk 2020). Although no consensus definition has been agreed upon, PEM is characterized by symptom exacerbation across a host of domains (e.g., fatigue, pain, cognition) following physical and/or mental exertion and is considered one of the most incapacitating aspects of these diseases. The presence of PEM is required for a diagnosis of ME/CFS, but not for either GWI or Long-COVID. Patient-based survey data indicate that between 78% and 99% (143/144) of people with ME/CFS experience symptom worsening triggered by physical exertion (Chu et al. 2018; Holtzman et al. 2019). Fatigue was the most frequently reported symptom (88%–94%), followed by poor concentration, difficulty thinking, poor memory, and cognitive exhaustion (63%–90%), sleep disturbance/unrefreshing sleep (68%–81%), muscle pain (69%–74%), and flu-like symptoms (58%–59%). The onset of symptoms has been reported to vary between hours and weeks, with the most frequent symptom onset time being 1–2 days (Holtzman et al. 2019). Prevalence of PEM in other CMIs is comparable to ME/CFS, ranging from 70%–96% in GWI (Khan et al. 2004; Fappiano and Baraniuk 2020) and 86% in Long-COVID (Davis et al. 2021), although characteristics of symptom type (e.g., fatigue, pain, brain fog), onset, and duration are less well understood.

When tested in controlled laboratory settings, the occurrence of fatigue worsening after exercise in CMIs is more variable, although still the most common symptom (Committee on the Diagnostic Criteria for Myalgic Encephalomyelitis et al. 2015). Measurement and design issues including symptom heterogeneity, limited measurement of associated symptoms, poor control over quantification of the exercise stimulus, symptom ceiling effects, and lack of control conditions have limited better understanding of the impact of exercise of various types, severities, and durations on PEM (Loy et al. 2016; Lindheimer et al. 2017b, 2020). Despite these limitations, meta-analytic estimates indicate moderate-to-large effects (0.73; 95% CI: 0.24 to 1.23) of acute exercise on ratings of fatigue in ME/CFS (Loy et al. 2016); effects that are opposite to that observed in non-ME/CFS conditions (Loy et al. 2013).

Acute exercise studies of PEM are typically conducted using case–control designs where the CMI sample of interest (i.e., ME/CFS, GWI, & Long-COVID) is compared to a control group. Symptom questionnaires (e.g., Fatigue Severity Scale (FSS; (Krupp et al. 1989)), POMS (McNair et al. 1981), SF-36 (Ware 2000) are administered at variable times (i.e., days to minutes) prior to exercise and then again at variable times during (e.g., mid-point and peak) and after (i.e., immediately, minutes, hours, days) exercise (Loy et al. 2016; Lindheimer et al. 2020). Most studies use aerobic exercise on a cycle ergometer for a duration of 25–30 min at 70–75% of age-predicted maximal heart rate or heart rate reserve (HRR) for submaximal exercise, or until volitional exhaustion for maximal exercise (Loy et al. 2016; Lindheimer et al. 2017b, 2020; Boruch et al. 2021). To date, this research has been conducted in ME/CFS and GWI; however, exercise research for people with Long-COVID is emerging (Durstenfeld et al. 2022; Joseph et al. 2021, 2023; Mancini et al. 2021; Singh et al. 2022). Future studies that incorporate symptom measurement before and after acute exercise in Long-COVID samples are needed.

Exercise intensity and duration have variable effects on fatigue in CMIs. Thirty minutes of intermittent treadmill exercise (i.e., 3 min of exercise and 3 min of rest) at a self-selected pace did not worsen fatigue in ME/CFS; however, this study did not have a control group (Clapp et al. 1999). In contrast, both 30 min of treadmill walking at 1-mph and serial six-minute walk tests have been shown to worsen fatigue for ME/CFS but not controls (Peterson et al. 1994; Friedberg et al. 2023). Prolonged increases in fatigue have been reported for acute exercise protocols ranging from moderate-to-high intensity in both ME/CFS and GWI (Lindheimer et al. 2017b, 2020; Boruch et al. 2021, 2023; Light et al. 2009, 2012; White et al. 2010; Meyer et al. 2013). Twenty-five minutes of combined leg- and arm-ergometry at 70% maximal heart rate resulted in large fatigue increases at 30 min, 8-, 24- and 48-h post-exercise in ME/CFS compared to controls (White et al. 2010). Thirty minutes of daytime cycling at 70% of HRR resulted induced large symptom increases in GWI; however, the most frequently reported symptoms were sleep related (i.e., "difficulty getting to sleep" and "unrefreshing sleep": 18/43, 41.9%), followed by musculoskeletal pain (9/43, 20.9%), headache (8/43, 18.6%), difficulty with memory (4/43, 9.3%), fatigue (3/43 6.9%), and nausea (1/43, 2.3%) (Boruch et al. 2021). Maximal exercise to volitional exhaustion has been reported to worsen fatigue ranging from 24 h to over 10 days in ME/CFS (Lindheimer et al. 2017b; Meyer et al. 2013; Bazelmans et al. 2005; Van Ness et al. 2010; Moore et al. 2023). When two maximal exercise tests were conducted 24-h apart (i.e., 2-Day cardiopulmonary exercise test (CPET)), ME/CFS participants reported greater symptom exacerbation compared to controls and took an average of two weeks to recover compared to two days for controls (Moore et al. 2023). A meta-analysis of PEM in ME/CFS found that six out of seven studies showed that post-exercise fatigue increases were larger in ME/CFS compared to healthy controls (Loy et al. 2016). Significant univariate effect sizes were found for moderate (0.955; 95% CI: 0.795 to 1.115) and high (0.618; 95% CI: 0.305 to 0.931) intensity exercise, but not light intensity exercise (0.081; 95% CI: −0.402 to 0.565). Fatigue differences were found at a minimum of 4- and up to 96-h post-exercise suggestive of a delayed and variable response.

The variability of PEM has been examined by a few studies. In ME/CFS, focus group interviews involving 43 patients revealed PEM responses that covered six categories (i.e., General, Cognitive, Neuromuscular Complaints, Sensory, Affect, & Ear, Nose, Throat) and 24 symptoms. Exhaustion was the most reported symptom response to exercise with variable onsets (2 to 48 h) and durations (up to 14 days) (Stussman et al. 2020). One study comparing maximal and submaximal exercise responses in ME/CFS found that pain, but not fatigue, was worsened at 24-h post submaximal exercise (Lindheimer et al. 2017b). Maximal exercise symptom responses differed from submaximal showing large increases in fatigue at 72 h, but not 48 h, post-exercise. Moreover, dot-density plots indicated high within-group variability for both maximal and submaximal symptom responses (Lindheimer et al. 2017b). Baseline symptom severity may also affect post-exertional fatigue. In ME/CFS, symptom exacerbation following a 1- or 2-Day CPET test post-exercise was related to baseline symptom severity (Stussman et al. 2020). Conversely, recovery time following a 2-Day CPET test was not predicted by baseline symptom severity (Moore et al. 2023). In GWI, multivariate modeling of PEM responses to 30 min of submaximal exercise at 70% of HRR did not show symptom worsening. However, subgroup analyses involving only those Veterans who endorsed "*feeling unwell after physical exercise or exertion*" at study intake revealed exacerbation of pain- and GWI-related symptoms (Lindheimer et al. 2020). A follow-up study reported that peak symptom responses for this group of Gulf War Veterans occurred at 48 h post-exercise for most GWI participants yet ranged from 24 h to 7 days (Boruch et al. 2021).

Sex differences in post-exertional fatigue ratings are currently understudied primarily due to the make-up of the population's studied (i.e., ME/CFS predominantly female, GWI predominantly male) and resultant difficulties in recruiting adequate samples of both female and male participants. In one ME/CFS study, sex differences in post-exercise fatigue and function following a 6-min walk test were found only for female ME/CFS participants (Friedberg et al. 2023). There are currently no studies investigating sex differences on post-exertional fatigue in GWI, yet baseline symptoms are more variable and severe for female Veterans compared to male Veterans (Dursa et al. 2019; Heboyan et al. 2019; Sullivan et al. 2020). Currently, there are no Long-COVID studies of PEM. A global meta-analysis estimated pooled post-COVID-19 condition prevalence rates of 49% (95% CI: 35% to 63%) for females and 37% (95% CI: 24% to 51%) for males (Chen et al. 2022). These data suggest that future PEM research testing for potential sex differences in Long-COVID may be more feasible than either ME/CFS or GWI.

Fatigue is one of many symptoms worsened with exercise in CMIs. Current data suggest acute exercise fatigue responses vary as a function of exercise intensity and duration, are highly variable with respect to symptom onset, intensity, and duration, and are somewhat dependent on whether PEM is endorsed at study intake. Experiments testing the consequences of different exercise doses and modes between male and female participants and compared to non-exercise control conditions are needed.

3.2 Evidence Regarding Plausible Mechanisms

The biological mechanisms that underlie PEM are not well understood; however, acute exercise has been used extensively to explore numerous physiological systems and biological markers. Both maximal and submaximal exercise protocols have been used to determine behavioral and physiological consequences of acute exercise challenge. These studies have demonstrated symptom exacerbation of variable intensity, type, and duration (White et al. 2010; Barhorst et al. 2021), inefficient/impaired cardiopulmonary responses to acute exercise (Joseph et al. 2023; Cook et al. 2022; Lindheimer et al. 2019), impaired pain regulation (Van Oosterwijck et al. 2010), altered immune function markers (e.g., cytokines, compliment c4, natural killer cells) (Light et al. 2012; Meyer et al. 2013; Smylie et al. 2013; Moneghetti et al. 2018); changes in gut microbiome interactions (Shukla et al. 2015), disruption of metabolites (Baraniuk et al. 2021; Germain et al. 2020), and altered brain function (Cook et al. 2017; Rayhan and Baraniuk 2021; Washington et al. 2020). It is clear from these studies that exercise differentially influences multiple physiological systems in CMIs. Although exercise has been used extensively to study disease pathophysiology, few studies have directly tested the associations between the physiological and behavioral manifestations of PEM. Thus, the extent that putative biomarkers of disease cause symptom exacerbation is largely unknown.

Blood-based biomarkers including those involving immune, metabolic, viral, and other agents have been tested in multiple CMI exercise studies, but with no clear pattern. In response to a 25-min submaximal exercise test, ME/CFS participants exhibited greater gene expression changes in leukocytes for metabolic detecting receptors (ASIC3, P2X4, and P2X5), adrenergic function (α-2A, β-1, β-2, & COMT), and immune function (IL-10, TLR4) compared to healthy controls (Light et al. 2009). These differences were observed at 30-min, 8-h, 24-h, and 48-h post-exercise and were significantly correlated with increased physical and mental fatigue. Similarly, maximal exercise was found to increase adrenergic α-2A receptor and nuclear receptor subfamily 3 (NR3C1) gene expression that were positively correlated with post-exercise fatigue (Meyer et al. 2013). These results suggest post-exertional fatigue may be associated with dysfunction of the hypothalamic-pituitary-adrenal axis, abnormal immune or stress response, and alterations of peripheral sensory signaling. Contrary to the results from ME/CFS samples, gene expression to submaximal exercise for adrenergic, metabolic, and immune-related receptors was found to decrease in GWI samples and was not predictive of post-exercise symptom exacerbation (Boruch et al. 2023).

Two-Day CPET protocols have been used to study transcriptomic, virome, and metabolomic composition/pathways (Bouquet et al. 2019; Germain et al. 2022). No differentially expressed genes related to immune signaling in response to exercise were found in ME/CFS compared to controls. Further, neither viral composition nor number of reads were associated with PEM (Bouquet et al. 2019). Numerous metabolite differences representing numerous pathways (e.g., fatty-acid metabolism, glycolysis, glutamate synthesis, carnitine synthesis) were found post-exercise in sample of 60 ME/CFS (45 female, 15 male) and 45 controls (30 female, 15 male); however, relationships with PEM symptoms were not directly tested (Germain et al. 2022).

Abnormal cardiopulmonary responses to exercise have been found for ME/CFS, GWI, and Long-COVID groups. These studies have found abnormal breathing patterns (Mancini et al. 2021; Lindheimer et al. 2019; Alexander et al. 2023), reduced ventilatory efficiency (Boruch et al. 2021; Joseph et al. 2023; Cook et al. 2022), impaired venous return and oxygen extraction (Joseph et al. 2021; Singh et al.

2022), and reduced exercise capacity (Durstenfeld et al. 2022; Lim et al. 2020b). Moreover, 2-Day CPET protocols have revealed an inability of ME/CFS samples to reproduce exercise outcomes such as power output at ventilatory threshold (Lim et al. 2020b). These abnormal responses could help explain why exercise worsens fatigue in CMIs; however, this has rarely been directly tested. One study of submaximal exercise reported that neither symptom nor cardiopulmonary responses during submaximal exercise were predictive of PEM in GWI (Boruch et al. 2021). PEM was defined as the peak symptom change from baseline in the 7 days following exercise. A general linear model including cumulative work (kJ), peak leg muscle pain during exercise, the Physical Component Score of the Veterans R and 36-item Health Survey, and the ventilatory equivalents for carbon dioxide (VE/VCO_2) was not predictive of symptom responses to exercise. Although the total model was not predictive of PEM, both VE/VCO2 (Pearson $r = 0.31$, 95% CI = 0.01, 0.56, $p = 0.05$) and leg muscle pain (Pearson $r = 0.31$, 95% CI = 0.01, 0.55, $p < 0.05$) were significantly correlated with peak symptom responses.

Dysfunction of the gut microbiota has also been proposed to contribute to PEM. Early studies reported baseline differences between ME/CFS and controls (Lakhan and Kirchgessner 2010), with irritable bowel syndrome (IBS) being a common comorbidity in ME/CFS (Riedl et al. 2008). Individuals with comorbid IBS and ME/CFS were found to report more severe fatigue compared to ME/CFS but not IBS (Lakhan and Kirchgessner 2010). The hypothesis that gut dysbiosis and bacterial translocation could contribute to PEM was examined in a small sample of ME/CFS ($n = 10$) and controls ($n = 10$) in response to maximal exercise (Shukla et al. 2015). Enteric (gut) bacterial concentrations and makeups in both stool and blood samples were measured before, 15 min, 48 and 72 h following exercise. ME/CFS showed greater increases for both stool and blood enteric bacterial concentration and makeups, which remained elevated at 72-h post-exercise suggestive of delayed blood clearance (Shukla et al., 2015). This co-occurred with large changes in fatigue over the same timeframe (i.e., 72-h post-exercise). These findings suggest the exercise-induced translocation of enteric bacteria may trigger systemic inflammation and fatigue worsening in ME/CFS.

Post-exercise worsening of fatigue could be emblematic of central nervous system dysfunction. Both self-paced and submaximal exercise were found to decrease pressure pain thresholds in ME/CFS compared to increases for controls (Van Oosterwijck et al. 2010). Changes in pressure pain thresholds were significantly and positively associated with fatigue changes. Functional brain responses during fatiguing cognition have been shown to be associated with symptom worsening in ME/CFS (Cook et al. 2017). ME/CFS patients and controls completed symptom assessments and functional magnetic resonance imaging while performing non-fatiguing (finger-tapping and simple number recognition) and fatiguing cognitive tasks (paced auditory serial addition task). Testing occurred at baseline and 24 h after 30 min of exercise at 70% of peak HR. Differential patterns of brain responses to the fatiguing cognitive task were found and were characterized by greater activity in bilateral hippocampus, left frontal, and cingulate cortices, and decreased activity in superior and inferior parietal cortices for ME/CFS compared to controls. No differential brain responses occurred during the non-fatiguing tasks. Symptoms and cognitive performance worsened for ME/CFS but improved for controls.

Symptom worsening with physical exertion in CMIs is opposite to the normal beneficial responses experienced by healthy people and therefore suggests physiological dysregulation. Although numerous physiological systems are differentially affected in ME/CFS, GWI, and Long-COVID, whether these differences cause symptom exacerbation (i.e., fatigue and other symptoms) is unknown. Research that determines whether select physiological systems are significantly predictive of symptom worsening is needed to begin pinpointing the pathophysiology of PEM, and perhaps CMIs. Moreover, efficacious treatments for CMIs, when discovered, could aid this effort by testing whether treatment mechanisms and symptom improvement reduce or eliminate PEM responses.

4 Chronic Physical Activity and Fatigue Symptoms in Neurotypical Adults

4.1 Summary of Chronic Exercise Training Studies with Neurotypical Participants

Numerous reviews have examined research aimed at understanding relationships between regular physical activity and symptoms of fatigue and/or energy (Puetz et al. 2006; Puetz 2006; O'Connor et al. 2010; O'Connor and Puetz 2005). Some of these reviews have focused on patients with specific chronic conditions such as arthritis (Rongen-van Dartel et al. 2015), cancer-related fatigue (Tomlinson et al. 2014), fibromyalgia (Estévez-López et al. 2020), multiple sclerosis (Pilutti et al. 2013; Razazian et al. 2020), or specific sub-types of cancers such as breast (Medeiros Torres et al. 2022) or gynecologic cancer (Ma et al. 2023). Most of these reviews show that exercise training results in small positive improvements in fatigue symptoms though there are some exceptions, often when the number of studies included is relatively small (Paneroni et al. 2020). There are few experiments directly comparing exercise training to an alternative method for potentially improving fatigue symptoms. One meta-analysis found that among cancer patients, regular exercise yielded larger improvements than the pharmacological therapies tested. Moreover, the exercise training effect on improving fatigue symptoms was equivalent to psychological treatments such as cognitive behavioral therapy (Mustian et al. 2017).

A meta-analysis published in 2022 represents a comprehensive review of randomized controlled trials, which are considered the strongest type of evidence (Wender et al. 2022). This investigation revealed that the influence of chronic exercise on symptoms of fatigue and energy has been examined in at least 81 randomized controlled trials. These experiments involved over 7000 participants who were predominantly mid-aged (mean of 50 years old) women (~70%) with a variety of health problems, most commonly cancer, heart disease, and neuropsychological concerns such as multiple sclerosis or fibromyalgia. The exercise interventions were typically about 3 months long and usually involved three sessions per week, each lasting about 45 min. About 82% of the investigations involved moderate intensity exercise bouts. About 36% of the investigations focused on aerobic exercise modes, about 10% used yoga, about 20% focused on resistance training, and 32% used a combination of both resistance and aerobic type exercise. Symptoms were categorized into feelings of energy or fatigue as well as vitality, which combines energy and fatigue questions into a single score, because the vitality scale of the SF-36 Health Survey was so frequently used as an outcome measure.

Across all of these trials, regular exercise resulted in moderate sized improvements in vitality; the mean standardized effect size g equaled = 0.54 and the 95% confidence interval did not overlap zero (0.40, 0.67). Regular exercise resulted in smaller, statistically significant, increases in feelings of energy ($g = 0.42$; 95% CI of 0.25 to 0.58) and reductions in feelings of fatigue ($g = -0.37$; 95% CI of −0.52 to −0.23). The authors concluded that "when groups adopt a moderate intensity exercise training program while participating in a randomized trial, compared to controls, this typically results in small-to-moderate average improvements in feelings of fatigue, energy, and vitality" (Wender et al. 2022).

This comprehensive quantitative review identified several meaningful moderators of the chronic exercise effects (Wender et al. 2022). For feelings of fatigue, larger improvements were found for moderate compared to low intensity exercise as well as for longer exercise sessions and greater weeks of exercise training. For feelings of energy, larger improvements were found for moderate compared to low intensity exercise as well as for resistance exercise mode whether performed alone or in combination with aerobic exercise. For vitality, larger improvements were found for low and high intensity compared to moderate intensity. The small number of high and low intensity trials means that more data are needed before treating this finding without caution. Also, for vitality, larger improvements were found for neurological/psychological samples compared to samples with cancer. Larger improvements in vitality also were found for exercise completed under supervision in a facility and when resistance exercise was combined with aerobic exercise modes. In summary, these observations are potentially useful for practitioners prescribing exercise programs to reduce symptoms of fatigue and enhance feelings of energy. To date, demographic and clinical characteristics have not been found to moderate the effects of exercise on fatigue (van Vulpen et al. 2020).

4.2 *Evidence Regarding Plausible Mechanisms*

There is a paucity of research testing specific mechanisms by which regular physical activity plausibly could cause improvements in feelings of energy and fatigue. Regular physical activity may cause adaptations in neural circuits that underlie feelings of energy and/or fatigue (Salamone and Correa 2024). Also, regular physical activity may have positive effects on behavior, such as improved sleep (Kline et al. 2021), or peripheral physiology, such as a favorable impact on hematocrit or hemoglobin in cancer patients (Courneya et al. 2008), that might help to improve fatigue or energy symptoms. Alternatively, because symptoms are self-reported exercise-induced placebo responses cannot be ruled out, and one review estimated that about half of the psychological response to exercise training appears to be attributable to placebo responses (Lindheimer et al. 2015). The absence of better data regarding mechanisms that could explain improvements in symptoms of energy and fatigue after the adoption of regular physical activity performed by neurotypical adults and patients with variety of medical illnesses is a major limitation in this literature and represents the greatest need regarding research in this area going forward.

5 Chronic Physical Activity and Fatigue Symptoms: ME/CFS, GWI, and Long-COVID

5.1 *Summary of Chronic Exercise Training Studies*

The occurrence of PEM complicates using regular physical activity as a fatigue treatment for CMIs. People with CMI report avoiding physical activity to limit symptom worsening and this avoidance behavior, sometimes termed kinesiophobia, has been shown to be associated with reduced self-reported physical activity in at least one study (Nijs et al. 2004). Physical activity pacing, a process whereby the person determines and adjusts the level of activity that they can tolerate without exacerbating symptoms, has been proposed as one strategy to avoid PEM in ME/CFS (Casson et al. 2022; Goudsmit et al. 2012). A meta-analysis of eleven pacing studies in ME/CFS reported moderate ($g = -0.52$; 95% CI of -0.73, -0.32) fatigue reductions (Casson et al. 2022). However, there is a paucity of pacing studies involving GWI and Long-COVID. Notably, this approach differs from graded exercise therapy (GET) studies where progression of the exercise training dose is typical; although pauses in progression are common in CMI studies when symptom worsening occurs. Regular physical activity research in CMIs largely almost uniformly uses the phrase graded exercise therapy (GET), and therefore this section will adopt that terminology for consistency with the literature.

One early randomized controlled trial (RCTs) for people with ME/CFS compared GET to flexibility training combined with relaxation therapy (Fulcher and White 1997). Twelve weeks of supervised exercise, prescribed at an intensity of 50%–60% of individual maximum heart rate, significantly improved fatigue compared to flexibility training plus relaxation therapy. Attrition due to symptom exacerbation occurred for only one participant from each treatment. More than a decade later, a meta-analysis compared the efficacy of GET and cognitive behavioral therapy (CBT) for fatigue and other health-related outcomes in ME/CFS (Castell et al. 2011). A total of $N = 21$ studies (GET: $N = 5$ studies, sample size $= 170$; CBT:

$N = 16$ studies, sample size = 1457) included both fatigue and functional impairment related outcomes. Both GET and CBT had small-to-moderate effects for reducing fatigue (GET: $g = 0.41$, 95% CI = 0.06, 0.77; CBT: $g = 0.36$, 95% CI = 0.25, 0.46) and functional impairment (GET: $g = 0.39$, 95% CI = 0.23, 0.56; CBT: $g = 0.36$, 95% CI = 0.26, 0.47). Heterogeneity was greater for GET ($I^2 = 62$) than CBT ($I^2 = 27$), warranting the need for additional studies that examine the characteristics of exercise that moderate fatigue and physical function in ME/CFS. Eight years later a meta-analysis of exercise therapy for ME/CFS (i.e., GET) examined studies directly comparing GET to other therapies (Larun et al. 2019). Eight GET trials, involving 1518 participants, and other therapies ranging from 12 to 26 weeks were included. Exercise intensities were variable across modes (e.g., walking, swimming, dancing, and cycling). Studies compared GET primarily to "passive" controls ($n = 8$ studies) but also to CBT ($n = 3$), adaptive pacing ($n = 1$), and antidepressants ($n = 1$). Fatigue reductions ($d = -0.66$, 95% CI = -1.01, -0.31) and physical function improvements (SF-36 physical functioning subscale (scale range 0–100) difference of 13.1 points) favored GET compared to the "passive" control conditions but not when compared to CBT or antidepressants for which there was much less evidence. The lone active pacing study reported reduced fatigue in favor of GET (1.99-point difference on the Chalder Fatigue Scale (scale range 0–33)).

Using a four-arm RCT, one large ME/CFS trial ($N = 641$ participants) compared the efficacy of three previously investigated treatment options (i.e., GET ($n = 160$), CBT ($n = 161$), and specialist medical care ($n = 160$)) with a fourth arm of physical activity pacing ($n = 160$) (White et al. 2011). At 52 weeks, all four treatment arms reduced fatigue (Chalder Fatigue Scale), though only GET (3.2 points, 95% CI = 1.7, 4.8) and CBT (3.4 points, 95% CI = 1.8, 5.0) resulted in significant fatigue reductions compared to specialist medical care ($p < 0.05$). Moreover, both GET and CBT resulted in improved physical function (SF-36 GET: 9.4 points, 95% CI = 4.4, 14.4; SF-36 CBT: 7.1 points, 95% CI = 2.0, 12.1). The synergistic effects of physical activity and CBT have also been estimated to produce moderate-to-large ($g = 0.72$, 95% CI = 0.51, 0.94) fatigue reductions and moderate ($g = 0.43$, 95% CI = 0.22, 0.64) depression improvements in ME/CFS (Marques et al. 2015).

The impact of patient engagement was tested in one RCT (Powell et al. 2001). The 12-week interventions consisted of: 1) a minimum intervention that included three hours of consultation and GET explanation; 2) a telephone intervention that included the minimum intervention, plus seven 30-min telephone consultations; and 3) a maximum intervention that included the minimum intervention plus seven 60-min in-person sessions. All three education-based treatment arms significantly improved fatigue in ME/CFS compared to a standard medical treatment control group. There were no significant differences between the intervention levels. These results are consistent with meta-analytic data indicating that minimal contact interventions (e.g., those minimizing face-to-face contact with the provider) have large effects ($g = 0.96$, 95% CI = 0.63, 1.28) on fatigue in ME/CFS samples (Marques et al. 2015). Fatigue may be influenced by various combinations of the physical, mental, and social demands of completing an exercise intervention. Moreover, minimal contact interventions may be more economical options and alleviate burdens (e.g., travel to and from appointments, scheduling conflicts, etc.) for both the patient and provider. Finally, placebo and/or expectancy responses have not been adequately studied in exercise training trials of CMIs. The placebo response was reported to be low in ME/CFS by one review (Cho et al. 2005). This study determined that among 28 RCTs, the pooled placebo response was 19.6% (95% CI: 15.4, 23.7).

Exercise interventions have been less well studied in GWI and Long-COVID. Two trials, one aerobic (Donta et al. 2003) and the other resistance exercise (Stegner et al. 2021) have been tested in GWI. One large RCT ($N = 1092$ participants) of Veterans with GWI investigated the efficacy of aerobic exercise training alone or as an adjuvant treatment (e.g., in tandem with either CBT or usual standard of care practices) for improving physical function and GWI symptoms. Participants completed one supervised exercise training session per week for 12 weeks at a self-selected intensity. Participants were also instructed to exercise at home two to three times per week throughout the intervention. Exercise alone and when combined with CBT improved fatigue, cognitive and mental health outcomes compared to usual care.

Sixteen weeks of resistance exercise training was found to be both safe and effective at improving strength and overall illness perceptions in Gulf War Veterans with widespread and chronic musculoskeletal pain (Stegner et al. 2021). However, neither fatigue nor pain was found to improve over the course of the intervention or at 6-month and 12-month follow-up periods. Despite the absence of symptom improvements, moderate-to-large ($g = 0.47$ to 0.78) increases in both maximal strength and training volume occurred. Moreover, there were no adverse events related to the RET intervention and adherence to training was high (87%, 27.9/32 sessions).

The COVID-19 pandemic drastically affected physical activity as well as physical and mental health in countries across the world (Aritake et al. 2015). Qualitative data based on semi-structured interviews of people with Long-COVID identified prolonged and varied challenges for physical activity engagement including reduced function, coping with brain fog, and fear of symptom exacerbation (Humphreys et al. 2021). We are aware of one exercise training intervention in Long-COVID comparing 8 weeks of multicomponent (i.e., both aerobic and resistance) exercise to standard rehabilitation and self-management recommendations of the World Health Organization (Jimeno-Almazán et al. 2022). Participants performed three sets of eight repetitions for squat, bench press, deadlift, and bench pull on two days per week starting at 50% of their one repetition maximum. This was combined with "moderate intensity variable training" ranging from 55% to 80% of heart rate reserve. The third day consisted of "light intensity continuous training" for 60 min at 65%–70% of heart rate reserve. Training progression was individualized and based on participant tolerance. Compared to the self-management control group, multicomponent exercise improved fatigue, quality of life, depression, cardiovascular fitness, and muscular strength. The fatigue improvements were large ($\eta_p^2 = 0.15$–0.23) and no adverse events were reported.

When considering physical activity for people with ME/CFS, clinicians face the challenge of helping patients avoid the negative effects of acute exercise (e.g., symptom exacerbation), while moving them toward experiencing the health benefits associated with a more physically active lifestyle. A logical approach is to develop exercise prescriptions which strike a balance between minimizing symptom exacerbation and maximizing physical function; however, there is limited information on the mode, duration, or intensity threshold at which this ideal balance occurs or guidance on how to establish this threshold for individual patients. It is noteworthy that organizations such as the American College of Sports Medicine, as part of their Exercise is Medicine campaign (https://www.exerciseismedicine.org/), have created exercise prescription materials for similar diseases such as Fibromyalgia (https://www.exerciseismedicine.org/wp-content/uploads/2021/04/EIM_Rx-for-Health_Fibromyalgia.pdf). These recommendations are logical in that they emphasize the individualized nature of exercise prescription and focus on light activity and slow intensity progression to minimize symptom exacerbation. More research is needed to determine for whom physical activity is both tolerable and effective for reducing symptoms, and for whom it may be contraindicated or unsafe.

5.2 Evidence Regarding Plausible Mechanisms of Chronic Exercise and Fatigue in Atypical Adults

The mechanisms that underlie the small-to-moderate fatigue improvements with exercise training in CMIs are unknown. Improvements in aerobic fitness and strength (Fulcher and White 1997; Stegner et al. 2021) suggest that the broad benefits of exercise training (e.g., more efficient cardiovascular function, increased aerobic capacity, and improved metabolism) may facilitate fatigue improvements. It is also plausible that exercise training, when appropriately structured, might positively affect multiple physiological systems (e.g., central, immune, autonomic, and gut microbiome) that have been shown to be associated with CMIs. Such responses might include adaptations to neural circuitry, peripheral and central inflammation, heart rate, blood pressure, and brain blood flow regulation, and gene methylation. However, these potential adaptations and their associations with feelings of fatigue have not been tested.

6 Conclusion

In neurotypical adults and compared to a resting control condition, dozens of experiments support that a single bout of low-to-moderate intensity exercise for 20 to 40 min almost always results in increased feelings of energy and common, but less so, are concomitant reductions in feelings of fatigue. Larger effects might be realized depending on the characteristics of the individual (e.g., presence of a psychiatric disorder) and the environment (e.g., exercising outdoors with a partner). High intensity and/or long duration exercise produces increased feelings of fatigue. For atypical adults (i.e., CMIs), acute exercise under similar circumstances can worsen fatigue and other associated symptoms. However, the exercise and symptom responses are highly variable with respect to exercise type, intensity, and duration. The mechanisms that underlie fatigue responses to acute exercise are not well understood and are largely untested for chronic exercise training. It is likely that multiple neural systems and their associated neurotransmitters interact to mediate feelings of energy and fatigue. In CMIs, dysregulation among multiple physiological systems, systemic inflammation, and skeletal muscle disuse issues could interact such that a single bout of exercise produces acute symptom worsening. The small number of RCT involving chronic physical activity have not reported acute symptom exacerbation in ME/CFS, GWI, or Long-COVID samples, and none have tested biological mechanisms of treatment responses. This suggests that acute exercise effects of symptom worsening in CMI should not be uniformly conflated with chronic exercise training which has not been shown to worsen symptoms. However, extreme care should be taken when prescribing any physical activity in CMI and should be tailored based on the individual's disease status and physical functional capacity.

References

Aaronson NK, Ahmedzai S, Bergman B, Bullinger M, Cull A, Duez NJ et al (1993) The European Organization for Research and Treatment of Cancer QLQ-C30: a quality-of-life instrument for use in international clinical trials in oncology. JNCI J Natl Cancer Inst 85(5):365–376

Al Maqbali M, Al Sinani M, Al Naamani Z, Al Badi K, MaI T (2021) Prevalence of fatigue in patients with cancer: a systematic review and meta-analysis. J Pain Symptom Manag 61(1): 167–89.e14

Alexander T, Watson MA, Klein-Adams JC, Ndirangu DS, Serrador JM, Falvo MJ et al (2023) Deployed veterans exhibit distinct respiratory patterns and greater dyspnea during maximal cardiopulmonary exercise: a case-control study. PLoS One 18(5):e0286015

Alghamdi I, Ariti C, Williams A, Wood E, Hewitt J (2021) Prevalence of fatigue after stroke: a systematic review and meta-analysis. Eur Stroke J 6(4):319–332

Arent SM, Alderman BL, Short EJ, Landers DM (2007) The impact of the testing environment on affective changes following acute resistance exercise. J Appl Sport Psychol 19(3):364–378

Aritake S, Kaneita Y, Ohtsu T, Uchiyama M, Mishima K, Akashiba T et al (2015) Prevalence of fatigue symptoms and correlations in the general adult population. Sleep Biol Rhythms 13(2): 146–154

Baraniuk JN, Kern G, Narayan V, Cheema A (2021) Exercise modifies glutamate and other metabolic biomarkers in cerebrospinal fluid from Gulf War illness and Myalgic encephalomyelitis/chronic fatigue syndrome. PLoS One 16(1):e0244116

Barhorst EE, Boruch AE, Cook DB, Lindheimer JB (2021) Pain-related post-exertional malaise in Myalgic encephalomyelitis/chronic fatigue syndrome (ME/CFS) and fibromyalgia: a systematic review and three-level meta-analysis. Pain Med 23:1144

Bartholomew JB, Morrison D, Ciccolo JT (2005) Effects of acute exercise on mood and well-being in patients with major depressive disorder. Med Sci Sports Exerc 37(12):2032

Basso JC, Suzuki WA (2017) The effects of acute exercise on mood, cognition, neurophysiology, and neurochemical pathways: a review. Brain Plasticity 2:127–152

Basu N, Yang X, Luben RN, Whibley D, Macfarlane GJ, Wareham NJ et al (2016) Fatigue is associated with excess mortality in the general population: results from the EPIC-Norfolk study. BMC Med 14(1):122

Bazelmans E, Bleijenberg G, Voeten MJ, van der Meer JW, Folgering H (2005) Impact of a maximal exercise test on symptoms and activity in chronic fatigue syndrome. J Psychosom Res 59(4):201–208

Boecker H, Drzezga A (2016) A perspective on the future role of brain pet imaging in exercise science. NeuroImage 131:73–80

Boolani A, Manierre M (2019) An exploratory multivariate study examining correlates of trait mental and physical fatigue and energy. Fatigue Biomed, Health Behav 7(1):29–40

Boolani A, O'Connor PJ, Reid J, Ma S, Mondal S (2019) Predictors of feelings of energy differ from predictors of fatigue. Fatigue Biomed, Health Behav 7(1):12–28

Boruch AE, Lindheimer JB, Klein-Adams JC, Stegner AJ, Wylie GR, Ninneman JV et al (2021) Predicting post-exertional malaise in Gulf War illness based on acute exercise responses. Life Sci 280:119701

Boruch AE, Lindheimer JB, Ninneman JV, Wylie GR, Alexander T, Klein-Adams JC et al (2023) Exercise-induced changes in gene expression do not mediate post exertional malaise in Gulf War illness. Brain Behav Immun Health 29:100612

Botsford C, Brellenthin AG, Cisler JM, Hillard CJ, Koltyn KF, Crombie KM (2023) Circulating endocannabinoids and psychological outcomes in women with PTSD. J Anxiety Disord 93: 102656

Bouquet J, Li T, Gardy JL, Kang X, Stevens S, Stevens J et al (2019) Whole blood human transcriptome and virome analysis of ME/CFS patients experiencing post-exertional malaise following cardiopulmonary exercise testing. PLoS One 14(3):e0212193

Brellenthin AG, Crombie KM, Hillard CJ, Koltyn KF (2017) Endocannabinoid and mood responses to exercise in adults with varying activity levels. Med Sci Sports Exerc 49(8):1688–1696

Brown A, Jason LA (2020) Meta-analysis investigating post-exertional malaise between patients and controls. J Health Psychol 25(13–14):2053–2071

Carruthers BM, Jain AK, De Meirleir KL, Peterson DL, Klimas NG, Lerner AM et al (2003) Myalgic encephalomyelitis/chronic fatigue syndrome: clinical working case definition, diagnostic and treatment protocols. J Chronic Fatigue Syndr 11(1):7–115

Casson S, Jones MD, Cassar J, Kwai N, Lloyd AR, Barry BK et al (2022) The effectiveness of activity pacing interventions for people with chronic fatigue syndrome: a systematic review and meta-analysis. Disabil Rehabil 1-15:3788

Castell BD, Kazantzis N, Moss-Morris RE (2011) Cognitive behavioral therapy and graded exercise for chronic fatigue syndrome: a meta-analysis. Clin Psychol Sci Pract 18(4):311

Cella D, Lai J-S, Jensen SE, Christodoulou C, Junghaenel DU, Reeve BB et al (2016) PROMIS fatigue item bank had clinical validity across diverse chronic conditions. J Clin Epidemiol 73: 128–134

Chen C, Haupert SR, Zimmermann L, Shi X, Fritsche LG, Mukherjee B (2022) Global prevalence of post-coronavirus disease 2019 (COVID-19) condition or long COVID: a meta-analysis and systematic review. J Infect Dis 226(9):1593–1607

Cho HJ, Hotopf M, Wessely S (2005) The placebo response in the treatment of chronic fatigue syndrome: a systematic review and meta-analysis. Psychosom Med 67(2):301–313

Chu L, Valencia IJ, Garvert DW, Montoya JG (2018) Deconstructing post-exertional malaise in myalgic encephalomyelitis/chronic fatigue syndrome: a patient-centered, cross-sectional survey. PLoS One 13(6):e0197811

Clapp LL, Richardson MT, Smith JF, Wang M, Clapp AJ, Pieroni RE (1999) Acute effects of thirty minutes of light-intensity, intermittent exercise on patients with chronic fatigue syndrome. Phys Ther 79(8):749–756

Clayton EW (2015) Beyond myalgic encephalomyelitis/chronic fatigue syndrome: an IOM report on redefining an illness. JAMA 313(11):1101–1102

Committee on the Diagnostic Criteria for Myalgic Encephalomyelitis, Chronic Fatigue S, Board on the Health of Select P, Institute of M (2015) The national academies collection: reports funded by National Institutes of Health. Beyond myalgic encephalomyelitis/chronic fatigue syndrome: redefining an illness. National Academies Press (US). Copyright 2015 by the National Academy of Sciences. All rights reserved, Washington

Cook DB, O'Connor PJ, Eubanks SA, Smith JC, Lee M (1997) Naturally occurring muscle pain during exercise: assessment and experimental evidence. Med Sci Sports Exerc 29(8):999–1012

Cook DB, Light AR, Light KC, Broderick G, Shields MR, Dougherty RJ et al (2017) Neural consequences of post-exertion malaise in myalgic encephalomyelitis/chronic fatigue syndrome. Brain Behav Immun 62:87

Cook DB, Van Riper S, Dougherty RJ, Lindheimer JB, Falvo MJ, Chen Y et al (2022) Cardiopulmonary, metabolic, and perceptual responses during exercise in Myalgic encephalomyelitis/chronic fatigue syndrome (ME/CFS): a multi-site clinical assessment of ME/CFS (MCAM) sub-study. PLoS One 17(3):e0265315

Cooper SL, Whitaker MM, Lydon GW (2022) Acute effects of continuous exercise, interval exercise, and sedentary colouring on anxiety, mood, and physical activity enjoyment. Int J Sport Exerc Psychol 1–18:951

Costello SE, O'Neill BV, Howatson G, van Someren K, Haskell-Ramsay CF (2022) Detrimental effects on executive function and mood following consecutive days of repeated high-intensity sprint interval exercise in trained male sports players. J Sports Sci 40(7):783–796

Courneya KS, Jones LW, Peddle CJ, Sellar CM, Reiman T, Joy AA et al (2008) Effects of aerobic exercise training in anemic cancer patients receiving Darbepoetin alfa: a randomized controlled trial. Oncologist 13(9):1012–1020

Crombie KM, Brellenthin AG, Hillard CJ, Koltyn KF (2018) Endocannabinoid and opioid system interactions in exercise-induced Hypoalgesia. Pain Med 19(1):118–123

Curran SL, Andrykowski MA, Studts JL (1995) Short form of the profile of mood states (POMS-SF): psychometric information. Psychol Assess 7(1):80–83

Cutler DM (2022) The economic cost of long COVID: an update. Published online July 2022

D'Silva A, Fox DE, Nasser Y, Vallance JK, Quinn RR, Ronksley PE et al (2022) Prevalence and risk factors for fatigue in adults with inflammatory bowel disease: a systematic review with meta-analysis. Clin Gastroenterol Hepatol 20(5):995–1009.e7

Davis HE, Assaf GS, McCorkell L, Wei H, Low RJ, Re'em Y et al (2021) Characterizing long COVID in an international cohort: 7 months of symptoms and their impact. eClinicalMedicine 38:202108. Available at SSRN 3820561.2021

Dishman RK, O'Connor PJ (2009) Lessons in exercise neurobiology: the case of endorphins. Ment Health Phys Act 2(1):4–9

Dishman RK, Thom NJ, Puetz TW, O'Connor PJ, Clementz BA (2010) Effects of cycling exercise on vigor, fatigue, and electroencephalographic activity among young adults who report persistent fatigue. Psychophysiology 47(6):1066–1074

Donta ST, Clauw DJ, Engel CC Jr, Guarino P, Peduzzi P, Williams DA et al (2003) Cognitive behavioral therapy and aerobic exercise for Gulf War veterans' illnesses: a randomized controlled trial. JAMA 289(11):1396–1404

Dursa EK, Barth SK, Porter BW, Schneiderman AI (2019) Health status of female and male Gulf War and gulf era veterans: a population-based study. Womens Health Issues 29:S39–S46

Durstenfeld MS, Sun K, Tahir P, Peluso MJ, Deeks SG, Aras MA et al (2022) Use of cardiopulmonary exercise testing to evaluate long COVID-19 symptoms in adults: a systematic review and meta-analysis. JAMA Netw Open 5(10):e2236057-e

Eignor DR (2013) The standards for educational and psychological testing. APA handbook of testing and assessment in psychology, Test theory and testing and assessment in industrial and organizational psychology. APA handbooks in psychology®, vol 1. American Psychological Association, Washington, pp 245–250

Estévez-López F, Mudie K, Wang-Steverding X, Bakken IJ, Ivanovs A, Castro-Marrero J et al (2020) Systematic review of the epidemiological burden of myalgic encephalomyelitis/chronic fatigue syndrome across Europe: current evidence and EUROMENE research recommendations for epidemiology. J Clin Med 9(5):1557

Fappiano CM, Baraniuk JN (2020) Gulf War illness symptom severity and onset: a cross-sectional survey. Mil Med 185(7–8):e1120–e11e7

Focht BC (2009) Brief walks in outdoor and laboratory environments: effects on affective responses, enjoyment, and intentions to walk for exercise. Res Q Exerc Sport 80(3):611–620

Friedberg F, Tintle N, Clark J, Bromet E (2015) Prolonged fatigue in Ukraine and the United States: prevalence and risk factors. Fatigue Biomed, Health Behav 3(1):33–46

Friedberg F, Adamowicz JL, Bruckenthal P, Milazzo M, Ramjan S, Zhang X et al (2023) Sex differences in post-exercise fatigue and function in myalgic encephalomyelitis/chronic fatigue syndrome. Sci Rep 13(1):5442

Fritz K, O'Connor PJ (2016) Acute exercise improves mood and motivation in young men with ADHD symptoms. Med Sci Sports Exerc 48(6):1153–1160

Fukuda K, Straus SE, Hickie I, Sharpe MC, Dobbins JG, Komaroff A (1994) The chronic fatigue syndrome: a comprehensive approach to its definition and study. International chronic fatigue syndrome study group. Ann Intern Med 121(12):953–959

Fukuda K, Nisenbaum R, Stewart G, Thompson WW, Robin L, Washko RM et al (1998) Chronic multisymptom illness affecting air force veterans of the Gulf War. JAMA 280(11):981–988

Fulcher KY, White PD (1997) Randomised controlled trial of graded exercise in patients with the chronic fatigue syndrome. BMJ 314(7095):1647

Germain A, Barupal DK, Levine SM, Hanson MR (2020) Comprehensive circulatory metabolomics in ME/CFS reveals disrupted metabolism of acyl lipids and steroids. Meta 10(1):34

Germain A, Giloteaux L, Moore GE, Levine SM, Chia JK, Keller BA et al (2022) Plasma metabolomics reveals disrupted response and recovery following maximal exercise in myalgic encephalomyelitis/chronic fatigue syndrome. JCI Insight 7(9):e157621

Goërtz YMJ, Braamse AMJ, Spruit MA, Janssen DJA, Ebadi Z, Van Herck M et al (2021) Fatigue in patients with chronic disease: results from the population-based lifelines cohort study. Sci Rep 11(1):20977

Goudsmit EM, Nijs J, Jason LA, Wallman KE (2012) Pacing as a strategy to improve energy management in myalgic encephalomyelitis/chronic fatigue syndrome: a consensus document. Disabil Rehabil 34(13):1140–1147

Greene DR, Greene CL, Petruzzello SJ (2023) Crank it up: Utilizing exercise to combat acute fatigue and pain in individuals with probable posttraumatic stress disorder. Sport Exerc Perform Psychol 12(4):229–240

Heboyan V, Krengel MH, Sullivan K, Iobst S, Klimas N, Wilson C et al (2019) Sex differences in Gulf War illness: a reanalysis of data from the CDC air force study using CDC and modified Kansas case definitions. J Occup Environ Med 61(7):610–616

Herring MP, O'Connor PJ (2009) The effect of acute resistance exercise on feelings of energy and fatigue. J Sports Sci 27(7):701–709

Holtzman CS, Bhatia S, Cotler J, Jason LA (2019) Assessment of post-exertional malaise (PEM) in patients with myalgic encephalomyelitis (ME) and chronic fatigue syndrome (CFS): a patient-driven survey. Diagnostics 9(1):26

Humphreys H, Kilby L, Kudiersky N, Copeland R (2021) Long COVID and the role of physical activity: a qualitative study. BMJ Open 11(3):e047632

Islam MF, Cotler J, Jason LA (2020) Post-viral fatigue and COVID-19: lessons from past epidemics. Fatigue Biomed, Health Behav 8(2):61–69

Jason LA, Mirin AA (2021) Updating the National Academy of medicine ME/CFS prevalence and economic impact figures to account for population growth and inflation. Fatigue Biomed, Health Behav 9(1):9–13

Jimeno-Almazán A, Franco-López F, Buendía-Romero Á, Martínez-Cava A, Sánchez-Agar JA, Sánchez-Alcaraz Martínez BJ et al (2022) Rehabilitation for post-COVID-19 condition through a supervised exercise intervention: a randomized controlled trial. Scand J Med Sci Sports 32(12):1791–1801

Joseph P, Arevalo C, Oliveira RKF, Faria-Urbina M, Felsenstein D, Oaklander AL et al (2021) Insights from invasive cardiopulmonary exercise testing of patients with myalgic encephalomyelitis/chronic fatigue syndrome. Chest 160(2):642–651

Joseph P, Singh I, Oliveira R, Capone CA, Mullen MP, Cook DB et al (2023) Exercise pathophysiology in Myalgic encephalomyelitis/chronic fatigue syndrome and post-acute sequelae of SARS-CoV-2: more in common than not? Chest 164:717

Khan F, Kennedy G, Spence VA, Newton DJ, Belch JJF (2004) Peripheral cholinergic function in humans with chronic fatigue syndrome, Gulf War syndrome and with illness following organophosphate exposure. Clin Sci 106(2):183–189

Kline CE, Hillman CH, Bloodgood Sheppard B, Tennant B, Conroy DE, Macko RF et al (2021) Physical activity and sleep: an updated umbrella review of the 2018 physical activity guidelines advisory committee report. Sleep Med Rev 58:101489

Krupp LB, LaRocca NG, Muir-Nash J, Steinberg AD (1989) The fatigue severity scale: application to patients with multiple sclerosis and systemic lupus erythematosus. Arch Neurol 46(10): 1121–1123

LaCaille RA, Masters KS, Heath EM (2004) Effects of cognitive strategy and exercise setting on running performance, perceived exertion, affect, and satisfaction. Psychol Sport Exerc 5(4): 461–476

Lakhan SE, Kirchgessner A (2010) Gut inflammation in chronic fatigue syndrome. Nutr Metab 7:1–10

Larun L, Brurberg KG, Odgaard-Jensen J, Price JR (2019) Exercise therapy for chronic fatigue syndrome. Cochrane Database Syst Rev 10:CD003200

Legrand FD, Race M, Herring MP (2018) Acute effects of outdoor and indoor exercise on feelings of energy and fatigue in people with depressive symptoms. J Environ Psychol 56:91–96

Light AR, White AT, Hughen RW, Light KC (2009) Moderate exercise increases expression for sensory, adrenergic, and immune genes in chronic fatigue syndrome patients but not in normal subjects. J Pain 10(10):1099–1112

Light AR, Bateman L, Jo D, Hughen RW, Vanhaitsma TA, White AT et al (2012) Gene expression alterations at baseline and following moderate exercise in patients with chronic fatigue syndrome and fibromyalgia syndrome. J Intern Med 271(1):64–81

Lim E-J, Ahn Y-C, Jang E-S, Lee S-W, Lee S-H, Son C-G (2020a) Systematic review and meta-analysis of the prevalence of chronic fatigue syndrome/myalgic encephalomyelitis (CFS/ME). J Transl Med 18(1):100

Lim E-J, Kang E-B, Jang E-S, Son C-G (2020b) The prospects of the two-day cardiopulmonary exercise test (CPET) in ME/CFS patients: a meta-analysis. J Clin Med 9(12):4040

Lindheimer JB, O'Connor PJ, Dishman RK (2015) Quantifying the placebo effect in psychological outcomes of exercise training: a meta-analysis of randomized trials. Sports Med 45(5):693–711

Lindheimer JB, O'Connor PJ, McCully KK, Dishman RK (2017a) The effect of light-intensity cycling on mood and working memory in response to a randomized, placebo-controlled design. Psychosom Med 79(2):243

Lindheimer JB, Meyer JD, Stegner AJ, Dougherty RJ, Van Riper SM, Shields M et al (2017b) Symptom variability following acute exercise in myalgic encephalomyelitis/chronic fatigue syndrome: a perspective on measuring post-exertion malaise. Fatigue Biomed, Health Behav 5(2):69–88

Lindheimer JB, Cook DB, Klein-Adams JC, Qian W, Hill HZ, Lange G et al (2019) Veterans with Gulf War illness exhibit distinct respiratory patterns during maximal cardiopulmonary exercise. PLoS One 14(11):e0224833

Lindheimer JB, Stegner AJ, Wylie GR, Klein-Adams JC, Almassi NE, Ninneman JV et al (2020) Post-exertional malaise in veterans with Gulf War illness. Int J Psychophysiol 147:202–212

Loy BD, O'Connor PJ (2016) The effect of histamine on changes in mental energy and fatigue after a single bout of exercise. Physiol Behav 153:7–18

Loy BD, O'Connor PJ, Dishman RK (2013) The effect of a single bout of exercise on energy and fatigue states: a systematic review and meta-analysis. Fatigue Biomed, Health Behav 1(4): 223–242

Loy BD, O'Connor PJ, Dishman RK (2016) Effect of acute exercise on fatigue in people with ME/CFS/SEID: a meta-analysis. Med Sci Sports Exerc 48(10):2003

Loy BD, Cameron MH, O'Connor PJ (2018) Perceived fatigue and energy are independent unipolar states: supporting evidence. Med Hypotheses 113:46–51

Ma RC, Li XJ, Liu R, Xu XX (2023) Effects of physical exercise interventions for individuals with gynecologic cancer: a systematic review and meta-analysis. Oncol Nurs Forum 50(4):531–543

Mancini DM, Brunjes DL, Lala A, Trivieri MG, Contreras JP, Natelson BH (2021) Use of cardiopulmonary stress testing for patients with unexplained dyspnea post–coronavirus disease. Heart Failure 9(12):927–937

Marques MM, De Gucht V, Gouveia MJ, Leal I, Maes S (2015) Differential effects of behavioral interventions with a graded physical activity component in patients suffering from chronic fatigue (syndrome): an updated systematic review and meta-analysis. Clin Psychol Rev 40:123–137

McDowell CP, Campbell MJ, Herring MP (2016) Sex-related differences in mood responses to acute aerobic exercise. Med Sci Sports Exerc 48(9):1798–1802

McNair DM, Lorr M, Droppleman LF (1981) Profile of mood states questionnaire. EDITS, San Diego

Medeiros Torres D, Jorge Koifman R, da Silva SS (2022) Impact on fatigue of different types of physical exercise during adjuvant chemotherapy and radiotherapy in breast cancer: systematic review and meta-analysis. Support Care Cancer 30(6):4651–4662

Meeusen R, Piacentini MF, Meirleir KD (2001) Brain microdialysis in exercise research. Sports Med 31(14):965–983

Meyer JD, Light AR, Shukla SK, Clevidence D, Yale S, Stegner AJ et al (2013) Post-exertion malaise in chronic fatigue syndrome: symptoms and gene expression. Fatigue Biomed, Health Behav 1(4):190–209

Meyer JD, Crombie KM, Cook DB, Hillard CJ, Koltyn KF (2019) Serum endocannabinoid and mood changes after exercise in major depressive disorder. Med Sci Sports Exerc 51(9): 1909–1917

Moneghetti KJ, Skhiri M, Contrepois K, Kobayashi Y, Maecker H, Davis M et al (2018) Value of circulating cytokine profiling during submaximal exercise testing in Myalgic encephalomyelitis/chronic fatigue syndrome. Sci Rep 8(1):2779

Monroe DC, Gist NH, Freese EC, O'Connor PJ, McCully KK, Dishman RK (2016) Effects of sprint interval cycling on fatigue, energy, and cerebral oxygenation. Med Sci Sports Exerc 48(4): 615–624

Moore GE, Keller BA, Stevens J, Mao X, Stevens SR, Chia JK et al (2023) Recovery from exercise in persons with Myalgic encephalomyelitis/chronic fatigue syndrome (ME/CFS). Medicina 59(3):571

Mustian KM, Alfano CM, Heckler C, Kleckner AS, Kleckner IR, Leach CR et al (2017) Comparison of pharmaceutical, psychological, and exercise treatments for cancer-related fatigue: a meta-analysis. JAMA Oncol 3(7):961–968

Nehme M, Chappuis F, Kaiser L, Assal F, Guessous I (2023) The prevalence, severity, and impact of post-COVID persistent fatigue, post-exertional malaise, and chronic fatigue syndrome. J Gen Intern Med 38(3):835–839

Nijs J, De Meirleir K, Duquet W (2004) Kinesiophobia in chronic fatigue syndrome: assessment and associations with disability. Arch Phys Med Rehabil 85(10):1586–1592

O'Brien PM, O'Conner PJ (2000) Effect of bright light on cycling performance. Med Sci Sports Exerc 32(2):439–447

O'Connor PJ (2004) Evaluation of four highly cited energy and fatigue mood measures. J Psychosom Res 57(5):435–441

O'Connor PJ, Puetz TW (2005) Chronic physical activity and feelings of energy and fatigue. Med Sci Sports Exerc 37(2):299–305

O'Connor PJ, Herring MP, Caravalho A (2010) Mental health benefits of strength training in adults. Am J Lifestyle Med 4(5):377–396

Oliva Ramirez A, Keenan A, Kalau O, Worthington E, Cohen L, Singh S (2021) Prevalence and burden of multiple sclerosis-related fatigue: a systematic literature review. BMC Neurol 21(1): 468

Paneroni M, Vitacca M, Venturelli M, Simonelli C, Bertacchini L, Scalvini S et al (2020) The impact of exercise training on fatigue in patients with chronic obstructive pulmonary disease: a systematic review and meta-analysis. Pulmonology 26(5):304–313

Peterson PK, Sirr SA, Grammith FC, Schenck CH, Pheley AM, Hu S et al (1994) Effects of mild exercise on cytokines and cerebral blood flow in chronic fatigue syndrome patients. Clin Diagn Lab Immunol 1(2):222–226

Petruzzello SJ, Hall EE, Ekkekakis P (2001) Regional brain activation as a biological marker of affective responsivity to acute exercise: influence of fitness. Psychophysiology 38(1):99–106

Pilutti LA, Greenlee TA, Motl RW, Nickrent MS, Petruzzello SJ (2013) Effects of exercise training on fatigue in multiple sclerosis: a meta-analysis. Psychosom Med 75(6):575

Powell P, Bentall RP, Nye FJ, Edwards RHT (2001) Randomised controlled trial of patient education to encourage graded exercise in chronic fatigue syndrome. BMJ 322(7283):387

Proffitt M (2023) Oxford English dictionary. OED.com. Oxford University Press

Proost M, Habay J, De Wachter J, De Pauw K, Rattray B, Meeusen R et al (2022) How to tackle mental fatigue: a systematic review of potential countermeasures and their underlying mechanisms. Sports Med 52(9):2129–2158

Puetz TW (2006) Physical activity and feelings of energy and fatigue. Sports Med 36(9):767–780

Puetz TW, O'Connor PJ, Dishman RK (2006) Effects of chronic exercise on feelings of energy and fatigue: a quantitative synthesis. Psychol Bull 132(6):866–876

Qu D, Zhang Z, Yu X, Zhao J, Qiu F, Huang J (2016) Psychotropic drugs for the management of cancer-related fatigue: a systematic review and meta-analysis. Eur J Cancer Care 25(6):970–979

Raichlen DA, Foster AD, Gerdeman GL, Seillier A, Giuffrida A (2012) Wired to run: exercise-induced endocannabinoid signaling in humans and cursorial mammals with implications for the 'runner's high. J Exp Biol 215(Pt 8):1331–1336

Rayhan RU, Baraniuk JN (2021) Submaximal exercise provokes increased activation of the anterior default mode network during the resting state as a biomarker of postexertional malaise in myalgic encephalomyelitis/chronic fatigue syndrome. Front Neurosci 15:748426

Razazian N, Kazeminia M, Moayedi H, Daneshkhah A, Shohaimi S, Mohammadi M et al (2020) The impact of physical exercise on the fatigue symptoms in patients with multiple sclerosis: a systematic review and meta-analysis. BMC Neurol 20(1):93

Reeves WC, Jones JF, Maloney E, Heim C, Hoaglin DC, Boneva RS et al (2007) Prevalence of chronic fatigue syndrome in metropolitan, urban, and rural Georgia. Popul Health Metrics 5(1):5

Regier DA, Kuhl EA, Kupfer DJ (2013) The DSM-5: classification and criteria changes. World Psychiatry 12(2):92–98

Riedl A, Schmidtmann M, Stengel A, Goebel M, Wisser A-S, Klapp BF et al (2008) Somatic comorbidities of irritable bowel syndrome: a systematic analysis. J Psychosom Res 64(6): 573–582

Rongen-van Dartel SAA, Repping-Wuts H, Flendrie M, Bleijenberg G, Metsios GS, van den Hout WB et al (2015) Effect of aerobic exercise training on fatigue in rheumatoid arthritis: a meta-analysis. Arthritis Care Res 67(8):1054–1062

Salamone JD, Correa M (2024) The neurobiology of activational aspects of motivation: exertion of effort, effort-based decision making, and the role of dopamine. Annu Rev Psychol 75(1):null

Shine K, Bloom FE, Cook KF, Cory-Slechta DA, Friedberg F, Grossblatt N (2014) Chronic multisymptom illness in Gulf War veterans: case definitions reexamined. National Academies Press, Washington

Shukla SK, Cook D, Meyer J, Vernon SD, Le T, Clevidence D et al (2015) Changes in gut and plasma microbiome following exercise challenge in myalgic encephalomyelitis/chronic fatigue syndrome (ME/CFS). PLoS One 10(12):e0145453

Singh I, Joseph P, Heerdt PM, Cullinan M, Lutchmansingh DD, Gulati M et al (2022) Persistent exertional intolerance after COVID-19: insights from invasive cardiopulmonary exercise testing. Chest 161(1):54–63

Smylie AL, Broderick G, Fernandes H, Razdan S, Barnes Z, Collado F et al (2013) A comparison of sex-specific immune signatures in Gulf War illness and chronic fatigue syndrome. BMC Immunol 14:29

Stahl SM (2002) The psychopharmacology of energy and fatigue. J Clin Psychiatry 63(1):7–8

Stavem K, Ghanima W, Olsen MK, Gilboe HM, Einvik G (2021) Prevalence and determinants of fatigue after COVID-19 in non-hospitalized subjects: a population-based study. Int J Environ Res Public Health 18(4):2030

Steele L (2000) Prevalence and patterns of Gulf War illness in Kansas veterans: association of symptoms with characteristics of person, place, and time of military service. Am J Epidemiol 152(10):992–1002

Stegner AJ, Almassi NE, Dougherty RJ, Ellingson LD, Gretzon NP, Lindheimer JB et al (2021) Safety and efficacy of short-term structured resistance exercise in Gulf War veterans with chronic unexplained muscle pain: a randomized controlled trial. Life Sci 282:119810

Stussman B, Williams A, Snow J, Gavin A, Scott R, Nath A et al (2020) Characterization of post–exertional malaise in patients with myalgic encephalomyelitis/chronic fatigue syndrome. Front Neurol 11:1025

Sullivan K, Krengel M, Heboyan V, Schildroth S, Wilson CC, Iobst S et al (2020) Prevalence and patterns of symptoms among female veterans of the 1991 Gulf War era: 25 years later. J Women's Health 29(6):819–826

Szabo A (2013) Acute psychological benefits of exercise: reconsideration of the placebo effect. J Ment Health 22(5):449–455

Thaweethai T, Jolley SE, Karlson EW, Levitan EB, Levy B, McComsey GA et al (2023) Development of a definition of postacute sequelae of SARS-CoV-2 infection. JAMA 329(22): 1934–1946

Tomlinson D, Diorio C, Beyene J, Sung L (2014) Effect of exercise on cancer-related fatigue: a meta-analysis. Am J Phys Med Rehabil 93(8):675

United States Department of Veterans Affairs (2008) Research advisory committee on Gulf War veterans I. Gulf War illness and the health of Gulf War veterans: scientific findings and recommendations. US Government Printing Office

Van Ness JM, Stevens SR, Bateman L, Stiles TL, Snell CR (2010) Postexertional malaise in women with chronic fatigue syndrome. J Women's Health (Larchmt) 19(2):239–244

Van Oosterwijck J, Nijs J, Meeus M, Lefever I, Huybrechts L, Lambrecht L et al (2010) Pain inhibition and postexertional malaise in myalgic encephalomyelitis/chronic fatigue syndrome: an experimental study. J Intern Med 268(3):265–278

van Vulpen JK, Sweegers MG, Peeters PHM, Courneya KS, Newton RU, Aaronson NK et al (2020) Moderators of exercise effects on cancer-related fatigue: a meta-analysis of individual patient data. Med Sci Sports Exerc 52(2):303–314

Wang T, Yin J, Miller AH, Xiao C (2017) A systematic review of the association between fatigue and genetic polymorphisms. Brain Behav Immun 62:230–244

Ward-Ritacco C, Poudevigne MS, O'Connor PJ (2016) Muscle strengthening exercises during pregnancy are associated with increased energy and reduced fatigue. J Psychosom Obstet Gynaecol 37(2):68–72

Ward-Ritacco CL, Wilson AR, O'Connor PJ (2021) An apple extract beverage combined with caffeine can improve alertness, mental fatigue, and information processing speed. J Cogn Enhanc 5(3):267–279

Ware JEJ (2000) SF-36 health survey update. Spine 25(24):3130–3139

Washington SD, Rayhan RU, Garner R, Provenzano D, Zajur K, Addiego FM et al (2020) Exercise alters brain activation in Gulf War illness and Myalgic encephalomyelitis/chronic fatigue syndrome. Brain Commun 2(2):fcaa070

Weinstein AA, Deuster PA, Francis JL, Beadling C, Kop WJ (2010) The role of depression in short-term mood and fatigue responses to acute exercise. Int J Behav Med 17(1):51–57

Wender CLA, Manninen M, O'Connor PJ (2022) The effect of chronic exercise on energy and fatigue states: a systematic review and meta-analysis of randomized trials. Front Psychol 13: 907637

White AT, Light AR, Hughen RW, Bateman L, Martins TB, Hill HR et al (2010) Severity of symptom flare after moderate exercise is linked to cytokine activity in chronic fatigue syndrome. Psychophysiology 47(4):615–624

White PD, Goldsmith KA, Johnson AL, Potts L, Walwyn R, DeCesare JC et al (2011) Comparison of adaptive pacing therapy, cognitive behaviour therapy, graded exercise therapy, and specialist medical care for chronic fatigue syndrome (PACE): a randomised trial. Lancet 377(9768): 823–836

Woo M, Kim S, Kim J, Petruzzello SJ, Hatfield BD (2009) Examining the exercise-affect dose–response relationship: does duration influence frontal EEG asymmetry? Int J Psychophysiol 72(2):166–172

Xiao P, Ding S, Duan Y, Li L, Zhou Y, Luo X et al (2022) Effect of light therapy on cancer-related fatigue: a systematic review and meta-analysis. J Pain Symptom Manag 63(2):e188–e202

Yoon JH, Park NH, Kang YE, Ahn YC, Lee EJ, Son CG (2023) The demographic features of fatigue in the general population worldwide: a systematic review and meta-analysis. Front Public Health 11:1192121

Yoshida T (1984) Effect of exercise duration during incremental exercise on the determination of anaerobic threshold and the onset of blood lactate accumulation. Eur J Appl Physiol Occup Physiol 53(3):196–199

Zou S, Kumar U (2018) Cannabinoid receptors and the endocannabinoid system: signaling and function in the central nervous system. Int J Mol Sci 19(3):833

Influence of Regular Physical Activity on Sleep

Melissa J. McGranahan and Patrick J. O'Connor

Contents

Abstract Good sleep and adequate physical activity are essential to health. Yet, large numbers of people are chronically deficient in sleep and physical activity. About 1 in 3 Americans get less than 7 h of sleep per night and only 1 of 4 adults regularly complete weekly physical activity in amounts recommended for good health. This chapter reviews research that has examined relationships between regular physical activity and sleep. The overall weight of evidence supports that regular physical activity is associated with better sleep quality among healthy adults, with epidemiological studies showing moderate-sized effects and more well-controlled randomized controlled trial experiments often showing larger effects. Large epidemiology studies suggest that the relationship between regular physical activity and better sleep quality may partially mediate the well-established associations between physical activity and reduced risk of mortality, cardiovascular diseases, and dementia. There is evidence that the completion of regular physical activity also is associated with better sleep quality among those with certain sleep

M. J. McGranahan and P. J. O'Connor (✉)
Department of Kinesiology, University of Georgia, Athens, GA, USA
e-mail: poconnor@uga.edu

Curr Topics Behav Neurosci (2024) 67: 309–328
https://doi.org/10.1007/7854_2024_503
Published Online: 31 July 2024

disorders (i.e., insomnia, obstructive sleep apnea, and restless legs syndrome), mental health disorders (i.e., depression and posttraumatic stress disorder), and medical illnesses (i.e., breast cancer survivors). The evidence is inadequate to support that regular physical activity substantially improves sleep quality either (i) in children, adolescents, and older adults, (ii) in those with cancers except for breast cancer, (iii) in those with fibromyalgia, or (iv) among those with chronic kidney disease. Also, there is inadequate evidence to conclude that sleep quality is disrupted during weeks when competitive athletes engage in periods of overtraining.

Keywords Exercise · Sleep Initiation and Maintenance Disorders · Sleep Quality · Strength Training

1 Introduction

The primary purpose of this chapter is to provide an overview of research that has examined the influence of regular physical activity on sleep.

Physical activity refers to any force produced by skeletal muscle action (Caspersen et al. 1985). Physical activity is a complex exposure variable that can meaningfully be categorized in multiple ways, including the purpose (e.g., occupational, transport, recreational, sport), frequency per week, the duration in minutes of a single bout, and the intensity from low to maximal (often expressed relative to a measure of an individual's capacity such as a percentage of maximal strength or peak aerobic capacity) (Beenackers et al. 2012). Regular physical activity (aka, chronic) refers to daily activity bouts that are repeated regularly such as two, three, or five times per week for multiple weeks (see Chapter "Introduction" for comprehensive overview). Complicating physical activity as an exposure variable with regard to sleep outcomes are factors that plausibly could interact with physical activity to influence sleep such as time of day, or the time during a circadian phase, that the activity occurs as well as if it occurs concomitantly with sunlight or heat exposure or as part of competitive or non-competitive social interactions (Youngstedt et al. 2019, 2016).

There are large changes in sleep across the lifespan. Sleep duration generally decreases from birth to the fifth decade of life and then increases around the age of 50. Clock time at sleep onset is earlier among children and older adults while the delay of sleep onset into the early morning hours is most apparent around 20 years of age (Su et al. 2022). Adequate amounts of physical activity and sleep are associated with better health status while inadequate or excessive levels of physical activity and sleep are associated with worse health status (Hale et al. 2020; Warburton and Bredin 2017). Sleep disorders are more common among people who are unable to engage in physical activity for extended periods of time either because of a movement-related pathology (Lélis et al. 2016), an injury (Giannoccaro et al. 2013) or environmental restrictions such as during space flight (Pandi-Perumal and

Gonfalone 2016). When sleep is prevented chronically, such as among individuals with a rare genetic disorder that prevents sleep, it is fatal (Llorens et al. 2017).

Chronic short sleep, often defined as less than 6 h per night, is significantly associated with obesity (mean relative risk of 1.38 and 95% Confidence Interval [CI] 1.25, 1.53), type II diabetes (1.37, 95% CI 1.22, 1.53), heart disease (1.26, 95% CI 1.15, 1.38), hypertension (1.17, 95% CI 1.09, 1.26), and mortality (1.12, 95% CI 1.08, 1.16) (Itani et al. 2017) as is chronic long sleep (> 9 h per night) with associated increased risks established for stroke (1.46, 95% CI 1.26, 1.69), mortality (1.39, 95% CI 1.31, 1.47), type II diabetes (1.26, 95% CI 1.11, 1.43), cardiovascular disease (1.25, 95% CI 1.14, 1.37), and coronary heart disease (1.24, 95% CI 1.13, 1.37) (Jike et al. 2018). Regular physical activity reduces the risk of the medical conditions listed above (see Chapter "Classification and Pharmacological Properties") and a limited amount of evidence suggests that the associations between sleep and mortality/adverse health outcomes are driven, in part, by physical inactivity.

Beyond the physical health problems associated with poor sleep, there are consistent links between sleep and mental health during both stable historical periods and especially stressful times driven by major social or economic upheavals. Psychological and health threats, such as occurred during the COVID pandemic, were associated with concomitant changes in physical activity, sleep, and mental health. On average, participation in physical activity was reduced during the COVID pandemic based on 57 worldwide studies of over 119,000 participants (Wunsch et al. 2022). People not meeting physical activity guidelines reported worse loneliness, depression, and stress which may have contributed to sleep problems (Meyer et al. 2020). Fear of acquiring COVID was associated with greater anxiety and stress which plausibly contributed to sleep problems (Alimoradi et al. 2022). Four of every 10 people reported a sleep problem during the pandemic with the prevalence highest among infected patients, children and adolescents, and hospital workers (Jahrami et al. 2022).

The next section will summarize research exploring the influence of regular physical activity on sleep that has been conducted among samples without major health problems. Subsequent sections will focus on the extent to which regular physical activity alters sleep in samples suffering from selected physical or psychiatric illness.

2 Effects of Regular Physical Activity on Sleep in Apparently Healthy Samples

In contrast to the common parental hope that children will fall asleep quickly and stay asleep longer after a physically active day, the evidence shows that physical activity is quite weakly associated with sleep duration in children aged 3 to 13 ($r = 0.02$; 95% CI −0.03, 0.07) based on a meta-analysis of 47 studies (Antczak et al. 2020). A similar weak, non-significant relationship also was observed in university students (averaged age between 18.9 and 23.1) based on a meta-analysis of 29 studies involving data from 141,035 students (Memon et al. 2021). Results of studies examining the effect of exercise on sleep are more mixed among samples of older adults. Studies of older adults can be especially challenging because of the frequent presence of co-morbid conditions that can influence sleep and the changes in sleep that occur with normal aging, including advanced sleep timing, increased wakefulness after sleep onset, less total sleep duration, decreased slow wave sleep, and more daytime naps (Li et al. 2022). A review of 21 prospective studies of older adults (aged >50 years) concluded that the most consistent risk factors for future sleep disturbances were: female sex, depressed mood, and having a physical illness while the evidence was less robust for low physical activity levels as a risk factor (Smagula et al. 2016). In contrast, meta-analysis of 13 randomized controlled trials with older adults (> 65 years old) showed small, significant improvements in sleep quality after adopting a program of regular exercise (Solis-Navarro et al. 2023). This small effect may be moderated by brain health given that a meta-analysis of 11 studies examining the influence of aerobic exercise on both cognitive function and sleep in older adults (aged ≥60 years) with mild cognitive impairment found that global cognitive function was improved but sleep was not (Ahn and Kim 2023).

Regular physical activity appears to be more consistently associated with sleep in samples of adults; however, the magnitude of the effect appears to be small-to-moderate (Atoui et al. 2021; Kline et al. 2021) and cross-sectional studies showing better sleep among physically active people may reflect confounding variables such as better health among the physically active (O'Connor and Wender 2021). In a study of over 50,000 British older adults (aged ≥60 years old), self-reported physical activity significantly moderated relationships between health-related quality of life and sleep duration (Hu et al. 2022).

Much of the exercise and sleep literature has been based on self-reports of both physical activity and sleep. There is more error in self-reports of sleep than more objective measures such as polysomnography or activity-monitor derived assessments (e.g., accelerometers), so it is perhaps not surprising that effects of exercise on sleep are typically smaller with the more objective measures. The effects of regular physical activity on sleep onset, total sleep time, and sleep efficiency are small when sleep has been measured objectively such as via polysomnography or actigraphy (Kline et al. 2021). In one analysis of over 3,000 adults aged 18 to 85 years from the National Health and Nutrition Examination Survey (NHANES) study during a year when participants wore an accelerometer on the hip to assess physical activity, after controlling for several confounders including age, it was found that the relative risk of excessive daytime sleepiness (an index of poor sleep) was reduced by a factor of 0.65 among those participants who met the U.S. federal government recommended amount of physical activity for health (Loprinzi and Cardinal 2011). Information providing a biological basis for potential relationships between physical activity and sleep is unfortunately rare but it is important and needed for making a strong case that physical activity per se meaningfully influences sleep. Thus, it is notable that one genome-wide association study of over 91,000 adults aged 45 to 79 years whose sleep and physical activity was documented from wearing a waist-worn accelerometer for one week found that as much as 18% of the variation in physical activity and sleep duration might be accounted for by common central nervous system genetic variations (Doherty et al. 2018).

The timing of exercise might influence sleep. Avoiding evening exercise so as to not negatively impact sleep has long been generally recommended based on clinical wisdom. The current evidence, based on 23 studies, supports the opposite; that is, small benefits of late night exercise for sleep except for high intensity exercise completed less than one hour prior to sleep which can lead to sleep disruptions (Stutz et al. 2019). One small study with insomniacs used reading before bed as a control condition and compared it to a separate night involving the completion of 30 min of moderate intensity treadmill exercise and 15 min of moderate muscle strengthening exercise 2 h prior to bedtime. Large sleep disruptions were observed in 16% of the sample, leading to the conclusion that people with insomnia should be cautious about completing a bout of moderate and higher intensity exercise within two hours of bedtime (Youngstedt et al. 2021).

The short-term adoption of regular exercise (typically for a few months) by groups of healthy adults (mean age of 42 years) results in large improvements in self-reported sleep quality (the effect size is about ¾ standard deviation) based on a meta-analysis of randomized controlled trials in predominantly community samples without sleep complaints (Kredlow et al. 2015). One systematic review of 13 studies focused on resistance exercise concluded that chronic strength training has moderate-to-large benefits for sleep quality and noted that the magnitude of the benefit appears to exceed the small-to-moderate improvement in sleep quality that accrues from sleep medications (Kovacevic et al. 2018). The effects of resistance exercise on the more objective sleep measures were small and inconsistent across the studies.

There is a small body of evidence regarding relationships among physical activity, sleep, and mortality and most of it has been limited to sleep duration and focused on older adult samples. Early epidemiological research found a U-shaped relationship between sleep duration and an increased risk of mortality, with reduced risk of mortality associated with a sleep duration of about 7 h per night (Bellavia et al. 2013; Youngstedt et al. 2017). The evidence is stronger for the negative effects of sleep that is too long and prospective cohort studies support a J-shaped relationship between sleep duration and mortality (Liu et al. 2017). For sleep durations between 4 and 6 h, the relative risk of mortality ranged between 1.02 and 1.09 while the relative risk ranged from 1.21 to 1.28 for a sleep duration of 10 or more hours per night. One review of prospective cohort studies concluded that high levels of physical activity plausibly could offset the increased risk of all-cause mortality associated with short sleep duration while long sleep duration was associated with increased mortality risk regardless of physical activity level (Duncan et al. 2023).

3 Effects of Regular Physical Activity on Sleep in People with Selected Sleep Disorders

Insomnia – About 20% of adults occasionally have insomnia symptoms while about 10% suffer from an insomnia disorder (Morin and Jarrin 2022). Those with an insomnia disorder have sleep problems for at least 3 months (Vahia 2013). They often have difficulties falling asleep, waking up during sleep, and having trouble going back to sleep, or they wake too early in the morning and are unable to get back to sleep. This often leads to daytime symptoms such as feelings of low energy and irritability as well as cognitive impairments such as problems with perceived or objectively measured attention and memory (Wardle-Pinkston et al. 2019).

At least five experiments have tested the influence of regular physical activity on sleep in adults with a diagnosed insomnia disorder. One study involved yoga (Afonso et al. 2012), a second used Tai Chi (Irwin et al. 2014), and several experiments involved aerobic exercise (Guilleminault et al. 1995; Hartescu et al. 2015; Reid et al. 2010). Four of the five experiments found significant improvements in sleep quality after exercise, one experiment likely lacked the statistical power to detect a true effect given there were only 10 participants in the exercise condition (Kline et al. 2021; Guilleminault et al. 1995; Passos et al. 2012; Banno et al. 2018; Yang et al. 2012).

The small body of research (<20 studies) examining the influence of exercise training on sleep in samples with insomnia symptoms has been reviewed at least five times (Kline et al. 2021; Passos et al. 2012; Banno et al. 2018; Yang et al. 2012; Lowe et al. 2019). Despite the methodological limitations in the studies (e.g., small samples), the consistency of the effect has led reviewers to conclude that there is a moderate amount of evidence that physical activity reduces the severity of insomnia among those with elevated insomnia symptoms.

Sleep apnea – Obstructive sleep apnea (OSA) results when the upper airway collapses during sleep resulting in transient cessation or attenuation of ventilation. By mid-age, OSA affects about 10% of men and 3% of women; yet, many people are unaware they have the problem (Peppard et al. 2013). At least 14 randomized controlled trial experiments have tested the influence of exercise training on some aspect of OSA, typically the apnea-hypopnea index (AHI), sleep quality, or a measure of daytime functioning (Kline et al. 2021). These investigations included a variety of exercise modes including cycling, walking, resistance exercise, and Tai Chi but there were not enough studies to test for moderator variables such as exercise mode. Overall, the studies show that exercise training, in the absence of meaningful loss of body mass, reduces OSA severity by a moderate amount (~0.5 standard deviations, equating to a range of 4 to 13 fewer events per hour) compared to controls (Kline et al. 2021; Aiello et al. 2016; Peng et al. 2022). Exercise training is not as effective in reducing OSA severity as treatment with continuous positive airway pressure (CPAP) or mandibular advancement devices but was more effective in improving sleep quality (Iftikhar et al. 2014). Subsequent research with OSA samples showed that exercise training improved sleep quality, daytime sleepiness and polysomnographic measures of sleep efficiency, sleep duration, and stage N3 sleep (Peng et al. 2022; Mendelson et al. 2018; Lins-Filho et al. 2020, 2023).

Restless legs syndrome – Restless legs syndrome (RLS) is an uncomfortable feeling in the legs while awake that can disrupt sleep onset. Most people with RLS also have periodic limb movements during sleep which reduce sleep continuity. RLS is more common in women, during pregnancy and for those with chronic pain and kidney disease. Randomized controlled trials consistently have found a reduced severity of RLS symptoms in response to exercise training primarily involving cycling but also including yoga, stretching, walking, resistance exercise, and a mixture of modes (Aukerman et al. 2006; Mortazavi et al. 2013; Çoban et al. 2023; Giannaki et al. 2013). A meta-analysis of four randomized controlled trials focused on patients undergoing hemodialysis and reported that the adoption of regular exercise training induced moderate-to-large reductions in RLS symptoms (Song et al. 2018).

4 Effects of Regular Physical Activity on Sleep in Patients with Selected Mental Illnesses

Depression – Depression is characterized by a depressed mood or loss of interest/pleasure in daily activities. Both insomnia and hypersomnia are symptoms included in the diagnostic and statistical manual of mental disorders (DSM-5) criteria for major depressive disorder (MDD). Insomnia symptoms in depression are more common (85–88%) compared to hypersomnia (27–48%) (Yates et al. 2004; Geoffroy et al. 2018). Both cross-sectional and longitudinal data support that poor sleep quality increases the odds of elevated depression symptoms (Furihata et al. 2017; Pan et al. 2022). Sleep architecture features also may vary based on MDD sub-types (Solelhac et al. 2023). In a notable meta-analysis ($k = 17$), patients with MDD had significantly higher subjective sleep disturbances (Hedges' $g = 2.1$, 95% CI 1.72, 2.47) compared to healthy controls (Cox and Olatunji 2020).

Physical activity and exercise-based interventions improve depression symptoms (Gordon et al. 2018) (see Chapter "Management of 'Challenging Experiences' and 'Bad Trips' in Psychedelic-assisted Therapy and Recreational Use") and recent studies suggest that the effects of physical activity on depressive symptoms may be mediated through changes in sleep. In a cross-sectional study of 1,576 participants, higher levels of self-reported physical activity significantly predicted fewer depression symptoms ($\beta = -0.114$). Mediation analysis indicated that sleep quality accounted for 19% of the effect of physical activity on depressive symptoms (Barham et al. 2022). In a similar analysis from the longitudinal Cardiovascular Risk in Young Finns Study, physical activity was associated with better sleep quality and lower depression symptoms, and sleep quality mediated the relation between physical activity and depression symptoms (Kaseva et al. 2019). However, after adjusting for baseline depressive symptoms, the mediation was no longer significant. In a large study evaluating 402,290 participants in the U.K. Biobank study, the risk of developing depressive disorder was lowest (hazard ratio = 0.39, 95% CI 0.33, 0.47) for individuals categorized as having healthy sleep (based on appropriate sleep duration, fewer insomnia symptoms, and less snoring and daytime sleepiness) and moderate-to-high levels of physical activity (Pan et al. 2022). However, whether physical activity levels lowered the risk of developing depression in those with poor sleep behavior was not evaluated (Pan et al. 2022).

Although the impact of exercise training on depression has been frequently studied, only a smaller number of studies have included sleep quality metrics (Gerber et al. 2019; Cheung and Lee 2018). In the most extensive meta-analysis on the topic ($n = 17$), vigorous aerobic exercise, moderate aerobic exercise plus treatment as usual (TAU), light strength training, and vigorous aerobic exercise plus TAU all had a moderate effect on improving sleep quality compared to passive controls. Large improvements in sleep quality based on effect size differences were found for moderate aerobic plus health education, mind-body exercises plus TAU, and vigorous strength training compared to passive control (Brupbacher et al. 2021). When compared to TAU, mind-body plus TAU (standardized mean difference (SMD) = −0.46, 95% CI −0.80, −0.12), and vigorous strength training (SMD = −0.61, 95% CI −1.12, −0.10) were the only interventions to significantly improve sleep quality (Brupbacher et al. 2021). This meta-analysis provides evidence that exercise interventions have potential for improving sleep quality in patients with diagnosed depression or those with elevated depressive symptoms. However, what factors may moderate the anti-depressant effect were not evaluated due to the low heterogeneity of the effects ($I^2 = 4.3\%$). In a sub-analysis, larger improvements in sleep quality were observed in studies with patients with diagnosed depressive disorder. Overall, the evidence suggests a beneficial effect of exercise training on sleep quality for individuals with depression.

Posttraumatic stress disorder (PTSD) – PTSD is a condition that develops following exposure to a traumatic event. Sleep disturbances prior to and following trauma exposure are associated with an elevated risk of developing PTSD (Neylan et al. 2020; Pace-Schott et al. 2015). Nightmares related to the trauma, an impaired ability to fall asleep, and frequent awakenings in the night are a part of two of the four symptom clusters in the DSM-5 PTSD criteria (American Psychiatric Association 2013). Self-reported sleep disturbances (e.g., nightmares, insomnia, restless sleep) range from 70 to 92% in samples of individuals with PTSD (Ohayon and Shapiro 2000; Seelig et al. 2010). Sleep disturbance often remains a residual problem following treatment such as after standardized cognitive behavioral therapy or exposure therapy (Brownlow et al. 2016; Taylor et al. 2020).

A few observational studies have evaluated the evidence between physical activity and sleep in patients with PTSD. In a one-year prospective study ($n = 736$), veterans who developed PTSD had significantly decreased physical activity levels at the end of the study compared to veterans without PTSD (Talbot et al. 2014). Poor sleep quality significantly mediated the relationship between baseline PTSD and the lower physical activity levels one year later (Talbot et al. 2014). While this suggests that in samples with PTSD, sleep quality may be an important predictor for physical activity levels, likewise increasing physical activity can have a beneficial impact on PTSD symptoms through improvements in sleep quality. In a three-month long internet-based study of 182 individuals with probable PTSD, higher exercise intensity was associated with decreases in common symptoms of PTSD such as hyperarousal, avoidance, and emotional numbing (Whitworth et al. 2017). Additionally, the effects of physical activity on PTSD symptoms were mediated by changes in sleep quality ($\beta = -2.54$) (Whitworth et al. 2017).

Currently, there are only a small number of high-quality studies evaluating the effects of exercise training on sleep quality in patients with PTSD. In a meta-analysis ($k = 5$) evaluating exercise training on sleep quality in patients with PTSD, the overall effect size was small (Hedges' $d = -0.47$, 95% CI −0.75, −0.18) (McGranahan and O'Connor 2021). In a recent randomized controlled trial evaluating 8 sessions of various exercise activities compared to 8 sessions of creative tasks, as adjunctive care during intensive trauma focused treatment, no significant difference in insomnia severity index occurred between the groups (Hedges' $d = -0.19$, CI −0.57, 0.19) (Voorendonk et al. 2023). There were several limitations to this study: (1) a lack of reporting of exercise intensity, (2) the actual duration participants were physically active was not measured, and (3) without a treatment alone comparison group, it is not possible to determine whether exercise or creative tasks had an impact on insomnia symptoms. The overall evidence, while limited, suggests that exercise training may help improve sleep for patients with PTSD. Whether exercise training is superior to other treatments for sleep outcomes in adults PTSD requires future well-designed studies.

5 Effects of Regular Physical Activity on Sleep in People with Selected Medical Illnesses

Cancer – Early small studies suggested that yoga has promising beneficial effects on sleep for cancer patients with insomnia and that the activity was well tolerated (Mustian 2013). This stimulated additional research investigating the impact of regular physical activity of various types on sleep in cancer survivors and this has yielded mixed results. A quantitative review of 17 randomized trials found no statistically significant effect of physical activity on subjective or objective measures of sleep in cancer survivors (Mercier et al. 2017). Several other meta-analyses, each with different inclusion and exclusion criteria, found small improvements in some aspect of sleep, most often self-reported sleep quality (Fang et al. 2019; Kreutz et al. 2019). A smaller body of randomized trials ($n = 6$) tested the influence of regular exercise on sleep problems among breast cancer survivors as assessed using the PSQI (Yang et al. 2021). It was found that increased physical activity improved sleep quality by a small mean difference (SMD = 0.22, 95% CI 0.04, 0.40) and there were smaller non-significant effects on sleep duration, latency, efficiency, and daytime functioning. A subsequent meta-analysis of 12 randomized controlled trials of the influence of yoga on sleep quality among breast cancer survivors found a significant, small mean improvement in sleep quality favoring yoga (SMD = −0.40, 95% CI −0.71, −0.09) compared to control conditions (Zhu et al. 2023). Another meta-analysis concluded that walking was more effective in improving sleep among cancer survivors than was yoga (Tang et al. 2019). Finally, a meta-analysis using individual patient data, which can be used to resolve study-level from individual-level effects in randomized trials, found small decrease in sleep disturbances after exercise but no effect on sleep quality even though sleep quality is the variable most frequently improved with the adoption of regular physical activity (Bernard et al. 2019).

Cardiovascular Disease – It is well established that regular leisure-time physical activity is associated with a reduced risk of cardiovascular disease; 70% of the possible risk reduction occurs by engaging in 150 min of brisk walking per week (Kraus et al. 2019). There is a U-shaped relationship between sleep duration and increased risk of heart disease and stroke; for example, compared to 7 h of sleep both 6 h or less and 8 h or more are associated with an increased risk of heart disease and stroke (Wang et al. 2022, 2016). The results of one large prospective cohort study of 502,616 adults based on self-reported physical activity concluded that the adverse associations of sleep with cardiovascular disease mortality were exacerbated by low levels of physical activity – suggesting synergistic effects of poor sleep and low levels of physical activity (Huang et al. 2022a). A cross-sectional analysis of 5,312 suburbanites living near Shanghai, China, found that activity intensity, duration, and frequency all significantly mediated the relationship between longer sleep duration and increased risk of stroke (Liu et al. 2022).

Dementia – Emerging evidence suggests bidirectional relationships between sleep disturbances and Alzheimer's disease and other dementias. That is, Alzheimer's disease can cause sleep problems and sleep disruptions can contribute to an increased risk of developing Alzheimer's disease (Lucey 2020). A small number of cross-sectional studies suggest that some of the negative effect of poor sleep on cognition might be attenuated when people are adequately physically active (Sewell et al. 2021). A similar small body of evidence suggests that the effects of physical activity on improving sleep might mediate the improved cognition associated with regular physical activity (Won et al. 2019). One large cross-sectional study (n = 431,924) examined relationships among sleep, physical activity, sedentary behavior, and dementia risk. The lowest dementia risk (Relative risk = 0.59, 95% CI 0.50, 0.69) was found for those reporting 7 h of sleep per day as well as moderate-to-high levels of leisure-time physical activity (LTPA) and low-to-moderate levels of sedentary behavior compared to the reference group (longer or shorter sleep/low LTPA/high sedentary behavior) (Huang et al. 2022b).

Fibromyalgia – Fibromyalgia (FM) is a disorder characterized by chronic widespread pain. Other symptoms include fatigue, cognitive difficulties, and sleep-related disruptions (Segura-Jiménez et al. 2015; Häuser et al. 2015). The estimated prevalence rate of FM in the United States ranges from 1.1 to 6.4% (Wolfe et al. 1995; Vincent et al. 2013). Similar rates of FM have been reported in European countries, specifically 1.4 to 3.7% (Branco et al. 2010). Prevalence rates of FM are higher in older compared to younger adults and in women compared to men (Branco et al. 2010; Jones et al. 2015).

Sleep related issues, including nonrestorative sleep and sleep disturbances, are common in patients with FM. In a sample of 2,596 people with FM, nonrestorative sleep was indicated as one of the top three most frequent symptoms, and sleep problems were in the top three problems listed to exacerbate FM symptoms (Bennett et al. 2007). Objective sleep metrics, as measured by polysomnography, indicate that there are distinct differences in the sleep architecture of patients with FM compared to healthy controls (Wu et al. 2017). Similarly, large differences in subjective sleep quality, as measured via the Pittsburgh Sleep Quality index (PSQI), were observed between patients with FM and healthy controls. Sleep was worse in patients with FM compared to healthy controls based on global PSQI scores (Hedges' $g = 2.19$, 95% CI 1.58, 2.79) and sleep onset latency scores (Hedges' $g = 1.75$, 95% CI 0.80, 2.70) that were higher and lower sleep efficiency scores (Hedges' $g = -1.08$, 95% CI –1.65, –0.51) (Wu et al. 2017).

Non-pharmacological interventions, such as physical activity, are currently recommended to help manage FM symptoms (Macfarlane et al. 2017). A study of 409 women with FM found that individuals with greater light and total physical activity minutes, assessed via wrist accelerometry, had significantly better PSQI scores (Borges-Cosic et al. 2019). While statistically significant, whether the changes were clinically meaningful is questionable because the raw change score was small (~1.2 points).

A recent systematic review evaluating the effects of resistance training on sleep quality in patients with FM stated that four of the six studies showed improvements (Bastos et al. 2023). With the exception of one study (Jones et al. 2002), when the effects sizes were calculated all were small (Hedges' d ranging from –0.06, –0.57) and the 95% CI's crossed zero (Häkkinen et al. 2001; Bircan et al. 2008; Andrade et al. 2017, 2019). The Jones et al. study showed large improvements (Hedges' $d = -3.73$, 95% CI, –4.68, –2.78) with the sleep visual analog scale (VAS) item from the Fibromyalgia Impact Questionnaire (FIQ). The sleep VAS item from the FIQ asks participants about how they felt in the morning upon waking with anchors of "felt tired" or "well-rested." Andrade, Sieczkowska, and Vilarino 2019 showed a moderate-sized improvement in the sleep VAS from the FIQ following four weeks of resistance training (Hedges' $d = -0.65$, 95% CI –1.25, –0.05), but no improvement was found for PSQI scores (Hedges' $d = -0.06$, 95% CI –0.64, 0.52,) (Andrade et al. 2019). Resistance training may improve some aspects of sleep quality, but more studies are needed to better understand which components of sleep quality are most impacted when FM patients complete regular resistance training.

One meta-analysis comparing usual care to exercise training ($k = 14$) in people with FM found a small, significant effect of exercise training for improving sleep quality (SMD = –0.17, 95% CI –0.32, –0.01) (Estévez-López et al. 2021). Studies included in the analysis involved a wide range of exercise modalities including yoga, qigong, tai chi, resistance training, and water aerobics. The length of the interventions ranged from 7 to 21 weeks. Due to the lack of heterogeneity ($I^2 < 30\%$), a moderation analysis was not conducted. In a secondary analysis of a small number of mind–body interventions (e.g., tai chi, yoga, and qigong) compared to other exercise interventions, large sized improvements in sleep quality were observed ($k = 5$, SMD = – 0.80; 95% CI −1.57, −0.02) (Estévez-López et al. 2021). The available evidence suggests the exercise training may improve sleep difficulties in patients with FM. Understanding the variables that moderate these effects, and the specific aspects of sleep quality which exercise may target, requires larger well-designed studies that more fully capture and report different sleep quality metrics.

Kidney Disease – Compared to the general population, sleep disorders are more prevalent in patients with chronic kidney disease and especially those with end-stage renal disease (Maung et al. 2016). While the reasons for this are likely multi-factorial, one hypothesis is that hyperactivity of the sympathetic nervous system and decreased vagal tone contribute to sleep disruptions (Maung et al. 2016; Hildreth 2011). Well-established adaptations to regular exercise, including increased vagal tone and reductions in sympathetic activity (Martinez Matthew et al. 2021), plausibly could contribute to improved sleep by changing the autonomic imbalance that is present among those with kidney disease. A meta-analysis of 4 experiments with a total of 130 people with chronic kidney disease reported significant improvements in sleep quality and a mean difference of about 5 points on the PSQI (Valera et al. 2023). A quantitative review of 20 randomized controlled trials involving 989 patients with chronic kidney disease found that the adoption of moderate intensity aerobic exercise resulted in a small improvement in sleep quality (SMD = −0.16) (Zhang et al. 2023). Large confidence intervals calculated in this larger meta-analysis (CI range = – 0.62 to 0.31) indicated that the true effect of exercise could either improve or worsen sleep quality (Zhang et al. 2023).

6 Overtraining in Athletes

Some endurance athletes periodically engage in a high volumes and intensities of physical activity and at times this amount of physical activity is extreme. For example, it was reported that the current male marathon world record holder, Kevin Kiptum, ran about 186 miles per week for three weeks preceding the 2023 London marathon which he won in a time of 2 h, 1 min, and 25 s. It is unclear what effect, if any, this level of physical activity had on Kiptum's sleep. His coach, Gervais Hakizimana, told a reporter that "All he does is run, eat and sleep" (McAlister 2023).

If there is a dose-dependent relationship between physical activity and sleep, then extremes of physical activity might result in very large changes in sleep. Only a few small sample studies have measured sleep objectively using wrist actigraphy before and during periods of intensified training. In these investigations generally small decreases in sleep duration and sleep efficiency of 1.6 to 7.9% were reported (Hausswirth et al. 2014; Killer et al. 2017; Schaal et al. 2015). The modest changes in sleep suggest that extremes of physical activity do not necessarily lead to dramatic changes in sleep. If these findings are true, then they discount the hypothesis that one purpose of sleep is for recovery from the prior day's metabolic processes (Shapiro 1982). Reviewers have concluded that more research is needed to understand this topic fully (Neil et al. 2021).

7 Conclusions and Future Directions

Overall, the evidence supports that regular physical activity is associated with better sleep quality, among healthy adults (Kredlow et al. 2015), individuals with sleep disorders (i.e., insomnia (Kline et al. 2021)), OSA (Aiello et al. 2016), and RLS (Song et al. 2018), individuals with mental health disorders (i.e., depression (Brupbacher et al. 2021)) and PTSD (McGranahan and O'Connor 2021), and among people with certain medical illnesses (i.e., breast cancer survivors (Zhu et al. 2023)). As well, evidence suggests that the positive benefit of physical activity on disease prevention (i.e., decreased risks of all-cause mortality, cardiovascular disease (Wang et al. 2022; Wang et al. 2016)), and dementia (Huang et al. 2022b) and mental health (i.e., odds of developing depression (Pan et al. 2022; Barham et al. 2022)) may be mediated by improved sleep. The evidence is inadequate to support that regular physical activity consistently improves sleep quality either (i) in children, adolescents (Antczak et al. 2020) and older adults (Smagula et al. 2016), (ii) in those with cancers (Mercier et al. 2017), (iii) those with fibromyalgia (Bastos et al. 2023; Estévez-López et al. 2021), or (iv) those with chronic kidney disease (Zhang et al. 2023).

To date there are a limited number of trials that evaluated sleep objectively using the gold standard of polysomnography. Notably, studies measuring sleep metrics with polysomnography following exercise training have primarily been in older adults (King et al. 2008), those with OSA (Lins-Filho et al. 2023), and patients with insomnia (Ferreira et al. 2022). In many of the conditions discussed in the chapter, the impact on polysomnography measured sleep has not been studied. The specific aspects of sleep architecture that exercise may impact and how that relates to the prevention of diseases and improvements in mental health should increasingly be considered as vital to the health of our society. Due to advances in technology, the quality and cost of at-home polysomnography are becoming more feasible and increasingly should be incorporated into research aimed at understanding the relationship between physical activity and sleep. Moreover, data are emerging regarding the accuracy of widely used smart watches to measure aspects of sleep (Roomkham et al. 2019). Thus, it is likely that future large studies will provide useful objective information about sleep and physical activity with this relatively inexpensive and palatable technology.

The wide ranges in 95% confidence intervals in many of the studies suggest there is a range of responses to the impact exercise training has on sleep. As stated in the beginning of this chapter, physical activity is a complex exposure variable which will require novel techniques to improve our understanding. One possible future direction is to leverage large databases with sleep and physical activity measures and implement machine-based learning algorithms to identify patterns and characteristics in those who report improved sleep after increased participation in physical activity. More studies are needed to understand aspects of physical activity that have the largest impact on sleep and what moderates the response across different conditions.

Another opportunity with regard to research design is the use of Mendelian randomization. This approach allows for potential causal effects to be estimated based on genetic variants previously associated with sleep and physical activity. One study using this approach failed to find evidence that sleep duration was causally related to the risk of cardiovascular disease while confirming that vigorous exercise does decrease the risk (Zhuang et al. 2020). More studies, and associated insights, are expected in the future using a Mendelian randomization design.

References

Afonso RF, Hachul H, Kozasa EH, de Souza OD, Goto V, Rodrigues D et al (2012) Yoga decreases insomnia in postmenopausal women: a randomized clinical trial. Menopause 19(2):186

Ahn J, Kim M (2023) Effects of aerobic exercise on global cognitive function and sleep in older adults with mild cognitive impairment: a systematic review and meta-analysis. Geriatr Nurs 51: 9–16

Aiello KD, Caughey WG, Nelluri B, Sharma A, Mookadam F, Mookadam M (2016) Effect of exercise training on sleep apnea: a systematic review and meta-analysis. Respir Med 116:85–92

Alimoradi Z, Ohayon MM, Griffiths MD, Lin C-Y, Pakpour AH (2022) Fear of COVID-19 and its association with mental health-related factors: systematic review and meta-analysis. BJPsych Open 8(2):e73

American Psychiatric Association (2013) Diagnostic and statistical manual of mental disorders (DSM-5®). American Psychiatric Pub

Andrade A, Vilarino GT, Bevilacqua GG (2017) What is the effect of strength training on pain and sleep in patients with fibromyalgia? Am J Phys Med Rehabil 96(12):889–893

Andrade A, Sieczkowska SM, Vilarino GT (2019) Resistance training improves quality of life and associated factors in patients with fibromyalgia syndrome. PM&R 11(7):703–709

Antczak D, Lonsdale C, Lee J, Hilland T, Duncan MJ, del Pozo CB et al (2020) Physical activity and sleep are inconsistently related in healthy children: a systematic review and meta-analysis. Sleep Med Rev 51:101278

Atoui S, Chevance G, Romain A-J, Kingsbury C, Lachance J-P, Bernard P (2021) Daily associations between sleep and physical activity: a systematic review and meta-analysis. Sleep Med Rev 57:101426

Aukerman M, Douglas A, Max B, Fred T, Lydia T, Beth B (2006) Exercise and restless legs syndrome: a randomized controlled trial. J Am Board Fam Med 19(5):487

Banno M, Harada Y, Taniguchi M, Tobita R, Tsujimoto H, Tsujimoto Y et al (2018) Exercise can improve sleep quality: a systematic review and meta-analysis. PeerJ 6:e 5172

Barham WT, Buysse DJ, Kline CE, Kubala AG, Brindle RC (2022) Sleep health mediates the relationship between physical activity and depression symptoms. Sleep Breath 26(3):1341–1349

Bastos ACRF, Vilarino GT, de Souza LC, Dominski FH, Branco JHL, Andrade A (2023) Effects of resistance training on sleep of patients with fibromyalgia: a systematic review. J Health Psychol 28(11):1072–1084

Beenackers MA, Kamphuis CBM, Giskes K, Brug J, Kunst AE, Burdorf A et al (2012) Socioeconomic inequalities in occupational, leisure-time, and transport related physical activity among European adults: a systematic review. Int J Behav Nutr Phys Act 9(1):116

Bellavia A, Åkerstedt T, Bottai M, Wolk A, Orsini N (2013) Sleep duration and survival percentiles across categories of physical activity. Am J Epidemiol 179(4):484–491

Bennett RM, Jones J, Turk DC, Russell IJ, Matallana L (2007) An internet survey of 2, 596 people with fibromyalgia. BMC Musculoskelet Disord 8(1):27

Bernard P, Savard J, Steindorf K, Sweegers MG, Courneya KS, Newton RU et al (2019) Effects and moderators of exercise on sleep in adults with cancer: individual patient data and aggregated meta-analyses. J Psychosom Res 124:109746

Bircan Ç, Karasel SA, Akgün B, El Ö, Alper S (2008) Effects of muscle strengthening versus aerobic exercise program in fibromyalgia. Rheumatol Int 28:527–532

Borges-Cosic M, Aparicio VA, Estévez-López F, Soriano-Maldonado A, Acosta-Manzano P, Gavilán-Carrera B et al (2019) Sedentary time, physical activity, and sleep quality in fibromyalgia: the al-Ándalus project. Scand J Med Sci Sports 29(2):266–274

Branco JC, Bannwarth B, Failde I, Abello Carbonell J, Blotman F, Spaeth M et al (2010) Prevalence of fibromyalgia: a survey in five European countries. Semin Arthritis Rheum 39(6):448–453

Brownlow JA, McLean CP, Gehrman PR, Harb GC, Ross RJ, Foa EB (2016) Influence of sleep disturbance on global functioning after posttraumatic stress disorder treatment. J Trauma Stress 29(6):515–521

Brupbacher G, Gerger H, Zander-Schellenberg T, Straus D, Porschke H, Gerber M et al (2021) The effects of exercise on sleep in unipolar depression: a systematic review and network meta-analysis. Sleep Med Rev 59:101452

Caspersen CJ, Powell KE, Christenson GM (1985) Physical activity, exercise, and physical fitness: definitions and distinctions for health-related research. Public Health Rep 100(2):126–131

Cheung LK, Lee S (2018) A randomized controlled trial on an aerobic exercise programme for depression outpatients. Sport Sci Health 14(1):173–181

Çoban Ö, Ün Yıldırım N, Yaşa ME, Sonkaya AR (2023) Effects of different exercise programs on symptoms, sleep, and quality of life in patients with primary restless legs syndrome. Mov Disord Clin Pract 10(9):1349–1359

Cox RC, Olatunji BO (2020) Sleep in the anxiety-related disorders: a meta-analysis of subjective and objective research. Sleep Med Rev 51:101282

Doherty A, Smith-Byrne K, Ferreira T, Holmes MV, Holmes C, Pulit SL et al (2018) GWAS identifies 14 loci for device-measured physical activity and sleep duration. Nat Commun 9(1):5257

Duncan MJ, Murphy L, Oftedal S, Fenwick MJ, Vincent GE, Fenton S (2023) The associations between physical activity, sedentary behaviour, and sleep with mortality and incident cardiovascular disease, cancer, diabetes and mental health in adults: a systematic review and meta-analysis of prospective cohort studies. J Act Sedent Sleep Behav 2(1):19

Estévez-López F, Maestre-Cascales C, Russell D, Álvarez-Gallardo IC, Rodriguez-Ayllon M, Hughes CM et al (2021) Effectiveness of exercise on fatigue and sleep quality in fibromyalgia: a systematic review and meta-analysis of randomized trials. Arch Phys Med Rehabil 102(4):752–761

Fang Y-Y, Hung C-T, Chan J-C, Huang S-M, Lee Y-H (2019) Meta-analysis: exercise intervention for sleep problems in cancer patients. Eur J Cancer Care 28(5):e13131

Ferreira WS, Santana MG, Youngstedt SD, de Assis DE, de Assis BP, de Cerqueira DP et al (2022) Effects of exercise training and exercise plus acupuncture on chronic insomnia: a feasibility study. Sleep Sci 15(3):288–296

Furihata R, Hall MH, Stone KL, Ancoli-Israel S, Smagula SF, Cauley JA et al (2017) An aggregate measure of sleep health is associated with prevalent and incident clinically significant depression symptoms among community-dwelling older women. Sleep 40(3):zsw 075

Geoffroy PA, Hoertel N, Etain B, Bellivier F, Delorme R, Limosin F et al (2018) Insomnia and hypersomnia in major depressive episode: prevalence, sociodemographic characteristics and psychiatric comorbidity in a population-based study. J Affect Disord 226:132–141

Gerber M, Minghetti A, Beck J, Zahner L, Donath L (2019) Is improved fitness following a 12-week exercise program associated with decreased symptom severity, better wellbeing, and fewer sleep complaints in patients with major depressive disorders? A secondary analysis of a randomized controlled trial. J Psychiatr Res 113:58–64

Giannaki CD, Sakkas GK, Karatzaferi C, Hadjigeorgiou GM, Lavdas E, Kyriakides T et al (2013) Effect of exercise training and dopamine agonists in patients with uremic restless legs syndrome: a six-month randomized, partially double-blind, placebo-controlled comparative study. BMC Nephrol 14:194

Giannoccaro MP, Moghadam KK, Pizza F, Boriani S, Maraldi NM, Avoni P et al (2013) Sleep disorders in patients with spinal cord injury. Sleep Med Rev 17(6):399–409

Gordon BR, McDowell CP, Hallgren M, Meyer JD, Lyons M, Herring MP (2018) Association of efficacy of resistance exercise training with depressive symptoms: meta-analysis and meta-regression analysis of randomized clinical trials. JAMA Psychiatry 75(6):566–576

Guilleminault C, Clerk A, Black J, Labanowski M, Pelayo R, Claman D (1995) Nondrug treatment trials in psychophysiologic insomnia. Arch Intern Med 155(8):838–844

Häkkinen A, Häkkinen K, Hannonen P, Alen M (2001) Strength training induced adaptations in neuromuscular function of premenopausal women with fibromyalgia: comparison with healthy women. Ann Rheum Dis 60(1):21–26

Hale L, Troxel W, Buysse DJ (2020) Sleep health: an opportunity for public health to address health equity. Annu Rev Public Health 41(1):81–99

Hartescu I, Morgan K, Stevinson CD (2015) Increased physical activity improves sleep and mood outcomes in inactive people with insomnia: a randomized controlled trial. J Sleep Res 24(5):526–534

Häuser W, Ablin J, Fitzcharles M-A, Littlejohn G, Luciano JV, Usui C et al (2015) Fibromyalgia. Nat Rev Dis Prim 1(1):15022

Hausswirth C, Louis J, Aubry A, Bonnet G, Duffield ROB, Le Meur Y (2014) Evidence of disturbed sleep and increased illness in overreached endurance athletes. Med Sci Sports Exerc 46(5):1036

Hildreth CM (2011) Prognostic indicators of cardiovascular risk in renal disease. Front Physiol 2: 121

Hu W, Chu J, Chen X, Liu S, Sun N, Han Q et al (2022) The role of depression and physical activity in the association of between sleep quality, and duration with and health-related quality of life among the elderly: a UK biobank cross-sectional study. BMC Geriatr 22(1):338

Huang B-H, Duncan MJ, Cistulli PA, Nassar N, Hamer M, Stamatakis E (2022a) Sleep and physical activity in relation to all-cause, cardiovascular disease and cancer mortality risk. Br J Sports Med 56(13):718–724

Huang S-Y, Li Y-Z, Zhang Y-R, Huang Y-Y, Wu B-S, Zhang W et al (2022b) Sleep, physical activity, sedentary behavior, and risk of incident dementia: a prospective cohort study of 431,924 UK biobank participants. Mol Psychiatry 27(10):4343–4354

Iftikhar IH, Kline CE, Youngstedt SD (2014) Effects of exercise training on sleep apnea: a meta-analysis. Lung 192(1):175–184

Irwin MR, Olmstead R, Carrillo C, Sadeghi N, Breen EC, Witarama T et al (2014) Cognitive behavioral therapy vs. Tai Chi for late life insomnia and inflammatory risk: a randomized controlled comparative efficacy trial. Sleep 37(9):1543–1552

Itani O, Jike M, Watanabe N, Kaneita Y (2017) Short sleep duration and health outcomes: a systematic review, meta-analysis, and meta-regression. Sleep Med 32:246–256

Jahrami HA, Alhaj OA, Humood AM, Alenezi AF, Fekih-Romdhane F, AlRasheed MM et al (2022) Sleep disturbances during the COVID-19 pandemic: a systematic review, meta-analysis, and meta-regression. Sleep Med Rev 62:101591

Jike M, Itani O, Watanabe N, Buysse DJ, Kaneita Y (2018) Long sleep duration and health outcomes: a systematic review, meta-analysis and meta-regression. Sleep Med Rev 39:25–36

Jones KD, Burckhardt CS, Clark SR, Bennett RM, Potempa KM (2002) A randomized controlled trial of muscle strengthening versus flexibility training in fibromyalgia. J Rheumatol 29(5):1041–1048

Jones GT, Atzeni F, Beasley M, Flüß E, Sarzi-Puttini P, Macfarlane GJ (2015) The prevalence of fibromyalgia in the general population: a comparison of the American College of Rheumatology 1990, 2010, and modified 2010 classification criteria. Arthritis Rheum 67(2):568–575

Kaseva K, Dobewall H, Yang X, Pulkki-Råback L, Lipsanen J, Hintsa T et al (2019) Physical activity, sleep, and symptoms of depression in adults-testing for mediation. Med Sci Sports Exerc 51:1162

Killer SC, Svendsen IS, Jeukendrup AE, Gleeson M (2017) Evidence of disturbed sleep and mood state in well-trained athletes during short-term intensified training with and without a high carbohydrate nutritional intervention. J Sports Sci 35(14):1402–1410

King AC, Pruitt LA, Woo S, Castro CM, Ahn DK, Vitiello MV et al (2008) Effects of moderate-intensity exercise on polysomnographic and subjective sleep quality in older adults with mild to moderate sleep complaints. J Gerontol A Biol Sci Med Sci 63(9):997–1004

Kline CE, Hillman CH, Bloodgood Sheppard B, Tennant B, Conroy DE, Macko RF et al (2021) Physical activity and sleep: an updated umbrella review of the 2018 physical activity guidelines advisory committee report. Sleep Med Rev 58:101489

Kovacevic A, Mavros Y, Heisz JJ, Fiatarone Singh MA (2018) The effect of resistance exercise on sleep: a systematic review of randomized controlled trials. Sleep Med Rev 39:52–68

Kraus WE, Powell KE, Haskell WL, Janz KF, Campbell WW, Jakicic JM et al (2019) Physical activity, all-cause and cardiovascular mortality, and cardiovascular disease. Med Sci Sports Exerc 51(6):1270–1281

Kredlow MA, Capozzoli MC, Hearon BA, Calkins AW, Otto MW (2015) The effects of physical activity on sleep: a meta-analytic review. J Behav Med 38(3):427–449

Kreutz C, Schmidt ME, Steindorf K (2019) Effects of physical and mind–body exercise on sleep problems during and after breast cancer treatment: a systematic review and meta-analysis. Breast Cancer Res Treat 176(1):1–15

Lélis ALPA, Cardoso MVLM, Hall WA (2016) Sleep disorders in children with cerebral palsy: an integrative review. Sleep Med Rev 30:63–71

Li J, Vitiello MV, Gooneratne NS (2022) Sleep in Normal Aging. Sleep Med Clin 17(2):161–171

Lins-Filho OL, Pedrosa RP, Gomes JML, Dantas Moraes SL, Vasconcelos BCE, Lemos CAA et al (2020) Effect of exercise training on subjective parameters in patients with obstructive sleep apnea: a systematic review and meta-analysis. Sleep Med 69:1–7

Lins-Filho O, Germano-Soares AH, Porto Aguiar JL, Vieira de Almedia JR, Felinto EC, Lyra MJ et al (2023) Effect of high-intensity interval training on obstructive sleep apnea severity: a randomized controlled trial. Sleep Med 112:316

Liu T-Z, Xu C, Rota M, Cai H, Zhang C, Shi M-J et al (2017) Sleep duration and risk of all-cause mortality: a flexible, non-linear, meta-regression of 40 prospective cohort studies. Sleep Med Rev 32:28–36

Liu X, Zhang J, Wang Y, Lu C, Gu X, Wan G et al (2022) The relationship between sleep duration and stroke risk: the mediating role of physical activity. Brain Sci 12(5):601

Llorens F, Zarranz J-J, Fischer A, Zerr I, Ferrer I (2017) Fatal familial insomnia: clinical aspects and molecular alterations. Curr Neurol Neurosci Rep 17(4):30

Loprinzi PD, Cardinal BJ (2011) Association between objectively-measured physical activity and sleep, NHANES 2005–2006. Ment Health Phys Act 4(2):65–69

Lowe H, Haddock G, Mulligan LD, Gregg L, Fuzellier-Hart A, Carter L-A et al (2019) Does exercise improve sleep for adults with insomnia? A systematic review with quality appraisal. Clin Psychol Rev 68:1–12

Lucey BP (2020) It's complicated: the relationship between sleep and Alzheimer's disease in humans. Neurobiol Dis 144:105031

Macfarlane GJ, Kronisch C, Dean LE, Atzeni F, Häuser W, Fluß E et al (2017) EULAR revised recommendations for the management of fibromyalgia. Ann Rheum Dis 76(2):318–328
Martinez Matthew W, Kim Jonathan H, Shah Ankit B, Phelan D, Emery Michael S, Wasfy Meagan M et al (2021) Exercise-induced cardiovascular adaptations and approach to exercise and cardiovascular disease. J Am Coll Cardiol 78(14):1453–1470
Maung SC, El Sara A, Chapman C, Cohen D, Cukor D (2016) Sleep disorders and chronic kidney disease. World J Nephrol 5(3):224–232
McAlister S (2023) Kelvin Kiptum: the remarkable 300km per week training regime of the marathon world record holder 2023. Available from https://olympics.com/en/news/kelvin-kiptum-incredible-300km-per-week-training-regime
McGranahan MJ, O'Connor PJ (2021) Exercise training effects on sleep quality and symptoms of anxiety and depression in post-traumatic stress disorder: a systematic review and meta-analysis of randomized control trials. Ment Health Phys Act 20:100385
Memon AR, Gupta CC, Crowther ME, Ferguson SA, Tuckwell GA, Vincent GE (2021) Sleep and physical activity in university students: a systematic review and meta-analysis. Sleep Med Rev 58:101482
Mendelson M, Bailly S, Marillier M, Flore P, Borel JC, Vivodtzev I et al (2018) Obstructive Sleep Apnea Syndrome, objectively measured physical activity and exercise training interventions: a systematic review and meta-analysis. Front Neurol 9:9
Mercier J, Savard J, Bernard P (2017) Exercise interventions to improve sleep in cancer patients: a systematic review and meta-analysis. Sleep Med Rev 36:43–56
Meyer J, McDowell C, Lansing J, Brower C, Smith L, Tully M et al (2020) Changes in physical activity and sedentary behavior in response to COVID-19 and their associations with mental health in 3052 US adults. Int J Environ Res Public Health 17(18):6469
Morin CM, Jarrin DC (2022) Epidemiology of insomnia: prevalence, course, risk factors, and public health burden. Sleep Med Clin 17(2):173–191
Mortazavi M, Vahdatpour B, Ghasempour A, Taheri D, Shahidi S, Moeinzadeh F et al (2013) Aerobic exercise improves signs of restless leg syndrome in end stage renal disease patients suffering chronic hemodialysis. Sci World J 2013(1):628142
Mustian KM (2013) Yoga as treatment for insomnia among cancer patients and survivors: a systematic review. Eur Med J Oncol 1:106–115
Neil PW, Shona LH, Charli S, Gregory DR, Mathieu N, Luke G et al (2021) Sleep and the athlete: narrative review and 2021 expert consensus recommendations. Br J Sports Med 55(7):356
Neylan TC, Kessler RC, Ressler KJ, Clifford G, Beaudoin FL, An X et al (2020) Prior sleep problems and adverse post-traumatic neuropsychiatric sequelae of motor vehicle collision in the AURORA study. Sleep 44(3):zsaa200
O'Connor PJ, Wender CLA (2021) Physical activity, sleep, and sleep disorders. In: Cacho V, Lum E, Cacho V, Lum E (eds) Integrative sleep medicine. Oxford University Press
Ohayon MM, Shapiro CM (2000) Sleep disturbances and psychiatric disorders associated with posttraumatic stress disorder in the general population. Compr Psychiatry 41(6):469–478
Pace-Schott EF, Germain A, Milad MR (2015) Sleep and REM sleep disturbance in the pathophysiology of PTSD: the role of extinction memory. Biol Mood Anxiety Disord 5(1):3
Pan C, Ye J, Wen Y, Chu X, Jia Y, Cheng B et al (2022) The associations between sleep behaviors, lifestyle factors, genetic risk and mental disorders: a cohort study of 402 290 UK biobank participants. Psychiatry Res 311:114488
Pandi-Perumal SR, Gonfalone AA (2016) Sleep in space as a new medical frontier: the challenge of preserving normal sleep in the abnormal environment of space missions. Sleep Sci 9(1):1–4
Passos GS, Poyares DL, Santana MG, Tufik S, Mello MT (2012) Is exercise an alternative treatment for chronic insomnia? Clinics (Sao Paulo) 67(6):653–660
Peng J, Yuan Y, Zhao Y, Ren H (2022) Effects of exercise on patients with obstructive sleep apnea: a systematic review and meta-analysis. Int J Environ Res Public Health 19(17):10845
Peppard PE, Young T, Barnet JH, Palta M, Hagen EW, Hla KM (2013) Increased prevalence of sleep-disordered breathing in adults. Am J Epidemiol 177(9):1006–1014

Reid KJ, Baron KG, Lu B, Naylor E, Wolfe L, Zee PC (2010) Aerobic exercise improves self-reported sleep and quality of life in older adults with insomnia. Sleep Med 11(9):934–940

Roomkham S, Hittle M, Cheung J, Lovell D, Mignot E, Perrin D (2019) Sleep monitoring with the Apple Watch: comparison to a clinically validated actigraph [version 1; peer review: 2 approved with reservations, 1 not approved]. F1000Res 8:754

Schaal K, Le Meur Y, Louis J, Filliard J-R, Hellard P, Casazza G et al (2015) Whole-body Cryostimulation limits overreaching in elite synchronized swimmers. Med Sci Sports Exerc 47(7):1416

Seelig AD, Jacobson IG, Smith B, Hooper TI, Boyko EJ, Gackstetter GD et al (2010) Sleep patterns before, during, and after deployment to Iraq and Afghanistan. Sleep 33(12):1615–1622

Segura-Jiménez V, Álvarez-Gallardo IC, Carbonell-Baeza A, Aparicio VA, Ortega FB, Casimiro AJ et al (2015) Fibromyalgia has a larger impact on physical health than on psychological health, yet both are markedly affected: the al-Ándalus project. Semin Arthritis Rheum 44(5):563–570

Sewell KR, Erickson KI, Rainey-Smith SR, Peiffer JJ, Sohrabi HR, Brown BM (2021) Relationships between physical activity, sleep and cognitive function: a narrative review. Neurosci Biobehav Rev 130:369–378

Shapiro CM (1982) Energy expenditure and restorative sleep. Biol Psychol 15(3–4):229–239

Smagula SF, Stone KL, Fabio A, Cauley JA (2016) Risk factors for sleep disturbances in older adults: evidence from prospective studies. Sleep Med Rev 25:21–30

Solelhac G, Berger M, Strippoli M-PF, Marchi NA, Stephan A, Petit J-M et al (2023) Objective polysomnography-based sleep features and major depressive disorder subtypes in the general population. Psychiatry Res 324:115213

Solis-Navarro L, Masot O, Torres-Castro R, Otto-Yáñez M, Fernández-Jané C, Solà-Madurell M et al (2023) Effects on sleep quality of physical exercise programs in older adults: a systematic review and meta-analysis. Clocks Sleep 5(2):152–166

Song Y-y, Hu R-j, Diao Y-s, Chen L, Jiang X-l (2018) Effects of exercise training on restless legs syndrome, depression, sleep quality, and fatigue among hemodialysis patients: a systematic review and meta-analysis. J Pain Symptom Manag 55(4):1184–1195

Stutz J, Eiholzer R, Spengler CM (2019) Effects of evening exercise on sleep in healthy participants: a systematic review and meta-analysis. Sports Med 49(2):269–287

Su S, Li X, Xu Y, McCall WV, Wang X (2022) Epidemiology of accelerometer-based sleep parameters in US school-aged children and adults: NHANES 2011–2014. Sci Rep 12(1):7680

Talbot LS, Neylan TC, Metzler TJ, Cohen BE (2014) The mediating effect of sleep quality on the relationship between PTSD and physical activity. J Clin Sleep Med 10(7):795–801

Tang M-F, Chiu H-Y, Xu X, Kwok JY, Cheung DST, Chen C-Y et al (2019) Walking is more effective than yoga at reducing sleep disturbance in cancer patients: a systematic review and meta-analysis of randomized controlled trials. Sleep Med Rev 47:1–8

Taylor DJ, Pruiksma KE, Hale W, McLean CP, Zandberg LJ, Brown L et al (2020) Sleep problems in active duty military personnel seeking treatment for posttraumatic stress disorder: presence, change, and impact on outcomes. Sleep 43(10):zsaa065

Vahia VN (2013) Diagnostic and statistical manual of mental disorders 5: a quick glance. Indian J Psychiatry 55(3):220–223

Valera S, Diz JC, Rey-Fernández B, González-Devesa D, García-Fresneda A, Ayán C (2023) Efficacy of physical exercise on sleep quality in patients with chronic kidney disease: a systematic review and meta-analysis. Sleep Breath 28:381

Vincent A, Lahr BD, Wolfe F, Clauw DJ, Whipple MO, Oh TH et al (2013) Prevalence of fibromyalgia: a population-based study in Olmsted County, Minnesota, utilizing the Rochester epidemiology project. Arthritis Care Res 65(5):786–792

Voorendonk EM, Sanches SA, Tollenaar MS, Hoogendoorn EA, de Jongh A, van Minnen A (2023) Adding physical activity to intensive trauma-focused treatment for post-traumatic stress disorder: results of a randomized controlled trial. Front Psychol 14:1215250

Wang D, Li W, Cui X, Meng Y, Zhou M, Xiao L et al (2016) Sleep duration and risk of coronary heart disease: a systematic review and meta-analysis of prospective cohort studies. Int J Cardiol 219:231–239

Wang H, Sun J, Sun M, Liu N, Wang M (2022) Relationship of sleep duration with the risk of stroke incidence and stroke mortality: an updated systematic review and dose–response meta-analysis of prospective cohort studies. Sleep Med 90:267–278

Warburton DER, Bredin SSD (2017) Health benefits of physical activity: a systematic review of current systematic reviews. Curr Opin Cardiol 32(5):541–556

Wardle-Pinkston S, Slavish DC, Taylor DJ (2019) Insomnia and cognitive performance: a systematic review and meta-analysis. Sleep Med Rev 48:101205

Whitworth JW, Craft LL, Dunsiger SI, Ciccolo JT (2017) Direct and indirect effects of exercise on posttraumatic stress disorder symptoms: a longitudinal study. Gen Hosp Psychiatry 49:56–62

Wolfe F, Ross K, Anderson J, Russell IJ, Hebert L (1995) The prevalence and characteristics of fibromyalgia in the general population. Arthritis Rheum 38(1):19–28

Won J, Alfini AJ, Weiss LR, Nyhuis CC, Spira AP, Callow DD et al (2019) Caudate volume mediates the interaction between total sleep time and executive function after acute exercise in healthy older adults. Brain Plasticity 5:69–82

Wu Y-L, Chang L-Y, Lee H-C, Fang S-C, Tsai P-S (2017) Sleep disturbances in fibromyalgia: a meta-analysis of case-control studies. J Psychosom Res 96:89–97

Wunsch K, Kienberger K, Niessner C (2022) Changes in physical activity patterns due to the Covid-19 pandemic: a systematic review and meta-analysis. Int J Environ Res Public Health 19(4):2250

Yang P-Y, Ho K-H, Chen H-C, Chien M-Y (2012) Exercise training improves sleep quality in middle-aged and older adults with sleep problems: a systematic review. J Physiother 58(3):157–163

Yang H, Yang Z, Pan H, Zhou Q (2021) Effects of physical activity on sleep problems in breast cancer survivors: a meta-analysis. Support Care Cancer 29(7):4023–4032

Yates WR, Mitchell J, Rush AJ, Trivedi MH, Wisniewski SR, Warden D et al (2004) Clinical features of depressed outpatients with and without co-occurring general medical conditions in STAR*D. Gen Hosp Psychiatry 26(6):421–429

Youngstedt SD, Kline CE, Elliott JA, Zielinski MR, Devlin TM, Moore TA (2016) Circadian phase-shifting effects of bright light, exercise, and bright light + exercise. J Circadian Rhythms 14:2

Youngstedt SD, Kripke DF, James S (2017) Sleep duration, morbidity, and mortality. In: Chokroverty S (ed) Sleep disorders medicine: basic science, technical considerations and clinical aspects. Springer, New York, pp 559–566

Youngstedt SD, Elliott JA, Kripke DF (2019) Human circadian phase–response curves for exercise. J Physiol 597(8):2253–2268

Youngstedt SD, Ito W, Passos GS, Santana MG, Youngstedt JM (2021) Testing the sleep hygiene recommendation against nighttime exercise. Sleep Breath 25(4):2189–2196

Zhang F, Wang H, Huang L, Bai Y, Wang W, Zhang H (2023) Effect of exercise interventions for sleep quality in patients with chronic kidney disease: a systematic review and meta-analysis. Int Urol Nephrol 55(5):1193–1204

Zhu J, Chen X, Zhen X, Zheng H, Chen H, Chen H et al (2023) Meta-analysis of effects of yoga exercise intervention on sleep quality in breast cancer patients. Front Oncol 13:13

Zhuang Z, Gao M, Yang R, Li N, Liu Z, Cao W et al (2020) Association of physical activity, sedentary behaviours and sleep duration with cardiovascular diseases and lipid profiles: a Mendelian randomization analysis. Lipids Health Dis 19(1):86

Exercise Training for Chronic Pain: Available Evidence, Current Recommendations, and Potential Mechanisms

Jacob V. Ninneman, Gunnar A. Roberge, Aaron J. Stegner, and Dane B. Cook

Contents

J. V. Ninneman, G. A. Roberge, A. J. Stegner, and D. B. Cook (✉)
Research Service, William S. Middleton Memorial Veterans Hospital, Madison, WI, USA

Department of Kinesiology, University of Wisconsin–Madison, Madison, WI, USA
e-mail: dane.cook@wisc.edu

Curr Topics Behav Neurosci (2024) 67: 329–366
https://doi.org/10.1007/7854_2024_504
Published Online: 10 August 2024

Abstract Chronic pain conditions pose a significant global burden of disability, with epidemiological data indicating a rising incidence. Exercise training is commonly recommended as a standalone or complementary approach for managing various chronic pain conditions like low back pain, osteoarthritis, rheumatoid arthritis, fibromyalgia syndrome, and neuropathic pain. Regardless of the specific condition or underlying cause (e.g., autoimmune disease, chronic inflammation), exercise training consistently leads to moderate to large reductions in pain. Moreover, exercise yields numerous benefits beyond pain alleviation, including small-to-moderate improvements in disability, quality of life, and physical function. Despite its efficacy, there is a lack of comprehensive research delineating the optimal intensity, duration, and type of exercise for maximal benefits; however, evidence suggests that sustained engagement in regular exercise or physical activity is necessary to achieve and maintain reductions in both clinical pain intensity ratings and the level that pain interferes with activities of daily living. Additionally, the precise mechanisms through which exercise mitigates pain remain poorly understood and likely vary based on the pathophysiological mechanisms underlying each condition.

Keywords Physical Activity · Inflammation · Guidelines

1 Introduction

1.1 Defining Chronic Pain

The International Association for the Study of Pain (IASP) defines pain as an unpleasant sensory and emotional experience associated with, or resembling that associated with, actual or potential tissue damage (Raja et al. 2020). Like any perception, pain is an inherently subjective experience that is influenced by multiple biological, psychological, social, and contextual factors. When pain persists or recurs for longer than 3 months it is considered chronic. The World Health Organization recently collaborated with the IASP to develop a classification system for chronic pain diagnoses in the 11th edition of International Classification of Diseases (ICD-11) (International Classification of Diseases, Eleventh Revision 2019; Treede et al. 2019). The ICD-11 divides chronic pain conditions into the following seven subgroups: chronic primary pain (Nicholas et al. 2019), chronic cancer-related pain (Bennett et al. 2019), chronic postsurgical or post-traumatic pain (Schug et al. 2019), chronic secondary musculoskeletal pain (Perrot et al. 2019), chronic secondary visceral pain (Aziz et al. 2019), chronic neuropathic pain (Scholz et al. 2019), and chronic secondary headache or orofacial pain (Benoliel et al. 2019).

Chronic primary pain conditions are those whose symptoms are not better accounted for by another condition and can be considered an independent disease. The remaining subgroups are considered chronic secondary pain conditions in which the pain experienced is initially regarded as a symptom of an underlying condition such as cancer or osteoarthritis (OA). Contributing factors to pain in these conditions may include injury, persistent inflammation, or diseases of the nervous system. It should be noted that it is not always possible to delineate between primary and secondary pain from the literature, as many musculoskeletal pain conditions get grouped together by location (e.g., low back pain, LBP). The full details and nuances of this classification system are beyond the scope of this chapter, but more information can be found in the above citations and on the IASP website (https://www.iasp-pain.org/advocacy/icd-11-pain-classification/). We will use the updated ICD-11 terminology throughout the chapter when feasible.

1.2 *Prevalence and Burden of Chronic Pain*

Chronic musculoskeletal pain (CMP) conditions are estimated to be the most common cause of morbidity and disability worldwide (Rice et al. 2016; Vos et al. 2015). Point prevalence data indicate that chronic pain affects more than 100 million Americans and is more common among females (34.3%) than males (26.7%; (Johannes et al. 2010). In Europe, pain lasting for greater than 6 months affects 19% of people 18 years of age and older (Breivik et al. 2006). Low- and middle-income countries are disproportionately affected with musculoskeletal pain rates of 26%, 39%, and 86% for the general population, the elderly, and workers, respectively (Jackson et al. 2015). For most countries, LBP alone is estimated to be the leading cause of years lived with disability (YLDs; e.g., 146 million YLDs estimated in 2013) and the prevalence of several CMP conditions is on the rise (Rice et al. 2016; Vos et al. 2015). Overall, chronic LBP, chronic neck pain, migraine, and a category designated as "other musculoskeletal disorders" are estimated as being within the top ten causes of YLDs in every developed and developing country (Rice et al. 2016; Vos et al. 2015).

The costs of chronic pain are staggering and current treatment standards are suboptimal (Institute of Medicine (US) Committee on Advancing Pain Research, Care, and Education 2011). In the United States of America, the economic impact has been estimated to be between \$560 and \$650 billion annually in terms of healthcare costs, prescription drug use, and loss of work productivity (Gaskin and Richard 2012). By comparison, the costs of cardiovascular diseases, cancers, and respiratory diseases were \$309 billion, \$243 billion, and \$112 billion, respectively (Gaskin and Richard 2012). Chronic pain is estimated to cost between 1.5% and 3.0% of the European Gross Domestic Product (Phillips 2006). The rising costs, low quality of life (QoL), and rates of disability are an enduring public health crisis and make clear that current approaches for treating chronic pain are lacking along several fronts.

1.3 Physical Activity and Exercise for the Prevention and Treatment of Chronic Pain

In 2011, the National Academies of Sciences, Engineering, and Medicine published a Consensus Study Report, titled *Relieving Pain in America: A Blueprint for Transforming Prevention, Care, Education, and Research* (Institute of Medicine (US) Committee on Advancing Pain Research, Care, and Education 2011). Recognizing an urgent need and that pain is a public health problem, this report made numerous recommendations focused on "transforming the way pain is understood, assessed, treated, and prevented." Recommendations focused on the need for better pain measurement and reporting; population level strategies for prevention, treatment, management, and research; development of person-centered treatments and initiatives that reduce barriers to access; expansion and redesign of educational programs; and increased federal investment in pain research. The report includes the importance of physical activity for chronic pain prevention and the value of exercise training and physical therapy as part of a multidisciplinary approach to chronic pain management.

The reasons for the rising prevalence of chronic pain are not well understood but a rapidly aging population, rising rates of obesity, and physical inactivity are suggested as primary reasons (Rice et al. 2016). Physical inactivity is a risk factor for developing CMP conditions. A retrospective longitudinal cohort study of young adults (19–24 years) in Sweden reported that stress (Odds Ratio (OR) = 1.6; 95% Confidence Interval (CI): 1.0, 2.4), non-restorative sleep (OR = 1.5; 95% CI: 1.0, 2.3), and physical inactivity (OR = 1.8; 95% CI: 1.1, 3.0) increased risk for developing chronic pain 1 year, but not 4 years, after baseline (Lindell and Grimby-Ekman 2022). Among Norwegian adults, exercising for more than 3 days per week at a medium intensity reduced the risk of chronic widespread musculoskeletal complaints such as pain or stiffness (OR = 0.72; 95% CI: 0.59, 0.88) and there was a 53% lower prevalence of chronic widespread musculoskeletal complaints for those that "maintained a medium activity level" over the 11-year study period (Holth et al. 2008). The relationship between chronic pain and physical activity is bidirectional. People with widespread pain were found to be more likely (relative risk ratio = 4.5; 95% CI: 3.2, 6.2) to report that their physical activity levels were "less" or "much less" than their age-group peers (McBeth et al. 2010). Device-measured physical activity and cardiopulmonary fitness have also been shown to be lower among those with chronic pain (Duque et al. 2009; McLoughlin et al. 2011; Pianosi et al. 2017; Smeets et al. 2006; Suárez et al. 2010; Zambolin et al. 2022) and pain is a primary barrier to a physically active lifestyle (Vos et al. 2015; Boutevillain et al. 2017; Clark 1999). Poor physical function combined with low levels of physical activity and aerobic fitness in chronic pain patients suggest that exercise interventions may aid in reducing the overall burden of chronic pain.

This chapter provides an overview of the strength of the evidence regarding exercise treatment for chronic LBP, OA, rheumatoid arthritis (RA), chronic cervical pain (CCP), chronic widespread pain (CWP), fibromyalgia syndrome (FMS), and chronic neuropathic pain. Current recommendations regarding exercise prescription are reviewed, and although limited, some of the proposed mechanisms for exercise's treatment effects are presented. The chapter concludes with a discussion of current knowledge gaps and needs for future research. This chapter is not meant to be a comprehensive review of exercise treatment for all chronic pain conditions, but rather a representative overview of some of the most common chronic pain conditions and thus should provide the reader with a general understanding for the current state of the literature.

2 Chronic Low Back Pain

Presenting in the area between the lower edge of the ribs and the buttocks, LBP can occur with or without accompanying sciatic nerve pain (i.e., pain radiating from the lower back and down the legs). Chronic LBP can be classified as either a primary or secondary musculoskeletal pain condition (often labeled non-specific or specific, respectively). These conditions are defined by pain felt in the musculoskeletal tissues (e.g., bones, joints, muscles, tendons) and are grouped by their location. Specific LBP is considered a secondary chronic musculoskeletal pain associated with spondylosis or after musculoskeletal injury (Perrot et al. 2019). Chronic secondary LBP can result from issues such as spondylosis, musculoskeletal injury, tumors, or neurologic conditions (e.g., Parkinson's disease, multiple sclerosis) and comprises 5–15% of all chronic LBP cases (Hoy et al. 2010a). For secondary LBP, treatment options are generally focused on the assumed root cause of the pain. LBP of any kind, whether acute, subacute, or chronic, has a lifetime prevalence of ~40% and is a leading cause of YLDs (Maher et al. 2017; Manchikanti et al. 2014). Most cases of LBP resolve in less than 6 weeks, with only 5–10% of patients experiencing persistent pain; however, it is also estimated that 24–33% of those with acute LBP will have a recurrence within 12 months (Manchikanti et al. 2014). The incidence, risk of recurrent episodes, and recovery rates for LBP are poorly understood due to a lack of high-quality evidence (Costa et al. 2009; Da Silva et al. 2017). Treatment in both recurrent and chronic LBP is focused on reducing pain symptoms and preventing disability (Verkerk et al. 2012). Exercise is generally considered a frontline treatment for chronic primary LBP and is recommended by most major medical organizations (Oliveira et al. 2018).

2.1 Exercise in Chronic Low Back Pain

Exercise is a generally safe and efficacious intervention to reduce pain symptoms, improve physical function, and avoid disability in chronic primary LBP. Multiple modes of exercise (e.g., aerobic, resistance, yoga, Pilates, stabilization/coordination, motor control therapy) have demonstrated significantly greater pain relief than usual care, with resistance and stabilization exercises often exhibiting the greatest benefits (Hayden et al. 2005; Owen et al. 2020). A meta-analysis of 39 randomized controlled trials (RCTs) reported a small general effect (Standard Mean Difference (SMD) = −0.32, 95% CI: −0.44, −0.19) of exercise compared to usual care. Exploratory subgroup analyses found that both stabilization (SMD = −0.47, 95% CI: −0.77, −0.18) and resistance (SMD = −0.50, 95% CI: −0.77, −0.24) exercise had larger effects, while aerobic (SMD = 0.04, 95% CI:−0.31, 0.39) and "combined" (SMD = −0.16, 95% CI:−0.29, −0.04) exercise modalities had smaller effects (Searle et al. 2015). A more recent network meta-analysis reported that Pilates, McKenzie therapy (i.e., a physical therapy technique involving primarily spinal extension and flexion exercises), and strengthening exercises were superior at reducing pain intensity and functional limitations compared to minimal therapy with clinically meaningful changes ranging from 15 to 19 points for pain (0–100 VAS) and 10 to 12 points for functional limitations (0–100 Disability Index; (Hayden et al. 2021). Limited data suggest that pain reductions may last for up to 3 years following exercise therapy, and continued adherence to the exercise significantly reduced the likelihood of reporting disability due to pain (Cecchi et al. 2014; Rasmussen-Barr et al. 2009). Multiple types of exercise have also been shown to reduce the risk of work-related disability (Oesch et al. 2010). Although neither short-term (4 weeks: OR = 0.80, 95% CI 0.51, 1.25) nor intermediate-term (6 months: OR = 0.78, 95% CI 0.45, 1.34) effects were found, odds of disability were reduced by 34% for longer-term follow-ups (1 year: OR = 0.66, 95% CI: 0.48, 0.92).

The optimal dose (e.g., duration, frequency, intensity) of exercise needed to reduce various pain outcomes for chronic LBP patients has not been identified for most types of exercise. A recent review of stabilization exercise therapy for LBP recommended that patients should exercise for at least 20–30 min per session and only 3–5 times per week, with the benefits for pain diminishing for larger doses (Mueller and Niederer 2020). Importantly, exercise, particularly when supervised, is safe for individuals with chronic primary LBP. Reports of adverse events are rare, with a majority related to a transient increase in pain or stiffness (Hayden et al. 2005; Saragiotto et al. 2016).

2.2 *Guidelines for Exercise in Low Back Pain*

Several medical organizations endorse regular exercise as a component of standard care for acute and chronic LBP. The European Spine Journal releases periodic reviews of clinical treatment guidelines. A recent review highlighted that all guidelines, across 14 different countries, recommend the use of exercise in treatment of "non-specific" LBP–although the mode, intensity, and language used for prescription were not consistent across organizations (Oliveira et al. 2018). The American College of Physicians strongly recommends initial and continued non-pharmacological treatment of chronic LBP with exercise. They further advise that pharmacologic therapy should only be considered following an inadequate response to non-pharmacologic options (Qaseem et al. 2017). The Academy of Orthopedic Physical Therapy recommends the use of multimodal exercise interventions as therapy for LBP. They have specifically endorsed trunk muscle strengthening exercises as well as endurance, aerobic, aquatic, and general exercise for both acute and chronic LBP (George et al. 2021). The Departments of Veterans Affairs and Defense jointly produced a clinical practice guideline for the treatment of LBP and endorsed "structured, clinician-directed exercise" to manage pain and disability but offered few specifics for an optimal program (VA/DoD Clinical Practice Guideline 2022). The American College of Sports Medicine (ACSM) recommends meeting the public health guidelines for physical activity with a focus on resistance exercise to help maintain good postural strength (American College of Sports Medicine 2019). Multiple modes of exercise are beneficial for reducing pain, improving physical function, and avoiding disability; treatment guidelines reflect this evidence through their recommendations. Considering there is insufficient evidence to support the optimal dose of exercise, clinical guidelines have stopped short of recommending a specific duration, intensity, frequency, and length of exercise for the treatment of chronic LBP.

2.3 *Potential Mechanism of Action in Low Back Pain*

Numerous mechanisms of how exercise can improve chronic primary LBP have been proposed, but few have been directly tested. Thus, the mechanisms of why exercise improves pain and physical function are largely unknown. Moreover, the root cause of pain in non-specific LBP is not well characterized, and this makes identifying a potential therapeutic target more challenging (Müller-Schwefe et al. 2017). The mechanisms most often discussed in chronic primary LBP research involve neuromuscular adaptations that occur during exercise training such as increased activation and strengthening of lower back musculature, increasing flexibility, and improved lumbar kinematics and proprioception (Wun et al. 2021). However, exercise interventions focusing on neuromuscular adaptions resulting from motor control exercises and spinal manipulation have generally failed to see any meaningful associations between changes in strength, flexibility, or control of lumbar movement and changes in pain (Steiger et al. 2012; Taulaniemi et al. 2019). Psychosocial (e.g., improvements in fear avoidance, pain catastrophizing), neurophysiological (e.g., neuroplastic changes in brain activation), and cardiometabolic factors have all similarly been theorized to partially explain the benefits of exercise, without much supportive data (Wun et al. 2021; Roussel et al. 2013). There is a need for trials that directly test hypothesized mechanisms of chronic LBP–i.e., whether the resulting changes in pain are mediated by a theoretical biological or biomechanical adaptation.

One randomized, double-blind, placebo-controlled cross-over study directly tested whether endogenous opioid function may partially explain pain improvements resulting from exercise training ($n = 38$, 55.3% female) compared to usual activity ($n = 44$, 65.9% female) in chronic LBP (Bruehl et al. 2020). The exercise intervention consisted of 30-min sessions of aerobic exercise training three times per week for 6 weeks at an intensity ranging from 70–85% of heart rate reserve. Endogenous opioid activity was operationalized as the difference in heat pain ratings in response to thermally evoked pain tolerance testing following the administration of naloxone or placebo infusions, with greater differences (i.e., higher pain ratings under naloxone compared to placebo) indicative of greater endogenous opioid activity. Aerobic exercise training resulted in meaningful changes in clinical pain ratings for 40.5% of the exercise training group compared to 27.3% of controls and pain interference ratings (PROMIS Interference) were moderately (Cohen's $d = 0.53$) improved. Endogenous opioid activity was found to increase from pre- to post-intervention, but only for the female participants in the exercise condition. Greater increases in endogenous opioid activity were associated with greater improvements in clinical pain ratings and higher exercise doses (i.e., greater average minutes in the target heart rate zone). These results suggest that pain improvements in chronic LBP from aerobic exercise training may be partially mediated by improvements in endogenous opioid activity for a subset of patients.

3 Osteoarthritis

OA is a common form of arthritis and one of the most common causes of joint disorders. When OA is accompanied by significant pain symptoms, it can include an additional diagnosis of chronic secondary musculoskeletal pain (Nicholas et al. 2019; Perrot et al. 2019). This condition is characterized by the degeneration of cartilage in the joint space and inflammation of the synovial fluid surrounding the joint. The prevalence of OA varies greatly based on the body location, geographic location, and method of diagnosis (Pereira et al. 2011). Point prevalence estimates of knee OA range from 16–28%, hip OA estimates range from 10–25%, and rates of hand OA range from 7–26% (Vina and Kwoh 2018; Johnson and Hunter 2014). The prevalence of OA is related to modifiable (e.g., muscular strength, a history of joint injury) and non-modifiable risk factors, with age being one of the strongest predictors of OA (Shane Anderson and Loeser 2010). The symptoms associated with OA include pain, joint stiffness, joint instability, reduced function, and when pain persists, psychological distress (Hunter et al. 2008; Wieland et al. 2005). Moreover, OA can lead to difficulties with mobility, completing activities of daily living, and decreases in overall QoL (Clynes et al. 2019). Exercise is commonly recommended with the goal of reducing pain, joint stiffness, and body weight, while increasing flexibility and muscular strength around the joint (Holden et al. 2021).

3.1 Exercise in Osteoarthritis

As detailed below, the beneficial effects of exercise on OA pain have been well documented by multiple meta-analyses. Pain (assessed using various measures of pain intensity and impact) is reduced in OA immediately following exercise interventions and there is moderate evidence for pain reductions in the short-term (2–-6 months) post-intervention (Fransen et al. 2015a; Raposo et al. 2021; Wellsandt and Golightly 2018). OA pain is generally reduced by a moderate to large degree immediately following an exercise intervention (SMD range: 0.49–1.20), resulting in, on average, a decrease of 12 points on a 100-point scale (Fransen et al. 2015a). Notably, the magnitude of the effect of exercise on pain is greater than commonly prescribed pharmacologic treatments (e.g., acetaminophen, naproxen, celecoxib, etc., compared to an oral placebo; Hedges' *g* range: 0.18–0.33) and is similar to the most effective pharmacologic interventions (interarticular hyaluronic acid, compared to an oral placebo (Hedges' $g = 0.63$; (Bannuru et al. 2015; Beaudart et al. 2020). Similar to pain, physical function is improved with exercise by a moderate degree (SMD range: 0.44–0.52), and these effects are consistently larger than reported for pharmacological treatments (SMD range: 0.18–0.40; (Fransen et al. 2015a; Raposo et al. 2021; Wellsandt and Golightly 2018). Exercise has a small, but consistent, impact on QoL in patients with OA (SMD range: 0.15–0.44; (Fransen et al. 2015a; Raposo et al. 2021; Wellsandt and Golightly 2018).

The benefits of exercise are not isolated to a certain mode; several types of exercise (e.g., aerobic, resistance, Pilates, yoga, tai chi) have been shown to be beneficial in OA. Meta-analytic data demonstrate only small differences in the magnitude of improvement of pain outcomes among aerobic, resistance, and performance exercise (SMD: 0.67, 95% CI: 0.39, 0.94; 0.62, 95% CI: 0.45, 0.79; 0.48, 95% CI: 0.11, 0.85, respectively; (Juhl et al. 2014). This same review also reported a lack of evidence to support that engaging in multiple types of exercise yields greater benefits. Trials incorporating a combination of exercise types were not as effective (SMD = 0.16, 95% CI: −0.04, 0.37) compared to more concentrated prescriptions focusing on a singular type of exercise (SMD = 0.61, 95% CI: 0.48, 0.75). One trial comparing non-weight-bearing quadriceps strengthening exercises to functional exercises reported that pain and physical function were significantly improved following 12 weeks of exercise in both groups, with no significant difference in the magnitude of improvement (Bennell et al. 2020). Further, the benefits of exercise can be realized in a variety of settings (e.g., home-based, class-based, or individually), with greater pain reductions occurring with individually-based interventions (Fransen et al. 2015b). The benefits of exercise on pain persisting beyond 6 months post-intervention are less consistent; however, it is unclear from the cited literature whether participants maintained the same level of activity during the follow-up. Consistent participation in an exercise program is likely necessary to maintain the therapeutic benefits of exercise on OA pain (Marks 2012).

3.2 Guidelines for Exercise in Osteoarthritis

According to current clinical guidelines, exercise is the most consistently recommended frontline treatment for OA (Conley et al. 2023). The most recent set of American College of Rheumatology (ACR) guidelines strongly recommend the use of exercise for the treatment of knee, hip, and hand OA, but does not set forth specific recommendations (Kolasinski et al. 2020). Non-specific recommendations are primarily the result of two factors: the lack of sufficient evidence for a universally efficacious prescription and the beneficial effects of multiple exercise types, frequencies, and doses. The European Alliance of Associations for Rheumatology (EULAR) recommends physical activity and weight loss, if overweight, as an integral part of standard care for OA (Fernandes et al. 2013; Kloppenburg et al. 2019; Rausch Osthoff et al. 2018). The National Institute for Health and Care Excellence (NICE) states that exercise should be offered based on needs (e.g., local muscle strengthening or general aerobic fitness) and combined with an educational program or behavioral change approaches (NICE 2022). The NICE guidelines also state, "doing regular and consistent exercise, even though this may initially cause pain or discomfort, will be beneficial for their joints and that long-term adherence to an exercise plan increases its benefits by reducing pain and increasing functioning and quality of life" (NICE 2022). The Ottawa Panel clinical practice guidelines recommend the use of progressive aerobic and strength training programs to best treat OA (Brosseau et al. 2017a,b). The ACSM similarly recommends a general exercise program, matching the general population recommendations, to improve pain and function in OA. Although the specific frequency, intensity, time, and type parameters for the most effective exercise intervention have not been defined, exercise in some form is consistently recommended in the treatment of OA.

3.3 Potential Mechanism of Action in Osteoarthritis

Exercise elicits a systemic anti-inflammatory response, that may partially explain the hypoalgesic effects of exercise in conditions characterized by chronic inflammation (i.e., OA, RA). Acute exercise stimulates the release of Interleukin (IL)-6 from contracting skeletal muscle (Pedersen et al. 2001). IL-6 is often considered a pro-inflammatory cytokine; however, when released via muscle contraction, it can act as a myokine, producing local and systemic anti-inflammatory effects through the downregulation of other pro-inflammatory cytokines (e.g., tumor necrosis factor α, C-reactive protein) and upregulation of anti-inflammatory cytokines (e.g., IL-10; (Benatti and Pedersen 2015; Metsios and Kitas 2018). Exercise also stimulates the release of cortisol and catecholamines, reduces the amount of pro-inflammatory monocytes in the blood, and downregulates the expression of toll-like receptors, all of which have anti-inflammatory effects (Gleeson et al. 2011). These acute benefits may accumulate over time, resulting in the chronic reduction of inflammation levels and consequently, reduced pain.

The mechanisms by which exercise specifically benefits OA are not well characterized, but there are several theories with modest levels of support. Cartilage, bone, and synovium release pro-inflammatory signaling molecules in OA, generating and maintaining a pro-inflammatory state in the affected joints (Berenbaum 2013) and the evidence suggests these inflammatory signals contribute to the pain experienced in OA (O'Neill and Felson 2018). Conversely, the established anti-inflammatory effects of exercise, as described above, may act to reduce pain post-exertion (Gleeson et al. 2011). Exercise may mitigate OA pain by suppressing inflammation through the promotion of an anti-inflammatory state and a decrease in the levels of circulating pro-inflammatory signaling molecules (Griffin et al. 2012; Lu et al. 2021). However, support for this hypothesis derives primarily from research employing animal models of OA (e.g., treadmill exercise of mice following a high fat diet or pharmacological injections into the intra-articular space). Studies designed to test and confirm this mechanism in humans with OA have been mixed, with only a modest number of studies (30%) reporting a decrease in circulating biomarkers and the majority of studies (63%) reporting no change following an exercise intervention (Bricca et al. 2019). Muscular strength and range-of-motion adaptations from strength training have also been investigated as potential mediators for pain improvements following exercise training (Zeng et al. 2021). Multiple RCTs have found significant strength benefits, and when restricting analyses to studies that have found benefits on pain, over 70% of the trials reported significant strength increases. Moreover, one trial reported that strength gains were significantly associated with reductions in pain and disability (Bartholdy et al. 2017). Brain activity in response to experimental pain stimuli and resting connectivity may also be altered in response to an exercise program in OA, but more research is needed to understand whether and how these brain adaptations translate to changes in pain regulation and perception (Liu et al. 2019; Öztürk et al. 2021).

4 Rheumatoid Arthritis

RA is a chronic autoimmune inflammatory disease that can affect many body systems, but most commonly impacts musculoskeletal joints leading to pain, joint stiffness, impaired range of motion, fatigue, and difficulty with activities of daily living (Aletaha et al. 2010; Arnett et al. 1988). Chronic pain symptoms associated with RA are diagnosed as secondary musculoskeletal pain (Nicholas et al. 2019; Perrot et al. 2019). RA can have profound effects on the synovial joints, causing damage through the erosion of bony surfaces and degradation of cartilage. This dual impact contributes to the progression of joint destruction and may result in severe disability. Moreover, the consequences of RA extend beyond joint damage, as the condition can impact other organ systems and is associated with an increased risk of premature mortality (Van Der Woude and Van Der Helm-van Mil 2018). It is estimated that RA affects approximately 0.24% of the global population, with the prevalence rate for women being roughly twice that for men. However, rates vary greatly by geographical location (Van Der Woude and Van Der Helm-van Mil 2018; Cross et al. 2014). Individuals with RA are less physically active than healthy individuals, as quantified by both self-report and device-based measures (Verhoeven et al. 2016). Pain and fatigue are consistently reported as the greatest barriers to physical activity participation in this patient population followed by reduced mobility/functional ability, stiffness, and lack of knowledge pertaining to the design and potential benefits of exercise programs for RA (Verhoeven et al. 2016; Veldhuijzen Van Zanten et al. 2015). Further, those with RA who report high levels of sedentary behavior are more likely to report worse symptoms and disability than those who are physically active (Henchoz et al. 2012).

4.1 Exercise in Rheumatoid Arthritis

Exercise is a crucial component in the non-pharmacological management of RA. It is safe, complements the primary pharmacological treatments like disease-modifying antirheumatic drugs, and offers multiple additional benefits, including, but not limited to, pain reduction. Both regular aerobic and resistance exercise are safe for RA patients and reduce pain by a moderate degree (SMD range: 0.31–0.61) and reduce the amount of swollen and tender joints (i.e., "disease activity") by a small-to-moderate degree (SMD range: 0.14–0.69; (Baillet et al. 2010; Wen and Chai 2021; Ye et al. 2022). Exercise also reduces fatigue by a small degree (SMD ~0.35; (Cramp et al. 2013; Kelley et al. 2018). Function, as assessed through the Health Assessment Questionnaire, consistently shows a modest improvement (mean differences (MD) range: 0.22–0.30) following both aerobic and resistance exercise training (Ye et al. 2022; Baillet et al. 2012). Few direct comparisons of the effectiveness of resistance and aerobic exercise have been conducted in RA patients. However, the magnitude of changes reported in the cited literature for each exercise modality are comparable (Juhl et al. 2014).

The characteristics of exercise training that may impact the beneficial effects of exercise for RA pain have not yet been determined. Two meta-analyses, one focusing on aerobic exercise and the other on resistance exercise, reported that different exercise parameters, such as session duration, training program duration, frequency, and intensity, did not consistently influence the exercise responses (Baillet et al. 2010, 2012). Few studies have directly tested how these parameters affect outcomes. One early trial compared 12 weeks of either high (i.e., weight-bearing activities including knee-bends, step-ups, and fast walking) or low intensity (i.e., non-weight-bearing range of motion and isometric) exercise and did not report significant differences for pain, function (i.e., self-reported and timed walk and stair tests), disease activity, or inflammation (Van Den Ende et al. 1996). It is unclear if similar benefits occur when exercise treatment is delivered at home or under expert supervision (Baillet et al. 2010, 2012). Emerging evidence has demonstrated that aquatic exercise may provide similar benefits for disease activity while reducing the likelihood of exacerbating joint pain and swelling (Siqueira et al. 2017). It remains unclear what mode, duration, frequency, or intensity of exercise is required to optimize clinical improvements in pain, physical function, and QoL in those with RA.

4.2 Guidelines for Exercise in Rheumatoid Arthritis

The existing treatment guidelines for RA from multiple organizations list pharmacological interventions as a first-line treatment and support the use of general exercise as an adjunct therapy. The 2022 ACR guidelines for the treatment of RA strongly recommend consistent engagement in exercise, with the specific prescription to be individually tailored based on capabilities, disease trajectory, and comorbid conditions. The ACR guidelines also conditionally recommend engagement in aerobic, aquatic, and mind-body exercise. The conditional recommendations are attributed to the low certainty of evidence stemming from the small sample sizes of included studies, the presence of serious biases (e.g., a lack of participant blinding, uncertainty about assessor blinding), a recognition of varying patient preferences for different types of exercise, and various obstacles (e.g., cost, access to facilities) that often prevent RA patients from regularly engaging in those types of exercise (England et al. 2023). The 2018 NICE guidelines for the treatment of RA encourage regular exercise and recommend the use of physiotherapy (i.e., strengthening, general fitness, stretching, and proprioceptive exercises) to improve general fitness. In addition, exercises intended to enhance joint flexibility, increase muscle strength, and manage functional impairments (e.g., pain or dysfunction in the hands) are also recommended. The NICE guidelines, however, do not go into any greater detail on the optimal dose, intensity, or duration of exercise to achieve these outcomes (Allen et al. 2018). The Ottawa Panel stated, "good evidence exists... that therapeutic exercise including functional strengthening and low- or high-intensity exercises, should be included as an intervention for patients with RA. Therapeutic exercises reduce pain while improving periarticular muscle force, aerobic capacity, and joint mobility" (Ottawa Panel Evidence 2004). The 2018 EULAR guidelines do not provide specific prescriptions for exercise but do recommend patients with inflammatory arthritis, including RA, perform enough physical activity to meet public health guidelines (Rausch Osthoff et al. 2018). The ACSM also recommends that RA patients attempt to meet the physical activity guidelines for healthy adults, although they advise avoiding vigorous exercise during symptom flare-ups; avoiding vigorous, highly repetitive activities; and to expect some discomfort, but not pain, after workouts.

4.3 Potential Mechanism of Action in Rheumatoid Arthritis

Pharmacologic therapies primarily alleviate RA disease activity by targeting inflammation through various proposed mechanistic pathways. Elevated levels of peripheral inflammation are believed to amplify nociceptive signaling, serving as a key factor driving pain symptoms in RA (Guo et al. 2018). The most widely reported mechanism of action for RA drugs is through the increased release of adenosine into the extracellular space leading to lower levels of inflammation (Friedman and Cronstein 2019). Due to the widespread nature of RA disease activity (e.g., circulating inflammatory biochemicals), other potential mechanisms for disease modification and pain relief include decreased leukocyte recruitment, decreased bone degeneration, and reduced spinal and supraspinal nociceptive signaling (Borea et al. 2018). Further, there is evidence that pharmacologic treatment can act directly on immune cells to reduce pro-inflammatory cytokine production. The benefits of exercise training for RA may be partially related to its anti-inflammatory properties, detailed above (Walsh and McWilliams 2012).

The cytokine response following acute resistance exercise is similar between RA and otherwise healthy controls (Pereira Nunes Pinto et al. 2017). Both aerobic and resistance exercise training in RA have been shown to decrease inflammation and circulating pro-inflammatory cytokines by a moderate to large effect (Wen and Chai 2021; Azeez et al. 2020). Regular exercise can reduce visceral fat, which in turn may slow the release of pro-inflammatory cytokines, as adipose tissue is a potent IL-6 secretor (Metsios et al. 2015, 2020). Although reductions in body fat and increases in lean mass are consistently observed in RA following resistance exercise trials, the impact of aerobic exercise on body composition in RA is still inadequately understood (Yekini and Grace 2023). Exercise has many other beneficial effects for patients with RA, including reducing risk factors for cardiovascular disease, decreasing risk of falls, and improving mental health (Cooney et al. 2011; Rongen-van Dartel et al. 2015; Sul et al. 2020). Trials that establish the optimal dose, intensity, and mode of exercise for clinical benefits, while testing associations with inflammatory mediators of disease and pain symptoms, are needed to advance exercise prescription for this patient population.

5 Chronic Cervical Pain

Cervical (i.e., neck) pain occurs in the upper back region, defined superiorly by the superior nuchal line and laterally/inferiorly by the spine of the scapula (Guzman et al. 2009). CCP can be considered either a primary or secondary musculoskeletal pain condition, commonly referred to as non-specific or specific, respectively. Common causes of secondary cervical pain may include whiplash or other soft tissue injury, tumors, or RA (Nicholas et al. 2019). It is estimated that nearly half of all adults will encounter neck pain at some juncture in their lives, with women experiencing it at higher rates than men (Fejer et al. 2006; Hoy et al. 2010b). The hypothesis that weak or easily fatigued neck musculature serves as a potential underlying cause of primary CCP has been proposed, leading to the suggestion of exercise as a potential treatment.

5.1 *Chronic Cervical Pain and Exercise*

Exercise has been shown to be effective at reducing pain, decreasing disability, and increasing QoL in CCP (Bertozzi et al. 2013; O'Riordan et al. 2014). Most studies have assessed the effectiveness of various forms of neck strengthening and/or stretching, primarily considering the localized nature of the painful region and the underlying muscular weakness. Resistance exercise reduces pain and disability by a moderate to large effect size (Hedges g range: 0.53–1.53, and 0.39–0.75, respectively) in the short to intermediate term (0–6 months post-intervention; (Bertozzi et al. 2013; De Zoete et al. 2021). The magnitude of pain and disability reductions is similar to that achieved with cognitive behavioral therapy (Monticone et al. 2015). When exercise and cognitive behavioral therapy are directly compared, exercise produces a more rapid improvement in pain, with benefits occurring as early as 4 weeks into an exercise program (Vonk et al. 2009). Few studies have focused on aerobic exercise in isolation but have found generally positive effects on pain and disability (Paraskevopoulos et al. 2023). Adding aerobic exercise to a neck-strengthening program resulted in greater improvements in pain, disability, and function than the neck-strengthening program alone (Daher et al. 2020). One meta-analysis reviewed the existing exercise trials to determine the optimal frequency, intensity, time, and type of an exercise program for CCP. This review reported that an exercise program should consist of three multimodal training sessions per week, each 30–45 min in duration, at an individually tailored intensity for the best results (O'Riordan et al. 2014).

5.2 *Guidelines for Exercise in Chronic Cervical Pain*

The American Physical Therapy Association recommends, on a moderate evidence base, that clinicians should consider coordination, strengthening, and endurance exercises to reduce neck pain (Blanpied et al. 2017). They also recommend the use of stretching, nerve mobilization, and thoracic mobilization and manipulation to reduce pain and disability. Clinical practice guidelines published in the Journal of Manipulative and Physiological Therapeutics and the European Spine Journal recommended the use of supervised strengthening exercises alone or as part of a multimodal treatment plan that also includes stretching and joint manipulation for primary CCP (Bussières et al. 2016; Côté et al. 2016). However, in a review of 17 European clinical treatment guidelines, only six included specific guidelines for the treatment of CCP, and among these, only three recommended exercise for CCP (Corp et al. 2021).

5.3 *Potential Mechanism of Action in Chronic Cervical Pain*

Two main mechanisms have been proposed to explain why exercise can reduce cervical pain. One mechanism is that strength training of the neck muscles increases both muscular strength and endurance (Andersen et al. 2011; Ylinen et al. 2006) thereby improving cervical postural control (Chung and Jeong 2018). Greater strength and endurance of the neck musculature could allow for better head and neck control throughout the day and less cervical lordosis (i.e., spinal compression), thereby leading to less pain. Notably, one trial compared stretching exercises to strengthening coupled with stretching exercises and reported no differences in pain upon completion of the 12-month training program (Häkkinen et al. 2008). However, the strength and stretching group did not demonstrate consistent strength improvements relative to the stretching group and this may explain the lack of differences in pain outcomes between the two groups. Alternatively, exercise can influence the central processing and modulation of pain, providing an additional explanation for the pain-relieving benefits (Heredia-Rizo et al. 2019; Rittig-Rasmussen et al. 2014)

6 Chronic Widespread Pain and Fibromyalgia Syndrome

CWP is a primary pain condition with diffuse pain in at least four of five body regions (i.e., axial spine, above and below the waist, both sides of the body) and is associated with significant emotional distress and/or functional disability (International Classification of Diseases, Eleventh Revision 2019; Nicholas et al. 2019). FMS is a form of CWP associated with concomitant fatigue, sleep disturbance, cognitive dysfunction, stiffness, and tenderness and is diagnosed based on clinical case definition criteria (Nicholas et al. 2019; Wolfe et al. 1990, 2016). Clinical presentations of FMS are heterogeneous and multimorbidity is common. Estimates suggest over half of FMS patients have comorbid depression/major depressive disorder, over one-third have bipolar disorder, panic disorders, anxiety disorders, or post-traumatic stress disorder, and nearly half may have comorbid myalgic encephalomyelitis/chronic fatigue syndrome (Kleykamp et al. 2021; Ramírez-Morales et al. 2022). Comorbid chronic pain conditions such as headache or orofacial pain, temporomandibular disorder, and irritable bowel syndrome are also common (Kleykamp et al. 2021). The global prevalence of CWP has been estimated as 11.8% and is more common in females with a ratio of 2:1 (Mansfield et al. 2016). The global prevalence of FMS is lower at 2.7%, but with a higher female-to-male ratio of 3:1 (Nicholas et al. 2019; Queiroz 2013).

6.1 *Exercise for Chronic Widespread Pain and Fibromyalgia Syndrome*

A series of Cochrane systematic reviews and meta-analyses have investigated randomized controlled trials (RCTs) of different types of exercise interventions for FMS including: resistance training (Busch et al. 2013), aquatic training (Bidonde et al. 2014), aerobic training (Bidonde et al. 2017), flexibility training (Kim et al. 2019), mixed exercise training (Bidonde et al. 2019), and mind-body exercise training (Theadom et al. 2015). Primary outcomes included health-related QoL, pain intensity, stiffness, fatigue, physical function, physical fitness, withdrawals, and adverse events. Preferred measures for each outcome were guided by consensus statements from the Outcome Measures in Rheumatoid Arthritis Clinical Trials initiative for clinical trials in FMS (Choy and Mease 2009). The overall certainty of the evidence was based on the Grading of Recommendations Assessment, Development, and Evaluation (GRADE) approach. The clinical importance of each treatment effect was based on the Philadelphia Panel guidelines, which consider a minimally clinically important difference as a 15% absolute improvement in pain intensity or a 15% relative improvement for all other outcomes (Albright et al. 2001). The main findings from each Cochrane review are described below.

Resistance Training Busch et al. (2013) evaluated five RCTs (241 participants) that compared resistance training versus control, aerobic training, or flexibility training. Compared to control conditions, there was low certainty evidence that resistance training for 16 to 21 weeks (2–3 sessions/week) improved QoL, physical function, pain intensity, tenderness, fatigue, and strength. Effect sizes were moderate to large (SMD range: 0.50–1.89) and many outcomes reached the threshold to be considered clinically important differences. One included study provided low certainty evidence that improvements for QoL, fatigue, and pain intensity were retained 12 weeks after the intervention, but these were small and considered clinically unimportant, suggesting that persistent training may be required to sustain benefits. Relative to aerobic exercise interventions, resistance training elicited smaller improvements in pain intensity and sleep quality. When compared to flexibility training, resistance exercise appeared to induce greater improvements in QoL, pain intensity, fatigue, and sleep quality. The certainty of these findings ranged from low to very low based primarily on the small number of studies and the risks of bias attributed to issues regarding inadequate blinding, selective reporting, and allocation concealment.

Aquatic Training Bidonde et al. (2014) evaluated 16 RCTs (881 participants) that compared aquatic training to control, land-based training, or different aquatic training programs. Aquatic training consisted of aerobic, resistance, and/or flexibility exercises performed in a pool. Interventions lasted 3 to 32 weeks (1–4 sessions/ week; 30–70 min/session). There was low to moderate certainty evidence that aquatic training, compared to control, improved QoL, physical function, pain intensity, stiffness, tenderness, sleep quality, anxiety, cognitive performance, and fitness outcomes. The improvement in stiffness was large (SMD = −1.00, 95% CI: −1.91, −0.10) and considered clinically important. Effect sizes for the remaining outcomes were small to moderate (SMD range: −0.44 to −0.63). Protocols with longer durations, higher frequencies, or greater accumulated time spent in the pool generally resulted in larger effect sizes for QoL, pain intensity, and strength outcomes. Larger improvements in QoL and strength were also shown in studies reporting lower baseline pain among their participants.

Aerobic Training Bidonde et al. (2017) evaluated 13 RCTs (839 participants) that compared aerobic training to control, different aerobic training programs, or non-exercise interventions. The average length of intervention was 15 weeks with most sessions occurring three times per week for an average of 35 min. There was moderate certainty evidence that aerobic training improved QoL and low certainty evidence that it improved pain intensity, physical function, fatigue, and stiffness. Improvements in pain intensity and physical function were considered clinically important. Interventions with training volumes lower than those recommended by the ACSM still improved FMS symptoms. Three studies provided low certainty evidence that improvements in pain intensity and physical function remained 24 to 208 weeks after intervention, but these estimates were small and imprecise, suggesting that persistent training may be required to sustain benefits. Results from comparisons to other aerobic training programs or non-exercise interventions were unclear given the low to very low certainty evidence, small number of studies, and large heterogeneity between studies.

Flexibility Training Kim et al. (2019) evaluated 12 RCTs (743 participants) that compared flexibility training to control, aerobic training, resistance training, or other interventions. The average length of intervention was 12 weeks with one to three sessions per week. There was very low certainty evidence that flexibility training was more beneficial than aerobic training for stiffness (MD = −29.6, 95% CI: −51.47, −7.73, 0–100 scale, with lower scores for less stiffness), but this was based on one study with a small sample size ($N = 15$). No other significant differences were found when comparing flexibility to resistance training or to untreated controls. The certainty of the evidence for these outcomes was very low.

Mixed Exercise Training Bidonde et al. (2019) evaluated 29 RCTs (2088 participants) that compared mixed exercise training to control, other exercise, or non-exercise interventions (e.g., biofeedback, medications). Mixed exercise interventions included at least two types of exercise (i.e., aerobic, resistance, or flexibility). The average length of intervention was 13 weeks with an average of three sessions per week ranging from 45 to 60 min in duration. There was moderate certainty evidence that mixed exercise improved QoL, physical function, and fatigue compared to control with the confidence intervals including both clinically important and unimportant improvements for each outcome. Improvements for pain intensity (moderate certainty) were small and considered clinically unimportant. There was low certainty evidence for improved stiffness with the confidence interval including both clinically important and unimportant changes. Very low certainty evidence from a small number of heterogeneous studies made it unclear whether these improvements were retained following intervention or how they compared to other exercise or non-exercise interventions.

Mind-Body Exercise Training Theadom et al. (2015) evaluated 11 RCTs involving mind-body exercise including tai chi, yoga, qi-gong, dance, and Pilates. There was very low-quality evidence of improvements for physical function, pain intensity, and mood in some of the studies. However, differences between the interventions, control groups, and outcomes used in these studies made conclusions difficult.

Collectively, these Cochrane reviews identified small-to-moderate group-average improvements in multiple outcomes for patients with FMS following exercise. Aerobic, aquatic, and mixed exercise training interventions provided the strongest evidence, with mixed exercise having moderate certainty evidence of clinically significant improvements in QoL, physical function, and fatigue. These findings are likely attributable to the larger samples in these reviews and their inclusion of additional studies. Further studies of resistance, flexibility, and mind-body exercise training will allow for better comparisons between exercise modalities. The original studies also varied in which outcomes were investigated, with more studies providing data for QoL and pain intensity than for outcomes such as fatigue, stiffness, physical function, or physical fitness.

Similar withdrawal rates between interventions and control conditions were found, suggesting exercise may be generally tolerated by FMS patients. However, the collection and reporting of adverse events was inconsistent and often inadequate in the analyzed studies. Among those that did, low rates and no serious adverse events were reported. The most frequently mentioned adverse events included increased pain, soreness, or fatigue during and/or after exercise.

Few of the original studies focused on fitness outcomes and many did not measure activity levels outside the intervention, making it difficult to quantify the exact relationship between exercise and clinical outcomes. Additionally, the percentage of females in the original studies ranged from 89–100% and the Cochrane reviewers only included studies of patients meeting criteria for FMS. This makes it difficult to generalize these findings to males and to those with CWP who do not meet criteria for FMS. A recent study by Stegner et al. (2021) examined the effects of resistance training (16 weeks; 2 sessions/week) in a predominately male sample of Gulf War Veterans that included both individuals who met CWP criteria alone and those who met additional criteria for FMS (Stegner et al. 2021). Training improved group averages for global perception of change and muscular strength, but had no effect on pain severity, fatigue, sleep quality, or total mood disturbance compared to a waitlist control condition. Device-measured physical activity data indicated that physical activity levels were maintained throughout the trial. No serious adverse events were reported and adherence to the protocol was high (87% of training sessions attended) indicating that the resistance exercise training program was generally well-tolerated.

6.2 *Guidelines for Exercise in Chronic Widespread Pain and Fibromyalgia Syndrome*

Recent clinical practice guidelines have recommended exercise along with other non-pharmacological therapies as first-line treatments for managing FMS (NICE 2021; Macfarlane et al. 2017; Kundakci et al. 2022a). Note that the following guidelines do not differentiate CWP from FMS; this is likely attributable to historical ambiguities surrounding the vocabulary and diagnostic criteria used for these conditions (Butler et al. 2016). In 2021, the United Kingdom's National Institute for Health and Care Excellence (NICE) published guidance for the management of chronic primary pain conditions (NICE 2021). The guidelines include offering patients supervised group exercise programs for managing their symptoms and promoting long-term general health. These programs should also take a patient's individual needs, preferences, and abilities into account. These recommendations are based on their review of 91 RCTs showing that exercise reduced pain (23 studies) and improved quality of life (22 studies) compared to usual care. Most interventions included supervised group-based exercise and most examined either female patients with FMS (58 studies) or patients with chronic neck pain (31 studies). They found no evidence that treatment effectiveness differed between pain conditions, therefore the NICE guidelines suggest applying these recommendations across all chronic primary pain conditions.

In 2017, the EULAR updated their recommendations for the management of FMS, which suggest initial treatment should focus on patient education and non-pharmacological therapies (Macfarlane et al. 2017). Of all the therapies reviewed (both pharmacological and non-pharmacological) only aerobic and strengthening exercise were given a "strong-for" recommendation by the committee. The guidelines suggest following patient education with individualized and graded physical exercise, which can be combined with other therapies if needed. However, some have questioned how the committee determined the strength of their recommendations due to apparent inconsistencies in how the criteria were applied, suggesting that the evidence for cognitive behavioral therapy should have also been given a strong-for recommendation (Thieme et al. 2017).

In 2022, a multidisciplinary expert panel performed a Delphi consensus to provide recommendations for core and adjunctive non-pharmacological interventions for pain, fatigue, sleep, and depression in FMS patients (Kundakci et al. 2022a). Panelists were asked to consider evidence from a recent meta-analysis (Kundakci et al. 2022b) as well as their clinical experience. Aerobic exercise was considered a core component of treatment for all four symptoms, along with patient education, sleep hygiene, and cognitive behavioral therapy. Mind-body exercise was considered a core treatment for pain, fatigue, and sleep, but an adjunct for depression. Strengthening exercise was considered a core treatment for pain and fatigue but was not recommended for sleep or depression. Flexibility was recommended as a core treatment for pain but was not recommended for the other three symptoms.

Clinical practice guidelines for FMS report the likely benefits of increasing physical activity and usually emphasize the importance of tailoring exercise to the patient's needs, preferences, and abilities. However, there is often little information provided for the clinician or the patient regarding how to implement this advice. The ACSM, as part of their Exercise is Medicine efforts, created an infographic for "Being Active with Fibromyalgia" promoting individual strategies for engaging in various physical activities (https://www.exerciseismedicine.org/wp-content/uploads/2021/04/EIM_Rx-for-Health_Fibromyalgia.pdf). Guidance is focused on the individualized nature of FMS, emphasizing starting slow, keeping impact low, and adjusting activity based on daily symptoms. Information regarding frequency, intensity, and durations for aerobic and strength training activities is included, along with information on navigating symptom flares, recovery, and personal choices. Additional guidance is provided for flexibility, mixed exercise, yoga, tai chi, and dance.

6.3 Potential Mechanism of Action in Chronic Widespread Pain and Fibromyalgia Syndrome

Although their exact etiology and pathophysiology remain unclear, CWP and FMS are often attributed to abnormal neural signaling in the central nervous system (CNS) leading to an amplification of pain perception and contributing to the broader symptoms that are associated with FMS (Clauw et al. 2011). This is supported by evidence for structural, functional, and biochemical abnormalities in the CNS, heightened sensitivity to various stimuli, and less robust pain inhibition, along with limited evidence of peripheral abnormalities that may contribute to pain symptoms (Sluka and Clauw 2016). Neuroinflammation is one potential factor that may underlie CNS dysfunction and symptoms in FMS (Bains et al. 2023). Regular exercise training can reduce neuroinflammation in neurological diseases and promote neuroplasticity by increasing multiple neurotrophic factors (Allen et al. 2015; Scheffer and Latini 2020; Spielman et al. 2016); however, no exercise trials of FMS have tested this mechanism. Animal models of widespread hyperalgesia have demonstrated that regular physical activity is protective and, in some cases, blocks the development of hyperalgesia (Lesnak and Sluka 2020; Lima et al. 2017; Sluka et al. 2012, 2013). These studies use repeated (5 days apart) injections of low pH saline (e.g., pH 4 or pH 5) to produce widespread mechanical hyperalgesia–meant to mimic CWP and FMS (Sluka et al. 2013). These studies further demonstrate that the widespread hyperalgesia is 1) prevented when mice are given access to a running wheel for 8 weeks prior to the low pH saline injections, and 2) treated when they are given access to a running wheel for 3 weeks after the development of widespread hyperalgesia (Lesnak and Sluka 2020; Lima et al. 2017).

The protective and pain-relieving effects of exercise are associated with activation of the NR1 subunit of N-methyl-d-Aspartate receptors in the brainstem (e.g., nucleus raphe magus & rostral ventromedial medulla; (Sluka et al. 2012). These studies suggest that regular physical activity can prevent chronic pain and treat it by enhancing CNS nociceptive regulatory activity, but to the best of our knowledge these mechanisms have not been rigorously tested in human chronic pain populations. One scoping review of human exercise training trials in FMS reported that 29 different potential mechanisms underlying the pain-relieving effects of exercise were identified (Neelapala et al. 2023). Both physiological and psychological mechanisms were proposed, and the most frequently discussed effects of exercise were exercise-induced hypoalgesia, autonomic modulation, improved sleep, increased muscle oxygenation, increased self-efficacy, improved mental health, and the general benefits of an aquatic environment. In general, most of the proposed mechanisms were not directly supported and the hypothetical reasoning was largely based on limited research in healthy populations.

7 Neuropathic Pain

Defined as pain directly attributed to a somatosensory system lesion or disease (Treede et al. 2008), neuropathic pain is estimated to affect somewhere between 7–10% of the general population (van Hecke et al. 2014). According to the new ICD-11, chronic neuropathic pain is a primary diagnosis with the secondary classification being determined by the nervous area (i.e., central or peripheral) affected by the identified lesion or disease (Scholz et al. 2019). Although frequently characterized by "neuropathic-like" symptoms (e.g., allodynia, hyperalgesia, ongoing burning pain, paroxysmal electric shock-like pain), symptoms of this type are neither required nor sufficient for a diagnosis of neuropathic pain (Bannister et al. 2020). These symptoms often become chronic and generally are not well managed with standard analgesic therapies (Colloca et al. 2017). Patients with neuropathic pain will also regularly report numbness, tingling, gait and balance disturbance, and problems with fine motor skills (Dhawan et al. 2020). A diagnosis of neuropathic pain for treatment is complicated and requires, at a minimum, a clinical examination to verify that patient history, pain symptoms, and other sensory responses all support a neuroanatomically plausible explanation (Finnerup et al. 2016). Even so, a diagnosis cannot be confirmed without a test locating the lesion or identifying the disease that corresponds to the previously proposed neurological dysfunction. The neurological dysfunctions that may give rise to neuropathic pain can be induced through a variety of other conditions. Consequently, neuropathic pain is associated with a number of common conditions, including postherpetic neuralgia, diabetic peripheral neuropathy (DPN), spinal cord injury (SCI), chemotherapy-induced peripheral neuropathy (CIPN), HIV polyneuropathy, multiple sclerosis, amputation, peripheral nerve damage, and stroke (Treede et al. 2008; Colloca et al. 2017). First-line pharmacotherapy for neuropathic pain generally falls into the classes of tricyclic antidepressants, serotonin and norepinephrine reuptake inhibitors, pregabalin, and gabapentin. Many sufferers of neuropathic pain, however, are also managing nociceptive pain and thus may be prescribed two or three analgesic medications in combination to manage their symptoms (Bannister et al. 2020). Given the expense and side effects associated with these treatments, not to mention their somewhat modest success rate, alternative cost-effective non-pharmacologic therapies, like exercise and physical therapy, should be explored.

7.1 Exercise in Neuropathic Pain

Compared to the other chronic pain conditions covered in this chapter, the body of research investigating the effectiveness of exercise and physical therapy for neuropathic pain is relatively small (Leitzelar and Koltyn 2021). In addition, the trials designed to explore the impact of exercise often focus on neuropathic pain as the result of a singular condition (e.g., DPN, CIPN, SCI) rather than general neuropathic pain. As that is the case, we will cover the evidence in support of exercise and physical therapy as treatments for neuropathic pain within a few of the particular patient groups.

Diabetic Peripheral Neuropathy Diabetes mellitus affects a large portion of the world population and is a growing problem. It has been estimated that more than 46 million people in the United States alone are living with diabetes (Wang et al. 2021). The most common complication of diabetes is DPN, which is estimated to affect nearly half of all diabetic patients (Albers and Pop-Busui 2014). While exercise has become a standard component of treatment for diabetes, its role in the prevention or treatment of DPN is less established. For example, a meta-analysis of exercise interventions designed to reduce neuropathic pain in patients with type 2 diabetes by Tatkiola et al. (2022) concluded that there was moderate quality evidence to support a significant improvement in pain complaints following exercise interventions in patients with type 2 diabetes. Even so, given the small number of studies which met their criteria, and the heterogeneity of the interventions reviewed (i.e., aerobic exercise alone or in combination with resistance exercise, high-intensity interval training, and flexibility training), the authors felt they could not make specific recommendations beyond the suggestion that exercise prescriptions should include a combination of aerobic and resistance exercise and be of a duration of 8–12 weeks before expecting any improvement. In a separate systematic review of the literature on exercise for neuropathic pain, the authors identified only two RCTs designed to investigate exercise as a treatment for painful DPN (Zhang et al. 2021). Nonetheless, their judgment was that there is moderate-level evidence to support aerobic or combined exercise regimens as a method of pain reduction in diabetic patients who have pain. In contrast to Tatkiola et al., the expert consensus convened in the process of preparing this review recommended general exercise with an emphasis on the distal extremities or moderate-to-vigorous aerobic exercise as a possible treatment for diabetic-related pain.

Chemotherapy-Induced Peripheral Neuropathy Neurotoxic cancer therapies, such as taxanes, are generally considered to be responsible for CIPN. Across all forms of this type of treatment, the incidence rate of CIPN is estimated at 38%, although depending on the specific therapy it may be as high as 90% (Kerckhove et al. 2017). Exercise has gained prominence as a possible therapy for other cancer treatment-related symptoms (Mustian et al. 2009), but the evidence to support its use as a treatment for CIPN is less clear. Kanzawa-Lee and colleagues (Kanzawa-Lee et al. 2020) completed a systematic review of the literature to evaluate the effectiveness of exercise on CIPN symptoms. They identified seven RCTs and six quasi-experiments which met their criteria for inclusion. The authors noted consistent improvements in CIPN symptoms, including pain and instability, with the exercise interventions employed. All interventions evaluated included a combination of exercise types (e.g., aerobic, resistance, balance), but each of those that included an aerobic exercise component demonstrated a significant improvement at the study conclusion. Even so, given the wide heterogeneity in exercise prescriptions and the moderate-to-high risk of bias determined for each of the studies, the authors concluded that definitive evidence for the effectiveness of exercise on CIPN pain is still lacking. A more recent review focused on exercise programs initiated at the start of chemotherapy, prior to the onset of CIPN symptoms (Lopez-Garzon et al. 2022). The authors concluded generally that exercise may be helpful in preventing CIPN symptoms. However, only two of the studies included monitored the response of pain symptoms to exercise and just one of those exhibited a significant improvement in pain (Kleckner et al. 2018).

Spinal Cord Injury Although there is strong support for the prescription of exercise and physical therapy for patients recovering from or managing an SCI (Martin Ginis et al. 2018; Hoekstra et al. 2020; Tweedy et al. 2017), the evidence for the influence of exercise on neuropathic pain associated with an SCI is particularly limited. Case in point, Toloui et al. (2023) conducted a systematic review and meta-analysis of clinical and preclinical controlled trials designed to investigate the effectiveness of physical exercise interventions for neuropathic pain following SCI. Notably, the authors were unable to identify a single human subjects trial which met their inclusion criteria. Nonetheless, they conducted an exploratory analysis of five RCTs designed to assess the effectiveness of exercise on pain in SCI patients. Although they judged the trials included to have a high risk of bias, the authors concluded that exercise improved the general perception of pain in chronic SCI patients, but that shoulder pain in these same patients was not significantly improved. The authors evaluation of the preclinical literature was that exercise results in significant improvements in mechanical allodynia, thermal hyperalgesia, and cold allodynia in rodent models of SCI. In general, there appears to be a fair level of evidence across conditions that exercise may be beneficial for patients suffering from neuropathic pain, but the consensus seems to be that more research is necessary to confirm the optimal prescription (Leitzelar and Koltyn 2021; Zhang et al. 2021).

7.2 Guidelines on Exercise for Neuropathic Pain

We were unable to identify organizational guidelines for exercise specific to the prevention or management of neuropathic pain. In general, research organizations focused on exercise recommendations for the treatment of the underlying causes of neuropathic pain (e.g., diabetes, cancer, etc.). In some cases, suggestions for starting exercise with peripheral neuropathy are included, but none of the guidelines we reviewed offered prescriptions tailored to specifically address neuropathic pain. For instance, the American Diabetes Association recommended that type 1 and 2 diabetic patients adopt the standard physical activity guidelines for healthy adults with a few caveats (Colberg et al. 2016). Beyond the standard 150 min/week of aerobic exercise and 2–3 sessions/week of resistance training, balance training was specifically recommended (2–3 sessions/week) for those suffering from peripheral neuropathy. The authors went on to make the point that exercising with neuropathy may be difficult, given the condition's potential impacts on gait, balance, and sensation. Hence, they advised the consideration of non-weight bearing activities. Similarly, the ACSM guidelines for diabetic patients adds that exercisers with peripheral neuropathy should limit their participation in activities which may be hard on their feet (e.g., hiking, jogging) and avoid activities requiring a high degree of balance. Regardless, as noted by others, exercise guidelines for reducing neuropathic pain in diabetic patients have not yet been defined (Tatikola et al. 2022).

In regard to CIPN, the ACSM convened a multidisciplinary roundtable in order to develop a consensus statement on exercise guidelines for cancer patients (Campbell et al. 2019). The roundtable participants made recommendations for aerobic exercise (3 sessions/week of at least 30 min) and resistance exercise (2 sessions/week) for 8–12 weeks for patients in treatment to manage cancer-related health outcomes but did not extend those guidelines to address CIPN or cancer-related pain. They concluded that there simply was not enough evidence to endorse exercise as a preventative or treatment for either. The American Society of Clinical Oncology published treatment guidelines for CIPN and similarly concluded that the lack of published data on the prevention or treatment of CIPN with exercise prevented them from making recommendations for or against its use in this population (Derman and Davis 2021).

Guidelines for exercise in SCI were also centered around promoting cardiometabolic health, physical fitness, and function rather than managing pain, neuropathic, or otherwise. The International Spinal Cord Society advocated that SCI patients work their way up to a regimen of 20 min of moderate-to-vigorous exercise and three sets of resistance exercises for all major muscle groups, each twice per week, to improve and maintain fitness (Martin Ginis et al. 2018). No discussion of neuropathic pain, however, was included in the write-up for these guidelines. Likewise, Exercise and Sports Science Australia omitted any commentary on the influence of exercise on neuropathic pain with their rather more rigorous recommendations for SCI patients; 30 min or more of moderate aerobic activity on 5 or more days per week, 2 or more days per week of strength training, and 2 or more days per week of flexibility training. These guidelines, once more, were presented with a goal of improving and maintaining health and fitness, rather than reducing pain complaints.

7.3 Potential Mechanisms for Action in Neuropathic Pain

The evidence for the mechanisms by which exercise may have an impact on neuropathic pain comes primarily from animal research. In their comprehensive review of exercise and neuropathic pain, Leitzelar and Koltyn (2021) covered a wide range of human and animal studies and proposed that potential mechanisms for pain reduction as a result of exercise could include: 1) normalization of microglial activity in the dorsal root ganglion, 2) prevention of the upregulation of pro-inflammatory cytokines while stimulating anti-inflammatory cytokine release, and 3) enabling the regeneration of diseased or damaged neurons by regulating neurotrophin activity. A more recent review by Wang et al. (2023) focused exclusively on the possible role of exercise in normalizing/regulating microglial activity in spinal pathways as a possible mechanism for neuropathic pain reduction. The authors concluded that the literature supports a role for microglial activity in the regulation of neuropathic pain via exercise, but the clinical evidence on how the mode and/or intensity of exercise may influence this system is lacking.

Exercise has been described by many in the literature as having "anti-inflammatory" effects (Pedersen 2017) and it has been proposed that exercise may reduce neuropathic pain by influencing patients' inflammatory state. As an example, a multicenter RCT designed to evaluate the effect of 6 weeks of walking and resistance exercise on CIPN symptoms and inflammatory responses was conducted by Kleckner et al. (2018, 2019). The trial was somewhat unique in that participants were cancer patients starting a neurotoxic chemotherapy and they were instructed to begin the exercise program in conjunction with their treatment, rather than waiting for CIPN symptoms to develop. The exercise regimen was deemed to be beneficial for the prevention and treatment of CIPN-related pain (Kleckner et al. 2018) and, in a secondary analysis, the authors further demonstrated collective changes in cytokine concentrations which were judged to improve the participants' inflammatory profile (Kleckner et al. 2019). These studies, however, are just two among a relatively small number of clinical studies designed to address questions on the relationship between exercise and inflammation in neuropathic pain. More research is needed before a firm conclusion can be drawn on the role of exercise and inflammation in neuropathic pain.

The relationship between exercise and nerve regeneration and repair was explored in a systematic review of human and animal studies by Chiaramonte et al. (2023). The authors determined that the body of evidence in animal studies supported the conclusion that physical exercise conducted in the early stages following neuronal damage or injury can facilitate axonal regeneration in peripheral sensory and motor nerves. They also reported that exercise could help to reduce the likelihood of harmful or maladaptive growth and connections. Regarding the human subject literature, however, they were unable to make any firm conclusions given the lack of relevant studies.

8 Conclusions

In many primary and secondary chronic pain conditions, meta-analytic data support the use of multiple modes, intensities, frequencies, and durations of exercise training to improve pain, physical function, and QoL. Moderate to large pain reductions occur following exercise training and this effect is often comparable to other standard treatment options. Aerobic capacity and muscular strength are often improved, which are benefits that are unique to exercise training and not often observed with pharmacologic and behavioral therapies. The long-term benefits of exercise are likely related to continued participation in regular physical activity or exercise. Adverse events are rare, although limited by lack of uniform measurement and reporting standards, and are likely to be mitigated by an individualized exercise program that accounts for symptom severity and exercise tolerance. Guidelines for exercise in chronic pain conditions are often non-specific with respect to optimal dose, intensity, type, and frequency. One notable exception is that resistance exercises that are designed to target the musculoskeletal structures that surround localized pain such as in chronic LBP and CCP have been shown to be more effective than general exercise training. The mechanisms by which exercise improves pain are not well understood and have not been rigorously nor systematically tested in humans. Future research should focus on determining why exercise works and how interventions can be designed to optimally activate the mechanism.

Although a robust literature base exists for multiple chronic pain conditions, there are several limitations. Considerable heterogeneity among studies exists, including differences in the prescribed exercise parameters, reported outcomes, statistical analyses, and the clinical criteria used for inclusion/exclusion of participants; and the certainty of the evidence is often reported as low (although strong evidence also exists). Thus, it is not currently known what exercise prescription will work best for which condition or for any single individual. Many investigators do not appear to account for baseline pain or overall illness severity when designing a study or interpreting its findings. This is important as baseline differences in symptoms can influence the clinical meaningfulness of a change for a given outcome (Smith et al. 2020). Floor and ceiling effects can also contribute to a lack of improvement when baseline pain is either too low or too high. The results for most outcomes we reviewed were based on mean changes between groups. However, only reporting average results can be misleading since responses to treatment can vary substantially between individuals. Responder analyses are recommended as an additional method for analyzing RCT data, but these were rarely reported in the original studies. These analyses examine the percentage of patients whose improvements met certain thresholds and can be used with cumulative distribution functions to better demonstrate variability in treatment effects (Smith et al. 2020). Greater inclusion of these types of analyses will assist with determining for whom certain exercise prescriptions are most efficacious and ultimately may help clinicians tailor treatment plans for individual patients. Physical activity that occurs outside of the intervention is rarely measured and could have meaningful effects on results–i.e., accounting for changes in physical activity and sedentary behaviors that may occur across the trial and at follow-up. Lastly, studies often do not control for participant expectations, fear of movement, or other relevant behavioral factors that are known to influence pain outcomes. Despite these limitations, exercise training is consistently associated with moderate to large improvements in pain and associated beneficial physical and psychological outcomes, is generally safe for most people with chronic pain, and it is recommended by most major clinical organizations.

References

Albers JW, Pop-Busui R (2014) Diabetic neuropathy: mechanisms, emerging treatments, and subtypes. Curr Neurol Neurosci Rep 14(8):473

Albright J, Allman R, Bonfiglio RP, Conill A, Dobkin B, Guccione AA et al (2001) Philadelphia panel evidence–based clinical practice guidelines on selected rehabilitation interventions for knee pain. Phys Ther 81(10):1675–1700

Aletaha D, Neogi T, Silman AJ, Funovits J, Felson DT, Bingham CO et al (2010) 2010 rheumatoid arthritis classification criteria: an American College of Rheumatology/European league against rheumatism collaborative initiative. Arthritis Rheum 62(9):2569–2581

Allen NE, Moloney N, Van Vliet V, Canning CG (2015) The rationale for exercise in the management of pain in Parkinson's disease. J Parkinsons Dis 5(2):229–239

Allen A, Carville S, McKenna F (2018) Diagnosis and management of rheumatoid arthritis in adults: summary of updated NICE guidance. BMJ 3:k3015

American College of Sports Medicine (2019) Exercising with low back pain. American College of Sports Medicine

Andersen LL, Saervoll CA, Mortensen OS, Poulsen OM, Hannerz H, Zebis MK (2011) Effectiveness of small daily amounts of progressive resistance training for frequent neck/shoulder pain: randomised controlled trial. Pain 152(2):440–446

Arnett FC, Edworthy SM, Bloch DA, Mcshane DJ, Fries JF, Cooper NS et al (1988) The american rheumatism association 1987 revised criteria for the classification of rheumatoid arthritis. Arthritis Rheum 31(3):315–324

Azeez M, Clancy C, O'Dwyer T, Lahiff C, Wilson F, Cunnane G (2020) Benefits of exercise in patients with rheumatoid arthritis: a randomized controlled trial of a patient-specific exercise programme. Clin Rheumatol 39(6):1783–1792

Aziz Q, Giamberardino MA, Barke A, Korwisi B, Baranowski AP, Wesselmann U et al (2019) The IASP classification of chronic pain for ICD-11: chronic secondary visceral pain. Pain 160(1):69–76

Baillet A, Zeboulon N, Gossec L, Combescure C, Bodin LA, Juvin R et al (2010) Efficacy of cardiorespiratory aerobic exercise in rheumatoid arthritis: meta-analysis of randomized controlled trials. Arthritis Care Res 62(7):984–992

Baillet A, Vaillant M, Guinot M, Juvin R, Gaudin P (2012) Efficacy of resistance exercises in rheumatoid arthritis: meta-analysis of randomized controlled trials. Rheumatology 51(3):519–527

Bains A, Kohrman S, Punko D, Fricchione G (2023) A link between inflammatory mechanisms and fibromyalgia. In: Kim YK (ed) Neuroinflammation, gut-brain axis and immunity in neuropsychiatric disorders [Internet], Advances in experimental medicine and biology, vol 1411. Springer Nature Singapore, Singapore, pp 357–378. [cited 2024 Jan 24]. https://doi.org/10.1007/978-981-19-7376-5_16

Bannister K, Sachau J, Baron R, Dickenson AH (2020) Neuropathic pain: mechanism-based therapeutics. Annu Rev Pharmacol Toxicol 60:257–274

Bannuru RR, Schmid CH, Kent DM, Vaysbrot EE, Wong JB, McAlindon TE (2015) Comparative effectiveness of pharmacologic interventions for knee osteoarthritis: a systematic review and network meta-analysis. Ann Intern Med 162(1):46–54

Bartholdy C, Juhl C, Christensen R, Lund H, Zhang W, Henriksen M (2017) The role of muscle strengthening in exercise therapy for knee osteoarthritis: a systematic review and meta-regression analysis of randomized trials. Semin Arthritis Rheum 47(1):9–21

Beaudart C, Lengelé L, Leclercq V, Geerinck A, Sanchez-Rodriguez D, Bruyère O et al (2020) Symptomatic efficacy of pharmacological treatments for knee osteoarthritis: a systematic review and a network meta-analysis with a 6-month time horizon. Drugs 80(18):1947–1959

Benatti FB, Pedersen BK (2015) Exercise as an anti-inflammatory therapy for rheumatic diseases – myokine regulation. Nat Rev Rheumatol 11(2):86–97

Bennell KL, Nelligan RK, Kimp AJ, Schwartz S, Kasza J, Wrigley TV et al (2020) What type of exercise is most effective for people with knee osteoarthritis and co-morbid obesity?: the TARGET randomized controlled trial. Osteoarthr Cartil 28(6):755–765

Bennett MI, Kaasa S, Barke A, Korwisi B, Rief W, Treede RD et al (2019) The IASP classification of chronic pain for ICD-11: chronic cancer-related pain. Pain 160(1):38–44

Benoliel R, Svensson P, Evers S, Wang SJ, Barke A, Korwisi B et al (2019) The IASP classification of chronic pain for ICD-11: chronic secondary headache or orofacial pain. Pain 160(1):60–68

Berenbaum F (2013) Osteoarthritis as an inflammatory disease (osteoarthritis is not osteoarthrosis!). Osteoarthr Cartil 21(1):16–21

Bertozzi L, Gardenghi I, Turoni F, Villafañe JH, Capra F, Guccione AA et al (2013) Effect of therapeutic exercise on pain and disability in the management of chronic nonspecific neck pain: systematic review and meta-analysis of randomized trials. Phys Ther 93(8):1026–1036

Bidonde J, Busch AJ, Webber SC, Schachter CL, Danyliw A, Overend TJ et al (2014) Aquatic exercise training for fibromyalgia. Cochrane Database Syst Rev 10:CD011336

Bidonde J, Busch AJ, Schachter CL, Overend TJ, Kim SY, Góes SM et al (2017) Aerobic exercise training for adults with fibromyalgia. Cochrane Musculoskeletal Group, editor. Cochrane Database Syst Rev [Internet] 6. https://doi.org/10.1002/14651858.CD012700

Bidonde J, Busch AJ, Schachter CL, Webber SC, Musselman KE, Overend TJ et al (2019) Mixed exercise training for adults with fibromyalgia. Cochrane Musculoskeletal Group, editor; [cited 2024 Jan 24]. Cochrane Database Syst Rev [Internet] 5. https://doi.org/10.1002/14651858.CD013340

Blanpied PR, Gross AR, Elliott JM, Devaney LL, Clewley D, Walton DM et al (2017) Neck pain: revision 2017: clinical practice guidelines linked to the international classification of functioning, disability and health from the Orthopaedic section of the American Physical Therapy Association. J Orthop Sports Phys Ther 47(7):A1–A83

Borea PA, Gessi S, Merighi S, Vincenzi F, Varani K (2018) Pharmacology of adenosine receptors: the state of the art. Physiol Rev 98(3):1591–1625

Boutevillain L, Dupeyron A, Rouch C, Richard E, Coudeyre E (2017) Facilitators and barriers to physical activity in people with chronic low back pain: a qualitative study. Jepson R, editor. PLoS One 12(7):e0179826

Breivik H, Collett B, Ventafridda V, Cohen R, Gallacher D (2006) Survey of chronic pain in Europe: prevalence, impact on daily life, and treatment. Eur J Pain 10(4):287–287

Bricca A, Struglics A, Larsson S, Steultjens M, Juhl CB, Roos EM (2019) Impact of exercise therapy on molecular biomarkers related to cartilage and inflammation in individuals at risk of, or with established, knee osteoarthritis: a systematic review and meta-analysis of randomized controlled trials. Arthritis Care Res 71(11):1504–1515

Brosseau L, Taki J, Desjardins B, Thevenot O, Fransen M, Wells GA et al (2017a) The Ottawa panel clinical practice guidelines for the management of knee osteoarthritis. Part two: strengthening exercise programs. Clin Rehabil 31(5):596–611

Brosseau L, Taki J, Desjardins B, Thevenot O, Fransen M, Wells GA et al (2017b) The Ottawa panel clinical practice guidelines for the management of knee osteoarthritis. Part three: aerobic exercise programs. Clin Rehabil 31(5):612–624

Bruehl S, Burns JW, Koltyn K, Gupta R, Buvanendran A, Edwards D et al (2020) Are endogenous opioid mechanisms involved in the effects of aerobic exercise training on chronic low back pain? A randomized controlled trial. Pain 161(12):2887–2897

Busch AJ, Webber SC, Richards RS, Bidonde J, Schachter CL, Schafer LA et al (2013) Resistance exercise training for fibromyalgia. Cochrane Database Syst Rev 12:CD010884

Bussières AE, Stewart G, Al-Zoubi F, Decina P, Descarreaux M, Hayden J et al (2016) The treatment of neck pain–associated disorders and whiplash-associated disorders: a clinical practice guideline. J Manip Physiol Ther 39(8):523–564.e27

Butler S, Landmark T, Glette M, Borchgrevink P, Woodhouse A (2016) Chronic widespread pain – the need for a standard definition. Pain 157(3):541

Campbell KL, Winters-Stone KM, Wiskemann J, May AM, Schwartz AL, Courneya KS et al (2019) Exercise guidelines for cancer survivors: consensus statement from international multidisciplinary roundtable. Med Sci Sports Exerc 51(11):2375

Cecchi F, Pasquini G, Paperini A, Boni R, Castagnoli C, Pistritto S et al (2014) Predictors of response to exercise therapy for chronic low back pain: result of a prospective study with one year follow-up. Eur J Phys Rehabil Med 50(2):143–151

Chiaramonte R, Pavone V, Testa G, Pesce I, Scaturro D, Musumeci G et al (2023) The role of physical exercise and rehabilitative implications in the process of nerve repair in peripheral neuropathies: a systematic review. Diagnostics 13(3):364

Choy EH, Mease PJ (2009) Key symptom domains to be assessed in fibromyalgia (outcome measures in rheumatoid arthritis clinical trials). Rheum Dis Clin N Am 35(2):329–337

Chung S, Jeong YG (2018) Effects of the craniocervical flexion and isometric neck exercise compared in patients with chronic neck pain: a randomized controlled trial. Physiother Theory Pract 34(12):916–925

Clark DO (1999) J Clin Geropsychol 5(1):51–62

Clauw DJ, Arnold LM, McCarberg BH (2011) The science of fibromyalgia. Mayo Clin Proc 86(9):907–911

Clynes MA, Jameson KA, Edwards MH, Cooper C, Dennison EM (2019) Impact of osteoarthritis on activities of daily living: does joint site matter? Aging Clin Exp Res 31(8):1049–1056

Colberg SR, Sigal RJ, Yardley JE, Riddell MC, Dunstan DW, Dempsey PC et al (2016) Physical activity/exercise and diabetes: a position statement of the American Diabetes Association. Diabetes Care 39(11):2065–2079

Colloca L, Ludman T, Bouhassira D, Baron R, Dickenson AH, Yarnitsky D et al (2017) Neuropathic pain. Nat Rev Dis Primers 3(1):1–19

Conley B, Bunzli S, Bullen J, O'Brien P, Persaud J, Gunatillake T et al (2023) Core recommendations for osteoarthritis care: a systematic review of clinical practice guidelines. Arthritis Care Res 75(9):1897–1907

Cooney JK, Law RJ, Matschke V, Lemmey AB, Moore JP, Ahmad Y et al (2011) Benefits of exercise in rheumatoid arthritis. J Aging Res 2011:1–14

Corp N, Mansell G, Stynes S, Wynne-Jones G, Morsø L, Hill JC et al (2021 Feb) Evidence-based treatment recommendations for neck and low back pain across Europe: a systematic review of guidelines. Eur J Pain 25(2):275–295

Costa LDCM, Maher CG, McAuley JH, Hancock MJ, Herbert RD, Refshauge KM et al (2009) Prognosis for patients with chronic low back pain: inception cohort study. BMJ 339:b3829–b3829

Côté P, Wong JJ, Sutton D, Shearer HM, Mior S, Randhawa K et al (2016) Management of neck pain and associated disorders: a clinical practice guideline from the Ontario Protocol for Traffic Injury Management (OPTIMa) collaboration. Eur Spine J 25(7):2000–2022

Cramp F, Hewlett S, Almeida C, Kirwan JR, Choy EH, Chalder T et al (2013) Non-pharmacological interventions for fatigue in rheumatoid arthritis. Cochrane Musculoskeletal Group, editor; [cited 2024 Jan 11]. Cochrane Database Syst Rev [Internet]. https://doi.org/10.1002/14651858.CD008322.pub2

Cross M, Smith E, Hoy D, Carmona L, Wolfe F, Vos T et al (2014) The global burden of rheumatoid arthritis: estimates from the global burden of disease 2010 study. Ann Rheum Dis 73(7):1316–1322

Da Silva T, Mills K, Brown BT, Herbert RD, Maher CG, Hancock MJ (2017) Risk of recurrence of low back pain: a systematic review. J Orthop Sports Phys Ther 47(5):305–313

Daher A, Carel RS, Tzipi K, Esther H, Dar G (2020) The effectiveness of an aerobic exercise training on patients with neck pain during a short- and long-term follow-up: a prospective double-blind randomized controlled trial. Clin Rehabil 34(5):617–629

De Zoete RM, Armfield NR, McAuley JH, Chen K, Sterling M (2021) Comparative effectiveness of physical exercise interventions for chronic non-specific neck pain: a systematic review with network meta-analysis of 40 randomised controlled trials. Br J Sports Med 55(13):730–742

Derman BA, Davis AM (2021) Recommendations for prevention and management of chemotherapy-induced peripheral neuropathy. JAMA 326(11):1058–1059

Dhawan S, Andrews R, Kumar L, Wadhwa S, Shukla G (2020) A randomized controlled trial to assess the effectiveness of muscle strengthening and balancing exercises on chemotherapy-induced peripheral neuropathic pain and quality of life among cancer patients. Cancer Nurs 43(4):269

Duque IL, Parra JH, Duvallet A (2009) Aerobic fitness and limiting factors of maximal performance in chronic low back pain patients. J Back Musculoskelet Rehabil 22(2):113–119

England BR, Smith BJ, Baker NA, Barton JL, Oatis CA, Guyatt G et al (2023) 2022 American College of Rheumatology Guideline for exercise, rehabilitation, diet, and additional integrative interventions for rheumatoid arthritis. Arthritis Care Res 75(8):1603–1615

Fejer R, Kyvik KO, Hartvigsen J (2006) The prevalence of neck pain in the world population: a systematic critical review of the literature. Eur Spine J 15(6):834–848

Fernandes L, Hagen KB, Bijlsma JWJ, Andreassen O, Christensen P, Conaghan PG et al (2013) EULAR recommendations for the non-pharmacological core management of hip and knee osteoarthritis. Ann Rheum Dis 72(7):1125–1135

Finnerup NB, Haroutounian S, Kamerman P, Baron R, Bennett DLH, Bouhassira D et al (2016) Neuropathic pain: an updated grading system for research and clinical practice. Pain 157(8):1599–1606

Fransen M, McConnell S, Harmer AR, Van Der Esch M, Simic M, Bennell KL (2015a) Exercise for osteoarthritis of the knee: a Cochrane systematic review. Br J Sports Med 49(24):1554–1557

Fransen M, McConnell S, Harmer AR, Van Der Esch M, Simic M, Bennell KL (2015b) Exercise for osteoarthritis of the knee. Cochrane Musculoskeletal Group, editor; [Internet], [cited 2023 Oct 29]. Cochrane Database Syst Rev 1:CD004376. https://doi.org/10.1002/14651858.CD004376.pub3

Friedman B, Cronstein B (2019) Methotrexate mechanism in treatment of rheumatoid arthritis. Joint Bone Spine 86(3):301–307

Gaskin DJ, Richard P (2012) The economic costs of pain in the United States. J Pain 13(8):715–724

George SZ, Fritz JM, Silfies SP, Schneider MJ, Beneciuk JM, Lentz TA et al (2021) Interventions for the management of acute and chronic low back pain: revision 2021: clinical practice guidelines linked to the international classification of functioning, disability and health from the Academy of Orthopaedic Physical Therapy of the American Physical Therapy Association. J Orthop Sports Phys Ther 51(11):CPG1–CP60

Gleeson M, Bishop NC, Stensel DJ, Lindley MR, Mastana SS, Nimmo MA (2011) The anti-inflammatory effects of exercise: mechanisms and implications for the prevention and treatment of disease. Nat Rev Immunol 11(9):607–615
Griffin TM, Huebner JL, Kraus VB, Yan Z, Guilak F (2012) Induction of osteoarthritis and metabolic inflammation by a very high-fat diet in mice: effects of short-term exercise. Arthritis Rheum 64(2):443–453
Guo Q, Wang Y, Xu D, Nossent J, Pavlos NJ, Xu J (2018) Rheumatoid arthritis: pathological mechanisms and modern pharmacologic therapies. Bone Res 6(1):15
Guzman J, Hurwitz EL, Carroll LJ, Haldeman S, Côté P, Carragee EJ et al (2009) A new conceptual model of neck pain. J Manip Physiol Ther 32(2):S17–S28
Häkkinen A, Kautiainen H, Hannonen P, Ylinen J (2008) Strength training and stretching versus stretching only in the treatment of patients with chronic neck pain: a randomized one-year follow-up study. Clin Rehabil 22(7):592–600
Hayden JA, van Tulder MW, Malmivaara A, Koes BW (2005) Exercise therapy for treatment of non-specific low back pain. Cochrane Database Syst Rev 3:CD000335
Hayden JA, Ellis J, Ogilvie R, Stewart SA, Bagg MK, Stanojevic S et al (2021) Some types of exercise are more effective than others in people with chronic low back pain: a network meta-analysis. J Physiother 67(4):252–262
Henchoz Y, Bastardot F, Guessous I, Theler JM, Dudler J, Vollenweider P et al (2012) Physical activity and energy expenditure in rheumatoid arthritis patients and matched controls. Rheumatology 51(8):1500–1507
Heredia-Rizo AM, Petersen KK, Madeleine P, Arendt-Nielsen L (2019) Clinical outcomes and central pain mechanisms are improved after upper trapezius eccentric training in female computer users with chronic neck/shoulder pain. Clin J Pain 35(1):65–76
Hoekstra F, McBride CB, Borisoff J, Fetterly MJ, Ginis S, Latimer-Cheung AE et al (2020) Translating the international scientific spinal cord injury exercise guidelines into community and clinical practice guidelines: a Canadian evidence-informed resource. Spinal Cord 58(6):647–657
Holden MA, Button K, Collins NJ, Henrotin Y, Hinman RS, Larsen JB et al (2021) Guidance for implementing best practice therapeutic exercise for patients with knee and hip osteoarthritis: what does the current evidence base tell us? Arthritis Care Res 73(12):1746–1753
Holth HS, Werpen HK, Zwart JA, Hagen K (2008) Physical inactivity is associated with chronic musculoskeletal complaints 11 years later: results from the Nord-Trøndelag Health Study. BMC Musculoskelet Disord 9:159
Hoy D, Brooks P, Blyth F, Buchbinder R (2010a) The epidemiology of low back pain. Best Pract Res Clin Rheumatol 24(6):769–781
Hoy DG, Protani M, De R, Buchbinder R (2010b) The epidemiology of neck pain. Best Pract Res Clin Rheumatol 24(6):783–792
Hunter DJ, McDougall JJ, Keefe FJ (2008) The symptoms of osteoarthritis and the genesis of pain. Rheum Dis Clin N Am 34(3):623–643
Institute of Medicine (US) Committee on Advancing Pain Research C re and Education (2011) Relieving Pain in America: A Blueprint for Transforming Prevention, Care, Education, and Research. Available from https://www.ncbi.nlm.nih.gov/pubmed/22553896
International Classification of Diseases, Eleventh Revision (ICD-11) [Internet] (2019). Available from https://icd.who.int/browse11. Licensed under Creative Commons Attribution-NoDerivatives 3.0 IGO licence (CC BY-ND 3.0 IGO)
Jackson T, Thomas S, Stabile V, Han X, Shotwell M, McQueen K (2015) Prevalence of chronic pain in low-income and middle-income countries: a systematic review and meta-analysis. Lancet Lond Engl 385(Suppl 2):S10
Johannes CB, Le TK, Zhou X, Johnston JA, Dworkin RH (2010) The prevalence of chronic pain in United States adults: results of an internet-based survey. J Pain 11(11):1230–1239
Johnson VL, Hunter DJ (2014) The epidemiology of osteoarthritis. Best Pract Res Clin Rheumatol 28(1):5–15

Juhl C, Christensen R, Roos EM, Zhang W, Lund H (2014) Impact of exercise type and dose on pain and disability in knee osteoarthritis: a systematic review and meta-regression analysis of randomized controlled trials. Arthritis Rheum 66(3):622–636

Kanzawa-Lee GA, Larson JL, Resnicow K, Smith EML (2020) Exercise effects on chemotherapy-induced peripheral neuropathy: a comprehensive integrative review. Cancer Nurs 43(3):E172

Kelley GA, Kelley KS, Callahan LF (2018) Aerobic exercise and fatigue in rheumatoid arthritis participants: a meta-analysis using the minimal important difference approach. Arthritis Care Res 70(12):1735–1739

Kerckhove N, Collin A, Condé S, Chaleteix C, Pezet D, Balayssac D (2017) Long-term effects, pathophysiological mechanisms, and risk factors of chemotherapy-induced peripheral neuropathies: a comprehensive literature review. Front Pharmacol 8:86

Kim SY, Busch AJ, Overend TJ, Schachter CL, Van Der Spuy I, Boden C et al (2019) Flexibility exercise training for adults with fibromyalgia. Cochrane Musculoskeletal Group, editor; [cited 2024 Jan 24]. Cochrane Database Syst Rev [Internet] 9. https://doi.org/10.1002/14651858.CD013419

Kleckner IR, Kamen C, Gewandter JS, Mohile NA, Heckler CE, Culakova E et al (2018) Effects of exercise during chemotherapy on chemotherapy-induced peripheral neuropathy: a multicenter, randomized controlled trial. Support Care Cancer 26(4):1019–1028

Kleckner IR, Kamen C, Cole C, Fung C, Heckler CE, Guido JJ et al (2019) Effects of exercise on inflammation in patients receiving chemotherapy: a nationwide NCORP randomized clinical trial. Support Care Cancer Off J Multinatl Assoc Support Care Cancer 27(12):4615–4625

Kleykamp BA, Ferguson MC, McNicol E, Bixho I, Arnold LM, Edwards RR et al (2021) The prevalence of psychiatric and chronic pain comorbidities in fibromyalgia: an ACTTION systematic review. Semin Arthritis Rheum 51(1):166–174

Kloppenburg M, Kroon FP, Blanco FJ, Doherty M, Dziedzic KS, Greibrokk E et al (2019) 2018 update of the EULAR recommendations for the management of hand osteoarthritis. Ann Rheum Dis 78(1):16–24

Kolasinski SL, Neogi T, Hochberg MC, Oatis C, Guyatt G, Block J et al (2020) 2019 American College of Rheumatology/Arthritis Foundation guideline for the Management of Osteoarthritis of the hand, hip, and knee. Arthritis Care Res 72(2):149–162

Kundakci B, Hall M, Atzeni F, Branco J, Buskila D, Clauw D et al (2022a) International, multidisciplinary Delphi consensus recommendations on non-pharmacological interventions for fibromyalgia. Semin Arthritis Rheum 1(57):152101

Kundakci B, Kaur J, Goh SL, Hall M, Doherty M, Zhang W et al (2022b) Efficacy of nonpharmacological interventions for individual features of fibromyalgia: a systematic review and meta-analysis of randomised controlled trials. Pain 163(8):1432–1445

Leitzelar BN, Koltyn KF (2021) Exercise and neuropathic pain: a general overview of preclinical and clinical research. Sports Med - Open 7(1):21

Lesnak JB, Sluka KA (2020) Mechanism of exercise-induced analgesia: what we can learn from physically active animals. PAIN Rep 5(5):e850

Lima LV, Abner TSS, Sluka KA (2017) Does exercise increase or decrease pain? Central mechanisms underlying these two phenomena. J Physiol 595(13):4141–4150

Lindell M, Grimby-Ekman A (2022) Stress non-restorative sleep and physical inactivity as risk factors for chronic pain in young adults: A cohort study. PLoS One 17(1) e0262601. https://doi.org/10.1371/journal.pone.0262601

Liu J, Chen L, Tu Y, Chen X, Hu K, Tu Y et al (2019) Different exercise modalities relieve pain syndrome in patients with knee osteoarthritis and modulate the dorsolateral prefrontal cortex: a multiple mode MRI study. Brain Behav Immun 82:253–263

Lopez-Garzon M, Cantarero-Villanueva I, Postigo-Martin P, González-Santos Á, Lozano-Lozano-M, Galiano-Castillo N (2022) Can physical exercise prevent chemotherapy-induced peripheral neuropathy in patients with cancer? A systematic review and meta-analysis. Arch Phys Med Rehabil 103(11):2197–2208

Lu J, Feng X, Zhang H, Wei Y, Yang Y, Tian Y et al (2021) Maresin-1 suppresses IL-1β-induced MMP-13 secretion by activating the PI3K/AKT pathway and inhibiting the NF-κB pathway in synovioblasts of an osteoarthritis rat model with treadmill exercise. Connect Tissue Res 62(5): 508–518

Macfarlane GJ, Kronisch C, Dean LE, Atzeni F, Häuser W, Fluß E et al (2017) EULAR revised recommendations for the management of fibromyalgia. Ann Rheum Dis 76(2):318–328

Maher C, Underwood M, Buchbinder R (2017) Non-specific low back pain. Lancet 389(10070):736–747

Manchikanti L, Singh V, Falco FJE, Benyamin RM, Hirsch JA (2014) Epidemiology of low back pain in adults. Neuromodulation Technol Neural Interface 17:3–10

Mansfield KE, Sim J, Jordan JL, Jordan KP (2016) A systematic review and meta-analysis of the prevalence of chronic widespread pain in the general population. Pain 157(1):55–64

Marks R (2012) Knee osteoarthritis and exercise adherence: a review. Curr Aging Sci 5(1):72–83

Martin Ginis KA, van der Scheer JW, Latimer-Cheung AE, Barrow A, Bourne C, Carruthers P et al (2018) Evidence-based scientific exercise guidelines for adults with spinal cord injury: an update and a new guideline. Spinal Cord 56(4):308–321

McBeth J, Nicholl BI, Cordingley L, Davies KA, Macfarlane GJ (2010) Chronic widespread pain predicts physical inactivity: results from the prospective EPIFUND study. Eur J Pain 14(9):972–979

McLoughlin MJ, Colbert LH, Stegner AJ, Cook DB (2011) Are women with fibromyalgia less physically active than healthy women? Med Sci Sports Exerc 43(5):905–912

Metsios GS, Kitas GD (2018) Physical activity, exercise and rheumatoid arthritis: effectiveness, mechanisms and implementation. Best Pract Res Clin Rheumatol 32(5):669–682

Metsios GS, Stavropoulos-Kalinoglou A, Kitas GD (2015) The role of exercise in the management of rheumatoid arthritis. Expert Rev Clin Immunol 11(10):1121–1130

Metsios GS, Moe RH, Kitas GD (2020) Exercise and inflammation. Best Pract Res Clin Rheumatol 34(2):101504

Monticone M, Cedraschi C, Ambrosini E, Rocca B, Fiorentini R, Restelli M et al (2015) Cognitive-behavioural treatment for subacute and chronic neck pain. Cochrane Back and Neck Group, editor; [cited 2024 Jan 21]. Cochrane Database Syst Rev [Internet] 2016(5). https://doi.org/10.1002/14651858.CD010664.pub2

Mueller J, Niederer D (2020) Dose-response-relationship of stabilisation exercises in patients with chronic non-specific low back pain: a systematic review with meta-regression. Sci Rep 10(1):16921

Müller-Schwefe G, Morlion B, Ahlbeck K, Alon E, Coaccioli S, Coluzzi F et al (2017) Treatment for chronic low back pain: the focus should change to multimodal management that reflects the underlying pain mechanisms. Curr Med Res Opin 33(7):1199–1210

Mustian KM, Sprod LK, Palesh OG, Peppone LJ, Janelsins MC, Mohile SG et al (2009) Exercise for the management of side effects and quality of life among cancer survivors. Curr Sports Med Rep 8(6):325

Neelapala YVR, Mercuri D, Macedo L, Hanna S, Kobsar D, Carlesso L (2023) Mechanisms hypothesized for pain-relieving effects of exercise in fibromyalgia: a scoping review. Ther Adv Musculoskelet Dis 15:1759720X231182894

NICE (2021) Chronic pain (primary and secondary) in over 16s: assessment of all chronic pain and management of chronic primary pain [Internet]. National Institute for Health and Care Excellence (NICE), London. [cited 2024 Feb 29]. (National Institute for Health and Care Excellence: Guidelines). Available from http://www.ncbi.nlm.nih.gov/books/NBK569960/

NICE (2022) Osteoarthritis in over 16s diagnosis and management. National Institute for Health and Care Excellence (NICE), London

Nicholas M, Vlaeyen JWS, Rief W, Barke A, Aziz Q, Benoliel R et al (2019) The IASP classification of chronic pain for ICD-11: chronic primary pain. Pain 160(1):28–37

O'Neill TW, Felson DT (2018) Mechanisms of osteoarthritis (OA) pain. Curr Osteoporos Rep 16(5):611–616

O'Riordan C, Clifford A, Van De Ven P, Nelson J (2014) Chronic neck pain and exercise interventions: frequency, intensity, time, and type principle. Arch Phys Med Rehabil 95(4):770–783

Oesch P, Kool J, Hagen KB, Bachmann S (2010) Effectiveness of exercise on work disability in patients with non-acute non-specific low back pain: systematic review and meta-analysis of randomised controlled trials. J Rehabil Med 42(3):193–205

Oliveira CB, Maher CG, Pinto RZ, Traeger AC, Lin CWC, Chenot JF et al (2018) Clinical practice guidelines for the management of non-specific low back pain in primary care: an updated overview. Eur Spine J 27(11):2791–2803

Ottawa Panel Evidence (2004) Ottawa panel evidence-based clinical practice guidelines for therapeutic exercises in the management of rheumatoid arthritis in adults. Phys Ther [Internet]. [cited 2023 Oct 31]; Available from https://academic.oup.com/ptj/article/84/10/934/2857560/Ottawa-Panel-EvidenceBased-Clinical-Practice

Owen PJ, Miller CT, Mundell NL, Verswijveren SJJM, Tagliaferri SD, Brisby H et al (2020) Which specific modes of exercise training are most effective for treating low back pain? Network meta-analysis. Br J Sports Med 54(21):1279–1287

Öztürk Ö, Algun ZC, Bombacı H, Erdoğan SB (2021) Changes in prefrontal cortex activation with exercise in knee osteoarthritis patients with chronic pain: an fNIRS study. J Clin Neurosci 90: 144–151

Paraskevopoulos E, Koumantakis GA, Papandreou M (2023) A systematic review of the aerobic exercise program variables for patients with non-specific neck pain: effectiveness and clinical applications. Healthcare 11(3):339

Pedersen BK (2017) Anti-inflammatory effects of exercise: role in diabetes and cardiovascular disease. Eur J Clin Investig 47(8):600–611

Pedersen BK, Steensberg A, Schjerling P (2001) Exercise and interleukin-6. Curr Opin Hematol [Internet] 8(3) Available from https://journals.lww.com/co-hematology/fulltext/2001/05000/exercise_and_interleukin_6.2.aspx:137

Pereira Nunes Pinto AC, Natour J, De Moura Castro CH, Eloi M, Lombardi JI (2017) Acute effect of a resistance exercise session on markers of cartilage breakdown and inflammation in women with rheumatoid arthritis. Int J Rheum Dis 20(11):1704–1713

Pereira D, Peleteiro B, Araújo J, Branco J, Santos RA, Ramos E (2011) The effect of osteoarthritis definition on prevalence and incidence estimates: a systematic review. Osteoarthr Cartil 19(11):1270–1285

Perrot S, Cohen M, Barke A, Korwisi B, Rief W, Treede RD et al (2019) The IASP classification of chronic pain for ICD-11: chronic secondary musculoskeletal pain. Pain 160(1):77–82

Phillips CJ (2006) Economic burden of chronic pain. Expert Rev Pharmacoecon Outcomes Res 6(5):591–601

Pianosi P, Emerling E, Mara K, Weaver A, Fischer P (2017) Aerobic fitness in adolescents with chronic pain or chronic fatigue: parallels and mechanisms? J Rehabil Med 49(5):441–446

Qaseem A, Wilt TJ, McLean RM, Forciea MA, for the Clinical Guidelines Committee of the American College of Physicians (2017) Noninvasive treatments for acute, subacute, and chronic low back pain: a clinical practice guideline from the American College of Physicians. Ann Intern Med 166(7):514

Queiroz LP (2013) Worldwide epidemiology of fibromyalgia. Curr Pain Headache Rep 17(8):356

Raja SN, Carr DB, Cohen M, Finnerup NB, Flor H, Gibson S et al (2020) The revised International Association for the Study of Pain definition of pain: concepts, challenges, and compromises. Pain 161(9):1976

Ramírez-Morales R, Bermúdez-Benítez E, Martínez-Martínez LA, Martínez-Lavín M (2022) Clinical overlap between fibromyalgia and myalgic encephalomyelitis. A systematic review and meta-analysis. Autoimmun Rev 21(8):103129

Raposo F, Ramos M, Lúcia CA (2021) Effects of exercise on knee osteoarthritis: a systematic review. Musculoskeletal Care 19(4):399–435

Rasmussen-Barr E, Äng B, Arvidsson I, Nilsson-Wikmar L (2009) Graded exercise for recurrent low-back pain: a randomized, controlled trial with 6-, 12-, and 36-month follow-ups. Spine 34(3):221–228

Rausch Osthoff AK, Niedermann K, Braun J, Adams J, Brodin N, Dagfinrud H et al (2018) 2018 EULAR recommendations for physical activity in people with inflammatory arthritis and osteoarthritis. Ann Rheum Dis 77(9):1251–1260

Rice ASC, Smith BH, Blyth FM (2016) Pain and the global burden of disease. Pain 157(4):791–796

Rittig-Rasmussen B, Kasch H, Fuglsang-Frederiksen A, Svensson P, Jensen TS (2014) The role of neuroplasticity in experimental neck pain: a study of potential mechanisms impeding clinical outcomes of training. Man Ther 19(4):288–293

Rongen-van Dartel SAA, Repping-Wuts H, Flendrie M, Bleijenberg G, Metsios GS, Van Den Hout WB et al (2015) Effect of aerobic exercise training on fatigue in rheumatoid arthritis: a meta-analysis. Arthritis Care Res 67(8):1054–1062

Roussel NA, Nijs J, Meeus M, Mylius V, Fayt C, Oostendorp R (2013) Central sensitization and altered central pain processing in chronic low back pain: fact or myth? Clin J Pain 29(7):625–638

Saragiotto BT, Maher CG, Yamato TP, Costa LOP, Costa LCM, Ostelo RWJG et al (2016) Motor control exercise for nonspecific low back pain: a Cochrane review. Spine 41(16):1284–1295

Scheffer DDL, Latini A (2020) Exercise-induced immune system response: anti-inflammatory status on peripheral and central organs. Biochim Biophys Acta (BBA) - Mol Basis Dis 1866(10):165823

Scholz J, Finnerup NB, Attal N, Aziz Q, Baron R, Bennett MI et al (2019) The IASP classification of chronic pain for ICD-11: chronic neuropathic pain. Pain 160(1):53–59

Schug SA, Lavand'homme P, Barke A, Korwisi B, Rief W, Treede RD et al (2019) The IASP classification of chronic pain for ICD-11: chronic postsurgical or posttraumatic pain. Pain 160(1):45–52

Searle A, Spink M, Ho A, Chuter V (2015) Exercise interventions for the treatment of chronic low back pain: a systematic review and meta-analysis of randomised controlled trials. Clin Rehabil 29(12):1155–1167

Shane Anderson A, Loeser RF (2010) Why is osteoarthritis an age-related disease? Best Pract Res Clin Rheumatol 24(1):15–26

Siqueira US, Orsini Valente LG, De Mello MT, Szejnfeld VL, Pinheiro MM (2017) Effectiveness of aquatic exercises in women with rheumatoid arthritis: a randomized, controlled, 16-week intervention – the HydRA trial. Am J Phys Med Rehabil 96(3):167–175

Sluka KA, Clauw DJ (2016) Neurobiology of fibromyalgia and chronic widespread pain. Neuroscience 338:114–129

Sluka KA, Danielson J, Rasmussen L, Dasilva LF (2012) Exercise-induced pain requires NMDA receptor activation in the medullary raphe nuclei. Med Sci Sports Exerc 44(3):420–427

Sluka KA, O'Donnell JM, Danielson J, Rasmussen LA (2013) Regular physical activity prevents development of chronic pain and activation of central neurons. J Appl Physiol 1985 114(6):725–733

Smeets RJEM, Wittink H, Hidding A, Knottnerus JA (2006) Do patients with chronic low back pain have a lower level of aerobic fitness than healthy controls?: are pain, disability, fear of injury, working status, or level of leisure time activity associated with the difference in aerobic fitness level? Spine 31(1):90–97

Smith SM, Dworkin RH, Turk DC, McDermott MP, Eccleston C, Farrar JT et al (2020) Interpretation of chronic pain clinical trial outcomes: IMMPACT recommended considerations. Pain 161(11):2446–2461

Spielman LJ, Little JP, Klegeris A (2016) Physical activity and exercise attenuate neuroinflammation in neurological diseases. Brain Res Bull 125:19–29

Stegner AJ, Almassi NE, Dougherty RJ, Ellingson LD, Gretzon NP, Lindheimer JB et al (2021) Safety and efficacy of short-term structured resistance exercise in Gulf War veterans with chronic unexplained muscle pain: a randomized controlled trial. Life Sci 282:119810

Steiger F, Wirth B, De Bruin ED, Mannion AF (2012) Is a positive clinical outcome after exercise therapy for chronic non-specific low back pain contingent upon a corresponding improvement in the targeted aspect(s) of performance? A systematic review. Eur Spine J 21(4):575–598

Suárez A, Guillamó E, Roig T, Blázquez A, Alegre J, Bermúdez J et al (2010) Nitric oxide metabolite production during exercise in chronic fatigue syndrome: a case-control study. J Women's Health 19(6):1073–1077

Sul B, Lee KB, Joo YB, Hong BY, Kim JS, Kim KJ et al (2020) Twelve weeks of strengthening exercise for patients with rheumatoid arthritis: a prospective intervention study. J Clin Med 9(9):2792

Tatikola SP, Natarajan V, Desai VK, Asirvatham AR, Rajsekhar H (2022) Effect of various exercise protocols on neuropathic pain in individuals with type 2 diabetes with peripheral neuropathy: a systematic review and meta-analysis. Diabetes Metab Syndr Clin Res Rev 16(9):102603

Taulaniemi A, Kankaanpää M, Tokola K, Parkkari J, Suni JH (2019) Neuromuscular exercise reduces low back pain intensity and improves physical functioning in nursing duties among female healthcare workers; secondary analysis of a randomised controlled trial. BMC Musculoskelet Disord 20(1):328

Theadom A, Cropley M, Smith HE, Feigin VL, McPherson K (2015) Mind and body therapy for fibromyalgia. Cochrane Musculoskeletal Group, editor; [cited 2024 Jan 24]. Cochrane Database Syst Rev [Internet] 2015:CD001980. https://doi.org/10.1002/14651858.CD001980.pub3

Thieme K, Mathys M, Turk DC (2017) Evidenced-based guidelines on the treatment of fibromyalgia patients: are they consistent and if not, why not? Have effective psychological treatments been overlooked? J Pain 18(7):747–756

Toloui A, Ramawad HA, Gharin P, Vaccaro AR, Zarei H, Hosseini M et al (2023) The role of exercise in the alleviation of neuropathic pain following traumatic spinal cord injuries: a systematic review and meta-analysis. Neurospine 20(3):1073

Treede RD, Jensen TS, Campbell JN, Cruccu G, Dostrovsky JO, Griffin JW et al (2008) Neuropathic pain. Neurology 70(18):1630–1635

Treede RD, Rief W, Barke A, Aziz Q, Bennett MI, Benoliel R et al (2019) Chronic pain as a symptom or a disease: the IASP classification of chronic pain for the international classification of diseases (ICD-11). Pain 160(1):19–27

Tweedy SM, Beckman EM, Geraghty TJ, Theisen D, Perret C, Harvey LA et al (2017) Exercise and sports science Australia (ESSA) position statement on exercise and spinal cord injury. J Sci Med Sport 20(2):108–115

VA/DoD Clinical Practice Guideline (2022) The diagnosis and treatment of low back pain. U.-S. Government Printing Office, Washington

Van Den Ende CHM, Hazes JM, le Cessie S, Mulder WJ, Belfor DG, Breedveld FC et al (1996) Comparison of high and low intensity training in well controlled rheumatoid arthritis. Results of a randomised clinical trial. Ann Rheum Dis 55(11):798

Van Der Woude D, Van Der Helm-van Mil AHM (2018) Update on the epidemiology, risk factors, and disease outcomes of rheumatoid arthritis. Best Pract Res Clin Rheumatol 32(2):174–187

van Hecke O, Austin SK, Khan RA, Smith BH, Torrance N (2014) Neuropathic pain in the general population: a systematic review of epidemiological studies. Pain 155(4):654–662

Veldhuijzen Van Zanten JJCS, Rouse PC, Hale ED, Ntoumanis N, Metsios GS, Duda JL et al (2015) Perceived barriers, facilitators and benefits for regular physical activity and exercise in patients with rheumatoid arthritis: a review of the literature. Sports Med 45(10):1401–1412

Verhoeven F, Tordi N, Prati C, Demougeot C, Mougin F, Wendling D (2016) Physical activity in patients with rheumatoid arthritis. Joint Bone Spine 83(3):265–270

Verkerk K, Luijsterburg PAJ, Miedema HS, Pool-Goudzwaard A, Koes BW (2012) Prognostic factors for recovery in chronic nonspecific low back pain: a systematic review. Phys Ther 92(9):1093–1108

Vina ER, Kwoh CK (2018) Epidemiology of osteoarthritis: literature update. Curr Opin Rheumatol 30(2):160–167

Vonk F, Verhagen AP, Twisk JW, Köke AJA, Luiten MWCT, Koes BW (2009) Effectiveness of a behaviour graded activity program versus conventional exercise for chronic neck pain patients. Eur J Pain 13(5):533–541

Vos T, Barber RM, Bell B, Bertozzi-Villa A, Biryukov S, Bolliger I et al (2015) Global, regional, and national incidence, prevalence, and years lived with disability for 301 acute and chronic diseases and injuries in 188 countries, 1990–2013: a systematic analysis for the global burden of disease study 2013. Lancet 386(9995):743–800

Walsh DA, McWilliams DF (2012) Pain in rheumatoid arthritis. Curr Pain Headache Rep 16(6):509–517

Wang L, Li X, Wang Z, Bancks MP, Carnethon MR, Greenland P et al (2021) Trends in prevalence of diabetes and control of risk factors in diabetes among US adults, 1999–2018. JAMA 326(8):704–716

Wang MJ, Jing XY, Wang YZ, Yang BR, Lu Q, Hu H et al (2023. [cited 2023 Oct 19]) Exercise, spinal microglia and neuropathic pain: potential molecular mechanisms. Neurochem Res [Internet] 49:29. https://doi.org/10.1007/s11064-023-04025-4

Wellsandt E, Golightly Y (2018) Exercise in the management of knee and hip osteoarthritis. Curr Opin Rheumatol 30(2):151–159

Wen Z, Chai Y (2021) Effectiveness of resistance exercises in the treatment of rheumatoid arthritis: a meta-analysis. Medicine (Baltimore) 100(13):e25019

Wieland HA, Michaelis M, Kirschbaum BJ, Rudolphi KA (2005) Osteoarthritis – an untreatable disease? Nat Rev Drug Discov 4(4):331–344

Wolfe F, Smythe HA, Yunus MB, Bennett RM, Bombardier C, Goldenberg DL et al (1990) The American College of Rheumatology 1990 criteria for the classification of fibromyalgia. Report of the multicenter criteria committee. Arthritis Rheum 33(2):160–172

Wolfe F, Clauw DJ, Fitzcharles MA, Goldenberg DL, Häuser W, Katz RL et al (2016) 2016 revisions to the 2010/2011 fibromyalgia diagnostic criteria. Semin Arthritis Rheum 46(3):319–329

Wun A, Kollias P, Jeong H, RRN R, Cashin AG, Bagg MK et al (2021) Why is exercise prescribed for people with chronic low back pain? A review of the mechanisms of benefit proposed by clinical trialists. Musculoskelet Sci Pract 51:102307

Ye H, Weng H, Xu Y, Wang L, Wang Q, Xu G (2022) Effectiveness and safety of aerobic exercise for rheumatoid arthritis: a systematic review and meta-analysis of randomized controlled trials. BMC Sports Sci Med Rehabil 14(1):17

Yekini A, Grace JM (2023) Effects of exercise on body composition and physical function in rheumatoid arthritis patients: scoping review. Open Access Rheumatol Res Rev 15:113–123

Ylinen JJ, Takala EP, Nykänen MJ, Kautiainen HJ, Häkkinen AH, Airaksinen OVP (2006) Effects of twelve-month strength training subsequent to twelve-month stretching exercise in treatment of chronic neck pain. J Strength Cond Res 20(2):304

Zambolin F, Duro-Ocana P, Faisal A, Bagley L, Gregory WJ, Jones AW et al (2022) Fibromyalgia and chronic fatigue syndromes: a systematic review and meta-analysis of cardiorespiratory fitness and neuromuscular function compared with healthy individuals. Juhl CB, editor. PLoS One 17(10):e0276009

Zeng CY, Zhang ZR, Tang ZM, Hua FZ (2021) Benefits and mechanisms of exercise training for knee osteoarthritis. Front Physiol 16(12):794062

Zhang YH, Hu HY, Xiong YC, Peng C, Hu L, Kong YZ et al (2021) Exercise for neuropathic pain: a systematic review and expert consensus. Front Med 8:756940

Exercise and Schizophrenia

Fabiola González, Olga Montoya, and Simon Rosenbaum

Contents

Abstract Schizophrenia is a complex mental disorder that affects millions of people worldwide and has a profound impact on various aspects of life, including physical activity. The relationship between schizophrenia and physical activity is an area of growing interest in medical and health research from a physical, mental, and psychosocial health perspective. Physical activity and structured exercise have been identified as promising interventions to improve physical and psychological health outcomes of people living with schizophrenia. This chapter provides a brief overview that explores various aspects of the relationship between physical activity, exercise, and schizophrenia. The impact of schizophrenia on human movement is discussed, along with an overview of physical activity and cardiorespiratory fitness levels in adults with schizophrenia. Additionally, the influence of exercise interventions on physical and psychological outcomes will be discussed, along with current physical activity recommendations for those living with schizophrenia.

Keywords Exercise · Physical activity · Schizophrenia

F. González and O. Montoya
Escuela Colombiana de Rehabilitación, Bogotá, Colombia

S. Rosenbaum (✉)
Discipline of Psychiatry and Mental Health, University of New South Wales, Sydney, NSW, Australia
e-mail: s.rosenbaum@unsw.edu.au

Curr Topics Behav Neurosci (2024) 67: 367–380
https://doi.org/10.1007/7854_2024_505
Published Online: 10 August 2024

1 Prevalence, Etiology, and Background on Schizophrenia

The fifth edition of the Diagnostic and Statistical Manual of Mental Disorders (DSM-5) published by the American Psychiatric Association (APA) classifies schizophrenia as a serious mental disorder that affects a person's thoughts and behaviors (American Psychiatric Association 2013). There has been a debate as to the definition and classification, specifically whether schizophrenia is part of a disorder (a group of symptomatic dimensions that cluster in different combinations in different people and for which there may or may not be one or more underlying illnesses) or whether it is a disease (a nosological entity with specific causes, symptoms, treatment, and course) (van Os 2009). A diagnosis of schizophrenia is made when a person presents with at least two of the following symptoms for a significant period (usually longer than one month): delusions, hallucinations, disorganized speech, highly disorganized or catatonic behavior, and negative symptoms such as reduced emotional expression (American Psychiatric Association 2013).

Although its etiology is not fully understood, evidence exists for an underlying neurophysiological basis, and schizophrenia involves a complex interaction between genetic, biochemical, and environmental factors that affect brain functioning (Patel et al. 2014; McGrath et al. 2008). Schizophrenia is usually diagnosed between the ages of 16 and 30 years following a first episode of psychosis (Patel et al. 2014; McGrath et al. 2008). Schizophrenia is a complex mental disorder that affects approximately 1% of the world's population (McGrath et al. 2008). Psychosis and the diagnosis of schizophrenia are related to disturbances in perception, thinking, mood, and behavior (American Psychiatric Association 2013; McGrath et al. 2008). Symptoms are divided into "positive" symptoms, such as hallucinations and delusions, and "negative" symptoms, which include apathy, lack of drive, poverty of expression, social withdrawal, and poor self-care (American Psychiatric Association 2013). Adults living with schizophrenia experience clinical symptoms (e.g., delusions, hallucinations) that vary widely between and within individuals over time (Kuipers et al. 2014). These symptoms, such as delusions and hallucinations, are often integrated into broader disorders along with symptoms of mania and depression. Although characterized by these symptoms, they are likely to be the result of multiple etiopathogenic causes and pathways, converging on similar outcomes (Patel et al. 2014; McGrath et al. 2008; Kuipers et al. 2014).

Schizophrenia is also thought to lead to alterations in cognitive, emotional, and sensory processing, which has led researchers to speculate that it is possible that individuals with schizophrenia may exhibit alterations in brain regions/networks responsible for such processes (e.g., attention has been devoted to amygdala, hippocampus, and cerebellum) (Xavier and Vorderstrasse 2016). Alterations in functional connectivity may affect communication between different brain regions and contribute to the characteristic symptoms of schizophrenia (Luvsannyam et al. 2022). At the neurochemical level, schizophrenia is associated with alterations in the transmission of neurotransmitters, such as dopamine, serotonin, and glutamate (Luvsannyam et al. 2022; Ruiz et al. 2010). Increased dopaminergic activity has been observed in certain brain regions, which is believed to be related to the appearance of positive symptoms of schizophrenia, such as hallucinations and delusions. It has also been suggested that a decrease in glutamate function may be related to the negative and cognitive symptoms of the disease (Luvsannyam et al. 2022; Kasai et al. 2002; St Clair and Lang 2021). It is important to note that neurological involvement in schizophrenia can vary from individual to individual and can be influenced by numerous factors, such as age of onset, severity of symptoms, and response to treatment (Kasai et al. 2002). Clearly, schizophrenia is a complex and multifactorial disorder, and understanding its neurobiological basis is still under investigation (Kasai et al. 2002). Early recognition and intervention are critical in effectively treating and managing schizophrenia from both a mental and physical health perspective (Patel et al. 2014; Fusar-Poli et al. 2017). A 2014 meta-analysis of observational studies found that a prolonged duration of untreated psychosis (DUP) is associated with more severe symptoms, more intense positive and negative symptoms, difficulty achieving remission, lower social functioning, and an unfavorable global outcome (Penttilä et al. 2014).

2 Schizophrenia and Human Movement

Impacts on various aspects of human movement have been observed in adults with schizophrenia, including, but not limited to, specific postural patterns, gait cycles, and subjective bodily experiences (Bernard and Mittal 2014). People with schizophrenia often experience difficulties in planning and executing coordinated movements (Tonna et al. 2023). These difficulties may manifest as generalized clumsiness, uncoordinated or stiff movements, and decreased fluidity of movements (Tonna et al. 2023; Varlet et al. 2012). Individuals often exhibit alterations in gait, such as slower, less coordinated, or uneven gait patterns, which affect mobility. These gait abnormalities are often manifested in conjunction with alterations in posture (Tonna et al. 2023). Maintaining a balanced, upright posture often becomes a challenge for people with schizophrenia, possibly due to a combination of motor and cognitive dysfunctions. In addition, studies have reported on alterations in nonverbal behavior and movement coordination during social interactions (Varlet et al. 2012). Relatedly, an investigation examining pendulum swings revealed that patients with schizophrenia face challenges in social coordination, as they typically exhibit fewer stable patterns and a tendency not to lead in intentional coordination situations (Teng et al. 2016). This indicates that schizophrenia affects social coordination and the dynamics of interactions and that explicit processes may be more affected than implicit ones (Teng et al. 2016). Beyond observable movements, schizophrenia significantly influences the subjective experience of one's own body. Distorted perceptions of the body may result in unusual sensations or discomfort. Additionally, motor parameters such as increased roll area and reduced gait cadence differentiate early-onset patients from adults with schizophrenia, providing insight into the different phases of the disorder (Teng et al. 2016; Picard et al. 2008).

At the neurophysiological level, people with schizophrenia may have alterations in the connections of sensory-perceptual networks related to movement, such as the cerebellum, basal ganglia, and motor cortex (Picard et al. 2008). As such, these alterations may affect the integration of sensory information and the generation of appropriate motor patterns (Picard et al. 2008). Motor symptoms in schizophrenia can vary significantly among affected individuals; some people may experience excessive and uncontrolled movements, known as dyskinesias, while others may have a general decrease in motor activity, known as akinesia (Teng et al. 2016). These symptoms may be influenced by antipsychotic medication used in the treatment of schizophrenia, which may also have unwanted effects on the motor system (Mathews et al. 2005). Lastly, positive and negative symptoms of schizophrenia, such as hallucinations, delusions, apathy, and social withdrawal, may interfere with motivation and participation in motor learning activities (Marvel et al. 2004).

3 Physical Inactivity, Cardiorespiratory Fitness, and Exercise Interventions in People with Schizophrenia

People with schizophrenia spend increased time in sedentary behaviors and are at an elevated risk of physical inactivity (Taliercio et al. 2020; Kiwan et al. 2020; Stubbs et al. 2018a, 2016; Firth et al. 2015). For example, Stubbs et al. (2018a, b), in a meta-review and position statement by the European Psychiatric Association (EPA), found that less than 50% of patients with schizophrenia meet the internationally recommended physical activity levels of 150 min of moderate-intensity aerobic exercise per week (Stubbs et al. 2018a). Additionally, a systematic review with a pooled analysis by Stubbs et al. showed that adults with schizophrenia ($N = 3{,}453$; mean age $= 40$ years old, mean BMI $= 30.00$) engage in approximately 80.44 min of light physical activity per day (95% CI: 68.32–92.52 min/day) and were found to engage in less vigorous physical activity ($g = -0.39$, 95% CI: -0.60 to -0.18) compared to those without a diagnosis of schizophrenia (Stubbs et al. 2016). This study also highlighted that in the prodromal and early stages of schizophrenia, physical inactivity has been shown to be associated with an increased risk of developing the disease (Stubbs et al. 2016). Another review also revealed that greater depressive symptom severity and older age among adults with schizophrenia was associated with lower levels of physical activity (Firth et al. 2015).

Additionally, a study focused on analyzing physical activity levels in Nigerian patients with schizophrenia suggests that clinical and sociodemographic factors are linked to physical activity participation (Esan and Ephraim-Oluwanuga 2022). In this study, a total of 215 patients were recruited from psychiatric hospitals in Ibadan, Nigeria, and their level of physical activity was assessed using the International Physical Activity Questionnaire (IPAQ) short form. The results revealed that the majority of the participants (67.1%) had a low level of physical activity (i.e., did not meet physical activity guidelines). Physical activity was related to patients' remission status, with higher activity levels among those in remission. In addition, inverse correlations were found between physical activity and education, waist circumference, and severity of depression. Additionally, positive relationships were observed between physical activity and social and occupational functioning (Esan and Ephraim-Oluwanuga 2022). It is noteworthy that the level of physical inactivity in this study compares with that of high-income countries, despite these countries having extensive facilities for physical activities like gyms, soccer fields, basketball courts, and walking trails.

The low proportion of participants with an adequate level of physical activity (i.e., meeting guidelines) is consistently reported in studies of patients with schizophrenia (Daumit et al. 2005; Lee et al. 2018). Individuals without schizophrenia tend to be more active, lead less sedentary lifestyles, and face fewer social or cognitive barriers that limit their participation in physical activities (Stubbs et al. 2018b). Additionally, they often experience fewer mental health issues that might otherwise influence their ability to maintain a regular exercise routine (Stubbs et al. 2018b). This is unfortunate as physical activity appears to be beneficial for adults with schizophrenia (Stubbs et al. 2018a). Evidence from a recent systematic review revealed that greater participation in moderate- to vigorous-intensity exercise was associated with reduced cardiometabolic risk, reduced psychotic symptoms and comorbid psychiatric symptoms, and improved cognitive function (Firth et al. 2015).

It has also been demonstrated that individuals with schizophrenia typically exhibit lower levels of cardiorespiratory fitness compared to age-matched controls (Abdel-Baki et al. 2013; Vancampfort et al. 2015). Vancampfort et al. systematically examined seven different aerobic exercise interventions and found significant improvement in cardiorespiratory fitness ($g = 0.40$, 95% CI = 0.16–0.64) among adults with schizophrenia following the varying exercise interventions (intervention duration ranged from 6 to 32 weeks, with 2–3 weekly sessions of 30–60 min of either cycling, walking, or treadmill running) (Vancampfort et al. 2015). As a result of these findings, researchers and practitioners in this area have stressed the importance of cardiorespiratory fitness as a primary target of physical activity and exercise interventions among adults with schizophrenia (Abdel-Baki et al. 2013; Vancampfort et al. 2015). This makes intuitive sense based on some of the findings – specifically that low cardiorespiratory fitness is a significant predictor of cardiovascular disease and overall mortality (Lee et al. 2010).

In addition to cardiorespiratory fitness, several trials have examined the influence of exercise interventions on multiple disorder-specific and more general mental health outcomes (e.g., positive and negative symptoms, cognition, sleep, physical functioning). A systematic review by Firth et al. revealed that in people experiencing their first episode of psychosis, exercise combined with cognitive training significantly improved total psychiatric symptoms, negative symptoms, psychosocial functioning, and verbal short-term memory (Firth et al. 2015). An important finding from this review was that the benefits of exercise occurred independent of changes in body weight. However, it should be noted that weight loss and at a very minimum weight maintenance should still be considered an important component for adults with schizophrenia as poor physical health in adults with schizophrenia has been shown to be positively associated with the duration of the illness, greater likelihood to smoke cigarettes, increased incidence of metabolic syndrome, and greater schizophrenia-specific symptom severity (Firth et al. 2015). Additionally, a recent meta-analysis of 22 studies evaluated the effects of different types of exercise interventions on negative symptoms of schizophrenia (Vogel et al. 2019). Specifically, the analysis focused on studies with mind-body interventions (yoga, Wu-style Cheng, Tai chi Chuan, relaxation, and breath control), aerobic exercise (cycling, jogging, jumping, dancing, aerobic video games, elliptical and bench exercises) and resistance training (machine and functional training). Results revealed that mind-body exercise interventions and aerobic exercise interventions led to medium (Hedges $g = 0.46$) and small (Hedges $g = 0.34$) reductions in negative symptoms of schizophrenia, respectively, compared to a control condition (Vogel et al. 2019). Unfortunately, the effect of resistance training could not be examined, and the overall methodological quality of the included studies was deemed to be poor (Vogel et al. 2019).

The effects of exercise interventions on quality of life and sleep in patients with schizophrenia have also been recognized. Korkmaz Aslan and Ekinci (2022) investigated the effect of a structured exercise program on quality of life and sleep in subjects with schizophrenia using the Pittsburgh Sleep Quality Index and World Health Organization Quality of Life scales. The 12-week intervention involved 40-min sessions: 5 min were allocated to a warm-up, 30 min were allocated for stretching exercises, resistance training, and aerobic exercise, with a 5-min cool-down. Results from this trial revealed significant improvements in quality of life and sleep quality for the exercise intervention group compared to the control group (Korkmaz Aslan and Ekinci 2022).

Previous research has also found improvements in cognitive performance (in addition to schizophrenia-specific symptoms) following participation in exercise interventions. For example, Shimada and colleagues conducted a recent systematic review and meta-analysis ($k = 15$ studies, $N = 593$) and found that participation in aerobic exercise interventions (average duration of 50.3 min, 2.54 times per week for 12 weeks) significantly improved global cognition compared to a control condition ($SMD = 0.21$, 95% $CI = 0.12$–0.30) (Shimada et al. 2022). Additionally, a recent meta-analysis by Sabe and colleagues examined the influence of various different aerobic (including walking or high-intensity interval training) and/or resistance exercise interventions of varying intensities (moderate to vigorous) session durations (between 45 and 270 min per week) and intervention durations (8–24 weeks) on positive and negative schizophrenia symptoms in adults with schizophrenia (Sabe et al. 2020). All studies included in this meta-analysis administered exercise as an adjunctive intervention to standard antipsychotic pharmacotherapy. Results showed a significant improvement in positive symptoms and a large reduction in negative symptoms (as assessed via the Positive and Negative Syndrome Scale) after aerobic exercise but not after nonaerobic exercise interventions (e.g., resistance exercise) (Sabe et al. 2020).

Additionally, a study examined the relationship between functional exercise capacity and global functioning in patients with schizophrenia (Vancampfort et al.

2012). Sixty-eight male and twenty-five female inpatients were evaluated using the six-minute walk test and the Global Assessment of Functioning (GAF) scale, along with other symptom assessment tools. The results showed a moderate association (Pearson correlation coefficient of 0.59) between distance walked on the 6-min walk test and GAF score, suggesting that exercise capacity is related to global functioning in these patients (Vancampfort et al. 2012). In addition, distance walked was also found to correlate with negative, depressive, and cognitive symptoms, body mass index, smoking status, and antipsychotic medication dosage (Vancampfort et al. 2012). The findings suggest that, in addition to global assessment of functioning, inclusion of the 6-min walk test could be beneficial in inpatient settings to measure the functional consequences of schizophrenia and its treatment. This could provide a more complete understanding of the physical and functional status of patients with schizophrenia.

In addition to the aforementioned psychological outcomes, a recent study examined the effects of a 12-week aerobic exercise intervention on several functional and physiological outcomes (Szortyka et al. 2022). Results revealed that individuals that participated in the exercise intervention exhibited reductions in systemic inflammation, reductions in body mass index, and reductions in elevated blood pressure (Szortyka et al. 2022). There have also been recent attempts to understand the impact of exercise on hippocampal volumes in adults with schizophrenia. To date, there have been equivocal findings. For example, Joe Firth and colleagues conducted a meta-analysis ($k = 14$) that revealed that aerobic exercise had no significant effect on total hippocampal volumes (Firth et al. 2018). In contrast, Li and colleagues conducted a meta-analysis and reported increased right hippocampal volume for participants following participation in an aerobic exercise intervention, compared to nonexercising control conditions (Li et al. 2017). Additional research is needed among adults with schizophrenia to determine the precise role of aerobic exercise interventions in preventing age-related decline in the hippocampus.

Lastly, a systematic review analyzed the impact of exercise on the global functioning of adults (M age = 37.9) with schizophrenia, most of which were overweight or obese (Korman et al. 2023). Most participants maintained stable psychiatric medication for 4–12 weeks prior to the study. The exercise interventions examined as part of this review had a mean duration of 12 weeks and were performed primarily in a group setting at a frequency of 3 times per week. Types of exercise included aerobic exercise, resistance training, combined aerobic and resistance training, and combined standing and stretching postures with aerobic activities. Exercise sessions were supervised by trained physical therapists, exercise therapists, mental health ward staff, or researchers. Aerobic exercise of moderate to vigorous intensity had a positive effect on the overall functioning of patients with schizophrenia. Aerobic exercise had a greater impact on global functioning compared to other types of exercise. The study also explored the effects of exercise on daily living skills and social functioning. Preliminary evidence suggests that exercise may improve daily living skills (Korman et al. 2023).

4 Antipsychotic Use and Exercise in Schizophrenia

Antipsychotics are commonly prescribed for patients with schizophrenia. As a result, it is plausible that the lower observed activity rates in schizophrenia could be a byproduct of the treatment (and not simply having a diagnosis of schizophrenia). Additionally, antipsychotics are also typically associated with significant weight gain (Allison et al. 1999). For example, meta-analytic evidence examining weight gain in first- and second-generation antipsychotics revealed that several antipsychotics (e.g., olanzapine, clozapine) caused a significant increase in weight (Allison et al. 1999). Subsequent studies expanded upon these findings and indicated that olanzapine and clozapine are the drugs that contribute most to weight gain, whereas aripiprazole, cariprazine, lurasidone, and ziprasidone have lesser metabolic effects (Rummel-Kluge et al. 2010; Pillinger et al. 2020). However, clinical decisions about antipsychotics should consider both long-term clinical improvement and metabolic side effects (Rummel-Kluge et al. 2010; Pillinger et al. 2020). Given that antipsychotics generally result in weight gain, it is plausible to suggest that exercise may be an effective and readily accessible treatment to adverse health outcomes associated with weight gain. However, a recent pilot study investigated attendance and adherence to exercise in schizophrenia patients taking olanzapine and experiencing significant weight gain. Despite providing access to exercise facilities, only 5% of the participants adhered to the exercise program for six months. Most patients did not regularly attend exercise sessions, primarily citing lack of motivation (Archie et al. 2003). These results (along with the prior research demonstrating low physical activity levels) highlight that although exercise may be beneficial for this population, getting individuals to adhere long enough for exercise to exert a positive impact remains a difficult challenge. The low activity levels and adherence rates to exercise programs are unfortunate as physical exercise appears to influence some of the same neuroprotective factors that are considered targets of antipsychotic drugs and nutraceuticals. For example, as discussed throughout this volume, exercise reduces neuroinflammation, regulates neurotrophic growth factors promotes remyelination, and stimulates neurogenesis (Cotman et al. 2007; Chieffi et al. 2017).

5 Physical Activity Recommendations for Adults with Schizophrenia

Based on the aforementioned benefits, practitioners may want to consider prescribing physical activity for adults with schizophrenia. In doing so, practitioners should consider the most relevant outcomes of interest for their patients and consider the patient's baseline physical activity and fitness levels, as well as contextual factors that may influence motivation to initiate and maintain physical activity participation. In 2018, an expert group of the EPA issued a guideline on exercise interventions in the treatment of severe mental illness, including schizophrenia (Stubbs et al. 2018a).

Table 1 Example of how an exercise prescription may evolve for adults with schizophrenia based on findings derived from prior research

	Dosage
Modality	Aerobic exercise/cardiorespiratory training
Examples	Stationary cycling, jogging, walking, jumping, dance, elliptical, sport participation, aerobic exercise-based video games
Intensity	Moderate to vigorous intensity: –60–77% Vo2 Max or –40–75% of reserve heart rate –55–70% of maximum heart rate **The intensity should be adjusted according to individual characteristics (e.g., age, physical performance, preferences, contraindications to physical activity)* **It is recommended that the method for prescription is from % Vo2 Maximum or from % of reserve heart rate.*
Frequency	*Three to five times per week*
Duration of sessions	*Aerobic exercise session can have a minimum duration of 10 and a maximum of 60 min. Studies reporting weekly duration of ≥ 90 min/week demonstrate larger effects on aspects related to physical health (cardiometabolic alterations) and mental health (cognitive functions and positive/negative symptoms typical of schizophrenia).*
Length of program	*8 weeks up to 24 weeks, 12 weeks is most commonly used.*
Implementation of the program	*Individual or group* **Group exercise may be associated with greater adherence to exercise programs with one-on-one support offered where needed.*

The panel concluded that combining physical exercise with psychological and dietary interventions can improve parameters of metabolic syndrome, cardiovascular fitness, and cognitive performance (Stubbs et al. 2018a). To date, there are no specific physical activity recommendations for individuals with schizophrenia. It is recommended that, if feasible, individuals follow the federal guidelines outlined in Chapter "Introduction" (i.e., 150 min per week of moderate-intensity aerobic exercise or 75 min per week of vigorous-intensity aerobic exercise or an equal combination of both, in addition to muscle-strengthening exercises on two or more days per week) (Piercy et al. 2018). Additionally, it is recommended that exercise should be adapted and supervised according to the individual needs of each person, with consideration of physical condition and potential comorbidities or contraindications to exercise, personal preferences, and functional limitations (e.g., age, exercise capacity, and patient preferences). Given the risk profile of people with schizophrenia, collaboration with multidisciplinary teams including physicians and physical therapists is recommended (Faulkner and Biddle 1999). Table 1 provides an example of how specific recommendations for schizophrenia populations could evolve based on the available evidence from prior studies conducted in the area (Vogel et al. 2019; Sabe et al. 2020; Korman et al. 2023; Oliva et al. 2023; Nyboe et al. 2019; Sunhary De Verville et al. 2022; Gyllensten et al. 2020).

6 Conclusion

Physical activity and exercise are promising targets for intervention to improve physical and psychological health outcomes in people with schizophrenia. Evidence suggests that supportive and supervised moderate to vigorous-intensity exercise programs can have a positive impact if participants adhere to the program and as a result may be provided as part of evidence-based care. Implementation of exercise programs, preferably in a group setting, has been shown to be effective in improving physical and mental well-being of people with schizophrenia. Potential benefits include improvements in global functioning, cognition, mood, sleep, and both negative and positive symptoms of schizophrenia. For practitioners, it is crucial to consider the appropriate variables and doses of exercise, based on available evidence, for effective prescribing as part of the comprehensive management of schizophrenia symptoms. Although additional research is needed, the collective body of evidence to date underscores a promising role of exercise in promoting well-being and improving health in people living with schizophrenia.

References

Abdel-Baki A, Brazzini-Poisson V, Marois F, Letendre E, Karelis AD (2013) Effects of aerobic interval training on metabolic complications and cardiorespiratory fitness in young adults with psychotic disorders: a pilot study. Schizophr Res 149(1–3):112–115

Allison DB, Mentore JL, Heo M, Chandler LP, Cappelleri JC, Infante MC, Weiden PJ (1999) Antipsychotic-induced weight gain: a comprehensive research synthesis. Am J Psychiatry 156(11):1686–1696

American Psychiatric Association (2013) Diagnostic and statistical manual of mental disorders, fifth edition (DSM-5). American Psychiatric Publishing, Washington

Archie S, Wilson JH, Osborne S, Hobbs H, McNiven J (2003) Pilot study: access to fitness facility and exercise levels in olanzapine-treated patients. Can J Psychiatr 48(9):628–632

Bernard JA, Mittal VA (2014) Cerebellar-motor dysfunction in schizophrenia and psychosis-risk: the importance of regional cerebellar analysis approaches. Front Psychol 5:160

Chieffi S, Messina G, Villano I, Messina A, Valenzano A, Moscatelli F, Salerno M, Sullo A, Avola R, Monda V, Cibelli G, Monda M (2017) Neuroprotective effects of physical activity; evidence from human and animal studies. Front Neurol 8:188

Cotman CW, Berchtold NC, Christie LA (2007) Exercise builds brain health: key roles of growth factor cascades and inflammation. Trends Neurosci 30(9):464–472

Daumit GL, Goldberg RW, Anthony C, Dickerson F, Brown CH, Kreyenbuhl J, Wohlheiter K, Dixon LB (2005) Physical activity patterns in adults with severe mental illness. J Nerv Ment Dis Octubre de 193(10):641–646

Esan O, Ephraim-Oluwanuga OT (2022) Physical activity of patients with schizophrenia in Nigeria: distribution and correlates. Ment Health Phys Act Marzo 22:100445

Faulkner G, Biddle S (1999) Exercise as an adjunct treatment for schizophrenia: a review of the literature. J Ment Health 8(5):441–457

Firth J, Cotter J, Elliott R, French P, Yung AR (2015) A systematic review and meta-analysis of exercise interventions in schizophrenia patients. Psychol Med 45(7):1343–1361

Firth J, Stubbs B, Vancampfort D, Schuch F, Lagopoulos J, Rosenbaum S et al (2018) Effect of aerobic exercise on hippocampal volume in humans: a systematic review and meta-analysis. NeuroImage 166:230–238

Fusar-Poli P, McGorry PD, Kane JM (2017) Improving outcomes of first-episode psychosis: an overview. World Psychiatry Off J World Psychiatr Assoc WPA 16(3):251–265

Gyllensten AL, Ovesson MN, Hedlund L, Ambrus L, Tornberg Å (2020) To increase physical activity in sedentary patients with affective - or schizophrenia spectrum disorders - a clinical study of adjuvant physical therapy in mental health. Nord J Psychiatry 74(1):73–82

Kasai K, Iwanami A, Yamasue H, Kuroki N, Nakagome K, Fukuda M (2002) Neuroanatomy and neurophysiology in schizophrenia. Neurosci Res 43(2):93–110

Kiwan N, Mahfoud Z, Ghuloum S, Chamali R, Yehya A, Hammoudeh S et al (2020) Self-reported sleep and exercise patterns in patients with schizophrenia: a cross-sectional comparative study. Int J Behav Med 27(4):366–377

Korkmaz Aslan T, Ekinci M (2022) The effect of structured exercise program on life and sleep quality in patients with schizophrenia. Med J SDU 29(3):446–453

Korman N, Stanton R, Vecchio A, Chapman J, Parker S, Martland R et al (2023) The effect of exercise on global, social, daily living and occupational functioning in people living with schizophrenia: a systematic review and meta-analysis. Schizophr Res 256:98–111

Kuipers E, Yesufu-Udechuku A, Taylor C, Kendall T (2014) Management of psychosis and schizophrenia in adults: summary of updated NICE guidance. BMJ 348:g 1173

Lee DC, Artero EG, Sui X, Blair SN (2010) Mortality trends in the general population: the importance of cardiorespiratory fitness. J Psychopharmacol Oxf Engl 24(4 Suppl):27–35

Lee SH, Kim G, Kim CE, Ryu S (2018) Physical activity of patients with chronic schizophrenia and related clinical factors. Psychiatry Investig 15(8):811–817

Li MY, Huang MM, Li SZ, Tao J, Zheng GH, Chen LD (2017) The effects of aerobic exercise on the structure and function of DMN-related brain regions: a systematic review. Int J Neurosci 127(7):634–649

Luvsannyam E, Jain MS, Pormento MKL, Siddiqui H, Balagtas ARA, Emuze BO et al (2022) Neurobiology of schizophrenia: a comprehensive review. Cureus 14(4):e23959

Marvel CL, Schwartz BL, Rosse RB (2004) A quantitative measure of postural sway deficits in schizophrenia. Schizophr Res 68(2–3):363–372

Mathews M, Gratz S, Adetunji B, George V, Mathews M, Basil B (2005) Antipsychotic-induced movement disorders. Psychiatry (Edgmont) 2(3):36–41

McGrath J, Saha S, Chant D, Welham J (2008) Schizophrenia: a concise overview of incidence, prevalence, and mortality. Epidemiol Rev 30:67–76

Nyboe L, Lemcke S, Møller AV, Stubbs B (2019) Non-pharmacological interventions for preventing weight gain in patients with first episode schizophrenia or bipolar disorder: a systematic review. Psychiatry Res 281:112556

Oliva HNP, Monteiro-Junior RS, Oliva IO, Powers AR (2023) Effects of exercise intervention on psychotic symptoms: a meta-analysis and hypothetical model of neurobiological mechanisms. Prog Neuro-Psychopharmacol Biol Psychiatry 125:110771

Patel KR, Cherian J, Gohil K, Atkinson D (2014) Schizophrenia: overview and treatment options. P T Peer-Rev J Formul Manag 39(9):638–645

Penttilä M, Jääskeläinen E, Hirvonen N, Isohanni M, Miettunen J (2014) Duration of untreated psychosis as predictor of long-term outcome in schizophrenia: systematic review and meta-analysis. Br J Psychiatry J Ment Sci 205(2):88–94

Picard H, Amado I, Mouchet-Mages S, Olié JP, Krebs MO (2008) The role of the cerebellum in schizophrenia: an update of clinical, cognitive, and functional evidences. Schizophr Bull 34(1): 155–172

Piercy KL, Troiano RP, Ballard RM, Carlson SA, Fulton JE, Galuska DA et al (2018) The physical activity guidelines for Americans. JAMA 320(19):2020–2028

Pillinger T, McCutcheon RA, Vano L, Mizuno Y, Arumuham A, Hindley G, Beck K, Natesan S, Efthimiou O, Cipriani A, Howes OD (2020) Comparative effects of 18 antipsychotics on metabolic function in patients with schizophrenia, predictors of metabolic dysregulation, and association with psychopathology: a systematic review and network meta-analysis. Lancet Psychiatry Enero de 7(1):64–77

Ruiz JS, de la Vega Sánchez DC, Páez PS (2010) Bases neurobiológicas de la esquizofrenia. [The neurobiological basis of schizophrenia.]. Clínica Salud 21(3):235–254

Rummel-Kluge C, Komossa K, Schwarz S, Hunger H, Schmid F, Lobos CA, Kissling W, Davis JM, Leucht S (2010) Head-to-head comparisons of metabolic side effects of second generation antipsychotics in the treatment of schizophrenia: a systematic review and meta-analysis. Schizophr Res 123(2–3):225–233

Sabe M, Kaiser S, Sentissi O (2020) Physical exercise for negative symptoms of schizophrenia: systematic review of randomized controlled trials and meta-analysis. Gen Hosp Psychiatry 62: 13–20

Shimada T, Ito S, Makabe A, Yamanushi A, Takenaka A, Kawano K et al (2022) Aerobic exercise and cognitive functioning in schizophrenia: an updated systematic review and meta-analysis. Psychiatry Res 314:114656

St Clair D, Lang B (2021) Schizophrenia: a classic battle ground of nature versus nurture debate. Sci Bull 66(10):1037–1046

Stubbs B, Firth J, Berry A, Schuch FB, Rosenbaum S, Gaughran F et al (2016) How much physical activity do people with schizophrenia engage in? A systematic review, comparative meta-analysis and meta-regression. Schizophr Res 176(2–3):431–440

Stubbs B, Vancampfort D, Hallgren M, Firth J, Veronese N, Solmi M et al (2018a) EPA guidance on physical activity as a treatment for severe mental illness: a meta-review of the evidence and position statement from the European psychiatric association (EPA), supported by the International Organization of Physical Therapists in Mental Health (IOPTMH). Eur Psychiatry J Assoc Eur Psychiatr 54:124–144

Stubbs B, Vancampfort D, Firth J, Hallgren M, Schuch F, Veronese N, Solmi M, Gaughran F, Kahl KG, Rosenbaum S, Ward PB, Carvalho AF, Koyanagi A (2018b) Physical activity correlates among people with psychosis: data from 47 low- and middle-income countries. Schizophr Res 193:412–417

Sunhary De Verville PL, Stubbs B, Etchecopar-Etchart D, Godin O, Andrieu-Haller C, Berna F et al (2022) Recommendations of the schizophrenia expert center network for adequate physical activity in real-world schizophrenia (FACE-SZ). Eur Arch Psychiatry Clin Neurosci 272(7): 1273–1282

Szortyka MF, Cristiano VB, Schuch F, Belmonte-de-Abreu P (2022) Inflammatory markers and reduced response to physical intervention in schizophrenia: a controlled study with chronic outpatients. Clin Schizophr Relat Psychoses 16(2)

Taliercio J, Bonasera B, Portillo C, Ramjas E, Serper M (2020) Physical activity, sleep-related behaviors and severity of symptoms in schizophrenia. Psychiatry Res 294:113489

Teng YL, Chen CL, Lou SZ, Wang WT, Wu JY, Ma HI et al (2016) Postural stability of patients with schizophrenia during challenging sensory conditions: implication of sensory integration for postural control. PLoS One 11(6):e0158219

Tonna M, Lucarini V, Lucchese J, Presta V, Paraboschi F, Marsella F et al (2023) Posture, gait and self-disorders: an empirical study in individuals with schizophrenia. Early Interv Psychiatry 17(5):447–461

van Os J (2009) "Salience syndrome" replaces "schizophrenia" in DSM-V and ICD-11: psychiatry's evidence-based entry into the 21st century? Acta Psychiatr Scand 120(5):363–372

Vancampfort D, Probst M, Scheewe T, Knapen J, De Herdt A, De Hert M (2012) The functional exercise capacity is correlated with global functioning in patients with schizophrenia. Acta Psychiatr Scand 125(5):382–387

Vancampfort D, Rosenbaum S, Ward PB, Stubbs B (2015) Exercise improves cardiorespiratory fitness in people with schizophrenia: a systematic review and meta-analysis. Schizophr Res 169(1–3):453–457

Varlet M, Marin L, Raffard S, Schmidt RC, Capdevielle D, Boulenger JP et al (2012) Impairments of social motor coordination in schizophrenia. PLoS One 7(1):e29772
Vogel JS, van der Gaag M, Slofstra C, Knegtering H, Bruins J, Castelein S (2019) The effect of mind-body and aerobic exercise on negative symptoms in schizophrenia: a meta-analysis. Psychiatry Res 279:295–305
Xavier RM, Vorderstrasse A (2016) Neurobiological basis of insight in schizophrenia: a systematic review. Nurs Res 65(3):224–237

The Placebo Effect in Exercise and Mental Health Research

John S. Raglin and Jacob B. Lindheimer

Contents

Abstract This chapter describes the placebo effect in the context of physical activity and mental health. Following a brief historical overview, definitions for placebos, placebo effects, and related terminology are discussed. Next, three major methodological challenges posed from studying the placebo effect in the context of exercise are described. Additionally, the role of a primary psychological mechanism of placebo and nocebo effects – expectations – will be summarized through the lens of key terminology. Findings from relevant chronic and acute exercise studies that have measured or manipulated exercise-associated expectations will be presented. The chapter concludes with recommendations for controlling or quantifying the placebo effect in exercise and mental health research and its implications in clinical practice.

J. S. Raglin (✉)
Department of Kinesiology, School of Public Health, Indiana University, Bloomington, IN, USA
e-mail: raglinj@indiana.edu

J. B. Lindheimer
William S. Middleton Memorial Veterans Hospital, Madison, WI, USA

Department of Kinesiology, University of Wisconsin-Madison, Madison, WI, USA

Curr Topics Behav Neurosci (2024) 67: 381–394
https://doi.org/10.1007/7854_2024_506
Published Online: 24 July 2024

Keywords Anxiety · Depression · Expectation · Mood · Nocebo · Psychosocial

It is exercise alone that supports the spirits, and keeps the mind in vigor.
Marcus Tullius Cicero, Roman statesman (106-43 BC).

1 Introduction

Long before the advent of scientific investigation into the benefits of physical exercise, its impact on mental health was recognized. Research consistently indicates that physical activity can both improve a variety of psychological variables (e.g., mood, energy) in healthy individuals and provide benefits for persons suffering from clinical disorders such as depression or anxiety that rival those associated with medication or therapy. Yet, despite over a half century of study, the mechanisms that underly these psychological benefits remain undetermined (Mikkelsen et al. 2017). This information is not simply a matter of scientific curiosity; it has important clinical implications determining what psychological conditions are most or least likely to benefit from exercise and for designing and implementing exercise programs that optimize improvement. For example, hypothesized biological explanations such as the endorphin hypothesis would logically predict that more vigorous exercise prescriptions (i.e., increased intensity and/or duration) should evoke a greater production of endogenous opioids which should then result in larger psychological benefit. Yet, research has generally failed to find dose-response relationships between exercise intensity and changes in mood and affect (Raglin and Wilson 1996; Perkins et al. 2023), which suggests that biological mechanisms (e.g., endorphin hypothesis) are not solely responsible for exercise-induced improvements in various different psychological outcomes. Accordingly, even the earliest reviews of the exercise and mental health literature emphasized the importance of accounting for the potential influence of the placebo or the related Hawthorne effect (Morgan 1970), including the beliefs and expectations exercisers possess about "their ability to profit psychologically from fitness training" (Folkins and Sime 1981). Due to the potential "contamination" from personal beliefs or incidentally induced expectations from demand characteristics, some researchers have contended that it is essentially impossible to distinguish treatment effects from placebo effects in exercise and mental health studies (Sime 2022), largely because of the inability to blind participants to the intervention and develop a convincing placebo substitute for exercise (Ojanen 1994).

The absence of placebo controls in exercise and mental health research continues to be raised as a concern. In a paper on the acute psychological benefits of exercise literature, Szabo concluded that "researchers have neglected the investigation of the role of the placebo effect in the psychological changes resulting from exercise for almost two decades" (Szabo 2013). The potential influence of the placebo effect is particularly relevant in the case of exercise because both the psychological outcomes (e.g., anxiety, depression, self-esteem) and sensory (e.g., pain) or feeling states (e.g., fatigue, energy) assessed in the majority of studies are highly susceptible to placebo interventions (Lindheimer et al. 2019; Lindheimer et al. 2015).

This chapter will provide an overview of the placebo effect in the context of psychological responses to physical activity. Following a brief historical discussion, standardized definitions of the placebo effect and related terminology will be provided. An overview of the challenges of studying the placebo effect in the context of exercise and mental health will be presented, as well as findings from studies which have attempted to navigate these challenges by implementing exercise placebo conditions or by measuring and in some cases manipulating expectations, an established psychological mechanism of placebo effects. Lastly, recommendations for future research directions and clinical practice are provided.

2 The Placebo: A Brief History

The term placebo was not initially linked to science or medicine. It first appeared in the fourteenth century Latin bible and was mistranslated as "I will please." At this time, the Black plague raged across Europe and people hired to fill in as mourners at funerals became known as placebos. Soon the term became more broadly used as an insult describing false sincerity, but it was not until the eighteenth century that the term placebo became associated with medicine. Its most celebrated initial application to science and medicine was conducted in 1784 by Benjamin Franklin. He and several other leading scientists had been commissioned by King Louis XVI to investigate the controversial medical treatments of the physician Franz Mesmer. Foremost among their tests, they utilized a prototype of a placebo control by blindfolding patients who were then exposed to objects either treated ("mesmerized") with animal magnetism or left untreated. This manipulation revealed that patients exhibited similar symptoms of mesmerism whether exposed either to treated or inert objects. Based on these and other results, the commission concluded that the effects of animal magnetism were solely the result of "imagination" (Annoni 2020).

The first medical definitions of the placebo began to appear in print in the late sixteenth century. A surprisingly contemporary definition published in 1811 in a medical dictionary described a placebo as "an epithet given to any medicine adapted more to please than to benefit the patient" (Quincy 1811). A century and a half later, Beecher modestly updated this description by substituting "pharmacologically inert substances" for the word "epithet" (Beecher 1955). More importantly, in this paper, Beecher reviewed the findings of medical studies that employed placebo controls and concluded that they provide actual benefits for an average of 35.2% patients (Beecher 1955), a claim that spurred both the serious consideration of the placebo effect and an equally spirited backlash against it which played out in the scientific literature over the next half century.

Foremost among the critiques of Beecher's review and other unfavorable perspectives on the placebo is the problem of nonspecific effects (Kienle and Kiene

1997; Roberts 1995), a range of factors that contribute to a partial or complete remission of medical signs or symptoms but which are not aspects of either the placebo or medical treatment. Examples of nonspecific effects include regression toward the mean, spontaneous improvement, and the natural fluctuation of symptoms, as well as results invalidated by experimental biases such as the Rosenthal effect (Kienle and Kiene 1997). The inclusion of a nontreatment control is then necessary to determine if nonspecific effects are present, an addition that is often omitted because of the additional expense or because of ethical concerns about withholding treatment (Kienle and Kiene 1997).

This dim view about the influence or even the existence of the placebo effect has faded with advent of innovative research that utilized brain imaging technology during the administration of placebos. The results of this work have revealed that placebos activate the same neurobiological pathways of the medication they are associated with, yielding real benefits and increasing the efficacy of many evidence-based treatments (e.g., pain medication).

Another important innovation in placebo research is the use of novel experimental designs. For example, the traditional randomized controlled trial (RCT) constrains the placebo effect through the use of a conditional (Loehr et al. 2014) expectation of treatment to minimize its efficacy. In contrast, the balanced placebo design involves conditions that manipulate the strength of expectation to ascertain its impact. For example, in one arm of this design, participants who received the treatment are told they received a placebo (i.e., a 0% expectation of treatment), allowing researchers to determine the influence of the efficacy of the treatment without the potential added benefits from expectation. In another arm, participants are truthfully informed they received the treatment (i.e., a 100% expectation of treatment), a condition which mimics the clinical setting wherein accurate information is provided the treatment along with information about the expected outcome (e.g., reduced pain). Other innovative approaches include the use of open-hidden designs, deceptive conditioning, or intentionally induced expectations; each has provided unique insights into how conditioning and expectation can modulate placebo outcomes. This work also has important implications in how the placebo effect may add to the efficacy of medical treatments in the clinical setting.

3 Definitions for Placebos, Placebo Effects, and Related Terminology

The term *placebo* classically describes an inert agent or sham intervention that lacks any inherent capacity to produce a biological or psychological benefit (Kirsch 1985), whereas a *placebo effect* refers to the changes specifically attributable to placebo mechanisms (Evers et al. 2018). Importantly, because classical conditioning and expectation are also associated with the presentation of an established treatment (Benedetti et al. 2003), placebo effects can arise without administering a placebo, meaning that the placebo effect can contribute to the efficacy of active treatments such as pain medication (Benedetti et al. 2003) and behavioral interventions such as exercise (Lindheimer et al. 2015). The term *placebo response* has been applied synonymously with placebo effect, but it is now defined as all health changes that result after administration of an inactive treatment (i.e., differences in symptoms before and after treatment), thus including nonspecific effects (e.g., natural history and regression to the mean) and experimental artifacts such as demand characteristics (i.e., the Rosenthal effect) (Evers et al. 2018). To distinguish the placebo response from a true placebo effect, the addition of a nontreatment control would be required.

Subtypes of placebos have also been identified in the literature, some of which have relevance to physical exercise research. The *inert placebo* when used in medical research must lack any pharmacological and physiological effects. Examples would include a saline injection or pill filled with corn powder. Establishing a physical activity intervention that qualifies as an inert placebo would first require identifying precisely what biological process (e.g., endorphin production) or adaptation (e.g., increased cardiovascular capacity) modulates changes in mental health and then either designing an exercise mode that does not stimulate this mechanism or if possible, by adding an intervention that would disable this mechanism. For example, if it were unequivocally determined that endogenous opioid activity was responsible for acute improvements in mental health, then an inert placebo condition would involve either an exercise prescription both insufficient in either intensity and duration to stimulate an increase in opioid production or the administration of an opioid antagonist during more vigorous exercise. Because there continues to be a debate over the mechanism(s) responsible for exercise-related changes in mental health as well as the possibility that different mechanisms explain different outcomes (e.g., increased energy versus reduced anxiety), identifying an inert exercise placebo may not be feasible.

An *active placebo* is designed to be convincing by mimicking the physical appearance of the medication as well as its sensory characteristics (smell or taste) and when possible its side effects. Taking the extra steps to create an active placebo can be valuable because research indicates that active placebos often yield greater effects compared with generic placebos (Miller and Colloca 2009). However, for behavioral interventions such as psychotherapy or physical exercise, the question arises of whether convincing placebo treatments can be identified. In psychiatry placebo research options such as support therapy, relaxation training and discussion groups have been employed (Weimer et al. 2015) as placebo conditions. But it has been noted (Ojanen 1994; Lindheimer et al. 2015) that realistic placebo substitutes for exercise may be unconvincing or are even impossible. Moreover, experienced participants can distinguish convincingly designed placebos from treatments that lack obvious sensory cues, a phenomenon referred to as functional unblinding. In a study involving frequent tanning bed users (Feldman et al. 2004), the investigators allowed participants to choose between a UV bed or an identical placebo bed that did not emit UV rays. It was found that 95% of the participants selected the UV bed despite the absence of any overt clues or sensory differences between the conditions.

These results, along with research that has found greater benefits in participants who strongly believe they have received the active treatment (Beedie et al. 2006), reinforce the need to assess level of confidence participants have of receiving either the active treatment or placebo (O'Connor et al. 2018). This would allow for researchers to conduct post hoc analyses to determine if expectations in participants influenced their responses to the experimental treatment or placebo during the intervention.

The *impure placebo* is a medication or treatment with pharmacological or physiological outcomes that either have no documented benefit on the patient's medical condition or symptoms or are administered at a low dosage deemed to be ineffective for improving a patient's condition (Fent et al. 2011). Several studies have employed low-intensity exercise conditions (e.g., stretching exercise) as an impure or minimal treatment placebo under the presumption that improved mental health is a function of enhanced cardiorespiratory capacity (Martinsen et al. 1989). However, these studies did not find greater benefits resulting from the more vigorous exercise prescriptions, which suggests that improvements in cardiovascular capacity are not the primary driver of psychological improvements from physical exercise and support the supposition the placebo effect contributes substantially.

The *nocebo effect* ("I will harm") stems from the administration of an inert medication or intervention given along with the expectation of negative outcomes (e.g., anxiety, pain, fatigue) (Benedetti et al. 2003; Carlino et al. 2014). Additionally, inert agents or even beneficial medications presented without messages of harm can sometimes result in a nocebo effect. This is most likely in individuals who already possess negative expectations because of past experiences (i.e., habitual expectation) or who observe nocebo responses in others (i.e., social learning). Another means by which nocebo effects may occur is when individuals receiving a treatment or placebo paradoxically develop negative expectations during study participation following unintentional exposure to certain psychosocial cues (i.e., incidentally induced expectation).

These definitions not only reveal that the efficacy of placebos is influenced by their sensory characteristics but also by the specific features of the psychosocial context in which they are administered (Benedetti 2013). These include the role of the provider of the placebo (e.g., physician, therapist, scientist, exercise instructor) and the setting in which the placebo is provided (e.g., hospital, laboratory, fitness club). This has led researchers to broaden the definition of the placebo to acknowledge the influence of these factors, describing it as "simulation of an active therapy within a psychosocial context" (Price et al. 2008).

4 Navigating Methodological Challenges in Exercise and Mental Health

Exercise researchers have recognized the need to account for placebo effects in clinical trials for quite some time (McCann and Holmes 1984). However, this objective is difficult to achieve for three primary reasons. First, as is the case with other behavioral interventions, it is debatable whether participants in an exercise study can be blinded from their treatment assignment. Second, there is no consensus on what might constitute a valid exercise placebo (e.g., one reflecting every aspect of exercise except the "active ingredients"). This problem is further compounded by the lack of reproducible evidence identifying the mechanisms responsible for the psychological improvements associated with exercise. Third, studying placebo effects in clinical trials is resource intensive and cost prohibitive. That is, to precisely distinguish treatment effects from placebo effects, the study must include a no-treatment control arm in addition to the treatment and placebo arms (Kienle and Kiene 1997).

While the above challenges make it difficult to distinguish placebo effects from treatment effects in exercise interventions, a small number of randomized controlled studies used designs which allow some placebo-related inferences to be made. Seeking to quantify the placebo effect in psychological outcomes of exercise training (e.g., anxiety, depression, fatigue, cognition), Lindheimer and colleagues conducted a meta-analysis of nine randomized controlled trials that included an exercise treatment, control, and "placebo" arm (Lindheimer et al. 2015). In this case, a placebo condition was defined as "an intervention that was not generally recognized as efficacious, that lacked adequate evidence for efficacy, and that has no direct pharmacological, bio-chemical, or physical mechanism of action according to the current standard of knowledge" (Lindheimer et al. 2015). Notably, the authors found that approximately 50% of the treatment effect could be explained by placebo effects, a finding that bears a striking resemblance to a more recent meta-analytic finding that exercise training was more effective than standard of care or nothing for chronic pain but was not more effective than placebo (Miller et al. 2022).

5 Expectations: Key Terms and Practical Examples

While the approach taken by Lindheimer and colleagues provided one pathway for studying placebo effects (Lindheimer et al. 2015), they later recognized that traditional placebo conditions are not necessarily germane to studying placebo effects in exercise studies (Lindheimer et al. 2019). Researchers have recognized that established psychological mechanisms of placebo effects can be measured and manipulated to provide insight into the likelihood and extent to which treatment responses to exercise are augmented by placebo effects or suppressed by nocebo effects (Lindheimer et al. 2019). While expectations can be generally defined as the belief that a given behavior will lead to an outcome (Bandura 1977), we define four specific types of expectations below and provide relevant practical examples from the scientific literature to help distinguish among these various subtypes of expectation:

1. *Habitual expectations*: An expectation that reflects an individual's previous experiences or cultural beliefs. These expectations are developed prior to engaging in a given study.

   Example. A randomized controlled trial by Gordon and colleagues found that eight weeks of resistance exercise training reduced anxiety symptom severity in young adults with subclinical generalized anxiety disorder ($d = 0.71$) (Gordon et al. 2021). However, during the trial's screening phase, 82.3% of respondents reported that they *typically* experience decreases in anxiety following exercise (Herring 2022). Thus, it is possible that the reported anxiolytic effects of the intervention were influenced by placebo effects from habitual expectations about the positive effects of exercise on anxiety.

2. *Study-specific expectations:* An expectation that is formed from experiences occurring *during* one or more phases of a study (e.g., recruitment, enrollment, test administration). Two different types of study-specific expectations are introduced below.

   (a) *Experimentally induced expectation:* A study-specific expectation arising from the investigator's intentional use of psychosocial cues to illicit a placebo or nocebo effect (e.g., expectancy manipulation).

   Example. Wang and colleagues measured affective responses to cycling exercise in 95 healthy adults after giving participants positive, negative, or neutral information about how exercise would make them feel (Wang et al. 2022). Relative to the neutral information control group, participants who received positive information had more positive expectations and reported feeling better both during and following exercise. Thus, the investigators used the positive information to experimentally induce the expectation that exercising would lead to feeling better.

   (b) *Incidentally induced expectation:* A study-specific expectation arising from exposure to psychosocial cues that were unintended by the investigator but may nevertheless illicit a placebo or nocebo effect (e.g., demand characteristics). These may threaten the internal validity of the study.

   Example. Lindheimer and colleagues tested the acute effects of active and passive cycling on state anxiety in 60 young adults with elevated trait anxiety (Lindheimer et al. 2017). To minimize psychosocial cues which might lead participants to develop positive, study-specific expectations, the investigators avoided statements about testing the therapeutic effects of exercise in the study advertisement and consent form and trained test administrators to refrain from discussing the study purpose with participants. As shown in Fig. 1, effect sizes for reductions in state anxiety after active and passive cycling were smaller than those reported by a meta-analysis of 36 randomized controlled trials (Ensari et al. 2015), perhaps due to the careful consideration of incidentally induced expectations.

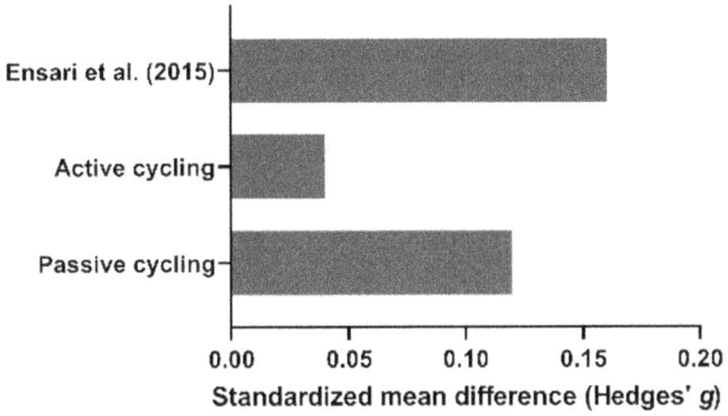

Fig. 1 Effect size comparison of state anxiety reductions between a study of active and passive cycling (Lindheimer et al. 2017) and a meta-analysis of 36 randomized controlled trials (Ensari et al. 2015)

6 The Influence of Expectation in Chronic Exercise Paradigms

Desharnais and colleagues randomly assigned 48 healthy adults to a ten-week aerobic exercise program (Desharnais et al. 1993). However, one group was informed that the exercise program was designed to enhance mental health, while the other group was only informed about the expected physical benefits. At the end of the intervention, both groups exhibited significant improvements in physical fitness but a significant increase for self-esteem occurred only in the group that received the mental health message, and the authors attributed this improvement to the placebo effect. In contrast, a recent study involving healthy adults participating in a seven-week aerobic exercise program failed to replicate these findings (Arbinaga et al. 2018). One group received a message that the exercise program was explicitly designed to enhance psychological well-being. Unlike the control group in the Desharnais et al.'s study in which the control group was not given any message about the psychological effects of exercise (Desharnais et al. 1993), Arbinaga et al. included not only an exercise with no expectation group but a third group in which participants were told the exercise program was not vigorous enough to improve psychological well-being (Arbinaga et al. 2018). Although the largest average improvement was found in the positive expectation group, comparable improvements were observed in the other conditions. In an effort to explain the equivocal findings between studies, the authors noted that repeated information sessions were employed in the Desharnais' study, which was in contrast to their study that involved a single information session, which may have been insufficient to affect participants' expectations. A third example stems a study of healthy, sedentary middle-aged adults who were randomized to a six-month walking/jogging intervention (King et al. 1989). In this case, expectations were not experimentally manipulated. However, the authors reported significant improvements in self-reported satisfaction with body image and fitness before and after accounting for habitual expectations in their statistical model. Thus, the effect of the intervention did not appear to be influenced by placebo effects.

7 Minimal Exercise ("Impure Placebo") Studies

Several studies have attempted to create an exercise placebo condition by using low-intensity activities such as stretching or passive movement. In an investigation of the influence of chronic exercise on the self-rated quality of life and other psychological measures (Daley et al. 2007), women undergoing breast cancer treatment were randomized to eight weeks of low-intensity aerobic exercise (<40% of HRR), a placebo-exercise program (light stretching and conditioning), or standard care. While quality of life improved more in the aerobic exercise condition, reductions in depression and other psychological variables were also found in the placebo condition, leading the authors to conclude: "We cannot rule out the possibility that light body conditioning and stretching exercises of minimal aerobic exertion also provide psychological benefit to cancer patients" (Daley et al. 2007). Indeed, Martinsen and colleagues found that a light stretching exercise program resulted in psychological benefits that were equivalent to a more vigorous aerobic exercise program in clinically anxious patients (Martinsen et al. 1989). On the other hand, Blumenthal et al. found only a modest correlation ($r = -0.25$) between improvements in aerobic capacity and reduced depression in clinically diagnosed adults following a 16-week exercise program and proposed that other biological mechanisms may also contribute to the psychological benefits of exercise (Blumenthal et al. 1999).

To summarize, studies of long-term exercise programs that have attempted to either alter expectations or utilize minimal intensity "placebo exercise" conditions have yielded mixed results. The failure of experimentally induced expectations to alter psychological responses in some studies (Arbinaga et al. 2018) has been attributed to the use of a single or infrequent communication exercise message. In the case of low-intensity exercise, prescriptions may evoke biological adaptations that are less dependent or unrelated to enhanced aerobic capacity, but additional research is needed to confirm this and other possible factors.

8 The Influence of Expectation in Acute Exercise Paradigms

A small number of studies have been conducted in which the psychological responses to a single bout of exercise were influenced by either habitual or experimentally induced expectations. In general, studies in which individuals possess positive habitual expectations about exercise are associated with greater psychological benefits, whether reductions in negative moods (e.g., state anxiety) or increases in positive mood states such as vigor or energy (Szabo 2013; Anderson and Brice 2011). This highlights the importance of measuring and accounting for habitual expectations in exercise studies. For instance, an early study by Tieman and colleagues found that the effect of light-intensity cycling on state anxiety was smaller after including habitual expectations as a covariate in their statistical model (Tieman et al. 2002). However, the results of studies involving experimentally induced expectations are less consistent. Mothes et al. found that habitual positive expectations were associated with both greater self-reports of psychological and neurophysiological benefits following 30 min of moderate intensity cycling, whereas experimentally induced positive expectations were associated with neurophysiological enhancements only (Mothes et al. 2016). Wang and colleagues found that experimentally induced, but not habitual, expectations were associated with greater affect during or after moderate-intensity cycling (Wang et al. 2022). Vaegter et al. found that negative study-specific expectations led to hyperalgesia (i.e., increased pain sensitivity) whereas neutral or positive study-specific expectations led to hypoalgesia (i.e., decreased pain sensitivity) following isometric squat exercise (Vaegter et al. 2020). Other researchers have observed greater psychological benefits (i.e., affect) following experimentally induced expectations of positive messages (Helfer et al. 2014; Kwan et al. 2017). The differences in the findings in these studies may be due to various factors including the strength of expectations or the psychological outcomes they are associated with, and further research is clearly needed.

9 General Summary and Recommendations

The mindset and emphasis of scientific inquiry has gradually shifted from lamenting the inability to distinguish treatment effects from placebo effects in psychological responses to exercise (Ojanen 1994), to recognizing the importance of studying mechanisms of placebo and nocebo effects, the psychosocial context in which they are activated, and how they can be leveraged to enhance the efficacy of exercise therapy on mental health (Lindheimer et al. 2019). A meta-analysis of the available research involving both a placebo and control arm has estimated this contribution to be approximately 50% of the overall benefit (Lindheimer et al. 2015), which is comparable to the contribution of the placebo effect in the benefits of psychotherapy for depression (Krogsbøll et al. 2009). Moreover, individuals who possess positive habitual expectations about exercise experience larger psychological benefits compared to those who have neutral or negative expectations, an important finding given that many people possess negative outcome expectations about exercise (e.g., fatigue, soreness) that can influence their decision to either begin or to continue an exercise program (Lindheimer et al. 2020; Loehr et al. 2014). The findings of research attempting to manipulate expectations are not entirely consistent, but it appears that presenting positive statements about the psychological benefits of exercise may sometimes result in greater psychological benefits, although it may require frequent reinforcement of such messages or elaboration of information to effect these changes (Wang et al. 2022).

These findings have implications for both research and practice. Given that habitual and experimentally induced expectations often, if not always, influence the experience of exercise-associated psychological effects, researchers should routinely assess expectations both at the outset of exercise intervention studies and throughout the experimental trial. This information would also help determine if unintentional expectations developed during study participation, even if efforts were made to avoid unintentional expectation through the use of single blind controls or deception.

Conversely, the finding that expectations can either enhance or diminish psychological responses to exercise has different implications for the clinician. As is the case in clinical medicine (Evers et al. 2018), it is important to educate exercise instructors about the potential influence that both the placebo and nocebo effects can have on the psychological outcomes of exercise. Communicating positive expectations about both the physical and psychological benefits of exercise to participants has both the potential to stimulate more positive psychological outcomes and improve long-term adherence. Moreover, determining whether a patient's negative expectations are associated with certain exercise stimulus characteristics (e.g., intensity, mode, duration) may help clinicians develop more effective exercise prescriptions. Indeed, these suggestions may help researchers and clinicians enhance the mental health benefits of exercise, but additional research is needed to further refine our understanding of the specific psychosocial conditions that optimize these benefits.

References

Anderson RJ, Brice S (2011) The mood-enhancing benefits of exercise: memory biases augment the effect. Psychol Sport Exerc 12:79–82

Annoni M (2020) Better than nothing: a historical account of placebos and placebo effects from modern to contemporary medicine. Int Rev Neurobiol 153:3–26

Arbinaga F, Fernández-Ozcorta E, Sáenz-López P, Carmona J (2018) The psychological effects of physical exercise: a controlled study of the placebo effect. Scand J Psychol 59(6):644–652

Bandura A (1977) Self-efficacy: toward a unifying theory of behavior change. Psychol Rev 84:191–215

Beecher HK (1955) The powerful placebo. JAMA J Am Med Assoc 159:1602–1606

Beedie CJ, Stuart EM, Coleman DA, Foad AJ (2006) Placebo effects of caffeine on cycling performance. Med Sci Sports Exerc 38(12):2159–2164

Benedetti F (2013) Placebo and the new physiology of the doctor-patient relationship. Physiol Rev 93:1207–1246

Benedetti F, Pollo A, Lopiano L, Lanotte M, Vighetti S, Rainero I (2003) Conscious expectation and unconscious conditioning in analgesic, motor, and hormonal placebo/nocebo responses. J Neurosci 23:4315–4323

Blumenthal JA, Babyak MA, Moore KA, Craighead WE, Herman S, Khatri P et al (1999) Effects of exercise training on older patients with major depression. Arch Intern Med 159:2349–2356

Carlino E, Piedimonte A, Frisaldi E (2014) The effects of placebos and nocebos on physical performance. Handb Exp Pharmacol 225:149–157

Daley AJ, Crank H, Saxton JM, Mutrie N, Coleman R, Roalfe A (2007) Randomized trial of exercise therapy in women treated for breast cancer. J Clin Oncol 25(13):1713–1721

Desharnais R, Jobin J, Cote C, Levesque L, Godin G (1993) Aerobic exercise and the placebo effect: a controlled study. Psychosom Med 55:149–154

Ensari I, Greenlee TA, Motl RW, Petruzzello SJ (2015) Meta-analysis of acute exercise effects on state anxiety: an update of randomized controlled trials over the past 25 years. Depress Anxiety 32:624–634

Evers AWM, Colloca L, Blease C, Annoni M, Atlas LY, Benedetti F et al (2018) Implications of placebo and nocebo effects for clinical practice: expert consensus. Psychother Psychosom 87:1–7

Feldman SR, Ligouri A, Kucenic M, Rapp SR, Fleischer AB, Lang W et al (2004) Ultraviolet exposure is a reinforcing stimulus in frequent indoor tanners. J Am Acad Dermatol 51:45–51

Fent R, Rosemann T, Fassler M, Senn O, Huber CA (2011) The use of pure and impure placebo interventions in primary care-a qualitative approach. BMC Fam Pract 12:1–7

Folkins CH, Sime WE (1981) Physical fitness training and mental health. Am Psychol 36:373

Gordon BR, McDowell CP, Lyons M, Herring M (2021) Resistance exercise training among young adults with analogue generalized anxiety disorder. J Affect Disord 281:153–159

Helfer SG, Elhai JD, Geers AL (2014) Affect and exercise: positive affective expectations can increase post-exercise mood and exercise intentions. Ann Behav Med 49:269–279

Herring MP (2022) The interplay between expected psychological responses to exercise and physical activity in analogue generalized anxiety disorder: a cross-sectional study. Int J Behav Med 30:221

Kienle GS, Kiene H (1997) The powerful placebo effect: fact or fiction? J Clin Epidemiol 50(12): 1311–1318

King A, Taylor C, Haskell W, DeBusk R (1989) Influence of regular aerobic exercise on psychological health: a randomized, controlled trial of healthy middle-aged adults. Health Psychol 8: 305–324

Kirsch I (1985) Response expectancy as a determinant of experience and behavior. Am Psychol 40: 1189–1202

Krogsbøll LT, Hróbjartsson A, Gøtzsche PC (2009) Spontaneous improvement in randomised clinical trials: meta-analysis of three-armed trials comparing no treatment, placebo and active intervention. BMC Med Res Methodol 9:1–10

Kwan BM, Stevens CJ, Bryan AD (2017) What to expect when you're exercising: an experimental test of the anticipated affect–exercise relationship. Health Psychol 36(4):309–319

Lindheimer JB, O'Connor PJ, Dishman RK (2015) Quantifying the placebo effect in psychological outcomes of exercise training: a meta-analysis of randomized trials. Sports Med 45(5):693–711

Lindheimer JB, O'Connor PJ, McCully KK, Dishman RK (2017) The effect of light-intensity cycling on mood and working memory in response to a randomized, placebo-controlled design. Psychosom Med 79:243–253

Lindheimer JB, Szabo A, Raglin JS, Beedie C (2019) Advancing the understanding of placebo effects in psychological outcomes of exercise: lessons learned and future directions. Eur J Sport Sci 20:1–12

Lindheimer JB, Szabo A, Raglin JS, Beedie C, Carmichael KE, O'Connor PJ (2020) Reconceptualizing the measurement of expectations to better understand placebo and nocebo effects in psychological responses to exercise. Eur J Sport Sci 20(3):338–346

Loehr VG, Baldwin AS, Rosenfield D, Smits JA (2014) Weekly variability in outcome expectations: examining associations with related physical activity experiences during physical activity initiation. J Health Psychol 19:1309–1319

Martinsen EW, Asle H, Solberg O (1989) Aerobic and non-aerobic forms of exercise in the treatment of anxiety disorders. Stress Med 5:115–120

McCann IL, Holmes DS (1984) Influence of aerobic exercise on depression. J Pers Soc Psychol 46: 1142–1147

Mikkelsen K, Stojanovska L, Polenakovic M, Bosevski M, Apostolopoulos V (2017) Exercise and mental health. Maturitas 106:48–56

Miller FG, Colloca L (2009) The legitimacy of placebo treatments in clinical practice: evidence and ethics. Am J Bioeth 9:39–47

Miller CT, Owen PJ, Than CA, Ball J, Sadler K, Piedimonte A et al (2022) Attempting to separate placebo effects from exercise in chronic pain: a systematic review and meta-analysis. Sports Med 52:789–816
Morgan WP (1970) Psychological concomitants of physical activity in the adult. In: Integrated development: relationships between motor and emotional domains. Indiana State Board of Health, Indianapolis, pp 39–45
Mothes H, Leukel C, Jo HG, Seelig H, Schmidt S, Fuchs R (2016) Expectations affect psychological and neurophysiological benefits even after a single bout of exercise. J Behav Med 40:293
O'Connor PJ, Loy BD, Lindheimer JB, Covert SF (2018) Adenosine A2A receptor gene polymorphisms (ADORA2A) are associated with maximal concentric contraction pain. Meta Gene 18:53–57
Ojanen M (1994) Can the true effects of exercise on psychological variables be separated from placebo effects? Int J Sport Psychol 25:63–80
Perkins S, Cook D, Herring M, Meyer J (2023) Dose-response effects of acute exercise intensity on state anxiety among women with depression. Front Psych 14:1090077
Price DD, Finniss DG, Benedetti F (2008) A comprehensive review of the placebo effect: recent advances and current thought. Annu Rev Psychol 59:565–590
Quincy J (1811) Quincy's lexicon-medicum: a new medical dictionary. Wellcome Collection
Raglin JS, Wilson M (1996) State anxiety following 20 minutes of bicycle ergometer exercise at selected intensities. Int J Sports Med 17:467–471
Roberts AH (1995) The powerful placebo revisited: magnitude of nonspecific effects. MindBody Med 1:1–10
Sime WE (2022) Psychological benefits of exercise training in the healthy individual. Behav Health:488–508
Szabo A (2013) Acute psychological benefits of exercise: reconsideration of the placebo effect. J Ment Health 22(5):449–455
Tieman JG, Peacock LJ, Cureton KJ, Dishman RK (2002) The influence of exercise intensity and physical activity. Int J Sports Psychol 33:155–166
Vaegter HB, Thinggaard P, Madsen CH, Hasenbring M, Thorlund JB (2020) Power of words: influence of Preexercise information on Hypoalgesia after exercise - randomized controlled trial. Med Sci Sports Exerc 52:2373–2379
Wang Y, Guo L, Fan J, Mao Z (2022) Expectations come true: the placebo effect of exercise on affective responses. Res Q Exerc Sport 94:1153
Weimer K, Colloca L, Enck P (2015) Placebo effects in psychiatry: mediators and moderators. Lancet Psychiatry 2:246–257

Printed in the USA
CPSIA information can be obtained
at www.ICGtesting.com
LVHW050239271124
797301LV00001BA/6
VPT0000109137